Applied Statistics

Applied Statistics

Theory and Problem Solutions with R

Dieter Rasch
Rostock
Germany

Rob Verdooren
Wageningen
The Netherlands

Jürgen Pilz
Klagenfurt
Austria

WILEY

This edition first published 2020

Registered Offices
John Wiley & Sons, Inc., 111 River Street, Hoboken, NJ 07030, USA
John Wiley & Sons Ltd, The Atrium, Southern Gate, Chichester, West Sussex, PO19 8SQ, UK

Editorial Office
9600 Garsington Road, Oxford, OX4 2DQ, UK

For details of our global editorial offices, customer services, and more information about Wiley products visit us at www.wiley.com.

Wiley also publishes its books in a variety of electronic formats and by print-on-demand. Some content that appears in standard print versions of this book may not be available in other formats.

Library of Congress Cataloging-in-Publication Data
Names: Rasch, Dieter, author. | Verdooren, L. R., author. | Pilz, Jürgen, 1951- author.
Title: Applied statistics : theory and problem solutions with R / Dieter Rasch (Rostock, GM), Rob Verdooren, Jürgen Pilz.
Description: Hoboken, NJ : Wiley, 2020. | Includes bibliographical references and index. |
Identifiers: LCCN 2019016568 (print) | LCCN 2019017761 (ebook) | ISBN 9781119551553 (Adobe PDF) | ISBN 9781119551546 (ePub) | ISBN 9781119551522 (hardcover)
Subjects: LCSH: Mathematical statistics–Problems, exercises, etc. | R (Computer program language)
Classification: LCC QA276 (ebook) | LCC QA276 .R3668 2019 (print) | DDC 519.5–dc23
LC record available at https://lccn.loc.gov/2019016568

Cover design: Wiley
Cover Images: © Dieter Rasch, © whiteMocca/Shutterstock

Set in 10/12pt WarnockPro by SPi Global, Chennai, India
Printed and bound in Singapore by Markono Print Media Pte Ltd

10 9 8 7 6 5 4 3 2 1

Contents

Preface

We wrote this book for people that have to apply statistical methods in their research but whose main interest is not in theorems and proofs. Because of such an approach, our aim is not to provide the detailed theoretical background of statistical procedures. While mathematical statistics as a branch of mathematics includes definitions as well as theorems and their proofs, applied statistics gives hints for the application of the results of mathematical statistics.

Sometimes applied statistics uses simulation results in place of results from theorems. An example is that the normality assumption needed for many theorems in mathematical statistics can be neglected in applications for location parameters such as the expectation, see for this Rasch and Tiku (1985). Nearly all statistical tests and confidence estimations for expectations have been shown by simulations to be very robust against the violation of the normality assumption needed to prove corresponding theorems.

We gave the present book an analogous structure to that of Rasch and Schott (2018) so that the reader can easily find the corresponding theoretical background there. Chapter 11 'Generalised Linear Models' and Chapter 12 'Spatial Statistics' of the present book have no prototype in Rasch and Schott (2018). Further, the present book contains no exercises; lecturers can either use the exercises (with solutions in the appendix) in Rasch and Schott (2018) or the exercises in the problems mentioned below.

Instead, our aim was to demonstrate the theory presented in Rasch and Schott (2018) and that underlying the new Chapters 11 and 12 using functions and procedures available in the statistical programming system R, which has become the golden standard when it comes to statistical computing.

Within the text, the reader finds often the sequence problem – solution – example with problems numbered within the chapters. Readers interested only in special applications in many cases may find the corresponding procedure in the list of problems in Appendix A.

We thank Alison Oliver (Wiley, Oxford) and Mustaq Ahamed (Wiley) for their assistance in publishing this book.

We are very interested in the comments of readers. Please contact:
d_rasch@t-online.de, l.r.verdooren@hetnet.nl, juergen.pilz@aau.at.
Rostock, Wageningen, and Klagenfurt, June 2019, the authors.

References

Rasch, D. and Tiku, M.L. (eds.) (1985). Robustness of statistical methods and nonparametric statistics. In: *Proceedings of the Conference on Robustness of Statistical Methods and Nonparametric Statistics, held at Schwerin (DDR), May 29-June 2, 1983.* Boston, Lancaster, Tokyo: Reidel Publ. Co. Dordrecht.

Rasch, D. and Schott, D. (2018). *Mathematical Statistics*. Oxford: Wiley.

1

The R-Package, Sampling Procedures, and Random Variables

1.1 Introduction

In this chapter we give an overview of the software package R and introduce basic knowledge about random variables and sampling procedures.

1.2 The Statistical Software Package R

In practical investigations, professional statistical software is used to design experiments or to analyse data already collected. We apply here the software package R. Anybody can extend the functionality of **R** without any restrictions using free software tools; moreover, it is also possible to implement special statistical methods as well as certain procedures of C and FORTRAN. Such tools are offered on the internet in standardised archives. The most popular archive is probably CRAN (Comprehensive R Archive Network), a server net that is supervised by the R Development Core Team. This net also offers the package OPDOE (optimal design of experiments), which was thoroughly described in Rasch *et al.* (2011). Further it offers the following packages used in this book: `car, lme4, DunnettTests, VCA, lmerTest, mvtnorm, seqtest, faraway, MASS, glm2, geoR, gstat.`

Apart from only a few exceptions, R contains implementations for all statistical methods concerning analysis, evaluation, and planning. We refer for details to Crawley (2013).

The software package R is available free of charge from http://cran.r-project.org for the operating systems Linux, MacOS X, and Windows. The installation under Microsoft Windows takes place via ‘Windows’. Choosing ‘base’ the installation platform is reached. Using ‘Download R 2.X.X for Windows’ (X stands for the required version number) the setup file can be downloaded. After this file is started the setup assistant runs through the installation steps. In this book, all standard settings are adopted. The interested reader will find more information about R at http://www.r-project.org or in Crawley (2013).

After starting R the input window will be opened, presenting the red coloured input request: ‘>’. Here commands can be written up and carried out by pressing the enter button. The output is given directly below the command line. However, the user can also realise line changes as well as line indents for increasing clarity. Not all this influences the functional procedure. A command to read for instance data $y = (1, 3, 8, 11)$ is as follows:

Applied Statistics: Theory and Problem Solutions with R, First Edition.
Dieter Rasch, Rob Verdooren, and Jürgen Pilz.

```
> y <- c(1,3,8,11)
```

The assignment operator in R is the two-character sequence '<-' or '='.

The Workspace is a special working environment in R. There, certain objects can be stored that were obtained during the current work with R. Such objects contain the results of computations and data sets. A Workspace is loaded using the menu

```
File - Load Workspace...
```

In this book the R-commands start with >. Readers who like to use R-commands must only type or copy the text after > into the R-window.

An advantage of R is that, as with other statistical packages like SAS and IBM-SPSS, we no longer need an appendix with tables in statistical books. Often tables of the density or distribution function of the standard normal distribution appear in such appendices. However, the values can be easily calculated using R.

The notation of this and the following chapters is just that of Rasch and Schott (2018).

Problem 1.1 Calculate the value $\varphi(z)$ of the density function of the standard normal distribution for a given value z.

Solution

Use the command > `dnorm(z, mean = 0, sd = 1)`. If the `mean` or `sd` is not specified they assume the default values of 0 and 1, respectively. Hence > `dnorm(z)` can be used in Problem 1.1.

Example

We calculate the value $\varphi(1)$ of the density function of the standard normal distribution using

```
> dnorm(1)
[1] 0.2419707
```

Problem 1.2 Calculate the value $\Phi(z)$ of the distribution function of the standard normal distribution for a given value z.

Solution

Use the command > `pnorm(z, mean = 0, sd = 1)`.

Example

We calculate the value $\Phi(1)$ of the distribution function of the standard normal distribution by > `pnorm(1, mean = 0, sd = 1)` or using the default values using > `pnorm(1)`.

```
> pnorm(1)
[1] 0.8413447
```

Also, for other continuous distributions, we obtain using `d` with the R-name of a distribution, the value of the density function and, using `p` with the R-name of a distribution, the value of the distribution function. We demonstrate this in the next problem for the lognormal distribution.

Problem 1.3 Calculate the value of the density function of the lognormal distribution whose logarithm has mean equal to `meanlog = 0` and standard deviation equal to `sdlog = 1` for a given value z.

Solution
Use the command `> dlnorm(z, meanlog = 0, sdlog = 1)` or use the default values `meanlog = 0` and `sdlog = 1` using `> dlnorm(z)`.

Example
We calculate the value of the density function of the lognormal distribution with `meanlog = 0` and `sdlog = 1` using

```
> dlnorm(1)
[1] 0.3989423
```

Problem 1.4 Calculate the value of the distribution function of the lognormal distribution whose logarithm has mean equal to `meanlog = 0` and standard deviation equal to `sdlog = 1` for a given value z.

Solution
Use the command `> plnorm(z, meanlog = 0, sdlog = 1)` or use the default values `meanlog = 0` and `sdlog = 1` using `> plnorm(z)`.

Example
We calculate the value of the distribution function for $z = 1$ of the lognormal distribution with `meanlog = 0` and `sdlog = 1` using

```
> plnorm(1)
[1] 0.5
```

From most of the other distributions we need the quantiles (or percentiles) $q_P = P(y \le P)$.
This can be done by writing q followed by the R-name of the distribution.

Problem 1.5 Calculate the P%-quantile of the t-distribution with df degrees of freedom and optional non-centrality parameter ncp.

Solution
Use the command `> qt(P,df, ncp)` and for a central t-distribution use the default by omitting `ncp`.

Example
Calculate the 95%-quantile of the central t-distribution with 10 degrees of freedom.

```
> qt(0.95,10)
[1] 1.812461
```

We demonstrate the procedure for the chi-square and the F-distribution.

Problem 1.6 Calculate the P%-quantile of the χ^2-distribution with df degrees of freedom and optional non-centrality parameter ncp.

Solution
Use the command > `qchisq(P,df, ncp)` and for the central χ^2-distribution with df degrees of freedom use > `qchisq(P,df)`.

Example
Calculate the 95%-quantile of the central χ^2-distribution with 10 degrees of freedom.

```
> qchisq(0.95,10)
[1] 18.30704
```

Problem 1.7 Calculate the P%-quantile of the F-distribution with df1 and df2 degrees of freedom and optional non-centrality parameter ncp.

Solution
Use the command > `qf(P,df1,df2, ncp)`, and for the central F-distribution with df1 and df2 degrees of freedom use > `qf(P,df1,df2)`.

Example
Calculate the 95%-quantile of the *central* F-distribution with 10 and 20 degrees of freedom!

```
> qf(0.95,10,20)
[1] 2.347878
```

For the calculation of further values of probability functions of discrete random variables or of distribution functions and quantiles the commands can be found by using the help function in the tool bar of R, and then you may call up the 'manual' or use Crawley (2013).

1.3 Sampling Procedures and Random Variables

Even if we, in this book, we mainly discuss how to plan experiments and to analyse observed data, we still need basic knowledge about random variables because, without this, we could not explain unbiased estimators or the expected length of a confidence interval or how to define the risks of a statistical tests.

Definition 1.1 A sampling procedure without replacement (wor) or with replacement (wr) is a rule of selecting a proper subset, named sample, from a well-defined finite basic set of objects (population, universe). It is said to be at random if each element of the basic set has the same probability p to be drawn into the sample. We also can say that in a random sampling procedure each possible sample has the same probability to be drawn.

A (concrete) sample is the result of a sampling procedure. Samples resulting from a random sampling procedure are said to be (concrete) random samples or shortly samples.

If we consider all possible samples from a given finite universe, then, from this definition, it follows that each possible sample has the same probability to be drawn.

There are several random sampling procedures that can be used in practice. Basic sets of objects are mostly called (statistical) populations or, synonymously, (statistical) universes.

Concerning random sampling procedures, we distinguish (among other cases):

- Simple (or pure) random sampling with replacement (wr) where each of the N elements of the population is selected with probability $\frac{1}{N}$.
- Simple random sampling without replacement (wor) where each unordered sample of n different objects has the same probability to be chosen.
- In cluster sampling, the population is divided into disjoint subclasses (clusters). Random sampling without replacement is done among these clusters. In the selected clusters, all objects are taken into the sample. This kind of selection is often used in area sampling. It is only random corresponding to Definition 1.1 if the clusters contain the same number of objects.
- In multi-stage sampling, sampling is done in several steps. We restrict ourselves to two stages of sampling where the population is decomposed into disjoint subsets (primary units). Part of the primary units is sampled randomly without replacement (wor) and within them pure random sampling without replacement (wor) is done with the secondary units. A multi-stage sampling is favourable if the population has a hierarchical structure (e.g. country, province, towns in the province). It is at random corresponding to Definition 1.1 if the primary units contain the same number of secondary units.
- Sequential sampling, where the sample size is not fixed at the beginning of the sampling procedure. At first, a small sample with replacement is taken and analysed. Then it is decided whether the obtained information is sufficient, e.g. to reject or to accept a given hypothesis (see Chapter 3), or if more information is needed by selecting a further unit.

When a cluster or in two-stage sampling the clusters or primary units have different sizes (number of elements or areas), more sophisticated methods are used (Rasch et al. 2008, Methods 1/31/2110, 1/31/3100).

Both a random sampling (procedure) and arbitrary sampling (procedure) can result in the same concrete sample. Hence, we cannot prove by inspecting the concrete sample itself whether or not the sample is randomly chosen. We have to check the sampling procedure used instead.

In mathematical statistics random sampling with a replacement procedure is modelled by a vector $\boldsymbol{Y} = (\boldsymbol{y}_1, \boldsymbol{y}_2, \dots, \boldsymbol{y}_n)^T$ of random variables $\boldsymbol{y}_i$, $i = 1, \dots, n$, which are independently distributed as a random variable $\boldsymbol{y}$, i.e. they all have the same distribution. The $\boldsymbol{y}_i$, $i = 1, \dots, n$ are said independently and identically distributed (i.i.d.). This leads to the following definition.

Definition 1.2 A random sample of size n is a vector $\boldsymbol{Y} = (\boldsymbol{y}_1, \boldsymbol{y}_2, \dots, \boldsymbol{y}_n)^T$ with n i.i.d. random variables $\boldsymbol{y}_i$, $i = 1, \dots, n$ as elements.

Random variables given in bold print (see Appendix A for motivation).

The vector $Y = (y_1, y_2, \dots, y_n)^T$ is called a realisation of $\boldsymbol{Y} = (\boldsymbol{y}_1, \boldsymbol{y}_2, \dots, \boldsymbol{y}_n)^T$ and is used as a model of a vector of observed values or values selected by a random selection procedure.

To explain this approach let us assume that we have a universe of 100 elements (the numbers 1–100). We like to draw a pure random sample without replacement (wor) of size $n = 10$ from this universe and model this by $\boldsymbol{Y} = (\boldsymbol{y}_1, \boldsymbol{y}_2, \ldots, \boldsymbol{y}_{10})^T$. When a random sample has been drawn it could be the vector $Y = (y_1, y_2, \ldots, y_{10})^T = (3, 98, 12, 37, 2, 67, 33, 21, 9, 56)^T = (2, 3, 9, 12, 21, 33, 37, 56, 67, 98)^T$. This means that it is only important which element has been selected and not at which place this has happened. All samples wor occur with probability $\frac{1}{\binom{100}{10}}$. The denominator $\binom{100}{10}$ can be calculated by R with the

> choose() command

```
> choose(100,10)
[1] 1.731031e+13
```

and from this the probability is $\frac{1}{1731031\times10^7}$.

We can now write

$$P\{(\boldsymbol{y}_1, \boldsymbol{y}_2, \ldots, \boldsymbol{y}_{10})^T = (2, 3, 9, 12, 21, 33, 37, 56, 67, 98)^T\} = \frac{1}{1731031 \times 10^7}.$$

In a probability statement, something must always be random. To write

$$P\{(y_1, y_2, \ldots, y_{10})^T = (2, 3, 9, 12, 21, 33, 37, 56, 67, 98)^T\}$$

is nonsense because $(y_1, y_2, \ldots, y_{10})^T$ as the vector on the right-hand-side is a vector of special numbers and it is nonsense to ask for the probability that 5 equals 7.

To explain the situation again we consider the problem of throwing a fair dice; this is a dice where we know that each of the numbers 1, …, 6 occurs with the same probability $\frac{1}{6}$. We ask for the probability that an even number is thrown. Because one half of the six numbers are even, this probability is $\frac{1}{2}$. Assume we throw the dice using a dice cup and let the result be hidden, than the probability is still $\frac{1}{2}$. However, if we take the dice cup away, a realisation occurs, let us say a 5. Now, it is stupid to ask, what is the probability that 5 is even or that an even number is even. Probability statements about realisations of random variables are senseless and not allowed. The reader of this book should only look at a probability statement in the form of a formula if something is in bold print; only in such a case is a probability statement possible.

We learn in Chapter 4 what a confidence interval is. It is defined as an interval with at least one random boundary and we can, for example, calculate with some small α the probability $1 - \alpha$ that the expectation of some random variable is covered by this interval. However, when we have realised boundaries, then the interval is fixed and it either covers or does not cover the expectation. In applied statistics, we work with observed data modelled by realised random variables. Then the calculated interval does not allow a probability statement. We do not know, by using R or otherwise, whether the calculated interval covers the expectation or not. Why did we fix this probability before starting the experiment when we cannot use it in interpreting the result?

The answer is not easy, but we will try to give some reasons. If a researcher has to carry out many similar experiments and in each of them calculates for some parameter a $(1 - \alpha)$ confidence interval, then he can say that in about $(1 - \alpha)100\%$ of all cases the interval has covered the parameter, but of course he does not know when this happened.

What should we do when only one experiment has to be done? Then we should choose $(1 - \alpha)$ so large (say 0.95 or 0.99) that we can take the risk of making an erroneous statement by saying that the interval covers the parameter. This is analogous to the situation of a person who has a severe disease and needs an operation in hospital. The person can choose between two hospitals and knows that in hospital A about 99% of people operated on survived a similar operation and in hospital B only about 80%. Of course (without further information) the person chooses A even without knowing whether she/he will survive. As in normal life, also in science; we have to take risks and to make decisions under uncertainty.

We now show how R can easily solve simple problems of sampling.

Problem 1.8 Draw a pure random sample without replacement of size $n < N$ from N given objects represented by numbers 1, …, N without replacing the drawn objects. There are $M = \binom{N}{n}$ possible unordered samples having the same probability $p = \frac{1}{M}$ to be selected.

Solution
Insert in R a data file y with N entries and continue in the next line with `>sample (y,n, replace = FALSE)` or `>sample (y,n, replace = F)` with $n < N$ to create a sample of size $n < N$ different elements from y; when we insert `replace = TRUE` we get random sampling with replacement. The default is `replace = FALSE`, hence for sampling without replacement we can use `>sample (y, n)`.

Example
We choose $N = 9$, and $n = 5$, with population values $y = (1,2,3,4,5,6,7,8,9)$

```
> y <- c(1,2,3,4,5,6,7,8,9)
> sample(y,5)
[1] 7 6 5 1 3
```

A pure random sampling with replacement also occurs if the random sample is obtained by replacing the objects immediately after drawing and each object has the same probability of coming into the sample using this procedure. Hence, the population always has the same number of objects before a new object is taken. This is only possible if the observation of objects works without destroying or changing them (examples are tensile breaking tests, medical examinations of killed animals, felling of trees, harvesting of food).

Problem 1.9 Draw with replacement a pure random sample of size n from N given objects represented by numbers 1, …, N with replacing the drawn objects. There are $M_{\text{rep}} = \binom{N+n-1}{n}$ possible unordered samples having the same probability $\frac{1}{M_{\text{rep}}}$ to be selected.

Solution
Insert in R a data file y with N entries and continue in the next line with `>sample (y, n, replace =TRUE)` or `>sample(y, n, replace=T)` to create a sample of size n not necessarily with different elements from y.

Examples
Example with $n < N$

```
> y<-c(1,2,3,4,5,6,7,8,9)
> sample(y,5,replace=T)
[1] 2 4 6 4 2
```

Example with $n > N$

```
> y<-c(1,2,3,4,5,6,7,8,9)
> sample(y,10,replace=T)
[1] 3 9 5 5 9 9 8 7 6 3
```

A method that can sometimes be realised more easily is systematic sampling with a random start. It is applicable if the objects of the finite sampling set are numbered from 1 to N, and the sequence is not related to the character considered. If the quotient $m = N/n$ is a natural number, a value i between 1 and m is chosen at random, and the sample is collected from objects with numbers $i, m + i, 2m + i, \dots, (n - 1)m + i$. Detailed information about this case and the case where the quotient m is not an integer can be found in Rasch et al. (2008, method 1/31/1210).

Problem 1.10 From a set of N objects systematic sampling with a random start should choose a random sample of size n.

Solution
We assume that in the sequence 1, 2, …, N there is no trend. Let assume that $m = \frac{N}{n}$ is an integer and select by pure random sampling a value $1 \le x \le m$ (sample of size 1) from the m numbers 1, …, m. Then the systematic sample with random start contains the numbers $x, x + m, x + 2m, \dots, x + (n - 1)m$.

Example
We choose $N = 500$ and $n = 20$, and the quotient $\frac{500}{20} = 25$ is an integer-valued. Analogous to Problem 1.1 we draw a random sample of size 1 from (1, 2, …, 25) using R.

```
> y<- c(1,2,3,4,5,6,7,8,9,10,11,12,13,14,15,
  16,17,18,19,20,21,22,23,24,25)
> sample(y,1)
[1] 9
```

The final systematic sample with random start of size $n = 20$ starts with number $x = 9$ and $m = 25$: (9, 34, 59, 84, 109, 134, 159, 184, 209, 234, 259, 284, 309, 334, 359, 384, 409, 434, 459, 484).

Problem 1.11 By cluster sampling, from a population of size N decomposed into s disjoint subpopulations, so-called clusters of sizes $N_1, N_2, \dots, N_s$, a random sample has to be drawn.

Solution
Partial samples of size n_i are collected from the ith stratum ($i = 1, 2, \dots, s$) where pure random sampling procedures without replacement are used in each stratum. This leads to a random sampling without replacement procedure for the whole population if the numbers n_i/n are chosen proportional to the numbers N_i/N. The final random sample contains $n = \sum_{i=1}^{s} n_i$ elements.

Example
Vienna, the capital of Austria, is subdivided into 23 municipalities. We repeat a table with the numbers of inhabitants N_i^* in these municipalities from Rasch et al. (2011) and round the numbers for demonstrating the example to values so that N_i/N is an integer, where $N = 1\,700\,000$.

Now we select by pure random sampling without replacement, as shown in Problem 1.8, from each municipality n_i from the N_i inhabitants to reach a total random sample of 1000 inhabitants from the 1 700 000 people in Vienna.

While for the stratified random sampling objects are selected without replacement from each subset, for two-stage sampling, subsets or objects are selected at random without replacement at each stage, as described below. Let the population consist of s disjoint subsets of size N_0, the primary units, in the two-stage case. Further, we suppose that the character values in the single primary units differ only at random, so that objects need not to be selected from all primary units. If the desired sample size is $n = r\,n_0$ with $r < s$, then in the first step, r of the s given primary units are selected using a pure random sampling procedure. In the second step, n_0 objects (secondary units) are chosen from each selected primary unit, again applying a pure random sampling. The number of possible samples is $\binom{s}{r} \cdot \binom{N_0}{n_0}$, and each object of the population has the same probability $p = \frac{r}{s} \cdot \frac{n_0}{N_0}$ to reach the sample corresponding to Definition 1.1.

Problem 1.12 Draw a random sample of size n in a two-stage procedure by selecting first from the s primary units having sizes N_{i} $(i = 1, \ldots, \mathrm{s})$ exactly r units.

Solution
To draw a random sample without replacement of size n we select a divisor r of n and from the s primary units we randomly select r proportional to the relative sizes $\frac{N_i}{N}$ with $N = \sum_{i=1}^{s} N_i$ $(\mathrm{i} = 1, \ldots, s)$. From each of the selected r primary units we select by pure random sampling without replacement $\frac{n}{r}$ elements as the total sample of secondary units.

Example
We take again the values of Table 1.1 and select $r = 5$ from the $s = 23$ municipalities to take an overall sample of $n = 1000$. For this we split the interval (0,1] into 23 subintervals $\left(1000\frac{N_{i-1}}{N}, 1000\frac{N_i}{N}\right]$ $i = 1, \ldots, 23$ with $N_0 = 0$ and generate five uniformly distributed random numbers in (0,1]. If a random number multiplied by 1000 falls in any of the 23 sub-intervals (which can be easily found by using the 'cum' column in Table 1.1) the corresponding municipality has to be selected. If a further random number falls into the same interval it is replaced by another uniformly distributed random number. We generate five such random numbers as follows:

```
> runif(5)
 [1] 0.18769112 0.78229430 0.09359499 0.46677904 0.51150546
```

The first number corresponds to Mariahilf, the second to Florisdorf, the third to Landstraße, the fourth to Hietzing, and the last one to Penzing. To obtain a random sample of size 1000 we take pure random samples of size 200 from people in Mariahilf, Florisdorf, Landstraße, Hietzing, and Penzing, respectively.

Table 1.1 Number $N_i^*, i = 1, \ldots, 23$ of inhabitants in 23 municipalities of Vienna.

Municipality	N_i^*	N_i	$n_i = 1000\frac{N_i}{N}$	cum
Innere Stadt	16 958	17 000	10	10
Leopoldstadt	94 595	102 000	60	70
Landstraße	83 737	85 000	50	120
Wieden	30 587	34 000	20	140
Margarethen	52 548	51 000	30	170
Mariahilf	29 371	34 000	20	190
Neubau	30 056	34 000	20	210
Josefstadt	23 912	34 000	20	230
Alsergrund	39 422	34 000	20	250
Favoriten	173 623	170 000	100	350
Simmering	88 102	85 000	50	400
Meidling	87 285	85 000	50	450
Hietzing	51 147	51 000	30	480
Penzing	84 187	85 000	50	530
Rudolfsheim	70 902	68 000	40	570
Ottakring	94 735	102 000	60	630
Hernals	52 701	51 000	30	660
Währing	47 861	51 000	30	690
Döbling	68 277	68 000	40	730
Brigittenau	82 369	85 000	50	780
Floridsdorf	139 729	136 000	80	860
Donaustadt	153 408	153 000	90	950
Liesing	91 759	85 000	50	1 000
Total	N^*=1 687 271	N= 1 700 000	n= 1 000	

Rounded numbers N_i, n_i, and cumulated n_i.
Source: From Statistik Austria (2009) Bevölkerungsstand inclusive Revision seit 1.1. 2002, Wien, Statistik Austria.

References

Crawley, M.J. (2013). *The* R *Book*, 2nd edition, Chichester: Wiley.

Rasch, D. and Schott, D. (2018). *Mathematical Statistics*. Oxford: Wiley.

Rasch, D., Herrendörfer, G., Bock, J., Victor, N., and Guiard, V. (2008). *Verfahrensbibliothek Versuchsplanung und - auswertung*, 2. verbesserte Auflage in einem Band mit CD. R. Oldenbourg Verlag München Wien.

Rasch, D., Pilz, J., Verdooren, R., and Gebhardt, A. (2011). *Optimal Experimental Design with* R. Boca Raton: Chapman and Hall.

2

Point Estimation

2.1 Introduction

The theory of point estimation is described in most books about mathematical statistics, and we refer here, as in other chapters, mainly to Rasch and Schott (2018).

We describe the problem as follows. Let the distribution P_θ of a random variable $\boldsymbol{y}$ depend on a parameter (vector) $\theta \in \Omega \subseteq R^p,\ p \geq 1$. With the help of a realisation, Y, of a random sample $\boldsymbol{Y} = (\boldsymbol{y}_1, \boldsymbol{y}_2, \ldots, \boldsymbol{y}_n)^T,\ n \geq 1$ we have to make a statement concerning the value of θ (or a function of it). The elements of a random sample $\boldsymbol{Y}$ are independently and identically distributed (i.i.d) like $\boldsymbol{y}$. Obviously the statement about θ should be as precise as possible. What this really means depends on the choice of the loss function defined in section 1.4 in Rasch and Schott (2018). We define an estimator $S(\boldsymbol{Y})$, i.e. a measurable mapping of R^n onto Ω taking the value $S(Y)$ for the realisation $Y=(y_1, y_2, \ldots, y_n)^T$ of $\boldsymbol{Y}$, where $S(Y)$ is called the estimate of θ. The estimate is thus the realisation of the estimator. In this chapter, data are assumed to be realisations $(y_1, y_2, \ldots, y_n)$ of one random sample where n is called the sample size; the case of more than one sample is discussed in the following chapters. The random sample, i.e. the random variable $\boldsymbol{y}$ stems from some distribution, which is described when the method of estimation depends on the distribution – like in the maximum likelihood estimation. For this distribution the rth central moment

$$\mu_r = E[(\boldsymbol{y} - \mu)^r] \tag{2.1}$$

is assumed to exist where $\mu = E(\boldsymbol{y})$ is the expectation and $\sigma^2 = E[(\boldsymbol{y} - \mu)^2]$ is the variance of $\boldsymbol{y}$. The rth central sample moment m_r is defined as

$$m_r = \frac{\sum_{i=1}^{n} (y_i - \bar{y})^r}{n} \tag{2.2}$$

with

$$\bar{y} = \frac{\sum_{i=1}^{n} y_i}{n}. \tag{2.3}$$

An estimator $S(\boldsymbol{Y})$ based on a random sample $\boldsymbol{Y} = (\boldsymbol{y}_1, \boldsymbol{y}_2, \ldots, \boldsymbol{y}_n)^T$ of size $n \geq 1$ is said to be unbiased with respect to θ if

$$E[S(\boldsymbol{Y})] = \theta \tag{2.4}$$

holds for all $\theta \in \Omega$.

The difference $b_n(\theta) = E[S(\boldsymbol{Y})] - \theta$ is called the bias of the estimator $S(\boldsymbol{Y})$.

Applied Statistics: Theory and Problem Solutions with R, First Edition.
Dieter Rasch, Rob Verdooren, and Jürgen Pilz.

We show here how R can easily calculate estimates of location and scale parameters as well as higher moments from a data set. We at first create a simple data set y in R. The following values are weights in kilograms and therefore non-negative.

```
> y <- c(5,7,1,7,8,9,13,9,10,10,18,10,15,10,10,11,8,11,12,13,15,
     22,10,25,11)
```

If we consider y as a sample, the sample size n can with R be determined via

```
> length(y)
  [1] 25
```

i.e. $n = 25$. We start with estimating the parameters of location.

In Sections 2.2, 2.3, and 2.4 we assume that we observe measurements in an interval scale or ratio scale; if they are in an ordinal or nominal scale we use the methods described in Section 2.5.

2.2 Estimating Location Parameters

When we estimate any parameter we assume that it exists, so speaking about expectations, skewness $\gamma_1 = \mu_3/\sigma^3$, kurtosis $\gamma_2 = [\mu_4/\sigma^4] - 3$ and so on we assume that the corresponding moments in the underlying distribution exist.

The arithmetic mean, or briefly, the mean

$$\bar{y} = \frac{1}{n}\sum_{i=1}^{n} y_i \tag{2.5}$$

is an estimate of the expectation μ of some distribution.

Problem 2.1 Calculate the arithmetic mean of a sample.

Solution

Use the command > `mean()`.

```
> mean(y)
```

Example

We use the sample Y already defined above and obtain

```
> y<- c(5,7,1,7,8,9,13,9,10,10,18,10,15,10,10,11,8,11,12,13,15,22,
    10,25,11)
> mean(y)
[1] 11.2
```

i.e. $\bar{y} = \frac{1}{25}\sum_{i=1}^{25} y_i = 11.2$.

The arithmetic mean is a least squares estimate of the expectation μ of $\boldsymbol{y}$.

The corresponding least squares estimator is $\bar{\boldsymbol{y}} = \frac{1}{n}\sum_{i=1}^{n} \boldsymbol{y}_i$ and is unbiased.

Problem 2.2 Calculate the extreme values $y_{(1)} = \min(y)$ and $y_{(n)} = \max(y)$ of a sample.

Solution
We receive the extreme values using the R commands `>min()` and `>max()`.

Example
Again, we use the sample y defined above and obtain

```
> min(y)
[1] 1
> max(y)
[1] 25
```

i.e. $y_{(1)} = 1$ and $y_{(25)} = 25$ if we denote the jth element of the ordered set of Y by $y_{(j)}$ such that $y_{(1)} \leq \ldots \leq y_{(n)}$ holds. Note: you can get both values using the command `> range(y)`.

Sometimes one or more elements of $\boldsymbol{Y} = (\boldsymbol{y}_1, \boldsymbol{y}_2, \ldots, \boldsymbol{y}_n)^T$ do not have the same distribution as the others and $\boldsymbol{Y} = (\boldsymbol{y}_1, \boldsymbol{y}_2, \ldots, \boldsymbol{y}_n)^T$ is not a random sample.

If only a few of the elements of $\boldsymbol{Y}$ have a different distribution we call them outliers. Often the minimum and the maximum values of y represent realisations of such outliers. If we conjecture the existence of such outliers we can use special L-estimators as the trimmed or the Winsorised mean. Outliers in observed values can occur even if the corresponding element of $\boldsymbol{Y}$ is not an outlier. This can happen by incorrectly writing down an observed number or by an error in the measuring instrument.

L-estimators are weighted means of order statistics (where L stands for linear combination). If we arrange the elements of the realisation Y of $\boldsymbol{Y}$ according to their magnitude, and if we denote the jth element of this ordered set by $y_{(j)}$ such that $y_{(1)} \leq \ldots \leq y_{(n)}$ holds, then

$$Y_{(.)} = (y_{(1)}, \ldots, y_{(n)})^T$$

is a function of the realisation of $\boldsymbol{Y}$, and $S(\boldsymbol{Y}) = \boldsymbol{Y}_{(.)} = (\boldsymbol{y}_{(1)}, \ldots, \boldsymbol{y}_{(n)})^T$ is said to be the order statistic vector, the component $\boldsymbol{y}_{(i)}$ is called the ith order statistic and

$$L(\boldsymbol{Y}) = \sum_{i=1}^{n} c_i \boldsymbol{y}_{(i)}; \quad c_i \geq 0, \quad \sum_{i=1}^{n} c_i = 1 \tag{2.6}$$

is said to be an L-estimator and $\sum_{i=1}^{n} c_i y_{(i)}$ is called an L-estimate.

If we put

$$c_1 = \cdots = c_t = c_{n-t+1} = \cdots = c_n = 0 \quad \text{and} \quad c_{t+1} = \cdots = c_{n-t} = \frac{1}{n-2t}$$

in (2.6) with $t < \frac{n}{2}$, then

$$L_{\mathrm{T}}(\boldsymbol{Y}) = \frac{1}{n-2t} \sum_{i=t+1}^{n-t} \boldsymbol{y}_{(i)} \tag{2.7}$$

is called the $\frac{t}{n}$ – trimmed mean.

If we do not suppress the t smallest and the t largest observations, but concentrate them in the values $y_{(t+1)}$ and $y_{(n-t)}$, respectively, then we get the so-called $\frac{t}{n}$ Winsorised mean

$$L_W(Y) = \frac{1}{n}\left[\sum_{i=t+1}^{n-t} y_{(i)} + ty_{(t+1)} + ty_{(n-t)}\right] \tag{2.8}$$

$$c_1 = \cdots = c_t = c_{n-t+1} = \cdots = c_n = 0 \quad \text{and} \quad c_{t+1} = \cdots = c_{n-t} = \frac{1}{n}.$$

The median in samples of even size $n = 2m$ can be defined as the 1/2 Winsorised mean

$$L_W(Y) = \frac{1}{2}(y_{(t+1)} + y_{(n-t)}). \tag{2.9}$$

To calculate the trimmed and Winsorised means using R we first order the samples of n observations by magnitude.

Problem 2.3 Order a vector of numbers by magnitude.

Solution
Use the vector y of numbers and the command `>sort()`.

```
> sortedy <- sort(y)
```

Example
We again use the sample

```
> y<- c(5,7,1,7,8,9, 13,9,10,10, 18,10, 15, 10,10, 11, 8,
      11,12,13, 15, 22, 10,25, 11)
```

and obtain

```
> sortedy <- sort(y)
> sortedy
 [1]  1  5  7  7  8  8  9  9 10 10 10 10 10 10 11 11 11 12 13 13 15 15 18
    22 25
```

Problem 2.4 Calculate the $\frac{1}{n}$ trimmed mean of a sample.

Solution
We at first order the sample Y using the command *sort*, as shown in Problem 2.3. Then we drop the smallest and the largest entry in y and denote the result as x. With `> mean(x)` we obtain the $\frac{1}{n}$ trimmed mean of a sample Y.

Example
We use `sortedy`, the ordered sample y from Problem 2.3 of the 25 observations.

```
[1]  1  5  7  7  8  8  9  9 10 10 10 10 10 10 11 11 11 12 13 13 15 15 18
 22 25
```

and drop manually the smallest and the largest entry and call the result x.

```
x<- c(5,7,7,8,8,9,9,10,10,10,10,10,10,11,11,11,12,13,13,15,15,18,22)
```

However, this can be done directly with R as follows

```
> x< - sortedy[-1]
> x <- x[-24]
> x
[1]  5  7  7  8  8  9  9 10 10 10 10 10 10 11 11 11 12 13 13 15 15 18 22
> length(x)
[1] 23
```

Then we calculate the mean of the entries in x.

```
> mean(x)
[1] 11.04348
```

and by rounding we obtain $L_T(Y) = \frac{1}{23}\sum_{i=2}^{24} y_{(i)} = 11.04$.

This is the $\frac{1}{25}$ – trimmed mean of y.

Note: you can directly find the trimmed mean using the command > `mean(y, trim=1/25)`.

Problem 2.5 Calculate the $\frac{1}{n}$ Winsorised mean of a sample of size n.

Solution

We at first order the sample Y using the command *sort*, as shown in Problem 2.3. Then we set $y_{(1)} = y_{(2)}$ and $y_{(n-1)} = y_{(n)}$ and call the result z.

Example

We calculate the $\frac{1}{25}$ Winsorised mean of y in

```
y<- c(5,7,1,7,8,9, 13,9,10,10, 18,10, 15, 10,
     10, 11, 8,11,12,13, 15, 22, 10,25, 11).
```

We at first calculate using sort the ordered sample

```
> sortedy <- sort(y)
 1  5  7  7  8  8  9  9 10 10 10 10 10 10 11 11 11 12 13 13 15 15 18 22 25
```

and shift manually 1 to 5 and 22 to 25. The result is

```
z<- c(5,5,7,7,8,8,9,10,10,10,10,10,10,11,11,11,12,13,13, 15,15,18,25,25)
```

Of course this can be done directly in R using

```
> sortedy[1] <- 5
> sortedy[24] <- 25
> z <- sortedy
> z
 [1]  5  5  7  7  8  8  9  9 10 10 10 10 10 10 11 11 11 12  13 13
  15 15 18 25 25
```

We get the $\frac{1}{25}$ Winsorised mean via

```
> mean(z)
[1] 11.48
```

or by rounding as $L_W(Y) = \frac{1}{25}\left[\sum_{i=2}^{24} y_{(i)} + y_{(2)} + y_{(24)}\right] = 11.5$.

Problem 2.6 Calculate the median of a sample.

Solution
We receive the median of a sample Y using R via

```
> median(y)
```

Example
We use the sample

```
y<- c(5,7,1,7,8,9, 13,9,10,10, 18,10, 15, 10,10, 11, 8,11,12,13, 15, 22, 10,25, 11)
```

and obtain

```
> median(y)
[1] 10
```

Further location measures are the quantiles of the empirical distribution of the sample Y. We denote by $q(P,Y)$ the P-quantile of Y. The P-quantile is the value in the range of y so that $100P\%$ of the values of Y are below and $100(1-P)\%$ are above P. The most important quantiles are the quartiles, which are three numbers $q(0.25, y) = Q_1(y)$, $q(0.50, y) = Q_2(y) = \text{median}(y)$ and $q(0.75, y) = Q_3(y)$ that divide the data set into four equal groups, each group comprising a quarter of the data.

Problem 2.7 Calculate the first and the third quartile of a sample.

Solution
Using the command >`summary( )` we get the minimum and the maximum, the first and the third quartile, and the median and the mean of the sample.

Example
We use again the sample y above and get

```
> summary(y)
   Min. 1st Qu.  Median    Mean 3rd Qu.    Max.
    1.0     9.0    10.0    11.2    13.0    25.0
```

The first and the third quartile are $Q_1(y) = 9$ and $Q_3(y) = 13$ respectively.

If the observed numbers are ratios it often is better to use the geometric mean in place of the arithmetic mean. The geometric mean G of a vector $Y=(y_1, y_2, \dots, y_n)^T$, $y_i > 0$ for $i = 1, \dots, n$ is defined as

$$G = \sqrt[n]{\prod_{i=1}^{n} y_i} = \left(\prod_{i=1}^{n} y_i\right)^{\frac{1}{n}}. \tag{2.10}$$

The geometric mean is less than the corresponding arithmetic mean if at least two of the elements of Y are different.

Problem 2.8 Calculate the geometric mean of a sample.

Solution

We rewrite (2.10) as $\ln G = \frac{1}{n}\sum_{i=1}^{n}\ln(y_i)$ and get G via

```
> exp(mean(log(y)))
```

Example

Using the 25 data sets in y of Problem 2.1, calculating

```
> exp(mean(log(y)))
[1] 9.892722
```

and rounding we get $G = 9.89$.

Another mean of $Y=(y_1, y_2, \dots, y_n)^T$ is the harmonic mean H.

The harmonic mean H of a vector $Y=(y_1, y_2, \dots, y_n)^T$, $y_i > 0$ is defined as

$$H = \frac{n}{\sum_{i=1}^{n}\frac{1}{y_i}}. \tag{2.11}$$

Problem 2.9 Calculate the harmonic mean of a sample.

Solution

We get H via

```
> length(y)/sum(1/y)
```

Example

Using the 25 data sets in y of Problem 2.1, calculating

```
> length(y)/sum(1/y)
[1] 7.480133
```

and rounding we get $H = 7.48$.

It can be shown that $H \le G \le \bar{y}$, as we can see from Problems 2.1, 2.8, and 2.9, where we got $7.48 < 9.89 < 11.2$.

2.2.1 Maximum Likelihood Estimation of Location Parameters

We now show how for location parameters of non-normal distributions maximum likelihood estimates are calculated. We start with the lognormal distribution.

Definition 2.1 A random variable $\boldsymbol{x}$ is lognormally distributed if its density function equals

$$f(x, \mu, \sigma) = \begin{cases} \frac{1}{\sigma x\sqrt{2\pi}} e^{-\frac{(\ln x-\mu)^2}{2\sigma^2}}, & x > 0 \\ 0, & x \le 0 \end{cases}, \quad \mu \in R, \sigma \in R^+. \tag{2.12}$$

Its expectation equals $\mu_x = e^{\mu+\frac{\sigma^2}{2}}$.

Problem 2.10 Calculate from n observations $(x_1, x_2, \dots, x_n)$ of a lognormal distributed random variable the maximum-likelihood (ML) estimate of the expectation $\mu_x = e^{\mu+\frac{\sigma^2}{2}}$.

Solution

Calculate the arithmetic mean $\widetilde{\mu} = \frac{1}{n}\sum_{i=1}^{n} \ln x_i$ of $\ln x_i$; it is the maximum likelihood estimate and a realisation of an unbiased estimator.

Example

Measures of the blood pressure of 15 male persons of ages between 25 and 40 are as follows: 132, 156, 128, 122, 130, 115, 123, 125, 128, 129, 132, 124, 127, 122, 141.

In R we write

```
> b<- c(132, 156, 128, 122, 130, 115, 123, 125, 128, 129, 132, 124, 127,
     122, 141)
```

and then

```
> mean(log(b))
[1] 4.856905
```

and $\widetilde{\mu} \approx 4.86$.

The binomial distribution with parameter p of an event (success) and n i.i.d random variables has the probability (likelihood) function

$$L(Y,p) = p^y(1-p)^{n-y}, \quad y = 0, 1, \dots, n; 0 < p < 1 \tag{2.13}$$

where y is the number of successes in n independent trials with probability p of a success. In the notation so far, we consider the random sample $\boldsymbol{Y} = (\boldsymbol{y}_1, \boldsymbol{y}_2, \dots, \boldsymbol{y}_n)^T$ as one, where $\boldsymbol{y}_i$ takes the value 1 with probability p and the value 0 with probability $1-p$.

By setting the derivative

$$\frac{\partial \ln L(Y,p)}{\partial p} = \frac{y}{p} - \frac{n-y}{1-p}$$

equal to 0 we get the solution $p = \frac{y}{n}$, which supplies a maximum of L, as the second derivative of $\ln L$ relative to p is negative. Therefore, the uniquely determined ML estimate is the relative frequency of successes.

$$\frac{y}{n} = \widehat{p}. \tag{2.14}$$

Problem 2.11 Estimate the parameter p of a binomial distribution.

Solution

We observe a sample of size n with zeros (for no success) and 1's (for success) and calculate its arithmetic mean.

Example

In a sample of 20 observations we kept the vector (0,0,1,0,1,1,0,1,0,0,1,0,0,0,1,0, 0,1,0,0).

Using R we obtain $\hat{p}$ via

```
> w<- c(0,0,1,0,1,1,0,1,0,0,1,0,0,0,1,0,0,1,0,0)
> mean(w)
[1] 0.35
```

Or by counting the numbers y of 1's in the observed sample and dividing this by $n = 20$, this also gives $\frac{7}{20} = \hat{p} = 0.35$.

In R we get the counting of 0 and 1 in w using the command:

```
> table(w)
w
 0  1
13  7
```

and the total number n using the command

```
> length(w)
[1] 20
```

Hence the calculation in R gives $\hat{p}$

```
> 7/20
[1] 0.35
```

Problem 2.12 Estimate the parameter λ of a Poisson distribution.

Solution

The Poisson distribution with a random sample $\boldsymbol{Y} = (\boldsymbol{y}_{1,\ldots,}\boldsymbol{y}_{\boldsymbol{n}})^T$ has the likelihood function

$$L(Y, \lambda) = \prod_{i=1}^{n} \frac{1}{y_i!} e^{[\ln \lambda \sum_{i=1}^{n} y_i - \lambda n]}, \quad y_i \in \{0, 1, 2, \ldots\}; \quad \lambda \epsilon R^+. \tag{2.15}$$

By setting the derivative

$$\frac{\partial \ln L(Y,p)}{\partial \lambda} = \frac{1}{\lambda} \sum_{i=1}^{n} y_i - n$$

equal to 0 we get the solution

$$\hat{\lambda} = \frac{\sum_{i=1}^{n} y_i}{n} \tag{2.16}$$

which supplies a maximum of L, as the second derivative of ln L relative to λ is negative.

Table 2.1 Number of noxious weed seeds.

Number of noxious seeds	0	1	2	3	4	5	6	7	8	9	10 or more
Observed frequency *f*	3	17	26	16	18	9	3	5	0	1	0

Example

The number of noxious weed seeds in 98 subsamples of *Phleum pratense* (meadow grass) is given in the frequency Table 2.1.

In R we do the calculation of $\hat{\lambda} = \frac{\sum_{i=1}^{n} y_i}{n}$ as follows:

```
> number    <-    c(0, 1, 2, 3, 4, 5, 6, 7, 8, 9, 10)
> frequency <- c(3, 17, 26, 16, 18, 9, 3, 5, 0, 1, 0)
> sum(number*frequency)
[1] 296
> sum(frequency)
[1] 98
> lambda <- sum(number*frequency)/sum(frequency)
> lambda
[1] 3.020408
```

Note: more advanced R-users can calculate maximum-likelihood estimates directly using the library "maxLik".

2.2.2 Estimating Expectations from Censored Samples and Truncated Distributions

We consider a random variable $\boldsymbol{y}$ that is normally distributed with expectation μ and variance σ^2. In animal breeding often selection means one-sided truncation of the distribution. All animals with a performance (birth weight for example) larger than a value a are excluded from further breeding. In general we can say that we only use such observations from a normally distributed random variable $\boldsymbol{y}$ that is larger than a. The left-sided truncated standard normal distribution is defined in the region $[a, \infty)$. Since the final area under the curve of a truncated distribution must be equal to 1, the new curve is stretched up to compensate for the lost truncated area over the region $(-\infty, a)$. Therefore the density function of the 'in a' truncated normal distribution is

$$\varphi_{\mathrm{T}}(y) = \frac{\varphi(y)}{\int_a^{\infty} \varphi(t)\mathrm{d}t}.$$

The expectation of $\boldsymbol{y}$ after truncation is

$$\mu_{\mathrm{T}} = \mu + \sigma \frac{\varphi\left(\frac{a-\mu}{\sigma}\right)}{1 - \Phi\left(\frac{a-\mu}{\sigma}\right)}. \tag{2.17}$$

The right-sided truncated distribution of the standard normal distribution is defined in the region $(-\infty, b]$. The density function of the 'in b' right-sided truncated normal distribution is

$$\varphi_T(y) = \frac{\varphi(y)}{\int_{-\infty}^{b} \varphi(t)dt}.$$

The expectation of y after truncation is in the left-sided case

$$\mu_T = \mu + \sigma \frac{\varphi\left(\frac{a-\mu}{\sigma}\right)}{1 - \Phi\left(\frac{a-\mu}{\sigma}\right)}$$

and in the right-sided case

$$\mu_T = \mu + \sigma \frac{\varphi\left(\frac{b-\mu}{\sigma}\right)}{\Phi\left(\frac{b-\mu}{\sigma}\right)}.$$

However, often after truncation (selection), the expectation μ of the initial distribution has to be estimated.

Problem 2.13 Estimate the expectation and the variance of the initial $N(\mu, \sigma^2)$ distribution after an 'in a' left-sided and after an 'in b' right-sided truncation.

Remark

The estimation of the variance actually belongs to Section 2.3 but we need an estimate of the variance to calculate an estimate of the expectation.

Solution

Truncation right-sided:

Calculate first the mean $\bar{y} = \hat{\mu}_1$ of the observations y_i $(i = 1, \ldots, n)$ and

$$m_{2,1} = \frac{\sum_{i=1}^{n} (y_i - \hat{\mu}_1)^2}{n}.$$

From this we calculate as initial values of an iteration $z_1 = \frac{a-\hat{\mu}_1}{\sqrt{m2,1}}$, $c_1 = -\frac{\varphi(z_1)}{1-\Phi(z_1)}$, and $v_1 = \frac{c_1}{c_1+z_1}$, and from this we obtain the next steps in the iteration

$$\hat{\mu}_{i+1} = \bar{y} - c_i(\hat{\mu}_i - a), m_{2,i+1} = \frac{\sum_{i=1}^{n} (y_i - \hat{\mu}_{i+1})^2}{n}, z_{i+1} = \frac{a - \hat{\mu}_{i+1}}{\sqrt{m2,\ i+1}},$$

$$c_{i+1} = -\frac{\varphi(z_{i+1})}{1 - \Phi(z_{i+1})}, v_{i+1} = \frac{c_{i+1}}{c_{i+1} + z_{i+1}}.$$

We stop the iteration with some small ε if $|\hat{\mu}_{i+1} - \hat{\mu}_i| < \varepsilon$.

If truncation is right-sided at b we use $c_1 = \frac{\varphi(z_1)}{\Phi(z_1)}$ and $\hat{\mu}_{i+1} = \bar{y} + c_i(\hat{\mu}_i - b)$ and proceed as above.

Example
From a cattle population 500 heifers are selected with a milk performance of at least 3000 kg milk performance during the first 300 days of milking period. Their mean milk performance was $\bar{y} = \hat{\mu}_1 = 4612$ kg, and the second sample moment $m_{2,1} = \frac{\sum_{i=1}^{n}(y_i - \hat{\mu}_1)^2}{n} = 971500\ \text{kg}^2$. We need estimates of the expectation and variance of all heifers of the population of heifers. Assuming that the milk performance can be modelled by a normal distribution, we can apply the iteration described above and obtain the results in Table 2.2.

If we choose $\varepsilon = 1$ kg iteration is stopped at step 20 because $|\hat{\mu}_{20} - \hat{\mu}_{19}| < 1$.

Now we estimate the expectation in the case of censoring. While truncation occurs in distributions, censoring occurs in samples. We have left- and right-sided censoring and two types of censored samples. Type I censoring occurs if an experiment has a set number of subjects or items and stops the experiment at a predetermined time, at which point any subjects remaining are right-censored. That is measurements y_i are only known if $y_i > y_0$ where y_0 is given by the experimenter before the experiment starts. We assume as our model a $N(\mu, \sigma^2)$-distributed random sample $\boldsymbol{Y} = (\boldsymbol{y}_1, \boldsymbol{y}_2, \dots, \boldsymbol{y}_N)^T$, $N \geq 1$ of size N. Left-sided censoring means, from $m = N - n > 0$ values of the realised sample $Y = (y_1, y_2, \dots, y_N)^T$ we only know that they are below y_0, and from n realisations we have a measured value. If on the other hand from $m = N - n$ values of the realised sample $Y = (y_1, y_2, \dots, y_N)^T$ we only know that they are above y_0 and from n realisations we have a measured value, we speak about right-sided censoring.

Type II censoring occurs, we speak about right-sided censoring, if an experiment has a set number N of subjects or items and stops the experiment when a predetermined number are observed to have failed; the remaining subjects are then right-censored. Analogously, left-censored can be defined. Here n and m are given before the experiment starts.

Problem 2.14 Estimate the expectation of a $N(\mu, \sigma^2)$-distribution based on a left-sided or a right-sided censored sample of type I.

Solution
Use the iteration described in the solution of Problem 2.13 using the following changes.

Table 2.2 Some results of the first 20 steps in the iteration of the Heifer example.

Step number, i	v_{i+1}	$\hat{\mu}_{i+1}$	$\sqrt{m_{2,\,i+1}}$
1	0.062 98	451 0.5	106 4.5
2	0.102 06	445 0.4	110 9.1
3	0.125 44	440 9.8	113 8.2
4	…	…	…
19	0.206 64	427 8.9	122 7.4
20	0.207 20	427 8.0	122 8.0

Replace in the scheme below the original in the first column by the entries in the second and third column respectively.

Original	Left-sided	Right-sided
Type I,II c	$-\dfrac{n}{N}\dfrac{\varphi(z)}{\Phi(z)}$	$\dfrac{n}{N}\dfrac{\varphi(z)}{[1-\Phi(z)]}$
Type II left a	$y_{(1)}$	
Type II right b		$y_{(N-n)}$

2.2.3 Estimating Location Parameters of Finite Populations

We assume that we have a finite population of size N. We first define the location parameters of such distributions and then show how to estimate them from a realised random sample of size n. It seems reasonable first to read Section 1.3. The usual procedure is sampling without replacement; when we sample with replacement the factor $\sqrt{1-\frac{n}{N}}$ in some of the formulae below is dropped. We write $Y_1, Y_2, \dots, Y_N$ for the N values in the finite population with expectation $\mu = \frac{1}{N}\sum_{j=1}^{N} Y_j$ and variance $\tilde{\sigma}^2 = \frac{1}{N-1}\sum_{j=1}^{N}(Y_j-\mu)^2$ for sampling without replacement or $\sigma^2 = \frac{1}{N}\sum_{i=1}^{N}(Y_j-\mu)^2 = \frac{N-1}{N}\tilde{\sigma}^2$ for sampling with replacement.

The quantity $\text{MSE}(\hat{\boldsymbol{\mu}}) = \text{var}(\hat{\boldsymbol{\mu}}) + B^2(\hat{\boldsymbol{\mu}})$ with the bias $B(\hat{\boldsymbol{\mu}}) = \mu - E(\hat{\boldsymbol{\mu}})$ of the estimator $\hat{\boldsymbol{\mu}}$ is called the mean square error (MSE) of $\hat{\boldsymbol{\mu}}$.

Problem 2.15 The expectation μ of a of finite population is to be estimated from the realisation of a pure random sample or a systematic sampling with random start. Give the estimates of the unbiased estimator for μ and of the estimator of the standard error of the estimator of μ.

Solution

The estimate (realisation of an unbiased estimator) is the arithmetic mean $\hat{\mu} = \bar{y}$ of the sample values. Its estimated standard deviation is $s_{\hat{\mu}} = \frac{s}{\sqrt{n}}\sqrt{1-\frac{n}{N}}$ for sampling without replacement and $s_{\hat{\mu}} = \frac{s}{\sqrt{n}}$ for sampling with replacement, and $s^2 = \frac{1}{n-1}\sum_{i=1}^{n}(y_i-\bar{y})^2$.

Problem 2.16 A universe of size N is subdivided into s disjoint clusters of size N_i ($i = 1, 2, \dots, s$). n_i sample units are drawn from stratum i by pure random sampling with a total sample size $n = \sum_{i=1}^{s} n_i$ in the universe. Estimate the expectation of the universe if the n_i/n are chosen proportional to N_i/N.

Solution

We estimate the expectation μ of the universe by the realisation of an unbiased estimator as

$$\hat{\mu} = \frac{1}{s}\sum_{i=1}^{s}\bar{y}_i. \tag{2.18}$$

Example

We refer to the data given in Table 1.1. To draw a sample of $n = 5000$ people from Vienna to estimate the average age of the population we multiply the values in the column n_i by 5 and receive the values n_i of persons from which we have to take their ages. For instance from the municipality 'Innere Stadt' we need the ages of 50 people selected by pure random sampling without replacement. Calculate the means of each municipality and use (2.18).

Problem 2.17 Estimate the expectation of a universe with N elements in s primary units (strata) having sizes N_i ($i = 1, \ldots$, s) by a two-stage sampling, drawing at first $r < s$ strata and then from each selected strata m elements.

Solution

Select in the first stage $r < s$ strata with probability $p_i = \frac{N_i}{N}$. Draw from the selected strata by pure random sampling without replacement a sample of size $m < N_i$ $(i = 1, \ldots, s)$.

Example

We will estimate the average age of the inhabitants of Vienna. We take the values of Table 1.1 and select $r = 5$ from the $s = 23$ municipalities to take an overall sample of $n = 1000$. For this we split the interval (0,1] into 23 subintervals $\left(1000\frac{N_{i-1}}{N}, 1000\frac{N_i}{N}\right]$ $i = 1, \ldots, 23$ with $N_0 = 0$ and generate five uniformly distributed random numbers in (0,1]. If a random number falls in any of the 23 sub-intervals (which can easily be found by using the 'cum' column in Table 1.1), the corresponding municipality has to be selected. If a further random number falls into the same interval it is dropped and replaced by another uniformly distributed random number. We generate five such random numbers as follows:

```
> runif(5)
 [1] 0.18769112 0.78229430 0.09359499 0.46677904 0.51150546
```

The first number corresponds to the municipality Mariahilf, the second to Florisdorf, the third to Landstraße, the fourth to Hietzing, and the last one to Penzing. To obtain a random sample of size 1000 we take pure random samples without replacement of size 200 from people in Mariahilf, Florisdorf, Landstraße, Hietzing, and Penzing respectively. Finally the mean of the ages of the 1000 selected inhabitants has to be calculated.

2.3 Estimating Scale Parameters

The most important scale parameters are the range, the interquartile range (IQR), and the standard deviation, variance. Except the variance all have the same dimensions as the observations.

The sample range R is a function of the order statistics of the sample its realisation is the difference between the largest and the smallest value of the sample, i.e. $R = (y_{(n)} - y_{(1)})$.

Problem 2.18 Calculate the range R of a sample.

Solution
Determine in a sample $Y = (y_1, y_2, \dots, y_n)$ the smallest value $y_{(1)} = \min(y_{(i)}), i = 1, \dots, n$ and the largest value $y_{(n)} = \max(y_{(i)}), i = 1, \dots, n$. Then $R = y_{(n)} - y_{(1)}$.

Example
In our vector y has already been used above in Problem 2.1 we found the minimum and the maximum values of y

```
  > min(y)
[1] 1
  > max(y)
[1] 25
```

and therefore the sample range is $R = 25{-}1 = 24$.

The interquartile range IQR(Y) of a sample Y is defined as the difference between the third and the first sample quartile.

$$\mathrm{IQR}(Y) = Q_3(Y) - Q_1(Y). \tag{2.19}$$

Problem 2.19 Calculate the interquartile range IQR(Y) of a sample Y.

Solution
Calculate the sample quartiles $Q_1(Y)$ and $Q_3(Y)$ of the sample Y and then IQR(Y) as the difference between the third quartile $Q_3(Y)$ and the first quartile $Q_1(Y)$.

Example
From the example of Problem 2.7 we know that $Q_1(Y) = 9$ and $Q_3(Y) = 13$, i.e.

$$\mathrm{IQR}(Y) = 13 - 9 = 4.$$

We can find directly Q_1 and Q_3 from a data vector y as follows and then calculate the IQR.

We use the data of Problem 2.1.

```
> y<- c(5,7,1,7,8,9, 13,9,10,10, 18,10, 15, 10, 10, 11, 8,11,12,13, 15,
        22,10,25, 11)

> summary(y)
   Min. 1st Qu.  Median    Mean 3rd Qu.    Max.
    1.0     9.0    10.0    11.2    13.0    25.0
```

However, if we are only interested in the interquartile range IQR we can find it directly in R:

```
> IQR(y, na.rm = TRUE, type = 7)
> IQR(y, na.rm = TRUE, type = 7)
[1] 4
```

A further scale parameter is the standard deviation. The standard deviation of the distribution of $\boldsymbol{y}$ is the positive square root σ of the variance

$$\sigma^2 = \text{var}(\boldsymbol{y}) = E[(\boldsymbol{y} - \mu)^2]. \tag{2.20}$$

An unbiased estimator of σ^2 from $\boldsymbol{Y} = (\boldsymbol{y}_1, \boldsymbol{y}_2, \ldots, \boldsymbol{y}_n)^T$ is

$$\boldsymbol{s}^2 = \frac{1}{n-1} \sum_{i=1}^{n} (\boldsymbol{y}_i - \overline{\boldsymbol{y}})^2. \tag{2.21}$$

Using R we can find the estimate s^2 of a data set y with the command `var(y)`.
The square root $\boldsymbol{s}$ of $\boldsymbol{s}^2$ is a biased estimator of σ.
An unbiased estimator of σ if the elements of $\boldsymbol{Y}$ are normally distributed is then

$$\hat{\sigma} = \boldsymbol{s} \frac{\Gamma\left(\frac{n-1}{2}\right)\sqrt{n-1}}{\sqrt{2}\Gamma\left(\frac{n}{2}\right)}. \tag{2.22}$$

In R the estimate sigmacap $\hat{\sigma} = s\frac{\Gamma\left(\frac{n-1}{2}\right)\sqrt{n-1}}{\sqrt{2}\Gamma\left(\frac{n}{2}\right)}$ is found by the commands for a data set y as follows:

```
> n <- length(y)
> s <- sqrt(var(y))
> sigmacap <- (s*gamma((n-1)/2)*sqrt(n-1)) / (sqrt(2)*gamma(n/2))
```

Problem 2.20 Calculate for an observed sample the estimate (realisation) of the square root $\boldsymbol{s}$ of $\boldsymbol{s}^2$ in (2.21).
Further give the estimate $\hat{\sigma}$.

Solution
Use in R `>sqrt(var(y))`. Alternatively, you can use directly `> sd(y)`.
For the commands of sigmacap see above.

Example
With y from Problem 2.1 we can calculate the realisation of $\boldsymbol{s}$ via

```
> sqrt(var(y))
[1] 5.024938
```

The estimate sigmacap is found as follows:

```
> n <- length(y)
> n
[1] 25
> s <- sqrt(var(y))
> s
[1] 5.024938
> sigmacap <- (s*gamma((n-1)/2)*sqrt(n-1)) / (sqrt(2)*gamma(n/2))
> sigmacap
[1] 5.077539
```

2.4 Estimating Higher Moments

In Section 2.2 the rth moment of a random variable was defined for any $r > 1$ and assumed that it exists if we discuss it. In (2.2) the rth sample moment was defined and for $r > 2$ we speak of higher moments. Usually for $r > 2$ the sample moments m_r are used as (biased) estimates of the corresponding moments μ_r of a random variable.

We consider here functions of the third and the fourth moment, the skewness and the kurtosis.

The skewness γ_1 is the standardised third moment

$$\gamma_1 = \frac{\mu_3}{\sigma^3}. \tag{2.23}$$

Sometimes it is estimated from a sample $Y = (y_1, y_2, \dots, y_n)^T$ by the sample skewness

$$g_1 = \frac{m_3}{s^3}$$

with s^2 defined in (2.21). The estimator

$$\boldsymbol{g}_1 = \frac{\boldsymbol{m}_3}{\boldsymbol{s}^3}$$

is biased.

In the statistical package SAS and IBM-SPSS Statistics with the weight 1 for all the sampled data the skewness is estimated as

$$g_1^* = \frac{m_3 \cdot n^2}{s^3(n-1)(n-2)}.$$

Problem 2.21 Calculate the sample skewness g_1 from a sample y for the data with weight 1.

Solution

Use the R -commands

```
> m <- mean(y)
> s <- sqrt(var(y))
> n <-length(y)
> devy <- y-m
> m3 <- sum(devy^3)/n
> g1 <- (m3*n^2) / [s^3*(n-1)*(n-2)]
```

Example

With Y from Problem 2.1

```
> y<- c(5,7,1,7,8,9, 13,9,10,10, 18,10, 15, 10,10, 11, 8,11,12,13, 15, 22,
       10,25, 11)
> m <- mean(y)
> m
[1] 11.2
> s <- sqrt(var(y))
> s
```

```
[1] 5.024938
> n <-length(y)
> n
[1] 25
> devy <- y-m
> devy
 [1]   -6.2   -4.2 -10.2   -4.2   -3.2   -2.2    1.8   -2.2   -1.2   -1.2   6.8
       -1.2    3.8  -1.2   -1.2   -0.2   -3.2   -0.2    0.8    1.8    3.8  10.8
       -1.2   13.8  -0.2
> m3 <- sum(devy^3)/n
> m3
[1] 111.168
> g1 <- (m3*n^2)/(s^3*(n-1)*(n-2))
> g1
[1] 0.9920388
```

The kurtosis γ_2 is the standardised fourth moment −3. The value 3 is subtracted because then the normal distribution has kurtosis 0.

$$\gamma_2 = \frac{\mu_4}{\sigma^4} - 3. \tag{2.24}$$

Sometimes it is estimated from a sample $Y = (y_1, y_2, \dots, y_n)^T$ by the sample kurtosis

$$g_2 = \frac{m_4}{s^4} - 3. \tag{2.25}$$

The estimator $\boldsymbol{g}_2$ is biased.

In the statistical packages SAS and SPSS with the weight 1 for all the sampled data the kurtosis is estimated as

$$g_2 = \{[(n+1)m_4 - 3(n-1)^3 s^4/n^2]n^2\}/[(n-1)(n-2)(n-3)s^4].$$

Problem 2.22 Calculate the sample kurtosis from a sample Y.

Solution

Use the R -commands

```
> m <- mean(y)
> s <- sqrt(var(y))
> n <-length(y)
> devy <- y-m
> m4 <- sum(devy^4)/n
> g2 <- (((n+1)*m4 -3*(n-1)^3*s^4/n^2)*n^2)/((n-1)*(n-2)*(n-3)*s^4)
```

Example

With Y from Problem 2.1

```
> y <- c(5,7,1,7,8,9, 13,9,10,10, 18,10, 15, 10, 10, 11, 8, 11,12,13, 15,
       22, 10,25, 11)
> m <- mean(y)
> m
[1] 11.2
> s <- sqrt(var(y))
> s
```

```
[1] 5.024938
> n <-length(y)
> n
[1] 25
> devy <- y-m
> devy
 [1]  -6.2  -4.2 -10.2  -4.2  -3.2  -2.2   1.8  -2.2  -1.2 -1.2   6.8
      -1.2   3.8  -1.2  -1.2  -0.2  -3.2  -0.2   0.8 1.8   3.8  10.8
      -1.2  13.8  -0.2
> m4 <- sum(devy^4)/n
> m4
[1] 2625.686
> g2 <- (((n+1)*m4 -3*(n-1)^3*s^4/n^2)*n^2)/((n-1)*(n-2)*(n-3)*s^4)
> g2
[1] 2.095743
```

2.5 Contingency Tables

Contingency tables are used when observations are nominally scaled. We describe here contingency tables in general, even if not only problems of estimation are handled by them. They will mainly be used in Chapter 3 but describing them here gives a unique approach. A k-dimensional contingency table with s_i levels of the ith of k factors F_i, ($i = 1, \ldots, k$) is given by $s_1 \cdot s_2 \cdot \ldots \cdot s_k$ classes, containing the number of observations from N investigated objects in a nominal scale with level s_i of the ith factor A_i. For such contingency tables there exist $k+1$ different models. The models depend on how many factors are observed by the experimenter (they are observation factors) and thus contain random results. The other factors are called fixed factors. We explain this by a two-dimensional contingency table.

2.5.1 Models of Two-Dimensional Contingency Tables

In two-dimensional contingency tables three models exist.

2.5.1.1 Model I

If we investigate N pupils and investigate whether they have blue eyes or not and if they are fair-haired or not, then we have $k = 2$ factors: A eye colour with $s_1 = 2$ levels and B hair colour with $s_2 = 2$ levels. The observations can be arranged in a contingency table like Table 2.3.

Here both factors are observation factors, the entries $\boldsymbol{n}_{ij}$, $i = 1, 2, j = 1, 2$ and the marginal sums $\boldsymbol{N}_{1.}, \boldsymbol{N}_{2.}, \boldsymbol{N}_{.1}$, and $\boldsymbol{N}_{.2}$ of the contingency Table 2.3 are random variables. Investigated is a random sample of size N. We call this situation model I of a contingency table.

2.5.1.2 Model II

If the marginal number of one of the factors, let's say A, are fixed in advance we obtain a contingency table like Table 2.4.

Such a situation occurs if N_1 female and N_2 male pupils are observed and it is counted how many have blue and how many do not have blue eyes. We call this model II of a contingency table.

Table 2.3 A two-by-two contingency table – model I.

	Factor *B*		
Factor *A*	B_1	B_2	**Sum**
A_1	n_{11}	n_{12}	$N_{1.}$
A_2	n_{21}	n_{22}	$N_{2.}$
Sum	$N_{.1}$	$N_{.2}$	N

Table 2.4 A two-by-two contingency table – model II.

	Factor *B*		
Factor *A*	B_1	B_2	**Sum**
A_1	n_{11}	n_{12}	$N_{1.}$
A_2	n_{21}	n_{22}	$N_{2.}$
Sum	$N_{.1}$	$N_{.2}$	N

2.5.1.3 Model III

The situation of model III with all marginal sums fixed in advance are of theoretical interest as in Fisher's 'problem of the lady tasting tea' reported in Fisher (1935, 1971). The lady in question (Muriel Bristol) claimed to be able to tell whether the tea or the milk was first added to the cup. Fisher proposed to give her eight cups, four of each variety, in random order. One could then ask what the probability was for her getting the specific number of cups she identified correctly, but just by chance. However, when the lady knows that for each variety four cups have been prepared she would make all marginal sums equal to four. That situation leads to Fisher's exact test in Chapter 3.

Here we describe two-dimensional contingency tables; three-dimensional tables are described in Rasch et al. (2008, Verfahren 4/31/3000).

In contingency tables, we can estimate measures but also test hypotheses. Here we only show how to calculate several measures from observed data in two-dimensional contingency tables. Tests of hypotheses can be found in Chapter 3.

The degree of association between the two variables (here factors) can be assessed by a number of coefficients, so-called association measures. The simplest, applicable only to the case of 2×2 contingency tables, are as follows.

2.5.2 Association Coefficients for 2 × 2 Tables

These coefficients do not depend on the marginal sums and are often calculated from a two-dimensional contingency table in the form of Table 2.5.

Example 2.1 In a random sample without replacement of size $n = 5375$ from a population of adult persons the factor A is tuberculosis (A_1 = lung tuberculosis, A_2 = other

Table 2.5 A two-by-two contingency table – for calculating association measures.

	Factor *B*		
Factor *A*	B_1	B_2	**Sum**
A_1	a	b	$a+b$
A_2	c	d	$c+d$
Sum	$a+c$	$b+d$	N

form of tuberculosis) and the factor B is gender (B_1 = male, B_2 = female) with $a = 3534$, $b = 1319$, $c = 270$, and $d = 250$. We call these data CT2x2.

From Yule (1900) we know the coefficient

$$\varphi = Q = \frac{ac - bd}{ac + bd}. \tag{2.26}$$

Problem 2.23 Calculate the association measure Q in (2.26) for data CT2x2.

Solution

Make for data CT2x2 a table from the two columns with the R command

```
> CT2x2 <- cbind( c(3534, 270), c(1319, 250) )
> a <- CT2x2[1,1]
> b <- CT2x2[1,2]
> c<- CT2x2[2,1]
> d<- CT2x2[2,2]
> numQ   <-   a*c - b*d  # num = numerator
> denomQ <-   a*c+b*d  # denom = denominator
> Q <- numQ/denomQ
```

Example

```
> CT2x2 <- cbind(c(3534, 270), c(1319, 250))
> CT2x2
     [,1] [,2]
[1,] 3534 1319
[2,]  270  250
> a<-CT2x2[1,1]
> a
[1] 3534
> b <- CT2x2[1,2]
> b
[1] 1319
> c<- CT2x2[2,1]
> c
[1] 270
```

```
> d<- CT2x2[2,2]
> d
[1] 250
> numQ <- a*c - b*d
> numQ
[1] 624430
> denomQ <- a*c + b*d
> denomQ
[1] 1283930
> Q <- numQ/denomQ
> Q
[1] 0.4863427
```

From Yule (1911) we know the coefficient

$$Y = \frac{\sqrt{ac} - \sqrt{bd}}{\sqrt{ac} + \sqrt{bd}}. \tag{2.27}$$

We have $|Y| \leq |Q|$.

Problem 2.24 Calculate the association measure Y in (2.27) for example CT2x2.

Solution
Using the same commands as in the solution of Problem 2.23 to get the values a, b, c, and d we have only to add the commands:

```
> numY <- (sqrt(a*c)) - (sqrt(b*d))
> denomY <- (sqrt(a*c)) + (sqrt(b*d))
> Y <- numY / denomY
```

Example

```
> numY <-  (sqrt(a*c)) -(sqrt(b*d))
> numY
[1] 402.5827
> denomY <- (sqrt(a*c)) + (sqrt(b*d))
> denomY
[1] 1551.06
> Y = numY/denomY
> Y <- numY/denomY
> Y
[1] 0.2595533
```

From Digby (1983) we know the coefficient

$$H = \frac{(ac)^{\frac{3}{4}} - (bd)^{\frac{3}{4}}}{(ac)^{\frac{3}{4}} + (bd)^{\frac{3}{4}}}. \tag{2.28}$$

Problem 2.25 Calculate the association measure H in (2.28) for example CT2x2.

Solution

Using the same commands as in the solution of Problem 2.23 to get the values *a*, *b*, *c*, and *d* we have only to add the commands:

```
> n1 <- ((a*c))^0.75
> n2 <- ((b*d))^0.75
> numH <- n1 - n2
> denomH <-  n1 + n2
> H <- numH/denomH
```

Example

```
> n1 <- ((a*c))^0.75
> n1
[1] 30529.71
> n2 <- ((b*d))^0.75
> n2
[1] 13760.64
> numH <- n1 - n2
> numH
[1] 16769.07
> denomH <- n1 + n2
> denomH
[1] 44290.35
> H <- numH/denomH
> H
[1] 0.3786169
```

If we have general s_1, s_2 – also written $a \times b$ – contingency tables where neither s_1 or s_2 is equal to 2, we use other measures, often depending on n_{ij} analogous to Table 2.3 via

$$\chi^2 = \sum_{i=1}^{s_1} \sum_{j=1}^{s_2} \frac{N\left(n_{ij} - \frac{N_{i\cdot} \cdot N_{\cdot j}}{N}\right)^2}{N_{i\cdot} \cdot N_{\cdot j}}. \tag{2.29}$$

One of these is the contingency coefficient C defined as

$$C = \sqrt{\frac{\chi^2}{\chi^2 + N}}. \tag{2.30}$$

This coefficient is smaller than 1 and therefore adjusted by

$$C_{\text{adj}} = \sqrt[4]{\frac{s_1 - 1}{s_1} \frac{s_2 - 1}{s_2}} C. \tag{2.31}$$

Example 2.2 From a population of children in Germany with the German language as their mother tongue (MTG) a random sample with replacement of size 50 was taken. Also from a population of children in Germany with the Turkish language as their

Table 2.6 Mother tongue and marital status of the mother of 50 children.

	MT, mother tongue	
MS, marital status	**MTG**	**MTT**
MSS	6	3
MSM	23	42
MSD	18	4
MSW	3	1

mother tongue (MTT) a random sample with replacement of size 50 was taken. From the children the marital status of their mother (MS) was determined, there were $s_I = 4$ classes of MS: MSS = single, MSM = married, MSD = divorced and MSW = widow. From the children the mother tongue (MT) has $s_2 = 2$ classes: MTG and MTT (see Table 2.6).

Problem 2.26 Calculate χ^2 and the association measure C in (2.30) for Example 2.2.

Solution
In R use the following commands:

```
> y <- c(6,3,23,42,18,4,3,1)
> CT4x2 <- matrix(y, nrow=4, byrow=T)
> colnames(CT4x2) <- c("MTG", "MTT")
> rownames(CT4x2) <- c("MSS", "MSM", "MSD", "MSW")
> CT4x2
> chisq.test(CT4x2)
> CHISQ <- 16.4629 # The value given by R as X-squared
> N <- sum(y)
> C2 <- CHISQ/(CHISQ + N)
> C <- sqrt(C2)
```

Example

```
> y <- c(6,3,23,42,18,4,3,1)
> CT4x2 <- matrix(y, nrow=4, byrow=T)
> colnames(CT4x2) <- c("MTG", "MTT")
> rownames(CT4x2) <- c("MSS", "MSM", "MSD", "MSW")
> CT4x2
    MTG MTT
MSS   6   3
MSM  23  42
MSD  18   4
MSW   3   1
> chisq.test(CT4x2)

        Pearson's Chi-squared test
```

```
data:  CT4x2
X-squared = 16.4629, df = 3, p-value = 0.0009112

Warning message:
In chisq.test(CT4x2) : Chi-squared approximation may be incorrect
> CHISQ <- 16.4629 # The value given by R as X-squared
> N <- sum(y)
> N
[1] 100
> C2 <- CHISQ/(CHISQ + N)
> C <- sqrt(C2)
> C
[1] 0.3759753
```

Problem 2.27 Calculate the association measure C_{adj} in (2.31) for Example 2.2.

Solution

Using the previous commands of Problem 2.26 we add the following R commands:

```
> s1 <-4
> s2 <- 2
> A <- (s1-1)*(s2-1)/(s1*s2)
> Cadj <- (A^0.25)*C
```

Example

```
> s1 <- 4
> s2 <- 2
> A <- (s1-1)*(s2-1)/(s1*s2)
> A
[1] 0.375
> Cadj <- (A^0.25)*C
> Cadj
[1] 0.2942166
```

If $s_1 = s_2 = k$ then we get the simple formula

$$C_{\text{adj}} = \sqrt{\frac{k-1}{k}}C.$$

Further measures are Tschuprow's coefficient

$$T = \sqrt{\frac{\chi^2}{N\sqrt{(s_1-1)(s_2-1)}}} \tag{2.32}$$

and Cramer's coefficient

$$V = \sqrt{\frac{\chi^2}{N[\min(s_1, s_2) - 1]}}. \tag{2.33}$$

If $s_1 = s_2$ then $T = V$, otherwise we have $T < V$. Between C and V the following relations exist:

$$C > V \text{ if } V < \sqrt[4]{\frac{k-1}{k}},$$

$$C = V \text{ if } V = \sqrt[4]{\frac{k-1}{k}},$$

$$C < V \text{ if } V > \sqrt[4]{\frac{k-1}{k}}.$$

Example 2.3 From the population of Germany a random sample with replacement of size $N = 2000$ was drawn. From each person the hair colour (factor A) was determined as A1 = 'blond', A2 = 'dark', and A3 = 'other hair colour'. Also from each person the eye colour (factor B) was determined as B1 = 'blue', B2 = 'dark', and B3 = 'other eye colour'.

The results are shown in Table 2.7.

Problem 2.28 Calculate the association measure T in (2.29) for Example 2.3.

Solution

Use the following R commands:

```
> y <- c(418, 362, 123, 153, 318, 164, 66, 131, 265)
> CT3x3 <- matrix(y, nrow=3, byrow=T)
> colnames(CT3x3) <- c("B1", "B2", "B3")
> rownames(CT3x3) <- c("A1", "A2", "A3")
> CT3x3
> chisq.test(CT3x3)
> CHISQ <- 359.9694 # The value given by R as X-squared
> N <- sum(y)
> s1 <- 3
> s2 <- 3
> prod <- (s1-1)*(s2-1)
> denom <- N*sqrt(prod)
>  T <- sqrt((CHISQ/denom))
> T
[1] 0.2999872
>
```

Table 2.7 Hair and eye colour of 2000 German persons.

	Eye colour, B		
Hair colour, A	B1	B2	B3
A1	418	362	123
A2	153	318	164
A3	66	131	265

Example

```
> y <- c(418, 362, 123, 153, 318, 164, 66, 131, 265)
> CT3x3 <- matrix(y, nrow=3, byrow=T)
> colnames(CT3x3) <- c("B1", "B2", "B3")
> rownames(CT3x3) <- c("A1", "A2", "A3")
> CT3x3
    B1  B2  B3
A1 418 362 123
A2 153 318 164
A3  66 131 265
> chisq.test(CT3x3)

        Pearson's Chi-squared test

data:  CT3x3
X-squared = 359.9694, df = 4, p-value < 2.2e-16
> CHISQ <- 359.9694 # The value given by R as X-squared
> N <- sum(y)
> N
[1] 2000
> s1 <- 3
> s2 <- 3
> prod <- (s1-1)*(s2-1)
> prod
[1] 4
> denom <- N*sqrt(prod)
> denom
[1] 4000
> T <- sqrt((CHISQ/denom))
> T
[1] 0.2999872
```

Problem 2.29 Calculate the association measure V in (2.33) for Example 2.3.

Solution

Using the previous commands of Problem 2.28 we add the following R commands:

```
> s <-c(s1,s2)
> mins <- min(s)
> V <- sqrt((CHISQ/(N*(mins-1)))
> V
```

Example

```
> s <-c(s1,s2)
> s
[1] 3 3
```

```
> mins <- min(s)
> mins
[1] 3
> V <- sqrt((CHISQ/(N*(mins-1))))
> V
[1] 0.06415392
```

References

Digby, P.G.N. (1983). Approximating the tetrachoric correlation coefficient. *Biometrics* 39: 753–757.

Fisher, R.A. (1971) [1935]. *The Design of Experiments*, 9e. New York: Macmillan. ISBN: 0-02-844690-9.

Rasch, D., Herrendörfer, G., Bock, J., Victor, N. and Guiard, V. (2008). *Verfahrensbibliothek Versuchsplanung und - auswertung*, 2. verbesserte Auflage in einem Band mit CD. R. Oldenbourg Verlag München Wien.

Rasch, D. and Schott, D. (2018) *Mathematical Statistics*. Wiley. Oxford.

Yule, G.U. (1900). On the association of attributes in statistics: with illustrations from the material of the Childhood Society, &c. *Philosophical Transactions of the Royal Society of London (A)* 194: 257–319.

Yule, G.U. (1911). *Introduction to the Theory of Statistics*. London Griffin.

3

Testing Hypotheses – One- and Two-Sample Problems

3.1 Introduction

In empirical research, a scientist often formulates conjectures about objects of his research. For instance, he may argue that the fat content in the milk of Jersey cows is higher than that of Holstein Friesians. To check conjectures, he will perform an experiment. Now statistics come into play.

Sometimes the aim of investigation is not to determine certain statistics (to estimate parameters), but to test or to examine carefully considered hypotheses (assumptions, suppositions) and often also wishful notions based on practical material. In addition, in this case we establish a mathematical model where the hypothesis is formulated in the form of model parameters.

We assume that we have one random sample or two random samples from special distributions. We begin with the one-sample problem and assume that the distribution of the components of the sample depends on a parameter (vector) θ. We would like to test a hypothesis about θ. First, we define what we have to understand by these terms.

A statistical test is a procedure that allows a decision for accepting or rejecting a hypothesis about the unknown parameter to occur in the distribution of a random variable. We shall suppose in the following that two hypotheses are possible. The first (or main) hypothesis is the null hypothesis H_0, the other one is the alternative hypothesis H_A. The hypothesis H_0 is right, if H_A is wrong, and vice versa. Hypotheses can be composite or simple. A simple hypothesis prescribes the parameter value θ uniquely, e.g. the hypothesis $H_0: \theta = \theta_0$ is simple. A composite hypothesis admits that the parameter θ can have several values.

We discuss hypotheses testing based on random samples. A special random variable, a so-called test statistic, is derived from the random sample. We reject the null hypothesis if the realisation of the test statistic has some relation to a real number; let us say it exceeds some quantile. This quantile is chosen so that the probability that the random test statistic exceeds it if the null hypothesis is correct is equal to a value $0 < \alpha < 1$; this is fixed in advance and is called the first kind risk or significance level. The first kind risk equals the probability to make an error of the first kind, i.e. to reject the null hypothesis if it is correct. Usually we choose α relatively small. From the time in which the quantiles mentioned above are calculated to produce tables – only for few values α-quantiles did exist. From this time often $\alpha = 0.05$ or 0.01 where used. However, even now it makes sense to use one of these values to make different experiments comparable.

Applied Statistics: Theory and Problem Solutions with R, First Edition.
Dieter Rasch, Rob Verdooren, and Jürgen Pilz.

Besides an error of the first kind, an error of the second kind may occur if we accept the null hypothesis although it is wrong; the probability that this occurs is the second kind risk.

Both errors have different consequences. Assume that the null hypothesis states that a special atomic power plant is safe. The alternative hypothesis is then that it is not safe. Of course, it is more harmful if the null hypothesis is erroneously accepted and the plant is assembled. Therefore, in this case the risk of the second kind is more important than the risk of the first kind. In many other cases, the risk of the first kind plays an important role and it is fixed in advance.

At the end of a classical statistical test we decide on one of two possibilities, namely accepting or rejecting the null hypothesis.

Tests for a given first kind risk α are called α-tests; usually we have many α-tests for a given pair of hypotheses. Amongst them, the ones that are preferable are those that have the smallest second kind risk β, or, equivalently, the largest power $1-\beta$. Such tests are called the most powerful α-tests. When $\alpha \leq \beta$ if H_A is valid, the test is called unbiased.

A special situation arises in sequential testing. Here three decisions are possible after each of a series of sequentially registered observations:

- accept H_0
- reject H_0
- make the next observation.

The value of the risk of the second kind depends on the distance between the parameter values of the null and the alternative hypothesis.

We need another term, the power function. This is the probability of rejecting H_0. Its value equals α if the null hypothesis is correct.

It is completely wrong to state, after rejecting a null hypothesis based on observations, that this decision is wrong with probability α. This decision is either right or wrong. On the other hand, it is correct to say that the decision is based on a procedure, which supplies wrong rejections with probability α.

However, often a decision must be made.

Then one may say H_0 is correct if accepting H_0, even if this may be wrong.

Therefore, the user is recommended to choose α (or β, respectively) small enough that a rejection (or acceptance) of H_0 allows the user to behave with a clear conscience, as H_0 would be wrong (or H_A right). However, there is also an important statistical consequence: if the user has to conclude many such decisions during his investigations, then he will wrongly decide in about 100α (and 100β, respectively) per cent of the cases. This is a realistic point of view, which essentially we have confirmed by experience. If we move in traffic, we should realise the risk of one's own and other people's incorrect actions or an accident (observe that in this case α is considerably smaller than 0.05), but we must participate, just as a researcher must derive a conclusion from an experiment, although he knows that he/she could be wrong. On the other hand, it is very important to control risks. Concerning risks of the second kind, this is only possible if the sample size is determined before the experiment. The user should take care not to transfer probability statements to single observed cases.

In this chapter, we discuss tests on expectations, variances, and other general parameters. Tests on special parameters like regression coefficients are found in the corresponding chapters.

Statistical tests and confidence estimations for expectations of normal distributions are extremely robust against the violation of the normality assumption – see Rasch and Tiku (1985).

3.2 The One-Sample Problem

We begin with the one-sample problem and assume that the distribution of the components of the sample depend on a parameter (vector) θ. We would like to test a hypothesis about θ. First, we define what we have to understand by these terms.

In this section we assume that a random sample $\boldsymbol{Y} = (\boldsymbol{y}_1, \boldsymbol{y}_2, \ldots, \boldsymbol{y}_n)^T$, $n \geq 1$ of size n is taken from a distribution with an unknown parameter. In Sections 3.2.1 and 3.2.3 we handle problems in which this distribution is a normal one with parameter vector $\theta = \binom{\mu}{\sigma^2}$.

3.2.1 Tests on an Expectation

We know simple and composite hypotheses.

In a simple hypothesis the value of $\theta = \mu$ is fixed at a real number μ_0.

A simple null hypothesis could be $H_0 : \mu = \mu_0$. In the pair $H_0 : \mu = \mu_0$; $H_A : \mu = \mu_1$ both hypotheses are simple.

Examples for composite null hypotheses are:

$$H_0 : \mu = \mu_0, \sigma^2 \text{ arbitrary}$$
$$H_0 : \mu = \mu_0 \text{ or } \mu = \mu_1$$
$$H_0 : \mu < \mu_0,$$
$$H_0 : \mu \neq \mu_0.$$

3.2.1.1 Testing the Hypothesis on the Expectation of a Normal Distribution with Known Variance

We first assume that any component of $\boldsymbol{Y} = (\boldsymbol{y}_1, \boldsymbol{y}_2, \ldots, \boldsymbol{y}_n)^T$ is independently distributed as $\boldsymbol{y}$, namely normally with unknown expectation μ and known variance σ^2.

We like to test the null hypothesis

$H_0 : \mu = \mu_0$ against

(a) $H_A : \mu = \mu_1 > \mu_0$ (one-sided alternative)

(b) $H_A : \mu = \mu_1 < \mu_0$ (one-sided alternative)

(c) $H_A : \mu = \mu_1 \neq \mu_0$ (two-sided alternative)

with a significance level α.

All hypotheses are simple hypotheses. In this case, usually the test statistic

$$\boldsymbol{z} = \frac{\bar{\boldsymbol{y}} - \mu_0}{\sigma}\sqrt{n} \tag{3.1}$$

is applied where $\bar{\boldsymbol{y}}$ is the mean taken from the random sample $\boldsymbol{Y}$ of size n.

Starting with the (random) sample mean $\bar{\boldsymbol{y}}$ first the value μ_0 of the null hypothesis is subtracted and then this difference is divided by the standard error $\frac{\sigma}{\sqrt{n}}$ of $\bar{\boldsymbol{y}}$.

We know that $\boldsymbol{z} = \frac{\bar{\boldsymbol{y}} - \mu_0}{\sigma}\sqrt{n}$ is normally distributed with variance 1 and, under the null hypothesis, the expectation of $\boldsymbol{z} = \frac{\bar{\boldsymbol{y}} - \mu_0}{\sigma}\sqrt{n}$ is 0. Under the alternative hypothesis the expectation of $\boldsymbol{z} = \frac{\bar{\boldsymbol{y}} - \mu_0}{\sigma}\sqrt{n}$ is $E(\boldsymbol{z}) = \frac{\mu_1 - \mu_0}{\sigma}\sqrt{n}$. The value

$$\lambda = \frac{\mu_1 - \mu_0}{\sigma}\sqrt{n} = \frac{\delta}{\sigma}\sqrt{n} \tag{3.2}$$

is called the non-centrality parameter (ncp) or, in applications, it is also called the relative effect size and $\delta = \mu_1 - \mu_0$ is called the effect size.

We calculate from a realised sample, i.e. from observations $Y = (y_1, y_2, \ldots, y_n)^T$, the sample mean $\bar{y}$ following Problem 2.1 and from this the realised test statistic

$$z = \frac{\bar{y} - \mu_0}{\sigma}\sqrt{n}. \tag{3.3}$$

Now we reject the null hypothesis for the alternative hypotheses (a), (b), and (c) with z from (3.3) and a fixed first kind risk or significance level α if:

(a) $z > Z(1-\alpha)$
(b) $z < Z(\alpha)$
(c) $z < Z\left(\frac{\alpha}{2}\right)$ or $z > Z\left(1 - \frac{\alpha}{2}\right)$.

Above $Z(P)$, $0 < P < 1$, is the P-quantile of $\boldsymbol{z}$, which has a standard normal distribution $N(0,1)$.

Problem 3.1 Determine the P-quantile $Z(P)$ of the standard normal distribution.

Solution
Use the R-command

```
> qnorm(P)
```

Example
Determine the 0.95-quantile $Z(0.95)$ of the standard normal distribution

```
> qnorm(0.95)
[1] 1.644854
```

Rounding gives us $Z(0.95) = 1.645$.

In hypothesis testing we often use $\alpha = 0.01, 0.05$, or 0.10. The corresponding quantiles for one-sided and two-sided alternative testing are given in Table 3.1.

A test seems to be better the smaller its risk of the first kind. Considering practical investigations, a risk of the first kind $\alpha = 0.05$ seems to be only just acceptable in most cases. Users may ask why the test is not designed in such a way that α has a very small value, say $\alpha = 0.000\,01$. The smaller α the larger will be the probability to make another error. Namely, if we calculate an estimate from the realisation of the sample, then we accept the null hypothesis, although this value would be also possible in the case that the alternative hypothesis is right and, consequently, the null hypothesis is wrong.

It is clear that α can only be reduced for a certain test and a fixed sample size if on the other hand a larger β is accepted. Hence, we cannot make the risks of first and second kind simultaneously arbitrarily small for a fixed sample size n. When we increase n, the

Table 3.1 *P*-quantiles *Z*(*P*) of the standard normal distribution.

P	$Z(P)$	$Z(1-P)$
0.01	−2.326	2.326
0.005	−2.576	2.576
0.05	−1.645	1.645
0.025	−1.960	1.960
0.1	−1.282	1.282

Table 3.2 Situations and decisions in hypotheses testing.

True situation	Decision	Result of the decision	Probability of the result
H_0 right (H_A wrong)	H_0 accepted (H_A rejected)	Right decision	Acceptance (or confidence) probability $1-\alpha$
	H_0 rejected (H_A accepted)	Error of the first kind	Significance or risk α of the first kind
H_0 wrong (H_A right)	H_0 accepted (H_A rejected)	Error of the second kind	Risk β of the second kind
	H_0 rejected (H_A accepted)	Right decision	Power $1-\beta$

variance $\frac{\sigma^2}{n}$ of $\overline{y}$ is smaller and by this also the risks. Because this is the case in all tests for location parameters, we can fix α and β before starting the experiment and by this calculate the sample size n.

Applying statistical tests, it is wrong but common to focus mainly on the risk of the first kind while the risk of the second kind is neglected. There are many examples where the wrong acceptance of the null hypothesis can produce serious consequences (consider 'genetic corn has no damaging side effects' or 'nuclear power stations are absolutely safe'). Therefore, it is advisable to control both risks, which is always possible by a suitably chosen sample size. In Table 3.2 the decisions performing a statistical test with respect to the facts (H_0 null hypothesis, H_A alternative hypothesis) are shown.

Actually, each difference of the parameters under the null hypothesis (μ_0) on the one hand and under the alternative hypothesis (μ_1) on the other hand can become significant as soon as the sample size is large enough. Hence, a significant result alone is not yet meaningful. It expresses nothing, because the difference could also be very small, for instance $|\mu_1 - \mu_0| = 0.000\,01$. Therefore, investigations have to be planned by fixing the difference of the parameter value from the null hypothesis (μ_0) to be practically relevant. For explaining the risk β of the second kind, we pretend the alternative hypothesis consists only of one single value μ_1. However, in most applications μ_1 can take all values apart from μ_0 for two-sided test problems, and all values smaller than or larger than μ_0 for one-sided test problems. The fact is that each value of μ_1 causes another value for the risk β of the second kind. More precisely, β becomes smaller the larger the difference $\mu_1 - \mu_0$ will be. The quantity $E = (\mu_1 - \mu_0)/\sigma$, i.e. the relative or standardised practically relevant difference is called the (relative) effect size.

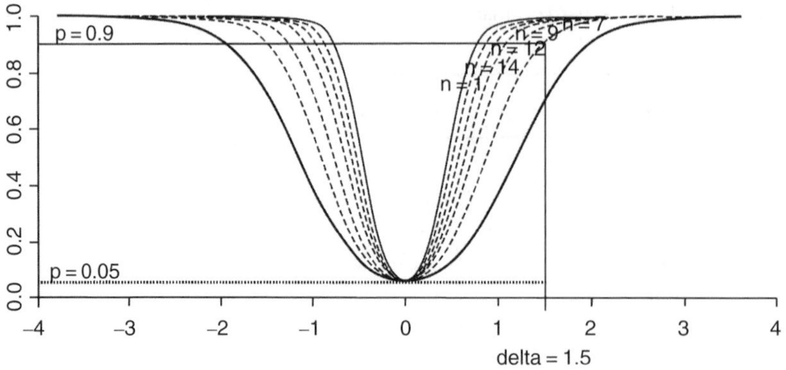

Figure 3.1 The power functions of the t-test testing the null hypothesis H_0: $\mu = \mu_0$ against H_A: $\mu \neq \mu_0$ for a risk $\alpha = 0.05$ of the first kind and a sample size $n = 5$ (bold plotted curve below) as well as other values of n (broken lined curves) up to $n = 20$ (bold plotted curve above).

Therefore, the fixing of the practically relevant minimal difference $\delta = \mu_1 - \mu_0$ is an essential step for planning an investigation. Namely, if δ is determined and if certain risks α of the first kind and β of the second kind are chosen, then the necessary sample size can be calculated. The fixing of α, β, and δ is called the precision requirement. Differences $\mu_1 - \mu_0$ equal to or larger than the prescribed δ should be overlooked only with a probability less than or equal to β.

The sample size, which fulfils the posed precision requirement, is obtained using the power function of the test. This function states the power for a given sample size and for all possible values of δ, i.e. the probability for rejecting the null hypothesis if indeed the alternative hypothesis holds. If the null hypothesis is true, the power function has the value α. It would not be fair to compare the power of a test with $\alpha = 0.01$ with that of a test with $\alpha = 0.05$, because larger α also means that the power is larger for all arguments referring to the alternative hypothesis. Hence, only tests with the same α should be compared.

For calculating the required sample size, we first look for all power functions related to all possible sample sizes, which have the probability α for μ_0, i.e. the parameter value under the null hypothesis. Now we look up the point of the minimum difference δ. Then we choose under all power functions that one which has the probability $1 - \beta$ at this point, i.e. the probability for the justified rejection of the null hypothesis; hence, at this point the probability of unjustified acceptance, i.e. of making an error of the second kind, is β. Finally, we have to choose the size n corresponding to this power function. For two-sided test problems the points $-\delta$ and $+\delta$ have to be fixed. Figure 3.1 illustrates that deviations larger than δ are overlooked with lower probability than the β chosen.

Problem 3.2 Calculate the power function of the one-sided test for (3.3) case (a).

Solution

$$z = \frac{\bar{y} - \mu_0}{\sigma}\sqrt{n}.$$

Power $= P(z > x; x \in R^1)$, i.e. for any x on the real line, where z has a normal distribution $N([(\mu_1 - \mu_0)\sqrt{n}]/\sigma, 1)$ or $z = ([(\mu_1 - \mu_0)\sqrt{n}]/\sigma + Z$ with Z being distributed as $N(0,1)$.

Table 3.3 Values of the power function $\pi(\delta) = P\left\{z > 1.6449 - \frac{\sqrt{n}\,\delta}{\sigma}\right\}$ for $n = 9, 16, 25, \sigma = 1$ and special δ.

δ	$\pi(\delta), n = 9$	$\pi(\delta), n = 16$	$\pi(\delta), n = 25$
0	0.05	0.05	0.05
0.1	0.0893	0.1066	0.1261
0.2	0.1480	0.1991	0.2595
0.3	0.2282	0.3282	0.4424
0.4	0.3282	0.4821	0.6387
0.5	0.4424	0.6387	0.8038
0.6	0.5616	0.7749	0.9123
0.7	0.6755	0.8760	0.9682
0.8	0.7749	0.9400	0.9907
0.9	0.8543	0.9747	0.9978
1.0	0.9123	0.9907	0.9996
1.1	0.9510	0.9971	0.9999
1.2	0.9747	0.9992	1.0000

Example
Let be $\delta = \mu_1 - \mu_0$ $(\delta \geq 0)$. Then the power function for $\alpha = 0.05$ is

$$\pi(\delta) = P\left\{Z > 1.6449 - \frac{\sqrt{n}\,\delta}{\sigma}\right\}.$$

Table 3.3 lists $\pi(\delta)$ for special δ and n.

In the applications, we call δ the practically interesting minimum difference to the value of the null hypothesis (also called the effect size). If we want to avoid such a difference with at most probability β, i.e. to discover it with probability at least $1 - \beta$, we have to prescribe a corresponding sample size n. Again we consider the general case that $\boldsymbol{Y}$ is a random sample of size n taken from a $N(\mu, \sigma^2)$ distribution.

In hypothesis testing we often use $\alpha = 0.01, 0.05$, or 0.10.

Problem 3.3 To test $H_0 : \mu = \mu_0$ for a given risk of the first kind α the sample size n has to be determined so that the second kind β is not larger than β_0 as long as $\mu_1 - \mu_0 \geq \delta$ in the one- and two-sided cases described above.

Solution
Cases (a) and (b) of (3.3) (one-sided alternatives). We restrict our attention on (a) because for (b) we receive the same sample size.

$H_0 : \mu = \mu_0$ is rejected if $z > z(1 - \alpha)$. For any $\lambda > 0$ our precision requirement in the least favourable case $\beta = \beta_0$ means that $\lambda + z(1 - \alpha) = z(\beta_0)$. From this it follows, with n^* for possibly non-integer values to calculate $n = \lceil n^* \rceil$,

$$\delta \frac{\sqrt{n^*}}{\sigma} + z(1 - \alpha) = z(\beta_0) \text{ or}$$

$$n^* = \frac{(z(\beta_0) - z(1-\alpha))^2\sigma^2}{\delta^2}.$$

The smallest integer larger or equal to x is written as $\lceil x \rceil$. Because we need integer numbers of observations, we use this operator $\lceil x \rceil$ and obtain

$$n = \left\lceil \frac{(z(\beta_0) - z(1-\alpha))^2\sigma^2}{\delta^2} \right\rceil. \tag{3.4}$$

If we have a two-sided alternative, analogously, we obtain

$$n = \left\lceil \frac{\left(z(\beta_0) - z\left(1-\frac{\alpha}{2}\right)\right)^2\sigma^2}{\delta^2} \right\rceil. \tag{3.5}$$

Often the variance σ^2 is not known. Then we can either use a relative effect size $\frac{\delta}{\sigma}$ so that we do not need a-priori information on σ or we use the value of a sample standard deviation from an analogous experiment in place of σ. A practical method is as follows: divide the expected range of the investigated character, that is the difference between the imaginably maximal and minimal realisation of the character, by six (assuming a normal distribution approximately 99% of the realisations lie between $\mu_0 - 3\sigma$ and $\mu_0 + 3\sigma$) and use the result as an estimate for σ.

If n and also β and α are given, we can use (3.4) or (3.5) to calculate δ.

Example

To test $H_0: \mu = \mu_0$ for a given risk of the first kind $\alpha = 0.05$ the sample n has to be determined so that the second kind risk β is not larger than 0.1 as long as $\mu_1 - \mu_0 \geq \delta{=}0.6\sigma$ in the one- and two-sided alternative cases.

One-sided alternative:

$$n = \left\lceil \frac{(z(\beta_0) - z(1-\alpha))^2\sigma^2}{\delta^2} \right\rceil = \left\lceil \frac{(-1.282 - 1.645)^2}{0.6^2} \right\rceil = \lceil 23.8 \rceil = 24.$$

Two-sided alternative:

$$n = \left\lceil \frac{\left(z(\beta_0) - z\left(1-\frac{\alpha}{2}\right)\right)^2\sigma^2}{\delta^2} \right\rceil = \left\lceil \frac{(-1.282 - 1.96)^2}{0.6^2} \right\rceil = \lceil 29.2 \rceil = 30.$$

Example 3.1 We take the data set y with $n = 25$ observations from Chapter 2 and test the null hypothesis $H_0: \mu = 10$. We further assume that y is a realised sample from a normal distribution with variance $\sigma^2 = 25$. Remembering that we found in Chapter 2

```
> mean(y)
[1] 11.2
```

we have $\bar{y} = 11.2$ and from this $z = \frac{\bar{y}-\mu_0}{\sigma}\sqrt{n} = \frac{11.2-10}{5} \cdot 5 = 1.2$.

When we use $\alpha = 0.05$ we have for a one-sided alternative to check whether $1.2 > z(0.9) = 1.645$. Because this is not the case, we accept $H_0: \mu = 10$. We have also to accept $H_0: \mu = 10$ for a two-sided alternative.

Because our sample size was 25, we know approximately that for a one-sided alternative the risk of the second kind is not larger than 0.1 as long as $\mu - 10 > 0.6 \cdot 5 = 3$.

3.2.1.2 Testing the Hypothesis on the Expectation of a Normal Distribution with Unknown Variance

We now assume that the components of a random sample $\boldsymbol{Y} = (\boldsymbol{y}_1, \boldsymbol{y}_2, \dots, \boldsymbol{y}_n)^T$ are distributed normally with unknown expectation μ and unknown variance σ^2.

We like to test the null hypothesis:

$H_0 : \mu = \mu_0, \sigma^2$ arbitrary against
(a) $H_A : \mu = \mu_1 > \mu_0, \sigma^2$ arbitrary (one-sided alternative)
(b) $H_A : \mu = \mu_1 < \mu_0, \sigma^2$ arbitrary (one-sided alternative)
(c) $H_A : \mu = \mu_1 \neq \mu_0, \sigma^2$ arbitrary (two-sided alternative).

All hypotheses are composite hypotheses because σ^2 is arbitrary. In this case, usually the test statistic (where $\boldsymbol{s}$ is the sample standard deviation):

$$t = \frac{\bar{y} - \mu_0}{s}\sqrt{n} \tag{3.6}$$

is used which is non-centrally t-distributed with $n - 1$ degrees of freedom and ncp

$$\lambda = \frac{\mu - \mu_0}{\sigma}\sqrt{n}.$$

Under the null hypothesis, the distribution is central t with $n - 1$ degrees of freedom.

If the type I error probability is α, H_0 will be rejected if:

in case (a), $t > t(n - 1;1 - \alpha)$,
in case (b), $t < -t(n - 1;1 - \alpha)$,
in case (c), $|t| > t(n - 1;1 - \alpha/2)$.

Our precision requirement is given by α and the risk of the second kind β if $\mu - \mu_0 = \delta$.

From this we get

$$t(n-1;1-\alpha) = t(n-1;\lambda;\beta)$$

where $t(n - 1;\lambda;\beta)$ is the β-quantile of the non-central t-distribution with $n - 1$ degrees of freedom and ncp

$$\lambda = \frac{\delta}{\sigma}\sqrt{n}.$$

For example, assuming a power of 0.9 the relative effect can be read on the abscissa; it is approximately 1.5 for $n = 7$.

From the precision requirement above the minimum sample size is iteratively determined as the integer solution of

$$t(n-1;P) = t\left(n-1;\frac{\delta}{\sigma}\sqrt{n};\beta\right) \tag{3.7}$$

(Rasch et al. 2011b) with $P = 1 - \alpha$ in the one-sided case and $P = 1 - \alpha/2$ in the two-sided case. In Problem 3.3 it is shown what to do when σ is unknown.

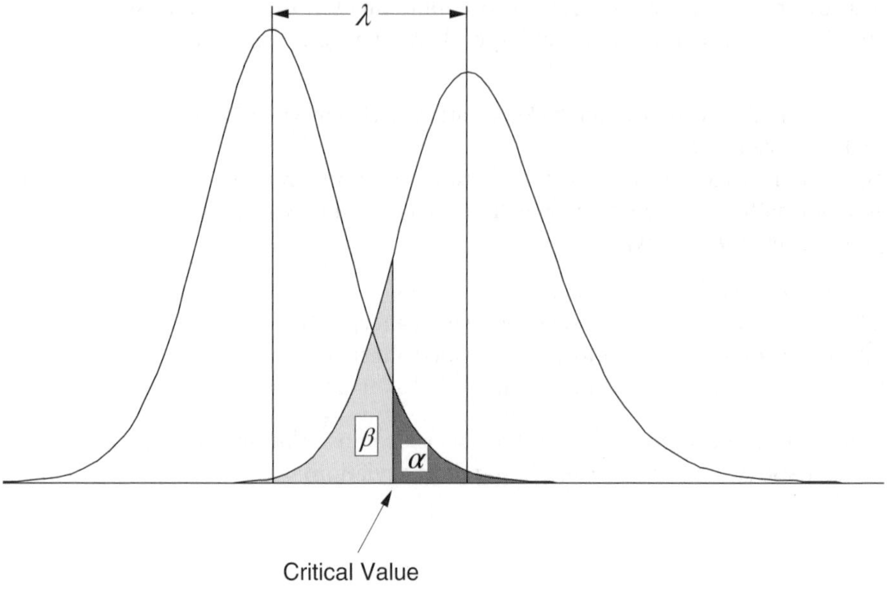

Figure 3.2 Graphical representation of the two risks α and β and the non-centrality parameter λ of a test based on (3.6).

The R-implementation reads as follows:

```
> size=function(p,rdelta,beta)
  {f=function(n,p,rdelta,beta)
  {qt(p,n-1,0)-qt(beta,n-1,rdelta*sqrt(n))}
  k=uniroot(f,c(2,1000),p=p,rdelta=rdelta,beta=beta)$root
  k0=ceiling(k)
  print(paste("optimum sample number: n =",k0), quote=F)}
```

In the command `rdelta` is equal to $\frac{\delta}{\sigma}$ and `p` equals $1 - \alpha$. As an example we calculate

```
> size(p=0.9,rdelta=0.5,beta=0.1)
[1] optimum sample number: n = 28.
```

The requirement (3.7) is illustrated in Figure 3.2 for the one-sided case.

Problem 3.4 Calculate the minimal sample size for testing the null hypothesis:

H_0: $\mu = \mu_0$ against one of the following alternative hypotheses:
(a) H_A: $\mu > \mu_0$ (one-sided alternative),
(b) H_A: $\mu \neq \mu_0$ (two-sided alternative).

Solution

We use the Optimal Design of Experiments (`OPDOE`) package in R where sd is used for σ.

We can use the function `>size`

```
> size(p=0.99,rdelta=0.6,beta=0.1)
[1] optimum sample number: n = 39
```

Or we use the OPDOE-command

```
> install.packages("OPDOE")
> library(OPDOE)
> size.t.test(type="one.sample",power=, delta= ,sd= ,
         sig.level= ,alternative="one.sided")
```

Example

The precision requirement is defined by $\delta = 0.6\sigma$; $\alpha = 0.01$ and $\beta = 0.1$.
For alternative (a)

```
> size.t.test(type="one.sample",power=0.9,delta=0.6,sd=1,
      sig.level=0.01,alternative = "one.sided")
[1] 39
```

For alternative (b)

```
> size.t.test(type="one.sample",power=0.9,delta=0.6,sd=1,
       sig.level=0.01,alternative = "two.sided")
[1] 45
```

We need a sample of size $n = 39$ in the one-sided alternative case and a sample of size $n = 45$ in the two-sided alternative case.

When we have a discrete distribution it may happen that there exists no test in the sense above so that a given α can be arrived at. Then we define more generally a statistical test via a critical function $k(Y)$ based on the likelihood function $L(Y, \theta)$ with the realisation $Y = (y_1, \dots, y_n)^T$ of the random sample $\boldsymbol{Y} = (\boldsymbol{y}_1, \dots, \boldsymbol{y}_n)^T$. The random critical function $k(\boldsymbol{Y})$ is the probability to reject a null hypothesis.

A test of the (simple) null hypothesis $H_0 : \theta = \theta_0$ against the (simple) alternative hypothesis $H_A : \theta = \theta_A$ is written with some constant $c_\alpha \geq 0$ as

$$k(Y) = \begin{cases} 1 & \text{for} \quad L(Y, \theta_A) > c_\alpha L(Y, \theta_0) \\ \gamma(Y) & \text{for} \quad L(Y, \theta_A) = c_\alpha L(Y, \theta_0) \\ 0 & \text{for} \quad L(Y, \theta_A) < c_\alpha L(Y, \theta_0) \end{cases}. \tag{3.8}$$

That such a test is a uniformly most powerful α-test is, in principle, the statement of the Neyman–Pearson lemma – the exact formulation is in theorem 3.1 of Rasch and Schott (2018). For more general problems and composite hypotheses two constants $c_{1\alpha}$ and $c_{2\alpha}$ can be needed. Such a test is called a randomised test because besides the acceptance of H_0 ($k(Y) = 0$) and the rejection of H_0 ($k(Y) = 1$) we have a third option: accept H_0 with probability $\gamma(Y)$. This means that after making observations we need a random decision if the likelihood function with both parameters fulfils $L(Y, \theta_A) = c_\alpha L(Y, \theta_0)$. This can be done by generating a uniformly distributed random variable in (0,1) and rejecting H_0 if its value is below $\gamma(Y)$. Experimenters certainly do not like such an approach but fortunately for continuous random variables this situation cannot occur because $P\{L(\boldsymbol{Y}, \theta_A) = c_\alpha L(\boldsymbol{Y}, \theta_0)\} = 0$. Contrary to Rasch and Schott (2018) a distribution function of a random variable $\boldsymbol{Y}$ is defined as $F(y) = P(\boldsymbol{Y} \leq y)$ and also so used in R.

Problem 3.5 Let y be binomially distributed (as $B(n,p)$). Describe an α-test for $H_0: p = p_0$ against $H_A: p = p_A < p_0$ and for $H_0: p = p_0$ against $H_A: p = p_A > p_0$.

Solution

$F(y, p)$ is the distribution function of $B(n, p)$. Use for $H_0: p = p_0$ against $H_A: p = p_A < p_0$

$$k^-(y) = \begin{cases} 1 & \text{for } y < y^- \\ \gamma_\alpha^- & \text{for } y = y^- \\ 0 & \text{for } y > y^- \end{cases} \qquad \text{with } \gamma_\alpha^- = \frac{\alpha - F(y^*, p_0)}{\binom{n}{y^*+1} p_0^{y^*+1} (1-p_0)^{n-y^*-1}}$$

and for $H_0: p = p_0$ against $H_A: p = p_A > p_0$

$$k^+(y) = \begin{cases} 1 & \text{for } y > y^+ \\ \gamma_\alpha^+ & \text{for } y = y^+ \\ 0 & \text{for } y < y^+ \end{cases} \qquad \text{with } \gamma_\alpha^+ = \frac{F(y^* + 1, p_0) - (1 - \alpha)}{\binom{n}{y^*+1} p_0^{y^*+1} (1-p_0)^{n-y^*-1}}$$

where y^- is determined by

$$F(y^-, p_0) \le \alpha < F(y^- + 1, \ p_0)$$

and y^+ by

$$F(y^+, p_0) < 1 - \alpha \le F(y^+ + 1, \ p_0).$$

Example

Let $n = 10$ and $H_0: p = 0.5$ is to be tested against $H_A: p = 0.1$, then the value $y^- = 1$ follows for $\alpha = 0.1$ where from R we find the binomial probabilities

```
> pbinom(1,10,0.5)
[1] 0.01074219
> pbinom(2,10,0.5)
[1] 0.0546875
```

and

```
> dbinom(2, 10, 0.5)
[1] 0.04394531
```

Hence we get

$$\gamma_{0.1} = \frac{0.05 - 0.010742}{0.04395} = 0.8933.$$

Then $k(Y)$ has the form

$$k(Y) = \begin{cases} 1 & \text{for } y < 2 \\ 0.8933 & \text{for } y = 2 \\ 0 & \text{for } y > 2 \end{cases}$$

i.e. for $y < 2$ the hypothesis $H_0: p = 0.5$ is rejected, for $y = 2$ H_0 is rejected with the probability 0.8933, and for $y > 2$ H_0 is accepted. The random trial in the case $y = 2$ can be simulated on a computer. Using a generator of random numbers supplying uniformly distributed pseudo-random numbers in the interval (0, 1) we obtain a value v. For $v < 0.8933$

the hypothesis H_0 is rejected and otherwise accepted. This test is a most powerful 0.05-test.

Problem 3.6 Let P be the family of Poisson distributions and $Y = (y_{1,\dots,}y_n)^T$ a realisation of a random sample $\boldsymbol{Y} = (\boldsymbol{y}_{1,\dots,}\boldsymbol{y}_n)^T$. The likelihood function is

$$L(Y,\lambda) = \prod_{i=1}^{n} \frac{1}{y_i!} e^{[\ln \lambda \sum_{i=1}^{n} y_i - \lambda n]}, \quad y_i = 0, 1, 2, \dots \; ; \quad \lambda \, \varepsilon \, R^+$$

How can we test the pair H_o: $\lambda = \lambda_o$, H_A: $\lambda \neq \lambda_o$ of hypotheses with a first kind risk α?

Solution

Solve with $\theta_0 = \ln \lambda_0$ and $\boldsymbol{T} = \sum_{i=1}^{n} \boldsymbol{y}_i$ the simultaneous equations

$$\alpha = P(\boldsymbol{T} < c_{1\alpha}|\theta_0) + P(\boldsymbol{T} > c_{2\alpha}|\theta_0) + \gamma'_{1\alpha} L_T(c_{1\alpha}, \theta_0) + \gamma'_{2\alpha} L_T(c_{2\alpha}, \theta_0),$$

$$\alpha = P(\boldsymbol{T} - 1 < c_{1\alpha}|\theta_0) + P(\boldsymbol{T} - 1 > c_{2\alpha}|\theta_0)$$

$$+ \gamma'_{1\alpha} L_T(c_{1\alpha} - 1, \theta_0) + \gamma'_{2\alpha} L_T(c_{2\alpha} - 1, \theta_0).$$

Example

Test $H_0 : \lambda = 10$ against $H_A : \lambda \neq 10$. R. can determine the values of the probability function and the likelihood function. We choose $\alpha = 0.1$ and look for possible pairs (c_1, c_2).

For $c_1 = 4$, $c_2 = 15$ we obtain the equations

$$0.006\,206 = 0.018\,917\gamma'_1 + 0.034\,718\gamma'_2$$

$$0.013\,773 = 0.007\,567\gamma'_1 + 0.052\,077\gamma'_2$$

supplying the improper solutions $\gamma'_1 = -0.215$, $\gamma'_2 = 0.296$. The pairs (4, 16) and (5, 15) also lead to improper values (γ'_1, γ'_2). Finally, we recognise that only the values $c_1 = 5$, $c_2 = 16$ and $\gamma'_1 = 0.697$, $\gamma'_2 = 0.799$ solve the problem. Hence, the test has the form

$$k(y) = \begin{cases} 1 & \text{for } y < 4 \text{ or } y > 15 \\ 0.697 & \text{for } y = 4 \\ 0.799 & \text{for } y = 15 \\ 0 & \text{else} \end{cases},$$

and $k(y)$ is the uniformly most powerful unbiased 0.1-test.

3.2.2 Test on the Median

Sometimes the population median m should be preferred to the expectation of a random variable $\boldsymbol{y}$. For any probability distribution of $\boldsymbol{y}$ on the real line R^1 with distribution function $F(y)$, regardless of whether it is a continuous distribution or a discrete one, a median is by definition any real number m that satisfies the inequalities

$$P(\boldsymbol{y} \leq m) \leq 0.5 \text{ and } P(\boldsymbol{y} \geq m) \geq 0.5.$$

If m is unique, the population median is the point where the distribution function equals 0.5.

It is better to use the median instead of the expectation if we expect serious outliers or if the distribution has an extremely skewness, so that in a sample the median better characterises the data than the mean. Tests on the population median are so-called non-parametric tests or distribution-free tests. These are tests without a specific assumption about the distribution.

It is notable that such an approach is not intrinsically justified by a suspected deviation from the normal distribution of the character in question. Particularly as Rasch and Guiard (2004) showed that all tests based on the t-distribution are extraordinarily robust against violations from the normal distribution. That is, even if the normal random variable is not a good model for the given character, the type I risk actually holds almost accurately. Therefore, if a test statistic results, on some occasions, in significance, we can be quite sure that an erroneous decision concerning the rejection of the null hypothesis has no larger risk than in the case of a normal distribution. We describe Wilcoxon's signed-ranks test useful for continuous distributions.

We test the hypothesis

H_0: $m = m_0$ against a one- or two-sided alternative:

H_A: $m < m_0$

H_A: $m > m_0$

H_A: $m \neq m_0$.

First, we calculate from each observation y_i of a sample $Y=(y_1, y_2, \dots, y_n)^T$ the differences $d_i = y_i - \text{med}(Y)$ between the observations and the median $=\text{med}(Y)$. If the value of d_i is equal to zero, we do not use these values. From the remaining data subset of size $n^* \leq n$ the rank of the absolute values $|d_i|$ is calculated. The sum,

$$V = \sum_{i=1}^{n*} \text{rank}(\mid d_i \mid)\,\text{sign}(d_i) \tag{3.9}$$

with $\text{sign}(d_i) = \begin{cases} -1 & if\ d_i < 0 \\ 0 & if\ d_i = 0 \\ 1 & if\ d_i > 0 \end{cases}$ is the test statistic. Wilcoxon's signed-ranks test is based on the distribution of the corresponding random variable, $\boldsymbol{V}$.

Planning a study is generally difficult for non-parametric tests, because it is hard to formulate a non-centrality parameter (however see, for instance, Munzel and Brunner 2002).

Problem 3.7 Perform a Wilcoxon's signed-ranks test for a sample Y.

Solution

We test the null hypothesis H_0: $m = 10$ against the alternative H_A: $m \neq 10$ with significance level $\alpha = 0.05$.

Use in R > `d <- y-m0` with `m0 = 10`.

```
> wilcox.test(d, alternative="two.sided",mu=0,correct = F,
     conf.level=0.95)
```

Example

We use the data y of Chapter 1, y = (5 7 1 7 8 9 13 9 10 10 18 10 15 10 10 11 8 11 12 13 15 22 10 25 11), and calculate first > `d <- y-10`.

```
> y = c(5,7,1,7,8,9,13,9,10,10,18,10,15,10,10,11,8,11,12,13,
        15,22,10,25,11)
> d <- y- 10
> wilcox.test(d, alternative="two.sided",mu=0, correct =
      F,conf.level=0.95)
        Wilcoxon signed rank test
data:  d
V = 118, p-value = 0.3528
alternative hypothesis: true location is not equal to 0

Warning messages:
1: In wilcox.test.default(d, alternative = "two.sided",
      mu = 0, correct = F,   :
  cannot compute exact p-value with ties
2: In wilcox.test.default(d, alternative = "two.sided",
      mu = 0, correct = F,   :
  cannot compute exact p-value with zeroes
```

The Wilcoxon signed rank test statistic V can be checked in R as follows:

```
> signd <- sign(d)
> absd <- abs(d)
> df0 <- data.frame(absd, signd)
> selnot0 <- df0$absd >0
> df1 <- data.frame(absd,signd,selnot0)
> df2 <- subset(df1,selnot0 == T)
> rabsd <- rank(df2$absd)
> df <- data.frame(rabsd,df2$signd)
> sel <- df2$signd >0
> wilcoxon <- df[sel, ]
> V <- sum(wilcoxon$rabsd)
> V
[1] 118
```

Problem 3.8 Perform a Wilcoxon's signed-rank test for two paired samples X and Y.

Solution

We test the null hypothesis H_0: $m = 0$ against the alternative H_A: $m \neq 0$ with significance level $\alpha = 0.05$.

Use in R

```
> wilcox.test(x,y, paired=T, alternative="two.sided",
     mu=0,conf.level=0.95)
```

Example

For seven people the reaction time under drugs X and Y are measured,

The reaction times under drug X are x = (0.223, 0.216, 0.211, 0.212, 0.209, 0.205, 0.201).

The reaction times under drug Y are $y = (0.208, 0.205, 0.202, 0.207, 0.206, 0.204, 0.203)$. The R commands are:

```
> x <- c(.223, .216, .211, .212, .209, .205, .201)
> y <- c(.208, .205, .202, .207, .206, .204, .203)
> wilcox.test(x,y, paired=T, alternative="two.sided",
     mu=0,conf.level=0.95)
   Wilcoxon signed rank test
data:  x and y
V = 26, p-value = 0.04688
alternative hypothesis: true location shift is not equal to 0
```

The Wilcoxon signed rank test statistic V can be checked in R as follows:

```
> dif <- (x-y)
> absdif <- abs(dif)
> selnot0  <- absdif >0
[1] TRUE TRUE TRUE TRUE TRUE TRUE TRUE
> # Because selnot0 is always TRUE we can find V directly as
     follows
> rabsdif<- rank(absdif)
> signdif <- sign(dif)
> df <- data.frame(rabsdif,signdif)
> sel <- df$signdif >0
> wilcoxon <- df[sel, ]
> V <- sum(wilcoxon$rabsdif)
> V
[1] 26
```

3.2.3 Test on the Variance of a Normal Distribution

Equation (2.21) shows an unbiased estimator s^2 of the variance σ^2 of a normal distribution.

We would like to test the null hypothesis

$H_0 : \sigma^2 = \sigma_0^2$ against one of the alternatives below:

- $H_A : \sigma^2 < \sigma_0^2$
- $H_A : \sigma^2 > \sigma_0^2$
- $H_A : \sigma^2 \neq \sigma_0^2$

with a first kind risk α.

We know that

$$\chi^2 = \frac{\boldsymbol{s}^2}{\sigma_0^2} \tag{3.10}$$

is centrally χ^2 distributed with $n - 1$ degrees of freedom.

We calculate from a sample of size n the realisation of (3.10) and compare the result with the P-quantile CS(P, $n-1$) of the central χ^2 distribution as follows:

$$\text{Reject}\begin{cases} \mathrm{H_A}: \sigma^2 < \sigma_0^2 \text{ if } \chi^2 < \mathrm{CS}(\alpha, n-1) \\ \mathrm{H_A}: \sigma^2 > \sigma_0^2 \text{ if } \chi^2 > \mathrm{CS}(1-\alpha, n-1) \\ \mathrm{H_A}: \sigma^2 \neq \sigma_0^2 \text{ if } \chi^2 < \mathrm{CS}\left(\frac{\alpha}{2}, n-1\right) \text{ or } \chi^2 > \mathrm{CS}\left(1-\frac{\alpha}{2}, n-1\right) \end{cases}. \tag{3.11}$$

Example 3.2 Intelligence tests are standardised to $\sigma = 15$. When in a sample from a class of pupils in some school the estimated variance of 28 observed intelligence quotient (IQ) is 120, is this a reason to reject the $H_0: \sigma^2 = 225$ against the two-sided alternative for $\alpha = 0.05$?

We calculate $\chi^2 = \frac{120}{225} = 0.5333$ and CS(0.025, 27); using R we calculate

```
> qchisq(0.025,27)
[1] 14.57338
```

and because $0.5333 < 14.573\,38$ we reject $H_0: \sigma^2 = 225$.

Problem 3.9 Show how, based on sample values Y, the null hypothesis $H_0: \sigma^2 = \sigma_0^2$ can be tested.

Solution
Input the values in an R-data file y and then calculate the sample variance of y. Divide this by σ_0^2 and compare the outcome with

```
> qchisq(P,n-1)
```

following (3.11).

Example
Take y from Chapter 2 in Problem 3.9 to test $H_0: \sigma^2 = 20$ against $H_A: \sigma^2 > 20$ with $\alpha = 0.05$.

```
> y
[1]  5  7  1  7  8  9 13  9 10 10 18 10 15 10 10 11  8 11
    12 13 15 22 10 25 11
```

and calculate

```
> var(y)
[1] 25.25
```

From this we obtain $\chi^2 = \frac{25.25}{20} = 1.26$ and this is smaller than

```
> qchisq(0.95,24)
[1] 36.41503
```

and $H_0: \sigma^2 = 20$ is accepted.

3.2.4 Test on a Probability

We already discussed the analysis problem in Problem 3.5. Here we show how to determine the sample size approximately.

Let p be the probability of some event. To test one of the pairs of hypotheses below:

(a) $H_0: p \le p_0, H_A: p > p_0$
(b) $H_0: p \ge p_0, H_A: p < p_0$
(c) $H_0: p = p_0, H_A: p \ne p_0$

with a risk of the first kind α, and if we want the risk of the second kind to be not larger than β as long as

(a) $p_1 > p_{0,}$ and $p_1 - p_0 \ge \delta$
(b) $p_1 < p_0$ and $p_0 - p_1 \ge \delta$
(c) $p_0 \ne p_1$ and $|p_0 - p_1| \ge \delta$.

Determine the minimum sample size from

$$n = \left\lceil \frac{[Z_{1-\alpha}\sqrt{p_0(1-p_0)} + Z_{1-\beta}\sqrt{p_1(1-p_1)}]^2}{(p_1 - p_0)^2} \right\rceil. \tag{3.12}$$

The same size is needed for the other one-sided test. In the two-sided case:

$$H_0 : p = p_0; H_A : p \ne p_0$$

we replace α by $\alpha/2$ and calculate (3.12) twice for $p_1 = p_0 - \delta > 0$ and for $p_1 = p_0 + \delta < 1$ if δ is the difference from p_0, which should not be overlooked with a probability of at least $1 - \beta$. From the two n-values, we then take the maximum.

Example 3.3 We would like to test that the probability $p = P(A)$ of the event A: 'an iron casting is faulty' equals $p_0 = 0.1$ against the alternative that it is larger. The risk of the first kind is planned to be $\alpha = 0.05$ and the power should be 0.9 as long as $p \ge 0.2$. How many castings should be tested?

From Table 3.1 we take $Z_{0.975} = 1.96$ and $Z_{0.9} = 1.282$ and get

$$n = \left\lceil \frac{[1.96\sqrt{0.1 \cdot 0.9} + 1.282\sqrt{0.2 \cdot 0.8}\,]^2}{0.1^2} \right\rceil = \lceil 121.18 \rceil = 122$$

(the other option is not possible because $p_1 = p_0 - \delta = 0$).

In R we can calculate n as follows:

```
> numerator <- (1.96*sqrt((0.1*0.9))
     + 1.282*sqrt((0.2*0.8)))^2
> denominator <- (0.1)^2
> n <- ceiling((numerator/denominator))
> n
[1] 122
```

We check the sample size $n = 122$ by first determining the right-sided critical value of $B(122, 0.10)$ for $\alpha = 0.05$ with R.

First, we calculate the 0.95-quantile of $B(122, 0.10)$:

```
> qbinom(p=0.95, size=122, prob=0.10)
[1] 18
```

Now we calculate Prob(B(122, 0.10) $\geq$ 18)

```
> 1-pbinom(q=17,size=122,prob=0.1)
[1] 0.06062245
```

and $0.060\,622\,45 > 0.05$.

And we calculate Prob(B(122,0.10) $\geq$ 19)

```
> 1-pbinom(q=18,size=122,prob=0.1)
[1] 0.03452349
```

and $0.034\,523\,49 < 0.05$.

Hence the right-sided critical value is 19 for $\alpha = 0.05$.

The power of $B(122, 0.20)$ is found as Prob($B(122, 0.20) \geq 19$)

```
> 1-pbinom(q=18,size=122,prob=0.2)
[1] 0.9125418
```

and $0.912\,541\,8 > 0.90$.

3.2.5 Paired Comparisons

If we measure two traits x and y from some individuals, we could say that we have a two-sample problem. However, from a mathematical point of view we can say that the two measurements are realisations of a two-dimensional random variable $\binom{x}{y}$ and from the corresponding universe one sample is drawn with measurements $\binom{x_i}{y_i}$ $i = 1, \dots, n$. We then speak about paired observations or statistical twins. In place of discussing the question of whether $\boldsymbol{x}_\mathrm{i}$ and $\boldsymbol{y}_\mathrm{i}$ have the same distribution we can ask whether the expectation of $\boldsymbol{d}_i = \boldsymbol{x}_i - \boldsymbol{y}_i$ is zero. The problem is reduced to that considered in Section 3.2.1 and must not be discussed again.

If the random variable $\boldsymbol{d}$ is extremely non-normal (skew for instance) with the median m, then based on n realisations d_i, $i = 1, \dots, n$ of the $\boldsymbol{d}_i$ null hypothesis, for the test below all $d_i = 0$ are excluded from the sample; the reduced sample size is denoted by n^*.

$H_0 : m = m_0$ against $H_A : m \neq m_0$ can be tested with Wilcoxon's signed rank test. For this the new differences $d_i^* = d_i - m_0, i = 1, \dots, n$ are calculated as well as their absolute values $|d_i^*|$ after deleting the values $d_i^* = 0$. The remaining data set has a size $n^* \leq n$. Finally, these values are to be ordered (ascending) and ranks R_i are to be assigned accordingly.

Ties (identical values) receive a rank equal to the average of the ranks they span. The sum of the so-called signed ranks $V = \sum_{i=1}^{n^*} \mathrm{sign}(\mathrm{rank} \mid d_i^* \mid \cdot R_i)$ is the test statistic of Wilcoxon's signed rank test. The corresponding random variable $\boldsymbol{V}$ has expectation $E(\boldsymbol{V}) = n^*(n^* + 1)/4$ and variance $var(\boldsymbol{V}) = \frac{n^*(n^*+1)(2n^*+1)}{24}$. The null hypothesis $H_0 : m = m_0$ is rejected if $V > V(n^*, 1 - \alpha)$, where α is the risk of the first kind. The critical values $V(n^*, 1 - \alpha)$ are calculated by the R program in the following example.

Table 3.4 The litter weights of mice (in grams) and the differences between the first and the second litter.

i	x_i	y_i	$d_i = x_i - y_i$
1	7.6	7.8	−0.2
2	13.2	11.1	2.1
3	9.1	16.4	−5.3
4	10.6	13.7	−3.1
5	8.7	10.7	−2
6	10.6	12.3	−1.9
7	6.8	14.0	−7.2
8	9.9	11.9	−2
9	7.3	8.8	−1.5
10	10.4	7.7	2.7
11	13.3	8.9	4.4
12	10.0	16.4	−6.4
13	9.5	10.2	−0.7

Example 3.4 Each of 13 mice gave birth to 2 litters. x_i is the weight of each of the first litter and y_i the weight of each of the second litter of these mice, which are considered to be a random sample from a population of mice. Test H_0: 'the first and second litter produce litter weights that are symmetric about a common axis' against the alternative H_A: 'the differences are not symmetric about a common axis' with significance level $\alpha = 0.05$.

Because no difference is zero, we have $n = n^*$. The analysis is done using R as follows. First we create data files from Table 3.4.

```
> x <- c(7.6, 13.2, 9.1, 10.6, 8.7, 10.6, 6.8, 9.9, 7.3,
        10.4, 13.3, 10.0, 9.5 )
> y <- c( 7.8, 11.1, 16.4, 13.7, 10.7, 12.3, 14.0, 11.9,
        8.8, 7.7, 8.9, 16.4, 10.2)
> d <- x-y
```

Then we apply Wilcoxon's signed rank test by typing

```
> wilcox.test( x, y, paired = TRUE, correct =
    FALSE,alternative= "two.sided",mu=0,conf.level=0.95)
        Wilcoxon signed rank test
data:  x and y
V = 25, p-value = 0.1518
alternative hypothesis: true location shift is not equal to 0
Warning message:
In wilcox.test.default(x, y, paired = TRUE, correct = FALSE,
     alternative = "two.sided",   :
  cannot compute exact p-value with ties
```

Of course we can also use the vector `d <- x-y` of the differences to test H_0:

```
> wilcox.test(d,correct = FALSE, alternative = "two.sided",
    mu=0, conf.level=0.95)
        Wilcoxon signed rank test
data:  d
V = 25, p-value = 0.1518
alternative hypothesis: true location is not equal to 0
Warning message:
In wilcox.test.default(d, correct = FALSE, alternative =
    "two.sided",  :
  cannot compute exact p-value with ties
```

Remark

The p-value is here calculated using R with the normal approximation where z is the standard normal variable:

```
 p-value = 2*Pr(z > |(25-[13*(13+1)/4])/
      √[13*(13+1)*(2*13+1)/24]|) = 2*Prob(z>|-1.433|) =
      2*Prob(z > 1.433)
> p_value <- 2*(1-pnorm(1.433))
> p_value
[1] 0.1518578
```

3.2.6 Sequential Tests

The technique of sequential testing offers the advantage that, given many studies, on average fewer research units need to be sampled as compared to the 'classic' approach of hypothesis testing in Section 3.2.1 with sample size fixed beforehand.

Nevertheless, we need also the precision requirements α, β, and δ, moreover a sequential test cannot be applied without this, which means that we cannot start observations without fixing β. Until now a sample of fixed size n was given; Stein (1945) proposed a method of realising a two-stage experiment. In the first stage a sample of size $n_0 > 1$ is drawn to estimate σ^2 using the variance s_0^2 of this sample and to calculate the sample size n of the method using (3.7). In the second stage $n-n_o$ further measurements are taken. Following the original method of Stein in the second stage at least one further measurement is necessary from a theoretical point of view. In this section we simplify this method by introducing the condition that no further measurements are to be taken for $n - n_o \leq 0$. Nevertheless, this yields an α-test of acceptable power.

Since both parts of the experiment are carried out one after the other, such experiments are called sequential. Sometimes it is even tenable to make all measurements step by step, where each measurement is followed by calculating a new test statistic. A sequential testing of this kind can be used if the observations of a random variable in an experiment take place successively in time. Typical examples are series of single experiments in a laboratory, psychological diagnostics in single sessions, medical treatments of patients in hospitals, consultations of clients of certain institutions, and certain procedures of statistical quality control, where the sequential approach was used the first time (Dodge and Romig 1929). The basic idea is to utilise the observations already made before the next are at hand.

For example, testing the hypothesis H_0: $\mu = \mu_0$ against H_A: $\mu > \mu_0$ (where the sample size can be determined by (3.7)) there are three possibilities in each step of evaluation, namely:

- accept H_0
- reject H_0
- continue the investigation.

The, up to now unsurpassed, textbook of Wald (1947) has since been reprinted and is therefore generally available (Wald 2004), and new results can be found in the books of Ghosh and Sen (1991) and DeGroot (2005). We do not recommend the application of this general theory, but we recommend closed plans, which end after a finite number of steps with certainty (and not only with probability 1). We first give a general definition useful in other sections independent of our one-sample problem.

Definition 3.1 A sequence $\boldsymbol{S} = \{\boldsymbol{y}_1, \boldsymbol{y}_2, \ldots\}$ of random variables is given where the components are identically and stochastically independent distributed with a parameter $P_\theta \in \Omega$. Let the parameter space Ω consist of the two different elements θ_0 and θ_A. A sequential test for H_0: $\theta = \theta_0$ against H_A: $\theta = \theta_A$, based on the ratio

$$\boldsymbol{LR}_n = \frac{L(\boldsymbol{Y}^{(n)}|\theta_A)}{L(\boldsymbol{Y}^{(n)}|\theta_0)}$$

of the likelihood functions $L(\boldsymbol{Y}^{(n)}|\theta)$ of both parameter values and on the first n elements $\boldsymbol{Y}^{(n)} = \{\boldsymbol{y}_1, \boldsymbol{y}_2, \ldots, \boldsymbol{y}_n\}$ of the sequence $\boldsymbol{S}$ is said to be the sequential likelihood ratio test (SLRT), if for certain numbers A and B with $0 < B < 1 < A$ the decomposition of $\{Y^{(n)}\}$ reads

$$M_0^n = \{Y^{(n)} : LR_n \leq B\},\quad M_A^n = \{Y^{(n)} : LR_n \geq A\},\quad M_F^n = \{Y^{(n)} : B < LR_n < A\}.$$

An SLRT that leads with probability 1 to a final decision with the strength (α, β) fulfils, with the numbers A and B from Definition 3.1, the conditions

$$A \leq \frac{1-\beta}{\alpha},$$

$$B \geq \frac{\beta}{1-\alpha}.$$

In the applications the equalities are often used to calculate approximately the bounds A and B. Such tests are called approximate tests.

It follows from the theory that SLRTs can hardly be recommended, since they end under certain assumptions only with probability 1. On the other hand, so far they are the most powerful tests for a given strength as the expectation of the sample size – the average sample number (ASN) – for such tests is minimal and smaller than the size for tests where the size is fixed. Since it is unknown for which maximal sample size the SLRT ends with certainty, it belongs to the class of open sequential tests. In comparison there are also closed sequential tests, i.e. tests with a secure maximal sample size, but this advantage is won by a bit larger ASN. We will concentrate on closed sequential triangular tests.

Their principle goes back to Whitehead (1997) and is as follows. Using the observed values y_i, $i = 1, 2, \ldots$, some cumulative ascertained ancillary values Z_v and V_v are calculated.

Let $\boldsymbol{S} = (\boldsymbol{y}_1, \boldsymbol{y}_2, \ldots)$ be a sequence with components distributed together with $\boldsymbol{y}$ as $N(\mu, \sigma^2)$ and $S = (y_1, y_2, \ldots)$ its realisation. We want to test

$$H_o : \mu = \mu_o, \sigma^2 \text{ arbitrary, against } H_A : \mu = \mu_1, \sigma^2 \text{ arbitrary.}$$

Then we get

$$z_n = \frac{\sum_{i=1}^{n} y_i}{\sqrt{\frac{\sum_{i=1}^{n} y_i^2}{n}}}, \qquad v_n = n - \frac{z_n^2}{2n}.$$

The boundary lines of the triangle follow from

$$a = \left(1 + \frac{z_{1-\beta}}{z_{1-\alpha}}\right) \frac{\ln \frac{\alpha}{2}}{\theta_1}, \tag{3.13}$$

$$b = \frac{\theta_1}{\left(1 + \frac{z_{1-\beta}}{z_{1-\alpha}}\right)}. \tag{3.14}$$

We put $\theta_1 = \frac{\mu_1 - \mu_0}{\sigma}$ and calculate from the n_1 and n_2 observations the maximum likelihood estimator

$$\tilde{\sigma}_n^2 = \frac{\sum_{i=1}^{n_1} (y_{1i} - \bar{y}_1)^2 + \sum_{i=1}^{n_2} (y_{2i} - \bar{y}_2)^2}{n_1 + n_2}.$$

Then we introduce

$$z_n = \frac{n_1 n_2}{n_1 + n_2} \frac{\bar{y}_1 - \bar{y}_2}{\tilde{\sigma}_n}, \qquad v_n = \frac{n_1 n_2}{n_1 + n_2} \frac{z_n^2}{2(n_1 + n_2)}.$$

The hypothesis $H_0 : \mu = \mu_0$ is accepted, if

$$z_n \geq a + bv_n \quad for \quad \mu_1 > 0$$

and if

$$z_n \leq -a + bv_n \quad \text{for} \quad \mu_1 < 0.$$

Both straight lines on the boundary of the triangle meet at the point

$$(v_{\max}; z_{\max}) = \left(\frac{a}{b}; 2a\right).$$

Problem 3.10 Show how a triangular sequential test can be calculated.

If the output

```
Test not finished, continue by adding single data via
     update() current sample size for x: 2
```

occurs, a further observation has to be made and input after

```
> conscient.tt <- update(conscient.tt, x =)
```

Solution

Use in the OPDOE-command $\mu_0 = \text{mu0}$ and $\mu_1 = \text{mu1}$

```
> conscient.tt <- triangular.test.norm(x =, mu0=mu1 =,
      sigma = 10, alpha =, beta = )
```

Example from Rasch et al. (2011c)

In a study with the aid of a personality questionnaire, the character 'conscientiousness' was tested. Conscientiousness is standardised to T-scores, i.e. modelled as a normally distributed random variable with mean $\mu_0 = 50$ and standard deviation $\sigma = 10$. The null hypothesis is H_0: $\mu = \mu_0 = 50$ and the alternative hypothesis is H_A: $\mu < 50$, if the sample does not stem from the population in question, then we expect the children to achieve lower test scores. We choose $\alpha = 0.05$, $\beta = 0.20$, and $\delta = 0.67$. We start with the first five children of our data set, who are tested with an intelligence test battery.

Their results are as follows: 50, 52, 53, 40, and 48.

In R, we type

```
> conscient.tt <- triangular.test.norm(x = c(50, 52), mu0 =
    50,mu1 = 43.3, sigma =  10,alpha = 0.05, beta = 0.2)
```

The outcome was

```
Triangular Test for normal distribution
Sigma known: 10
H0: mu = 50 versus H1: mu < 43.3 alpha: 0.05 beta:0.2
       Test not finished, continue by adding single data via
         update()
```

We now sequentially input further values like 38, 45, 56, 45, 50, 53, 68, 64, 55, 54; again the test could not be finished. However, after input 51 we obtain

```
> conscient.tt <- update(conscient.tt, x=51)
```

and obtain Figure 3.3.

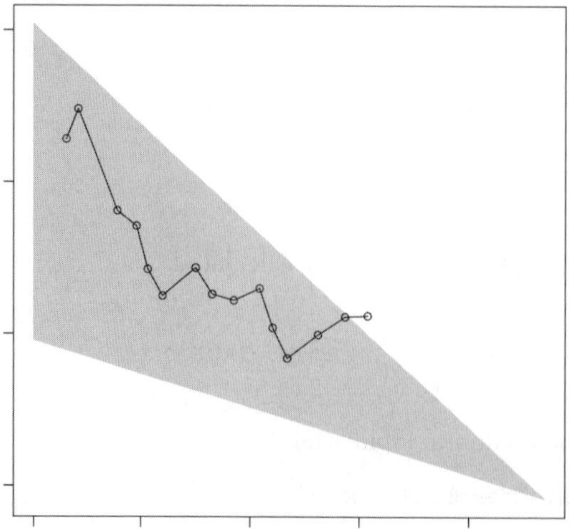

Figure 3.3 Result of the example.

3.3 The Two-Sample Problem

We assume that two independent random samples from each of two populations are investigated.

Independent means that each outcome in one sample, as a particular event, will be observed independently of all outcomes in the other sample. Once the outcomes of the particular character are given, (point) estimates are calculated separately for each sample, as shown in Section 3.2. However, we discuss here the problem whether the two samples really stem from two different populations.

We will first consider again the parameter μ; that is, the two means of the two populations underlying the two samples.

3.3.1 Tests on Two Expectations

We assume that the character of interest is in each population modelled by a normally distributed random variable. That is to say, independent random samples of size n_1 and n_2, respectively, are drawn from populations 1 and 2. The observations of the random variables $\boldsymbol{y}_{11}, \boldsymbol{y}_{12}, \dots \boldsymbol{y}_{1n_1}$ on the one hand and $\boldsymbol{y}_{21}, \boldsymbol{y}_{22}, \dots \boldsymbol{y}_{2n_2}$ on the other hand will be $y_{11}, y_{12}, \dots y_{1n_1}$ and $y_{21}, y_{22}, \dots y_{2n_2}$. We call the underlying parameters μ_1 and μ_2, and σ_1^2 and σ_2^2, respectively. The unbiased estimators are then used for the expectations $\hat{\boldsymbol{\mu}}_1 = \bar{\boldsymbol{y}}_1$ and $\hat{\boldsymbol{\mu}}_2 = \bar{\boldsymbol{y}}_2$, respectively, and for the variances $\hat{\boldsymbol{\sigma}}_1^2 = \boldsymbol{s}_1^2$ and $\hat{\boldsymbol{\sigma}}_2^2 = \boldsymbol{s}_2^2$, respectively (according to Section 3.2).

We test the null hypothesis

H_0: $\mu_1 = \mu_2 = \mu$; against one of the alternative hypotheses

(a) H_A: $\mu_1 < \mu_2$
(b) H_A: $\mu_1 > \mu_2$
(c) H_A: $\mu_1 \neq \mu_2$.

According to recent findings from simulation studies (Rasch et al. 2011a), the traditional approach is usually unsatisfactory because it is not theoretically regulated for all possibilities.

Nevertheless, we start with an introduction to the traditional approach because the current literature on applications of statistics refers to it, and methods for more complex issues are built on it. Further, there are situations in which it is certain that both populations have equal variances, especially if standardisation takes place, as in intelligence tests.

3.3.1.1 The Two-Sample *t*-Test

The traditional approach bases upon the test statistic in Formula (3.15). It examines the given null hypothesis for the case that the variances σ_1^2 and σ_2^2, respectively, of the relevant variables in the two populations, are not known; this is the usual case. However, it assumes further that these variances are equal in both populations. The test statistic

$$\mathbf{t} = \frac{\bar{\boldsymbol{y}}_1 - \bar{\boldsymbol{y}}_2}{\sqrt{\frac{(n_1 - 1) \cdot \boldsymbol{s}_1^2 + (n_2 - 1) \cdot \boldsymbol{s}_2^2}{n_1 + n_2 - 2}}} \cdot \sqrt{\frac{n_1 n_2}{n_1 + n_2}} \tag{3.15}$$

is (central) t-distributed with $n_1 + n_2 - 2$ degrees of freedom. This means, that we can decide in favour of, or against, the null hypothesis, analogous to Section 3.2, by using the $(1 - \alpha)$-quantile of the (central) t-distribution $t(n_1 + n_2 - 2, 1 - \alpha)$ and $t(n_1 + n_2 - 2, 1 - \frac{\alpha}{2})$, respectively, depending on whether it is a one- or two-sided question.

H_0: $\mu_1 = \mu_2 = \mu$; will be rejected if:

(a) $t < t(n_1 + n_2 - 2, \alpha)$
(b) $t > t(n_1 + n_2 - 2, 1 - \alpha)$
(c) $|t| > t\left(n_1 + n_2 - 2, 1 - \frac{\alpha}{2}\right)$.

If the null hypothesis is rejected, we say that the two expectations differ significantly.

Example 3.5 We consider the entries of Table 3.4 with the litter weights of mice (in grams) but interpret the entries differently. We assume that the x and y values are obtained independently from two strains of mice.

```
> x <- c(7.6, 13.2, 9.1, 10.6, 8.7, 10.6, 6.8, 9.9, 7.3,
        10.4, 13.3, 10.0, 9.5 )
> y <- c( 7.8, 11.1, 16.4, 13.7, 10.7, 12.3, 14.0, 11.9,
        8.8, 7.7, 8.9, 16.4, 10.2)
```

We obtain using R the t-value $t\ (24, 0.05) = -1.785$ with 24 degrees of freedom.

For $\alpha = 0.05$ and $f = 24$ we find using R

```
> qt(0.05,24)
[1] -1.710882
> qt(0.025,24)
[1] -2.063899
```

In a practical problem, exactly one of the three possible alternative hypotheses will be used. Here we show the procedures for all the possible choices.

Case 1: the experimenter tests H_0 against

$$H_A : \mu_1 > \mu_2$$

Decision: accept H_0 because

$$-1.78 < 1.711.$$

Case 2: the experimenter tests H_0 against

$$H_A : \mu_1 < \mu_2$$

Decision: reject H_0, because

$$-1.78 < -1.711.$$

Case 3: the experimenter tests H_0 against

$$H_A : \mu_1 \neq \mu_2$$

Decision: accept H_0, because

$$|-1.78| < 2.064.$$

With R we proceed as follows:

```
> t.test(x,y,alternative = "less", mu=0, conf.level =0.95,
     var.equal=TRUE)

        Two Sample t-test

data:  x and y
t = -1.7849, df = 24, p-value = 0.04346
alternative hypothesis: true difference in means is
     less than 0
95 percent confidence interval:
       -Inf -0.0730922
sample estimates:
mean of x mean of y
 9.769231 11.530769

> t.test(x,y,alternative = "greater", mu=0, conf.level=0.95,
     var.equal=TRUE)

        Two Sample t-test

data:  x and y
t = -1.7849, df = 24, p-value = 0.9565
alternative hypothesis: true difference in means is
     greater than 0
95 percent confidence interval:
 -3.449985       Inf
sample estimates:
mean of x mean of y
 9.769231 11.530769

> t.test(x,y,alternative = "two.sided",mu=0,conf.level=0.95,
     var.equal=TRUE)

        Two Sample t-test

data:  x and y
t = -1.7849, df = 24, p-value = 0.08692
alternative hypothesis: true difference in means is not
     equal to 0
95 percent confidence interval:
 -3.798372  0.275295
sample estimates:
mean of x mean of y
 9.769231 11.530769
```

Which sample size is needed for this two-sample t-test for a risk of the first kind α and a risk of the second kind not larger than β as long as $|\mu_1 - \mu_2| > \delta$?

Given the total number of observations $n_1 + n_2$ then $var(\bar{y}_1 - \bar{y}_2)$ is minimised for $n_1 = n_2 = n$, i.e. when the two sample sizes are equal.

The value of n can be determined from the demand that the $(1 - \alpha)$-quantile for the one-sided alternatives above or the $\left(1 - \frac{\alpha}{2}\right)$-quantile for the two-sided alternative of the central t-distribution with $2n - 2$ degrees of freedom equals the $(1 - \beta)$-quantile of the non-central t-distribution with ncp $\lambda = \frac{\mu_1-\mu_2}{\sigma}\sqrt{\frac{n_1 n_2}{n_1+n_2}} = \frac{\delta}{\sigma}\sqrt{\frac{n}{2}}$ and $2n - 2$ degrees of freedom. This illustrates Figure 3.1. In Problem 3.3 it is shown what to do when σ is unknown.

We show how to proceed with R in Problem 3.11 below.

Problem 3.11 How can we calculate the minimum sample size n per sample for the two-sample t-test for a risk of the first kind α and a risk of the second kind not larger than β as long as $|\mu_1 - \mu_2| > \delta$ with $\sigma = 1$?

Solution

Use in R the OPDOE command, e.g. for the one-sided test

```
> size.t.test(delta=0.5,sd=1,sig.level=0.01,power=0.9,
     type="two.sample", alternative="one.sided")
> size.t.test(delta=0.5,sd=1,sig.level=0.01,power=0.9,
     type="two.sample", alternative="two.sided")
```

Here sd is the standard deviation σ and power $= 1 - \beta$.

Example

We demonstrate the command above for a one- and a two-sided alternative $\alpha = 0.01$, $\delta = 0.5$; $\sigma = 1$, and $\beta = 0.1$.

```
> size.t.test(delta=0.5,sd=1,sig.level=0.01,power=0.9,
     type="two.sample", alternative="one.sided")
[1] 106
> size.t.test(delta=0.5,sd=1,sig.level=0.01,power=0.9,
     type="two.sample", alternative="two.sided")
[1] 121.
```

Thus, the minimum sample size for the one-sided case is 106, for the two-sided case we would need at least 121 observations each.

3.3.1.2 The Welch Test

We now discuss the case that either it is known that the variances σ_1^2 and σ_2^2 of the two populations are unequal or that we are not sure that they are equal – this is often the case. Therefore, we recommend always using the proposed Welch-test below in place of the two-sample t-test. The recommendation to use the Welch test instead of the t-test cannot be found in most textbooks of applied statistics and may therefore surprise experienced users. This recommendation is instead based on the results of current research

Table 3.5 A summary of a simulation study in Rasch et al. (2011a).

			$\delta = 0$					
			With pre-tests			Without pre-tests		
$\sqrt{(\sigma^2_1/\sigma^2_2)}$	n_1	n_2	t	Welch (Levene)	W (KS)	t	Welch	W
1	10	10	4.95	4.55 (4.84)	0.00 (0.02)	4.96	4.82	5.21
	30	30	4.97	4.82 (4.98)	11.11 (0.09)	4.96	4.95	4.92
	10	30	5.00	4.87 (4.31)	0.00 (0.06)	5.01	5.05	4.87
	30	10	4.97	5.09 (4.32)	0.00 (0.06)	4.96	5.15	5.00
	30	100	4.86	5.00 (4.80)	9.09 (0.11)	4.84	4.91	4.82
2	10	10	6.08	3.33 (32.99)	0.00 (0.03)	5.38	4.93	6.06
	30	30	7.37	4.78 (91.89)	10.00 (0.10)	5.15	4.98	5.88
	10	30	1.02	5.90 (30.18)	0.00 (0.07)	0.90	5.00	2.19
	30	10	19.38	2.98 (73.73)	16.67 (0.06)	15.51	5.19	10.09
	30	100	1.33	4.99 (98.38)	0.00 (0.11)	0.63	4.96	1.89

The entries give the percentage of rejecting the (final) hypothesis H_0: $\mu_1 = \mu_2$. t is for student's t-test and W for the Wilcoxon–Mann–Whitney-test both with and without pre-testing. Levene means the Levene test and KS the Kolmogorov–Smirnov test on normality.

from Rasch et al. (2011a). There it is also explained that pre-testing first for equality of variances and if the equality is accepted to continue with the two-sample t-test makes no sense. We show in Table 3.5 a part of table 2 from Rasch et al. (2011a).

Pre-tests of the assumptions are worthless, as shown by our simulation experiments. First testing the normal assumption per sample, testing then the null hypothesis for equality of variances, with the Levene test from Section 3.3.4, and finally, testing the null hypothesis for equality of means leads to terrible risks, i.e. the risk estimated using simulation is only in the case of the Welch test near the chosen value $\alpha = 0.05$ – it is exceeded by more than 20% in only a few cases. By contrast, the t-test rejects the null hypothesis largely. Moreover, the Wilcoxon W-test (see Section 3.3.1.3) does not maintain the type-I risk in any way; it is too large and it has to be carried out with a newly collected pair of samples.

The Welch test is based on the test statistic

$$\boldsymbol{t}^* = \frac{\overline{\boldsymbol{y}}_{1.} - \overline{\boldsymbol{y}}_{2.} - (\mu_1 - \mu_2)}{\sqrt{\frac{\boldsymbol{s}_1^2}{n_1} + \frac{\boldsymbol{s}_2^2}{n_2}}}, \quad \boldsymbol{s}_k^2 = \frac{1}{n_k - 1} \sum_{i=1}^{n_k} (\boldsymbol{y}_{ki} - \overline{\boldsymbol{y}_{k.}})^2, \quad k = 1, 2.$$

The distribution of the test statistic was derived by Welch (1947) and is presented in theorem 3.18 in *Mathematical Statistics* (Rasch and Schott, 2018).

If the pair of hypotheses

$$H_0 : \mu_1 = \mu_2 = \mu, \sigma_1^2, \sigma_2^2 \text{ arbitrary}$$

$$H_A : \mu_1 \neq \mu_2, \sigma_1^2, \sigma_2^2 \text{ arbitrary}$$

is to be tested, often the approximate test statistic

$$t^* = \frac{\bar{y}_{1.} - \bar{y}_{2.}}{\sqrt{\frac{s_1^2}{n_1} + \frac{s_2^2}{n_2}}}$$

is used. H_0 is rejected if $|t^*|$ is larger than the corresponding quantile of the central t-distribution with approximate degrees of freedom f, the so-called Satterthwaite's f,

$$f = \frac{\left(\frac{s_1^2}{n_1} + \frac{s_2^2}{n_2}\right)^2}{\frac{s_1^4}{{n_1}^2(n_1 - 1)} + \frac{s_2^4}{{n_2}^2(n_2 - 1)}}$$

degrees of freedom.

Referring to the t-test, the desired relative effect size is, for equal sample sizes

$$\frac{\mu_1 - \mu_2}{\sigma}.$$

For the Welch test, the desired relative effect size then reads:

$$\frac{\mu_1 - \mu_2}{\sqrt{\frac{\sigma_1^2 + \sigma_2^2}{2}}}.$$

Planning a study for the Welch test is carried out in a rather similar manner to that for two-sample t-test. A problem arises, however: in most cases it is not known whether the two variances in the relevant population are equal or not, and if not, to what extent they are unequal. If one knew this, then it would be possible to calculate the necessary sample size exactly. In the other case, one can use the largest appropriate size, which results from equal, realistically maximum expected variances.

Problem 3.12 Use the Welch test based on the approximate test statistic to test:

H_0: $\mu_1 = \mu_2 = \mu$; against one of the alternative hypotheses:

(a) H_A: $\mu_1 < \mu_2$
(b) H_A: $\mu_1 > \mu_2$
(c) H_A: $\mu_1 \neq \mu_2$.

Solution

In R use the commands respectively for (a), (b) and (c):

```
> t.test(x,y,alternative= "less", mu=0, conf.level =0.95,
    var.equal=FALSE)
> t.test(x,y,alternative= "greater", mu=0, conf.level =0.95,
    var.equal=FALSE)
```

```
> t.test(x,y,alternative= "two.sided", mu=0, conf.level
    =0.95, var.equal=FALSE)
```

Because the default option in the `>t.test( )` is `var.equal= FALSE` we do not use this option.

Example
We use the entries of Table 3.4 to test respectively for (a), (b), and (c) whether the strains of mice have the same expectations, which we have used in Example 3.6.

```
> t.test(x,y,alternative= "less", mu=0, conf.level =0.95)

        Welch Two Sample t-test

data:  x and y
t = -1.7849, df = 21.021, p-value = 0.04435
alternative hypothesis: true difference in means is
     less than 0
95 percent confidence interval:
        -Inf -0.06343762
sample estimates:
mean of x mean of y
 9.769231 11.530769

> t.test(x,y,alternative= "greater", mu=0, conf.level =0.95)

        Welch Two Sample t-test

data:  x and y
t = -1.7849, df = 21.021, p-value = 0.9556
alternative hypothesis: true difference in means is
     greater than 0
95 percent confidence interval:
 -3.459639       Inf
sample estimates:
mean of x mean of y
 9.769231 11.530769

> t.test(x,y,alternative= "two.sided", mu=0, conf.level
     =0.95)

        Welch Two Sample t-test

data:  x and y
t = -1.7849, df = 21.021, p-value = 0.08871
alternative hypothesis: true difference in means is not
     equal to 0
```

```
95 percent confidence interval:
 -3.8137583  0.2906814
sample estimates:
mean of x mean of y
 9.769231 11.530769
```

3.3.1.3 The Wilcoxon Rank Sum Test

Wilcoxon (1945) proposed for equal sample sizes, and later Mann and Whitney (1947) extended for unequal sample sizes, a two-sample distribution-free test based on the ranks of the observations. This test is not based on the normal assumption, in its exact form only two continuous distributions with all moments existing are assumed; we call it the Wilcoxon test. As is seen from the title of Mann and Whitney (1947), this test is testing whether one of the underlying random variables is stochastically larger than the other. It can be used for testing the equality of medians under additional assumptions: if the continuous distributions are symmetric, the medians are equal to the expectations. The null hypothesis tested by the Wilcoxon test corresponds to the hypothesis of equality of the medians m_1 and m_2, $H_0 : m_1 = m_2 = m$ if and only if all higher moments of the two populations exist and are equal. Otherwise a rejection of the Wilcoxon hypothesis says little about the rejection of $H_0 : m_1 = m_2 = m$. The test runs as follows based on the observations $y_{11}, y_{12}, \dots y_{1n_1}$ and $y_{21}, y_{22}, \dots y_{2n_2}$ of the random variables $\boldsymbol{y}_{11}, \boldsymbol{y}_{12}, \dots \boldsymbol{y}_{1n_1}$ and $\boldsymbol{y}_{21}, \boldsymbol{y}_{22}, \dots \boldsymbol{y}_{2n_2}$, respectively. We assume the sample size $n_1 \leq n_2$.

Calculate:

$$d_{ij} = \begin{cases} 1 & if\, y_{1i} > y_{2j} \\ 0 & \text{otherwise} \end{cases} \tag{3.16}$$

and

$$W = W_{12} = \frac{n_1(n_1+1)}{2} + \sum_{i=1}^{n_1} \sum_{j=1}^{n_2} d_{ij}. \tag{3.17}$$

We denote the P-quantiles of $\boldsymbol{W}$ based on $n_1 \leq n_2$ observations by $W(n_1, n_2, P)$.

Reject H_0:

in case (a), if $W > W(n_1, n_2, 1-\alpha)$,
in case (b), if $W < W(n_1, n_2, \alpha)$,
in case (c), if either $W < W\left(n_1, n_2, \frac{\alpha}{2}\right)$ or $W > W\left(n_1, n_2, 1 - \frac{\alpha}{2}\right)$,

and accept it otherwise (the three cases are those given in problem 3.13).

The expectation of $\boldsymbol{W}$ if H_0 is valid is $E(\boldsymbol{W}) = n_1\,(n_1 + n_2)/2$.

The variance of $\boldsymbol{W}$ if H_0 is valid is $\mathrm{var}(\boldsymbol{W}) = n_1\, n_2(n_1 + n_2 + 1)/12$.

Mann–Whitney used the test statistic $\boldsymbol{U}$, which is the number of times that an x observation is larger than a y observation. The relation between $\boldsymbol{U}$ and $\boldsymbol{W}$ is: $\boldsymbol{U} = \boldsymbol{W} - n_1(n_1+1)/2$. Hence the expectation of $\boldsymbol{U}$ is $E(\boldsymbol{U}) = E(\boldsymbol{W}) - n_1(n_1+1)/2$ and the variance of $\boldsymbol{U}$ is $\mathrm{var}(\boldsymbol{U}) = \mathrm{var}(\boldsymbol{W} - n_1(n_1+1)/2)$

The expressions for $\mathrm{var}(\boldsymbol{W})$ and $\mathrm{var}(\boldsymbol{U})$ must be corrected if there are ties (equal observations) in the data.

If there are ties of size t_i then var($\boldsymbol{W}$) must be subtracted by $n_1n_2[\Sigma(t_i^3 - t_i)]/[12(N^2 - N)]$ and with ties we have var($\boldsymbol{W}$) = var($\boldsymbol{U}$) = $\{n_1\ n_2\ (N^3 - N - \Sigma(t_i^3 - t_i)\}/[12(N^2 - N)]$.

The expectations of $E(W)$ and $E(\boldsymbol{U})$ remain the same in the case of ties.

Be aware that the quantiles of the random test statistic $\boldsymbol{U}$ are in R denoted by $\boldsymbol{W}$ and are given by the R-command

```
> qwilcox(P,n1,n2)
```

as an example take

```
> qwilcox(0.05,15,20)
[1] 101
>  qwilcox(0.95,15,20)
[1] 199
```

Problem 3.13 Test for two continuous distributions:

H_0: $m_1 = m_2 = m$; against one of the alternative hypotheses:
(a) H_A: $m_1 < m_2$
(b) H_A: $m_1 > m_2$
(c) H_A: $m_1 \neq m_2$.

Solution
In R use the commands with significance level $\alpha = 0.05$ respectively for (a), (b), and (c):

```
> wilcox.test(x,y,alternative="less", mu=0,exact=T,
     conf.level=0.95)
> wilcox.test(x,y,alternative="greater", mu=0,exact=T,
     conf.level=0.95)
> wilcox.test(x,y,alternative="two.sided", mu=0,exact=T,
     conf.level=0.95)
```

Example
We use the entries of Table 3.4 to test respectively for (a), (b), and (c) whether the strains of mice have the same expectations that we used in Example 3.6.

```
> wilcox.test(x,y,alternative="less", mu=0,exact=T,
     conf.level=0.95)

        Wilcoxon rank sum test with continuity correction

data:  x and y
W = 51, p-value = 0.04524
alternative hypothesis: true location shift is less than 0

Warning message:
In wilcox.test.default(x, y, alternative = "less", mu = 0,
     exact = T,  :
  cannot compute exact p-value with ties
```

```
> wilcox.test(x,y,alternative="greater", mu=0,exact=T,
     conf.level=0.95)

        Wilcoxon rank sum test with continuity correction

data:  x and y
W = 51, p-value = 0.9594
alternative hypothesis: true location shift is greater
     than 0

Warning message:
In wilcox.test.default(x, y, alternative = "greater",
     mu = 0, exact = T,  :
  cannot compute exact p-value with ties

> wilcox.test(x,y,alternative="two.sided", mu=0,exact=T,
     conf.level=0.95)

        Wilcoxon rank sum test with continuity correction

data:  x and y
W = 51, p-value = 0.09048
alternative hypothesis: true location shift is not
     equal to 0

Warning message:
In wilcox.test.default(x, y, alternative = "two.sided",
     mu = 0,  :
  cannot compute exact p-value with ties
```

The p-value is in this case of ties calculated in R by the normal approximation.

For the alternative (a), alternative = 'less', we calculate with the continuity correction of 0.5 as follows. There are two ties, the observation 10.6 and also the observation 16.4 occurs twice, all other 22 observations occur once.

```
N=13+13=26, t1 = 2, t2 = 2, The p-value =
Prob(Z<[51 - 13*13/2+ 0.5]/√[13*13*{(263 -26 -(23 -2)
     -(23 -2)-22*0}/{12*(262 -26}] =
Prob(Z < -33/√[2963922/7800]) = Prob (Z < - 33/19.4933) =
     Prob (Z < -1.6929)
> p_value <- pnorm(-1.6929) [1] 0.04523725
```

3.3.1.4 Definition of Robustness and Results of Comparing Tests by Simulation

We give here an introduction to methods of empirical statistics via simulations and methods, which will be used later in this chapter and also in other chapters (especially in Chapter 11).

The robustness of a statistical method means that the essential properties of this method are relatively insensitive to variations of the assumptions. We especially want to investigate the robustness of the methods of this section with respect to violating normality.

Definition 3.2 Let $\boldsymbol{k}_\alpha$ be an α-test $(0 < \alpha < 1)$ for the pair $\{H_0, H_A\}$ of hypotheses in the class G_1 of distributions of the random sample $\boldsymbol{Y}$ with size n. And let G_2 be a class of distributions containing G_1 and at least one distribution, which does not fulfil all assumptions for guaranteeing $\boldsymbol{k}_\alpha$ to be an α-test.

Finally, let $\alpha(g)$ be the risk of the first kind for $\boldsymbol{k}_\alpha$ concerning the element g of G_2 (estimated by simulation). Here and in the sequel we write $\alpha(g) = \alpha_{\text{act}}$, the actual α and the α pre-given nominal α written as α_{nom}.

Then $\boldsymbol{k}_\alpha$ is said to be $(1 - \varepsilon)$-robust in the class G_2 if

$$\max_{g \,\varepsilon\, G_2} |\alpha_{\text{act}} - \alpha_{\text{nom}}| \leq \varepsilon.$$

We call a statistical test acceptable if $100(1 - \varepsilon)\% \geq 80\%$.

We use for G_1 the family of univariate normal $(N(\mu, \sigma^2))$ distributions and for G_2 the Fleishman system of distributions.

Definition 3.3 A distribution belongs to the Fleishman system (Fleishman 1978) if its first four moments exist and if it is the distribution of the transform

$$\boldsymbol{y} = a + b\boldsymbol{x} + c\boldsymbol{x}^2 + d\boldsymbol{x}^3 \tag{3.18}$$

where $\boldsymbol{x}$ is a standard normal random variable (with mean 0 and variance 1).

By a proper choice of the coefficients a, b, c, and d, the random variable $\boldsymbol{y}$ will have any quadruple of first four moments $(\mu, \sigma^2, \gamma_1, \gamma_2)$. By γ_1 and γ_2 we denote the skewness (standardised third moment) and the kurtosis (standardised fourth moment) of any distribution, respectively. For instance, any normal distribution (i.e. any element of G_1) with mean μ and variance σ^2 can be represented as a member of the Fleishman system by choosing $a = \mu$, $b = \sigma$ and $c = d = 0$. This shows that we really have $G_2 \supset G_1$, as demanded in Definition 3.3.

It is known that all probability and empirical distributions (with existing fourth moment) fulfil the inequality

$$\gamma_2 \geq \gamma_1^2 - 2\,(g_2 \geq g_1^2 - 2). \tag{3.19}$$

The equality sign in (3.19) defines a parabola in the (γ_1,γ_2) – plane $\{(g_1,g_2)$ – plane for observations).

Rasch and Guiard (2004) selected seven (γ_1, γ_2) values in that parabola for the robustness investigations reported in this paper. The values, together with the coefficients a, b, c, and d of the elements in the Fleishman system for the case $\mu = 0$ (which means $a = -c$) and $\sigma = 1$ (which means $b^2 + 6bd + 2c^2 + 15d^2 = 1$), are given in Table 3.6.

Rasch and Guiard (2004) used in a simulation experiment 10 000 runs for comparing two means of independent random samples of size n_1 and n_2 drawn from two populations with parameters $\mu_1, \sigma_1^2, \gamma_{11}, \gamma_{12}$ and $\mu_2, \sigma_2^2, \gamma_{21}, \gamma_{22}$, respectively.

To test the null hypothesis:

H_o: $\mu_1 = \mu_2$ against one of the following one- or two-sided alternative hypotheses:

(a) H_A: $\mu_1 > \mu_2$
(b) H_A: $\mu_1 < \mu_2$
(c) H_A: $\mu_1 \neq \mu_2$

amongst others the two sample t-test, the Welch test, and the Wilcoxon (Mann–Whitney) test have been used. The variances (σ_1^2, σ_2^2) were chosen as (0.5, 1), (0.5, 0.5), and (0.5, 2), respectively. The simulation was performed under the null hypothesis (the relative frequency of rejecting H_0 was used as α_{act}) and for

Table 3.6 (γ_1,γ_2)-values and the corresponding coefficients used in the simulations.

No of distribution	γ_1	γ_2
1	0	0
2	0	1.5
3	0	3.75
4	0	7
5	1	1.5
6	1.5	3.75
7	2	7

$\frac{|\mu_i-\mu_3|\cdot\sqrt{n}}{\sqrt{\sigma_1{}^2+\sigma_2{}^2}} = \Delta_i$ the relative frequency of rejecting H_0 was used as an estimate of the power. Summarising, the results lead to the conclusion that for continuous distributions the Wilcoxon test, even for very skew distributions with large kurtosis, is no better than the Welch test, which is generally used. If we are not sure whether the distributions of the two populations are normal we can use the Welch test because Rasch and Guiard (2004) as well as Rasch et al. (2011a) have shown that the first kind risk is near to the nominal value even if the distributions are non-normal and the power is sufficient.

3.3.1.5 Sequential Two-Sample Tests

The technique of sequential testing offers the advantage that, given many studies, on average fewer research units need to be sampled as compared to the 'classic' approach of hypothesis testing in Sections 3.3.1.1–3.3.1.4 with sample sizes fixed beforehand. Nevertheless, we also need the precision requirements α, β, and δ. In the case of two samples from two populations and testing the null hypothesis H_0: $\mu_1 = \mu_2$, a sequential triangular test is preferable. Its application is completely analogous to Section 3.2.5. The only difference is that we use $\delta = 0 - (\mu_1 - \mu_2) = \mu_2 - \mu_1$ for the relevant (minimum difference) between the null and the alternative hypothesis, instead of $\delta = \mu_0 - \mu$. Again we keep on sampling data, i.e. outcomes, until we are able to make a terminal decision, namely to accept or to reject the null hypothesis.

Problem 3.14 Show how a triangular sequential two-sample test is calculated.

Solution

Use the OPDOE command in R:

```
> valid.tt <- triangular.test.norm(x= , y= , mu1= , mu2= ,
     sigma= , alpha= , beta= )
```

Example

In R, for now a sequential triangular test corresponding to the Welch test is not available. Hence we have to use, rather inappropriately, the one for the t-test, staying well aware that this leads, in the case of distinct variances in both populations, to falsely rejecting the null hypothesis more often than the nominal for the assumed α. We use the data of Problem 3.10.

```
> valid.tt <- triangular.test.norm(x = 48, y = 54,
            mu1 = 50, mu2 = 60, sigma = 10,
            alpha = 0.05, beta = 0.2)
```

i.e. we apply the function `triangular.test.norm()` and use the first observation value of group 1 (`x = 48`) and group 2 (`y = 54`) as the first two arguments; furthermore we set μ_0 (`mu1 = 50`) and μ_1 (`mu2 = 60`), and `sigma = 10` for σ. Finally, we set the appropriate precision requirements according to `alpha = 0.05` and `beta = 0.2`. All of this we assign to object `valid.tt`.

As a result, we get:

```
Triangular Test for normal distribution

Sigma known: 10

H0: mu1=mu2= 50  versus H1: mu1= 50  mu2>= 60
alpha: 0.05  beta: 0.2
Test not finished, continue by adding single data
     via update()
```

After the input of further data

```
> valid.tt <- update(valid.tt, x=48, y=54)
Triangular Test for normal distribution
Sigma known: 10
H0: mu1=mu2= 50  versus H1: mu1= 50  mu2>= 60
alpha: 0.05  beta: 0.2
Test not finished, continue by adding single data via
     update()
current sample size for x:  2
current sample size for y:  2
> valid.tt <- update(valid.tt, x=47, y=55)
Triangular Test for normal distribution
sigma known: 10
H0: mu1=mu2= 50  versus H1: mu1= 50  mu2>= 60
alpha: 0.05  beta: 0.2
Test not finished, continue by adding single data via
update()
current sample size for x:  3
current sample size for y:  3
> valid.tt <- update(valid.tt, x=49, y=61)
Triangular Test for normal distribution
Sigma known: 10
H0: mu1=mu2= 50  versus H1: mu1= 50  mu2>= 60
```

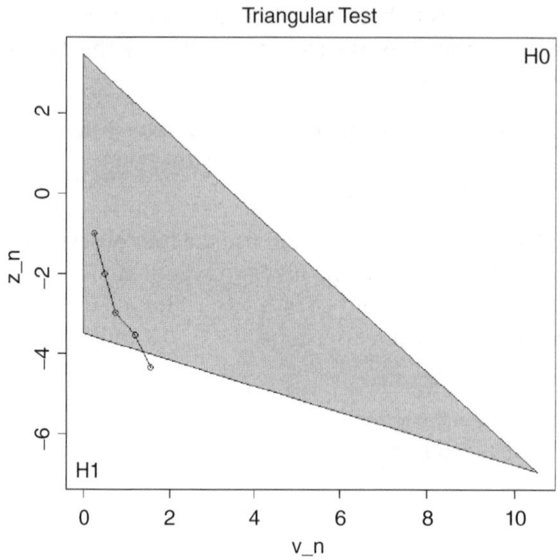

Figure 3.4 Graph of the triangular sequential two-sample test of Problem 3.14.

```
alpha: 0.05  beta: 0.2
Test not finished, continue by adding single data via
     update()
current sample size for x:  4
current sample size for y:  4
> valid.tt <- update(valid.tt, x=45, y=63)
Triangular Test for normal distribution
Sigma known: 10
H0: mu1=mu2= 50  versus H1: mu1= 50  mu2>= 60
alpha: 0.05  beta: 0.2
Test finished: accept H1
Sample size for x:  5
Sample size for y:  5
```

A graph of this test is shown in Figure 3.4.

3.3.2 Test on Two Medians

To test H_0 whether two populations have the same median, against the alternative hypothesis H_A that the medians are different, a random sample of size n_i $(i = 1,2)$ is drawn from each population. Be aware that the scale of the continuous measurement is at least ordinal, or else the term *median* is without meaning. A 2×2 contingency table is constructed, with rows '>median' and '≤median' and the columns for the sample $i = 1, 2$ of the two populations. The two entries in the ith column are the numbers of observations in the ith sample, which are above and below the combined sample grand median $\hat{M}$; this is the median of all observations combined.

The median test has three appealing properties from the practical standpoint. In the first place it is primarily sensitive to differences in location between cells and not to the shapes of the cell distributions. Thus, if the observations of some cells were symmetrically distributed while in other cells they were positively skewed, the rank test would be inclined to reject the null hypothesis even though all population medians were the same. The median test would not be much affected by such differences in the shapes of the cell distributions. In the second place the computations associated with the median test are quite simple and the test itself is nothing more than the familiar contingency table test. In the third place when we come to consider more complex experiments it will be found that the median tests are not much affected by differing cell sizes. Bradley (1968) gives the following rationale and example for the Westenberg–Mood median test (Westenberg 1948; Mood 1950, 1954).

3.3.2.1 Rationale

Suppose that populations I and II are both continuously distributed populations of variate-values, that the median variate-value of the combined sample has the value $\widehat{M}$, and that the two row categories are $<\widehat{M}$ and $\geq\widehat{M}$, respectively. Then (if the number of observations in the combined sample is odd) the single observation equal to $\widehat{M}$ is excluded from all frequencies in the 2×2 table so that N (the total number of remaining data) is even, Fisher's exact method for the 2×2 contingency table becomes a test of hypotheses that equal but inexactly known proportions of populations I and II lie below $\widehat{M}$ (or if the original combined sample contained an even number of observations, below some population value lying between the $[N/2]$th and $[(N/2)+1]$th observations in order of increasing size in the combined sample). Thus it tests the hypothesis that $\widehat{M}$ (or some value close to it) is the same, but unknown, quantile in the two populations; and if H_0 is true and N is large the common proportion of the two populations lying below $\widehat{M}$ will tend to be close to 0.5, so $\widehat{M}$ will tend to be a population quantile, which is nearly a median. For this reason the Westenberg–Mood test is sometimes stated to attest to identical population medians; however, it is quite possible for two populations to have unequal medians and equal '$\widehat{M}$ – quantiles' or vice versa, especially when both samples are small. Instead of a median test, therefore, it might be better described as a quasi-median test or a test for a common, probably more or less centrally located, quantile.

Example 3.6 An experimenter, content to let the exact nature of 'location' to be defined a posteriori, wishes a rough test (at the $\alpha = 0.05$ significance level) of whether or not two continuous populations have a common location. From population I he draws a random sample of size $n_1 = 13$ consisting of the values 2, 3, 5, 6, 8, 9, 13, 15, 16, 18, 31, 54, 89, and from population II, a random sample of size $n_2 = 15$, he obtains the sample observations 10, 11, 17, 19, 20, 22, 23, 27, 30, 33, 34, 35, 39, 40, 42. Of the $N = n_1 + n_2 = 28$ observations in both samples combined, the fourteenth and fifteenth in increasing order of size are 19 and 20, so $\widehat{M} = \frac{19+20}{2} = 19.5$.'

Using R we perform the test as follows:

```
> sampleI <- c(2, 3, 5, 6, 8, 9, 13, 15, 16, 18, 31, 54, 89)
> sampleII <- c(10, 11, 17, 19, 20, 22, 23, 27, 30, 33, 34,
                35, 39, 40, 42)
> M <-median(c(sampleI, sampleII))
> M   #  median of the combined samples
```

```
[1] 19.5
> combined <- c(sampleI, sampleII)
> group <- rep(1:2, c(length(sampleI), length(sampleII)))
> CT2x2  <-  table(combined <= M, group)
> CT2x2
             group
               1  2
  FALSE        3 11
  TRUE        10  4
> median.test<- function(y1,y2){
   z<-c(y1,y2)
   g <- rep(1:2, c(length(y1),length(y2)))
   m<-median(z)
   fisher.test(z  <=  m,g)$p. value
   }
> median.test(sampleI,sampleII)
[1] 0.02130091
```

We may reject, therefore, at the predesigned $\alpha = 0.05$ level, the null-hypothesis that equal proportions of the two populations lie below $\widehat{M} = 19.5$ (or at least below some point between 19 and 20) because the two-sided p-value $0.0213 < 0.05$.

3.3.3 Test on Two Probabilities

Let $\boldsymbol{y}$ be distributed as $B(n, p)$. Knowing one observation $\boldsymbol{y} = Y$ we want to test $H_0{:}\, p = p_0$ against $H_A{:}\, p \neq p_0, p_0 \in (0, 1)$. The natural parameter is $\eta = \ln \frac{p}{1-p}$, and $\boldsymbol{y}$ is sufficient with respect to the family of binomial distributions. Therefore the uniformly most powerful unbiased (UMPU)-α-test is given by (3.27) in Rasch and Schott (2018), where the $c_{i\alpha}$ and $\gamma'_{i\alpha}$ $(i = 1, 2)$ have to be determined from (3.28) and (3.29) in Rasch and Schott (2018). With

$$L_n(y|\,p) = \binom{n}{y} p^y (1-p)^{n-y}$$

Equation (3.28) in Rasch and Schott (2018) has the form

$$\sum_{y=0}^{c_{1\alpha}-1} L_n(y|p_0) + \sum_{y=c_{2\alpha+1}}^{n} L_n(y|p_0) + \gamma'_{1\alpha} L_n(c_{1\alpha}|p_0) + \gamma'_{2\alpha} L_n(c_{2\alpha}|p_0) = \alpha. \tag{3.20}$$

Regarding

$$\mathrm{y}\, L_n(y|\,p) = \mathrm{np}\, L_{n-1}(y-1|\,p) \quad \text{and} \quad E(\boldsymbol{y}|\,p_0) = np_0$$

the relation (3.29) in Rasch and Schott (2018) leads to

$$\sum_{y=0}^{c_{1\alpha}-1} L_{n-1}(y-1|p_0) + \sum_{y=c_{2\alpha}+1}^{n} L_{n-1}(y-1|p_0) + \gamma'_{1\alpha} L_{n-1}(c_{1\alpha}-1|p_0)$$
$$+ \gamma'_{2\alpha} L_{n-1}(c_{2\alpha}-1|p_0) = \alpha. \tag{3.21}$$

The solution of these two simultaneous equations can be obtained by statistical software, e.g. by R. Further results can be found in the book of Fleiss et al. (2003).

3.3.4 Tests on Two Variances

We consider two independent random samples $\boldsymbol{Y}_1 = (\boldsymbol{y}_{11}, \dots, \boldsymbol{y}_{1n_1})^T$, $\boldsymbol{Y}_2 = (\boldsymbol{y}_{21}, \dots, \boldsymbol{y}_{2n_2})^T$, where the components $\boldsymbol{y}_{ij}$ are supposed to be distributed as $N(\mu_i, \sigma_i^2)$. We intend to test the null hypothesis

$$H_0 : \sigma_1^2 = \sigma_2^2 = \sigma^2, \mu_1, \mu_2 \text{ arbitrary}$$

against the alternative

$$H_A : \sigma_1^2 \neq \sigma_2^2, \mu_1, \mu_2 \text{ arbitrary.}$$

Under H_0 we have $\frac{\sigma_1^2}{\sigma_2^2} = 1$, and the random variable

$$F = \frac{\sum_{i=1}^{n_1} (\boldsymbol{y}_{1i} - \overline{\boldsymbol{y}}_1)^2 \frac{n_2-1}{\sigma_1^2}}{\sum_{i=1}^{n_2} (\boldsymbol{y}_{2i} - \overline{\boldsymbol{y}}_2)^2 \frac{n_1-1}{\sigma_2^2}}$$

does not depend on μ_1 and μ_2, and $\sigma_1^2 = \sigma_2^2 = \sigma^2$. Since the random variable $\boldsymbol{F}$ is centrally distributed as $F(n_1 - 1, n_2 - 1)$ under H_0, the function

$$k(Y) = \begin{cases} 1 & \text{if } F < F\left(n_1 - 1, n_2 - 1 \,\middle|\, \frac{\alpha}{2}\right) \\ & \text{or } F > F\left(n_1 - 1, n_2 - 1 \,\middle|\, 1 - \frac{\alpha}{2}\right) \\ 0 & \text{else} \end{cases}$$

defines a UMPU-α-test, where $F(n_1 - 1, n_2 - 1| P)$ is the P-quantile of the F-distribution with $n_1 - 1$ and $n_2 - 1$ degrees of freedom. This test is very sensitive to deviations from the normal distribution. Therefore the following Levene test should be used instead of it in the applications. Therefore we present no example or problem.

Instead we propose the Levene test. Box (1953) has already mentioned the extreme non-robustness of the F-test comparing two variances. Rasch and Guiard (2004) have reported on extensive simulation experiments devoted to this problem. The results of Box show that non-robustness has to be expected already for relatively small deviations from the normal distribution. Hence, we generally suggest applying the test of Levene (1960) which is described now.

For $j = 1, 2$ we put

$$\boldsymbol{z}_{ij} = (\boldsymbol{y}_{ij} - \overline{\boldsymbol{y}}_{.j})^2; i = 1, \dots, n_j$$

and

$$\boldsymbol{SS}_{\text{between}} = \sum_{j=1}^{2} \sum_{i=1}^{n_j} (\overline{\boldsymbol{z}}_{.j} - \overline{\boldsymbol{z}}_{..})^2, \quad \boldsymbol{SS}_{\text{within}} = \sum_{j=1}^{2} \sum_{i=1}^{n_j} (\boldsymbol{z}_{ij} - \overline{\boldsymbol{z}}_{.j})^2$$

where $\overline{\boldsymbol{z}}_{.j} = \frac{1}{n_j} \sum_{i=1}^{n_j} \boldsymbol{z}_{ij}$, $\overline{\boldsymbol{z}}_{..} = \frac{1}{n_1+n_2} \sum_{j=1}^{2} \sum_{i=1}^{n_j} \boldsymbol{z}_{ij}$.

The null hypothesis H_0 is rejected if

$$F^* = \frac{\text{SS}_{\text{between}}}{\text{SS}_{\text{within}}} (n_1 + n_2 - 2) > F\left(1, n_1 + n_2 - 2 \,\middle|\, 1 - \frac{\alpha}{2}\right). \tag{3.22}$$

In R the Levene test is not available in the base package, but it is quite easy to program it yourself or it can be taken from the package `car v3.0.0` by John Fox, see https://cran.r-project.org/web/packages/car/car.pdf.

Example 3.7 We consider the entries of Table 3.4 with the litter weights of mice (in grams) but interpret the entries differently. We assume that the continuous x and y values are obtained independently from two strains of mice. Let $\text{var}(\boldsymbol{x}) = \sigma_1^2$ and $\text{var}(\mathbf{y}) = \sigma_2^2$. The continuous distributions are not assumed to be normally distributed. The null-hypothesis given above concerning the equality of variances is tested using the Levene test with significance level $\alpha = 0.05$.

The Levene test here is based on the squared deviations from the sample mean for x and y, respectively. We make in R the function `>levene_square(x,y)`.

```
> x <- c(7.6, 13.2, 9.1, 10.6, 8.7, 10.6, 6.8, 9.9, 7.3,
     10.4, 13.3, 10.0, 9.5 )
> y <- c( 7.8, 11.1, 16.4, 13.7, 10.7, 12.3, 14.0, 11.9,
     8.8, 7.7, 8.9, 16.4, 10.2)
> levene_square <- function(x,y){
   nx <- length(x)
   dfx <- nx -1
   mx <- mean(x)
   devx <- x- mx
   ny <- length(y)
   dfy <- ny -1
   my <- mean(y)
   devy <- y-my
   z1 <- devx^2
   mz1 <- mean(z1)
   SSz1 <- dfx*var(z1)
   z2 <- devy^2
   mz2 <- mean(z2)
   SSz2 <- dfy*var(z2)
   z <- c(z1,z2)
   mz <- mean(z)
   SSbetween <- nx*(mz1-mz)^2 + ny*(mz2-mz)^2
   SSwithin <-  SSz1 + SSz2
   Fstar <- SSbetween * (nx + ny -2)/ SSwithin
   p_value <- pf(Fstar, 1, nx+ny-2, lower.tail=FALSE)
   list(p_value)
   }
> levene_square (x,y)
[[1]]
[1] 0.1136212
```

Because the p-value $0.113\,621\,2 > 0.05$ we cannot reject the null hypothesis H_0 of equal variances at the level $\alpha = 0.05$.

Another version of the Levene test is based on the absolute deviations from the sample mean for x and y, respectively. We make in R the function `>levene_abs(x,y)`.

```
> x <- c(7.6, 13.2, 9.1, 10.6, 8.7, 10.6, 6.8, 9.9, 7.3,
     10.4, 13.3, 10.0, 9.5)
> y <- c( 7.8, 11.1, 16.4, 13.7, 10.7, 12.3, 14.0, 11.9,
     8.8, 7.7, 8.9, 16.4, 10.2)
> levene_abs <- function(x,y){
  nx <- length(x)
  dfx <- nx -1
  mx <- mean(x)
  devx <- x- mx
  ny <- length(y)
  dfy <- ny -1
  my <- mean(y)
  devy <- y-my
  z1 <- abs(devx)
  mz1 <- mean(z1)
  SSz1 <- dfx*var(z1)
  z2 <- abs(devy)
  mz2 <- mean(z2)
  SSz2 <- dfy*var(z2)
  z <- c(z1,z2)
  SSbetween <- nx*(mz1-mz)^2 + ny*(mz2-mz)^2
  SSwithin <-  SSz1 + SSz2
  Fstar <- SSbetween * (nx + ny -2)/ SSwithin
  p_value <- pf(Fstar, 1, nx+ny-2, lower.tail=FALSE)
  list( p_value)
  }
> levene_abs(x,y)
[[1]]
[1] 0.1198781
```

Because the p-value 0.119 878 1 > 0.05 we cannot reject the null hypothesis H_0 of equal variances at the level $\alpha = 0.05$.

References

Box, G.E.P. (1953). Non-normality and tests on variances. *Biometrika* 40: 318–335.

Bradley, J.V. (1968). *Distribution-Free Statistical Tests*. Englewood Cliffs, New Jersey: Prentice-Hall, Inc.

DeGroot, M.H. (2005). *Optimal Statistical Decisions*. New York: Wiley Online Library.

Dodge, H.F. and Romig, H.G. (1929). A method of sampling inspection. *Bell Syst. Tech. J.* 8: 613–631.

Fleishman, A.J. (1978). A method for simulating non-normal distributions. *Psychometrika* 43: 521–532.

Fleiss, J.L., Levin, B., and Paik, M.C. (2003). *Statistical Methods for Rates and Proportions*, 3e. Hoboken: Wiley.

Ghosh, M. and Sen, P.K. (1991). *Handbook of Sequential Analysis*. Boca Raton: CRC Press.

Levene, H. (1960). Robust tests for equality of variances. In: *Contributions to Probability and Statistics*, 278–292. Stanford: Stanford University Press.
Mann, H.H. and Whitney, D.R. (1947). On a test whether one of two random variables is stochastically larger than the other. *Ann. Math. Stat.* 18: 50–60.
Mood, A.M. (1950). *Introduction to the Theory of Statistics*. New York: McGraw-Hill.
Mood, A.M. (1954). On asymptotic efficiency of certain nonparametric two-sample tests. *Ann. Math. Stat.* 25: 514–533.
Munzel, U. and Brunner, E. (2002). An exact paired rank test. *Biom. J.* 44: 584–593.
Rasch, D. and Guiard, V. (2004). The robustness of parametric statistical methods. *Psychol. Sci.* 46: 175–208.
Rasch, D. and Schott, D. (2018). *Mathematical Statistics*. Oxford: Wiley.
Rasch, D. and Tiku, M.L. (eds. 1985). Robustness of statistical methods and nonparametric statistics Proc. Conf. on Robustness of Statistical Methods and Nonparametric Statistics, Schwerin (DDR), May 29–June 2, 1983. Reidel Publishing Company Dordrecht.
Rasch, D., Kubinger, K.D., and Moder, K. (2011a). The two-sample *t* test: pre-testing its assumptions does not pay-off. *Stat. Pap.* 52: 219–231.
Rasch, D., Pilz, J., Verdooren, R., and Gebhardt, A. (2011b). *Optimal Experimental Design with R*. Boca Raton: Chapman and Hall.
Rasch, D., Kubinger, K.D., and Yanagida, T. (2011c). *Statistics in Psychology Using R and SPSS*. Hoboken: Wiley.
Stein, C. (1945). A two sample test for a linear hypothesis whose power is independent of the variance. *Ann. Math. Stat.* 16: 243–258.
Wald, A. (1947). *Sequential Analysis*. New York: Dover Publ.
Wald, A. (2004). *Sequential Analysis Reprint*. New York: Wiley.
Welch, B.L. (1947). The generalization of students problem when several different population variances are involved. *Biometrika* 34: 28–35.
Westenberg, J. (1948). Significance test for median and interquartile range in samples from continuous populations of any form. *Proc. Koninklijke Nederlandse Akademie van Wetenschappen, (The Netherlands)* 51: 252–261.
Whitehead, J. (1997). *The Design and Analysis of Sequential Clinical Trials*, 2e Revised. New York: Wiley.
Wilcoxon, F. (1945). Individual comparisons by ranking methods. *Biom. Bull.* 1: 80–82.

4

Confidence Estimations – One- and Two-Sample Problems

4.1 Introduction

In confidence estimation, we construct random regions in the parameter space so that this region covers the unknown parameter with a given probability, the confidence coefficient. In this book we consider special regions, namely intervals called confidence intervals. We then speak about interval estimation. We will see that there are analogies to the test theory concerning the optimality of confidence intervals, which we exploit to simplify many considerations.

A confidence interval is a function of the random sample; that is to say, it is also random in the sense of depending on chance, hence we speak of a random interval. However, once calculated based on observations, the interval has, of course, non-random bounds. Sometimes only one of the bounds is of interest; the other is then fixed – this concerns the estimator as well as the estimate itself. At least one boundary of a confidence interval must be random; if both boundaries are random, the interval is two-sided, and if only one boundary is random, it is one-sided.

We make no difference between a random interval and its realisation with real boundaries but speak always about confidence intervals. What we mean in special cases will be easy to understand. When we speak about the expected length of a confidence interval, we of course mean a random interval.

Definition 4.1 Let $\boldsymbol{Y} = (\boldsymbol{y}_1, \boldsymbol{y}_2, \dots, \boldsymbol{y}_n)^T$ be a random sample with realisations $Y \in \{Y\}$, whose components are distributed with a parameter $\theta \in \Omega$. Let $K(\boldsymbol{Y})$ be a random set with realisations $K(Y)$ in Ω. $K(\boldsymbol{Y})$ is said to be a confidence region for θ with the corresponding confidence coefficient (confidence level) $1 - \alpha$ if

$$P[\theta \in K(\boldsymbol{Y})] \geq 1 - \alpha \text{ for all } \theta \in \Omega. \tag{4.1}$$

In condensed form $K(\boldsymbol{Y})$ is also said to be a $(1 - \alpha)$ confidence region. If $\Omega \subset R^1$ and $K(Y)$ is a connected set for all $Y \in \{Y\}$, then $K(\boldsymbol{Y})$ is a $(1 - \alpha)$ confidence interval. The realisation $K(Y)$ of a confidence region is called a realised confidence region.

Statistical tests and confidence estimations for normal distributions are extremely robust against the violation of the normality assumption, as was shown by simulation – see Rasch and Tiku (1985).

Applied Statistics: Theory and Problem Solutions with R, First Edition.
Dieter Rasch, Rob Verdooren, and Jürgen Pilz.

4.2 The One-Sample Case

We start with normal distributions and confidence intervals for the expectation.

4.2.1 A Confidence Interval for the Expectation of a Normal Distribution

Problem 4.1 Construct a one-sided $(1-\alpha)$ confidence interval for the expectation of a $N(\mu, \sigma^2)$ distribution if the variance is known.

Solution

Let the $n > 1$ components of a random sample $\boldsymbol{Y} = (\boldsymbol{y}_1, \boldsymbol{y}_2, \ldots, \boldsymbol{y}_n)^T$ be distributed as $N(\mu,\sigma^2)$, where σ^2 is known. The mean $\overline{\boldsymbol{y}}$ follows a $N(\mu, \frac{\sigma^2}{n})$ distribution. A $(1-\alpha)$ confidence interval $K(\boldsymbol{Y})$ with respect to μ has to satisfy $P[\mu \in K(\boldsymbol{Y})] = 1-\alpha$. This means (with $\hat{\boldsymbol{\mu}}_l$ as the random lower bound and $\hat{\boldsymbol{\mu}}_u$ as the random upper bound) in the case of a two-sided $(1-\alpha)$ confidence interval

$$P(\hat{\boldsymbol{\mu}}_l \le \mu \le \hat{\boldsymbol{\mu}}_u) = 1-\alpha.$$

Since $\overline{\boldsymbol{y}}$ is distributed as $N(\mu,\frac{\sigma^2}{n})$, it holds that

$$P\left\{Z_{\alpha_1} \le \frac{\overline{\boldsymbol{y}}-\mu}{\sigma}\sqrt{n} \le Z_{1-\alpha_2}\right\} = 1-\alpha_1-\alpha_2 = 1-\alpha$$

for $\alpha_1 + \alpha_2 = \alpha$, $\alpha_1 \ge 0$, $\alpha_2 \ge 0$. Consequently, we have

$$P\left\{\overline{\boldsymbol{y}} - \frac{\sigma}{\sqrt{n}}Z_{1-\alpha_2} \le \mu \le \overline{\boldsymbol{y}} - \frac{\sigma}{\sqrt{n}}Z_{\alpha_1}\right\} = 1-\alpha \tag{4.2}$$

so that $\hat{\boldsymbol{\mu}}_l = \overline{\boldsymbol{y}} - \frac{\sigma}{\sqrt{n}}Z_{1-\alpha_2}$ and $\hat{\boldsymbol{\mu}}_u = \overline{\boldsymbol{y}} - \frac{\sigma}{\sqrt{n}}Z_{\alpha_1} = \overline{\boldsymbol{y}} + \frac{\sigma}{\sqrt{n}}Z_{1-\alpha_1}$ is fulfilled. For $1-\alpha$ there are infinitely many confidence intervals according to the choice of α_1 and $\alpha_2 = \alpha - \alpha_1$. If $\alpha_1 = 0$ or $\alpha_2 = 0$, respectively, then the confidence intervals are one-sided (i.e. only one interval bound is random). The more the values α_1 and α_2 differ from each other, the larger the expected width $E(\hat{\boldsymbol{\mu}}_l - \hat{\boldsymbol{\mu}}_u) = \frac{\sigma}{\sqrt{n}}(Z_{1-\alpha_2} - Z_{\alpha_1})$. For example, the width becomes infinite for $\alpha_1 = 0$ or for $\alpha_2 = 0$. Finite confidence intervals result for $\alpha_1 > 0$, $\alpha_2 > 0$ and an optimal choice is $\alpha_1 = \alpha_2 = \frac{\alpha}{2}$.

In the case of a one-sided $(1-\alpha)$ confidence interval we have either

$P(\hat{\boldsymbol{\mu}}_l \le \mu < \infty) = 1-\alpha$ for the left-sided interval or
$P(-\infty < \mu \le \hat{\boldsymbol{\mu}}_u) = 1-\alpha$ for the right-sided interval.

From this, it follows either

$$P\left\{\overline{\boldsymbol{y}} - \frac{\sigma}{\sqrt{n}}Z_{1-\alpha} \le \mu < \infty\right\} = 1-\alpha \tag{4.3}$$

or

$$P\left\{-\infty < \mu \le \overline{\boldsymbol{y}} + \frac{\sigma}{\sqrt{n}}Z_{1-\alpha}\right\} = 1-\alpha. \tag{4.4}$$

Example

We construct in R a realised left-sided and right-sided 0.95 and a 0.99 confidence interval for the normally distributed random sample of x-data in Table 3.4 assuming that the variance is known as $\sigma^2 = 1$.

```
> x <- c(7.6, 13.2, 9.1, 10.6, 8.7, 10.6, 6.8,9.9,7.3, 10.4,
         13.3, 10.0, 9.5)
> xbar <- mean(x)
> var <- 1
> sdx <- sqrt(var/length(x))
> conf.level <- 0.95
> z <- qnorm(1-conf.level, lower.tail=FALSE)
> xl <- xbar -z*sdx  # lower 0.95-confidence limit
> xl
[1] 9.31303
> xu <- xbar + z*sdx  # upper 0.95-confidence limit
> xu
[1] 10.22543
> conf.level <- 0.99
> z <- qnorm(1-conf.level, lower.tail=FALSE)
> xl <- xbar -z*sdx  # lower 0.99-confidence limit
> xl
[1] 9.124018
> xu <- xbar + z*sdx  # upper 0.99-confidence limit
> xu
[1] 10.41444
```

Problem 4.2 Construct a two-sided $(1 - \alpha)$ confidence interval for the expectation of a $N(\mu, \sigma^2)$ distribution if the variance is known.

Solution

We see from the solution of Problem 4.1 the connection between a confidence interval and a statistical test.

In

$$P\left\{Z_{\alpha_1} \leq \frac{\overline{y} - \mu}{\sigma}\sqrt{n} \leq Z_{1-\alpha_2}\right\} = 1 - \alpha_1 - \alpha_2 = 1 - \alpha$$

$\frac{\overline{y}-\mu}{\sigma}\sqrt{n}$ is the test statistic (3.1) for testing the null hypothesis $H_0 : \mu = \mu_0$ against a two-sided alternative $H_A : \mu = \mu_1 \neq \mu_0$.

From (4.2) we find that the upper bound equals $\overline{y} - \frac{\sigma}{\sqrt{n}}Z_{\alpha_1}$ and the lower bound equals $\overline{y} - \frac{\sigma}{\sqrt{n}}Z_{1-\alpha_2}$; the difference is $\frac{\sigma}{\sqrt{n}}Z_{1-\alpha_2} - \frac{\sigma}{\sqrt{n}}Z_{\alpha_1}$. This difference is a minimum for $\alpha_1 = \alpha_2 = \frac{\alpha}{2}$ and then equals $\frac{\sigma}{\sqrt{n}}Z_{1-\frac{\alpha}{2}} - \frac{\sigma}{\sqrt{n}}Z_{\frac{\alpha}{2}}$.

Example

We construct a realised two-sided 0.95 and a 0.99 confidence interval using R. We assume that the sample of x-data in Table 3.4 stems from a normal distribution with known variance $\sigma^2 = 1$.

```
> x <- c(7.6, 13.2, 9.1, 10.6, 8.7, 10.6, 6.8, 9.9, 7.3,
      10.4, +13.3, 10.0, 9.5 )
> # we make a function to construct a normal two-sided +
      confidence-interval
> norm.confinterval = function(data, variance = var.data,
      conf.level = conf) {
     z = qnorm((1 - conf.level)/2, lower.tail = FALSE)
     xbar = mean(data)
    sdx = sqrt(variance/length(data))
    c(xbar - z * sdx, xbar + z * sdx)
 }
> # confidence interval with confidence-coefficient 0.95
> norm.confinterval(x, 1, 0.95)
[1]  9.225635 10.312827
> # confidence interval with confidence-coefficient 0.99
> norm.confinterval(x, 1, 0.99)
[1]  9.054824 10.483637
```

We have already used the term 'expected length' of a two-sided confidence interval as the expectation of the difference between the upper and the lower bound of the interval. We give now a definition.

Definition 4.2 The expected length L of a two-sided $(1-\alpha)$ confidence interval for a parameter θ is defined as the expectation of the difference between the (random) upper and the lower bound of this interval. The two bounds we denote by $\hat{\theta}_l$ for the lower bound and $\hat{\theta}_u$ for the upper bound. Then

$$L = E(\hat{\theta}_u - \hat{\theta}_l). \tag{4.5}$$

In the case of one-sided $(1-\alpha)$ confidence intervals we call the expectation of the distance between the finite bound and θ the expected length.

Problem 4.3 Determine the minimal sample size for constructing a two-sided $(1-\alpha)$ confidence interval for the expectation μ of a normally distributed random variable with known variance σ^2 so that the expected length L is below 2δ.

Solution

From (4.2) we find that the upper bound equals $\bar{y} - \frac{\sigma}{\sqrt{n}} Z_{\alpha_1}$ and the lower bound equals $\bar{y} - \frac{\sigma}{\sqrt{n}} Z_{1-\alpha_2}$; the difference is $\frac{\sigma}{\sqrt{n}} Z_{1-\alpha_2} - \frac{\sigma}{\sqrt{n}} Z_{\alpha_1}$. This difference is a minimum for $\alpha_1 = \alpha_2 = \frac{\alpha}{2}$ and then equals $\frac{\sigma}{\sqrt{n}} Z_{1-\frac{\alpha}{2}} - \frac{\sigma}{\sqrt{n}} Z_{\frac{\alpha}{2}}$. This is one of the very rare cases in which the length is not random and we must not determine its expectation. Because $Z_{\frac{\alpha}{2}} = -Z_{1-\frac{\alpha}{2}}$ we obtain $L = 2\frac{\sigma}{\sqrt{n}} Z_{1-\frac{\alpha}{2}}$.

Our precision requirement was $L = 2\frac{\sigma}{\sqrt{n}} Z_{1-\frac{\alpha}{2}} < 2\delta$ or, equivalently, $n > \frac{\sigma^2}{\delta^2} Z^2_{1-\frac{\alpha}{2}}$; i.e.

$$n = \left\lceil \frac{\sigma^2}{\delta^2} Z^2_{1-\frac{\alpha}{2}} \right\rceil. \tag{4.6}$$

If we compare with (3.5) $n = \left\lceil \sigma^2 \frac{\left(Z(\beta_0) - Z\left(1-\frac{\alpha}{2}\right)\right)^2}{\delta^2} \right\rceil$ we see that (4.6) equals (3.5) for $Z(\beta_0) = 0$ and this is the case for $\beta_0 = 0.5$ (the 50%-quantile for the standard normal distribution is 0). Therefore, we can also use formulae and R-commands for the sample size determination of tests for the confidence estimation.

Example
For a given confidence coefficient $1 - \alpha = 0.95$ the sample size n has to be determined so that the half-expected length is $\delta = 0.6\,\sigma$ in the one- and also in the two-sided case.

One-sided case:

$$n = \left\lceil \sigma^2 \frac{(Z(1-\alpha))^2}{\delta^2} \right\rceil = \left\lceil \frac{(1.645)^2}{0.6^2} \right\rceil = \lceil 7.51 \rceil = 8.$$

Two-sided case:

$$n = \left\lceil \sigma^2 \frac{\left(Z\left(1-\frac{\alpha}{2}\right)\right)^2}{\delta^2} \right\rceil = \left\lceil \frac{(1.96)^2}{0.6^2} \right\rceil = \lceil 10.67 \rceil = 11.$$

In R we use the commands:

```
> Z0.95 <- qnorm(0.95)
> Z0.95
[1] 1.644854
> n_onesided <- ceiling((Z0.95/0.6)^2)
> n_onesided
[1] 8
> Z0.975 <- qnorm(0.975)
> Z0.975
[1] 1.959964
> n_twosided <- ceiling((Z0.975/0.6)^2)
> n_twosided
[1] 11
```

Remark
When we construct confidence intervals for location parameters, the factor two always occurs in the formula of the expected length. We therefore replace the expected length by the half-expected length, which is better comparable with the precision requirement for one-sided intervals in Definition 4.2.

Problem 4.4 Determine the minimal sample size for constructing a one-sided $(1 - \alpha)$ confidence interval for the expectation μ of a normally distributed random variable with known variance σ^2 so that the distance between the finite (random) bound of the interval and 0 is below δ.

Solution

From (4.3) and (4.4) we find that the distance of the finite bound and μ equals $\bar{y} - \frac{\sigma}{\sqrt{n}} Z_{1-\alpha}$. The expectation of this distance is $E\left(\mu - \left(\bar{y} - \frac{\sigma}{\sqrt{n}} Z_{1-\alpha}\right)\right)$ is $\frac{\sigma}{\sqrt{n}} Z_{1-\alpha}$ and from $\frac{\sigma}{\sqrt{n}} Z_{1-\alpha} < \delta$ this gives

$$n = \left\lceil \frac{\sigma^2}{\delta^2} Z_{1-\alpha}^2 \right\rceil. \tag{4.7}$$

Example

For a given confidence coefficient $1 - \alpha = 0.95$ the sample size n has to be determined so that the expected distance between the upper bound and μ of a one-sided $(1 - \alpha)$ confidence interval is below $\delta = 0.6\sigma$.

From (4.7) we obtain

$$n = \left\lceil \frac{1}{0.6^2} 1.645^2 \right\rceil = \lceil 7.52 \rceil = 8.$$

In addition, this is exactly the solution of the one-sided part given in the example of Problem 4.3.

Now we discuss the practically more interesting case that the variance is unknown. We assume that the elements of the random sample are $N(\mu, \sigma^2)$ distributed but σ^2 is unknown. From (3.6) we know that

$$\boldsymbol{t} = \frac{\bar{\boldsymbol{y}} - \mu}{\boldsymbol{s}} \sqrt{n}$$

is non-centrally t-distributed with $n - 1$ degrees of freedom and non-centrality parameter $\lambda = \frac{\mu - \mu_0}{\sigma} \sqrt{n}$. We estimate σ by the estimator $\boldsymbol{s}$ or the realisation (the estimate) s. Analogously to (4.2) we obtain

$$P\left\{\bar{\boldsymbol{y}} - \frac{\boldsymbol{s}}{\sqrt{n}} t(1 - \alpha_2, n - 1) \leq \mu \leq \bar{\boldsymbol{y}} + \frac{\boldsymbol{s}}{\sqrt{n}} t(1 - \alpha_1, n - 1)\right\} = 1 - \alpha_1 - \alpha_2 = 1 - \alpha.$$

It can be shown (Rasch and Schott (2018)) that the expected length of such an interval is minimum if we put $\alpha_1 = \alpha_2 = \frac{\alpha}{2}$. From this an optimal two-sided $(1 - \alpha)$ confidence interval is given by

$$\left[\bar{\boldsymbol{y}} - \frac{\boldsymbol{s}}{\sqrt{n}} t\left(1 - \frac{\alpha}{2}, n - 1\right); \bar{\boldsymbol{y}} + \frac{\boldsymbol{s}}{\sqrt{n}} t\left(1 - \frac{\alpha}{2}, n - 1\right)\right]. \tag{4.8}$$

The one-sided $(1 - \alpha)$ confidence intervals are given by

$$\left[\bar{\boldsymbol{y}} - \frac{\boldsymbol{s}}{\sqrt{n}} t(1 - \alpha, n - 1); \infty\right) \tag{4.9}$$

and

$$\left(-\infty; \bar{\boldsymbol{y}} + \frac{\boldsymbol{s}}{\sqrt{n}} t(1 - \alpha, n - 1)\right]. \tag{4.10}$$

Problem 4.5 Construct a realised left-sided 0.95 confidence interval of μ for the x-data in Table 3.4.

Solution
Use the R command > `t.test(x, alternative = "less", conf.level = 0.95)`.

Example

```
> x <-c(7.6, 13.2, 9.1, 10.6, 8.7, 10.6, 6.8, 9.9, 7.3,
     10.4,13)
>  t.test(x, alternative = "less", conf.level = 0.95)
      One Sample t-test data:  x
t = 17.7289, df = 12, p-value = 1
alternative hypothesis: true mean is less than 0
95 percent confidence interval:
     -Inf 10.75133
sample estimates:
mean of x
 9.769231
     -Inf means -∞
```

Problem 4.6 Construct a realised right-sided 0.95 confidence interval for the x-data in Table 3.4.

Solution
Use the R command > `t.test(x, alternative = "greater", conf.level = 0.95)`.

Example

```
> t.test(x, alternative = "greater", conf.level = 0.95)
      One Sample t-test data:  x
t = 17.7289, df = 12, p-value = 2.835e-10
alternative hypothesis: true mean is greater than 0
 95 percent confidence interval:
 8.787129      Inf
sample estimates:
mean of x
 9.769231
```

Problem 4.7 Construct a realised two-sided 0.95 confidence interval for the x-data in Table 3.4.

Solution
Use the R command > `t.test(x, alternative = "two.sided", conf.level = 0.95)`.

Example

```
> t.test(x, alternative = "two.sided", conf.level = 0.95)
```

```
  One Sample t-test data:  x
t = 17.7289, df = 12, p-value = 5.67e-10
alternative hypothesis: true mean is not equal to 0
95 percent confidence interval:
  8.56863 10.96983
sample estimates:
mean of x
 9.769231
```

The expected length of the two-sided $(1 - \alpha)$ confidence interval reads

$$E\left\{\bar{y} + \frac{s}{\sqrt{n}}t\left(1 - \frac{\alpha}{2}, n - 1\right) - \left[\bar{y} - \frac{s}{\sqrt{n}}t\left(1 - \frac{\alpha}{2}, n - 1\right)\right]\right\}$$

$$= \frac{2E(s)}{\sqrt{n}}t\left(1 - \frac{\alpha}{2}, n - 1\right) = \frac{2\sigma\Gamma\left(\frac{n}{2}\right)\sqrt{2}}{\sqrt{n}\Gamma\left(\frac{n-1}{2}\right)\sqrt{n-1}}t\left(1 - \frac{\alpha}{2}, n - 1\right)$$

which follows from example 2.6 in Rasch and Schott (2018).

Analogously, we obtain for the one-sided $(1 - \alpha)$ confidence intervals

$$E\left\{\bar{y} + \frac{s}{\sqrt{n}}t(1 - a, n - 1)\right\} = \frac{\sigma\Gamma\left(\frac{n}{2}\right)\sqrt{2}}{\sqrt{n}\Gamma\left(\frac{n-1}{2}\right)\sqrt{n-1}}t(1 - \alpha, n - 1).$$

Now we define the following precision requirements for the construction of two- and one-sided $(1 - \alpha)$ confidence intervals:

- One-sided $(1 - \alpha)$ confidence interval: determine n so that

$$\frac{\sigma\Gamma\left(\frac{n}{2}\right)\sqrt{2}}{\sqrt{n}\Gamma\left(\frac{n-1}{2}\right)\sqrt{n-1}}t(1 - \alpha, n - 1) < \delta \text{ or, approximately, } \frac{\sigma}{\sqrt{n}}t(1-\alpha, n - 1) < \delta.$$

- Two-sided $(1 - \alpha)$ confidence intervals: determine n so that the half expected width fulfils

$$\frac{\sigma\Gamma\left(\frac{n}{2}\right)\sqrt{2}}{\sqrt{n}\Gamma\left(\frac{n-1}{2}\right)\sqrt{n-1}}t\left(1 - \frac{\alpha}{2}, n - 1\right) < \delta \text{ or, approximately,}$$

$$\frac{\sigma}{\sqrt{n}}t\left(1 - \frac{\alpha}{2}, n - 1\right) < \delta.$$

From the approximate formulae, we obtain the formulae for the minimum sample size as $n = \left\lceil \frac{\sigma^2}{\delta^2}t\left(1 - \frac{\alpha}{2}, n - 1\right)^2 \right\rceil$ and $\left\lceil \frac{\sigma^2}{\delta^2}t(1 - \alpha, n - 1)^2 \right\rceil$, respectively.

In Table 4.1 we show how good the approximate formulae are. When the values of the table are near 1, the approximation is good.

As we can see the approximation may be used for $n > 18$. The approximate formula is solved iteratively by starting with a value n_0 via

$$n_i = \left\lceil \frac{\sigma^2}{\delta^2}t^2\left(1 - \frac{\alpha}{2}, n_{i-1} - 1\right) \right\rceil.$$

Table 4.1 Values of $\frac{\sigma\Gamma\left(\frac{n}{2}\right)\sqrt{2}}{\Gamma\left(\frac{n-1}{2}\right)\sqrt{n-1}}$.

n	$\frac{\sigma\Gamma\left(\frac{n}{2}\right)\sqrt{2}}{\Gamma\left(\frac{n-1}{2}\right)\sqrt{n-1}}$
2	$0.797\,884\,6\sigma$
3	$0.886\,226\,9\sigma$
4	$0.921\,317\,7\sigma$
5	$0.939\,985\,6\sigma$
6	$0.951\,532\,9\sigma$
7	$0.959\,368\,8\sigma$
8	$0.965\,030\,5\sigma$
9	$0.972\,659\,3\sigma$
10	$0.972\,659\,3\sigma$
11	$0.975\,350\,1\sigma$
12	$0.977\,559\,4\sigma$
13	$0.979\,405\,6\sigma$
14	$0.980\,971\,4\sigma$
15	$0.982\,316\,2\sigma$
16	$0.983\,483\,5\sigma$
17	$0.984\,506\,4\sigma$
18	$0.985\,410\,0\sigma$
19	$0.986\,214\,1\sigma$
20	$0.986\,934\,3\sigma$

Example 4.1 For $i = 1$ and $n_0 = \infty$ we obtain with $\frac{\sigma^2}{\delta^2} = 1$ and $\alpha = 0.05$, $n_1 = \lceil 1.96^2 \rceil = 4$, $n_2 = \left\lceil t\left(1 - \frac{\alpha}{2}, 4 - 1\right)^2 \right\rceil = \lceil 3.1824^2 \rceil = 11$ and so on. Finally, we receive this also via the optimal design of experiments (OPDOE)-command in R:

```
> size.t.test(type="one.sample",power=0.5,delta=1,sd=1,
          sig.level=0.01,
          alternative="two.sided")
[1] 10
```

Thus, the final value from our iteration process is $n = 10$. Comparing this with the entries of Table 4.1

$$n = 10 \text{ gives } 0.972\,659\,3\sigma.$$

4.2.2 A Confidence Interval for the Variance of a Normal Distribution

We again assume that the components of the random sample $\boldsymbol{Y} = (\boldsymbol{y}_1, \dots, \boldsymbol{y}_n)^T$ are $N(\mu, \sigma^2)$ distributed with μ and σ^2 unknown. To construct a two-sided $(1 - \alpha)$ confidence interval for σ^2 we use its unbiased estimator s^2 and use as in Chapter 3 the test statistic $\chi^2 = \frac{s^2}{\sigma^2}$, which is centrally χ^2 -distributed with $n-1$ degrees of freedom. Therefore

$$P\left\{ \mathrm{CS}(\alpha_1, n-1) \le \frac{s^2}{\sigma^2} \le \mathrm{CS}(1-\alpha_2, n-1) \right\} = 1 - \alpha_1 - \alpha_2 = 1 - \alpha. \tag{4.11}$$

However, contrary to the tests about expectations it is not optimal to split α into equal parts as reasonable for the corresponding uniformly most powerful unbiased (UMPU) test. However, the unequal case does not always give a shorter expected length and therefore we use the split of α into equal parts. From (4.11) with $\alpha_1 = \alpha_2 = \alpha$ we obtain a $(1-\alpha)$ confidence interval for σ^2 as

$$\left[\frac{(n-1)s^2}{\mathrm{CS}\left(1-\frac{\alpha}{2}, n-1\right)}; \frac{(n-1)s^2}{\mathrm{CS}\left(\frac{\alpha}{2}, n-1\right)} \right]. \tag{4.12}$$

The half-expected length of this interval is $\frac{1}{2}(n-1)\sigma^2 \left\{ \frac{1}{\mathrm{CS}\left(\frac{\alpha}{2}, n-1\right)} - \frac{1}{\mathrm{CS}\left(1-\frac{\alpha}{2}, n-1\right)} \right\}$.

Problem 4.8 Construct a realised two-sided 0.95 confidence interval for σ^2 for the random sample of normally distributed x-data in Table 3.4.

Solution

Here we have a random sample X of size n from $N(\mu, \sigma^2)$ with unknown parameters μ and σ^2. We want to construct a two-sided $1-\alpha$ confidence interval for σ^2 with confidence level = clalpha. We get the solution in R using the following function `> var.interval`.

```
> var.interval = function(data, conf.level = clalpha) {
    df = length(data) - 1
    chilower = qchisq((1 - conf.level)/2, df)
    chiupper = qchisq((1 - conf.level)/2, df, lower.tail
               = FALSE)
    v = var(data)
    c(df * v/chiupper, df * v/chilower)
  }
```

Example

```
> x <- c(7.6, 13.2, 9.1,10.6,8.7, 10.6, 6.8, 9.9, 7.3, 10.4,
         13.3, 10.0, 9.5 )
> var.interval(x, 0.95)
[1]  2.029754 10.756123
```

Problem 4.9 Determine the sample size for constructing a $(1 - \alpha)$ confidence interval for the variance σ^2 of a normal distribution so that

(a) $$\frac{n-1}{2}\sigma^2\left\{\frac{1}{\mathrm{CS}\left(\frac{\alpha}{2}, n-1\right)} - \frac{1}{\mathrm{CS}\left(1-\frac{\alpha}{2}, n-1\right)}\right\} < \delta \text{ or}$$

(b) $$\frac{\mathrm{CS}\left(1-\frac{\alpha}{2}, n-1\right) - \mathrm{CS}\left(\frac{\alpha}{2}, n-1\right)}{\mathrm{CS}\left(1-\frac{\alpha}{2}, n-1\right) + \mathrm{CS}\left(\frac{\alpha}{2}, n-1\right)} < \delta_{\mathrm{rel}}, \delta_{\mathrm{rel}} = \frac{\delta}{\sigma}.$$

Solution
Use the `OPDOE`-command in R `>size.variance.confint(alpha=,delta=)` in case (a) or `>size.variance.confint(alpha=,deltarel=)` in case (b).

Example
We use $(1 - \alpha) = 0.9$ and $\delta_{\mathrm{rel}} = \delta = 0.3$ and get in case (a) with $\sigma^2 = 1$

```
> size.variance.confint(alpha=0.1,delta=0.3)$n
[1] 67
```

and in case (b)

```
> size.variance.confint(alpha=0.1,deltarel=0.3)$n
[1] 59
```

4.2.3 A Confidence Interval for a Probability

Let us presume n independent trials in each of which a certain event A occurs with the same probability p. A $(1 - \alpha)$ confidence interval for p can be calculated from the number of occurrences y of the event A under the n observations as $[l(n,y,\alpha); u(n,y,\alpha)]$, with the lower bound $l(n,y,\alpha)$ and the upper bound, $u(n,y,\alpha)$, respectively given by:

$$l(n, y, \alpha) = \frac{y}{y + (n - y + 1)F\left[2(n - y + 1), 2y, 1 - \frac{\alpha}{2}\right]} \tag{4.13}$$

$$u(n, y, \alpha) = \frac{(y + 1)F\left[2(y + 1), 2(n - y), 1 - \frac{\alpha}{2}\right]}{n - y + (y + 1)F\left[2(y + 1), 2(n - y), 1 - \frac{\alpha}{2}\right]}; \tag{4.14}$$

$F(f_1, f_2, P)$ is the P-quantile of an F-distribution with f_1 and f_2 degrees of freedom.

For other interval estimators of a binomial proportion see Pires and Amado (2008).

To determine the minimum sample size approximately, it seems better to use the half-expected width of a normal approximated confidence interval for p. Such an interval is given by

$$\left[\frac{\mathbf{y}}{n} - Z_{1-\frac{\alpha}{2}}\sqrt{\frac{1}{n}\frac{\mathbf{y}}{n}\left(1-\frac{\mathbf{y}}{n}\right)}; \frac{\mathbf{y}}{n} + Z_{1-\frac{\alpha}{2}}\sqrt{\frac{1}{n}\frac{\mathbf{y}}{n}\left(1-\frac{\mathbf{y}}{n}\right)}\right]. \tag{4.15}$$

This interval has an approximate half-expected width $Z_{1-\frac{\alpha}{2}}\sqrt{\frac{p(1-p)}{n}}$. The requirement that this is smaller than δ leads to the sample size

$$n = \left\lceil \frac{p(1-p)}{\delta^2} Z^2\left(1-\frac{\alpha}{2}\right)\right\rceil.$$

If nothing is known about p, we must take into account the least favourable case $p = 0.5$, which gives the maximum of the minimal sample size, the *maximum* size.

Problem 4.10 Compare the realised exact interval with the realised bounds (4.13) and (4.14) with the realised approximate interval (4.15).

Solution

Use the R-program

```
> binom.test(x,n, p=0.5, alternative = c("two.sided", "less",
     "greater"), conf.level  = 0.95)
  x           number of successes, or a vector of length 2 giving
              the numbers of successes and failures respectively.
  n           number of trials; ignored if x has length 2.
  p           hypothesized probability of success.
 alternative     indicates the alternative hypothesis and must be
                 one of "two.sided", "less" or "greater". You can
                 specify just the initial letter.
 conf.level  confidence level of the returned confidence interval.
```

Example

For $n = 20$, number of successes $y = 5$ and $\alpha = 0.05$.

Exact binomial test in R:

```
> binom.test(5,20,p =0.5, alternative = "two.sided" ,
       conf.level = 0.95)
       Exact binomial test
data:  5 and 20
number of successes = 5, number of trials = 20, p-value =
 0.04139
alternative hypothesis: true probability of success is not
 equal to 0.5
95 percent confidence interval:
 0.08657147 0.49104587
sample estimates:
probability of success
                  0.25
```

Normal approximate interval in R.

```
> n <- 20
> y <- 5
> est_p <- y/n
> est_p
[1] 0.25
> z_0.975 <- qnorm(0.975)
> z_0.975
[1] 1.959964
> n <- 20
> half_width <- z_0.975* sqrt((est_p*(1-est_p)/n))
> half_width
[1] 0.1897727
> lowerCL <- est_p - half_width
> upperCL <- est_p + half_width
> CL <- c(lowerCL, upperCL)
> CL
[1] 0.0602273 0.4397727
```

We get with the exact result for CL (0.086 571 47; 0.491 045 87) and with the normal approximation for CL (0.060 227 3; 0.439 772 7).

Problem 4.11 Determine the sample size for the approximate interval (4.15).

Solution
Use the OPDOE-command in R `>size.prop.confint(p=,delta=,alpha=)`.

Example
We require a confidence interval for the probability $p = P(A)$ of the event A: 'an iron casting is faulty'. The confidence coefficient is specified as 0.90. How many castings should be tested if the half-expected width of the interval is:

(a) Not greater than $\delta = 0.15$, and nothing is known about p?
(b) Not greater than $\delta = 0.15$, when we know that at most 10% of castings are faulty?
We use the command above and obtain:

```
(a) > size.prop.confint(p=0.5,delta=0.15,alpha=0.05)
[1] 43
(b) > size.prop.confint(p=0.1,delta=0.15,alpha=0.05)
[1] 16
```

As we can see, the maximum size is much larger than the size using prior information. Even if $p = 0.3$ we spare some observations.

```
> size.prop.confint(p=0.3,delta=0.15,alpha=0.05)
[1] 36
```

4.3 The Two-Sample Case

We discuss here only differences between location parameters of two populations; for variances we had to consider ratios but the reader can derive the corresponding intervals using the approach described for differences. We assume that the character of interest in each population is modelled by a normally distributed random variable. That is to say, we draw independent random samples of size n_1 and n_2, respectively, from populations 1 and 2. The observations of the random variables $\boldsymbol{y}_{11}, \boldsymbol{y}_{12}, \dots, \boldsymbol{y}_{1n_1}$ on the one hand and $\boldsymbol{y}_{21}, \boldsymbol{y}_{22}, \dots, \boldsymbol{y}_{2n_2}$ on the other hand will be $y_{11}, y_{12}, \dots, y_{1n_1}$ and $y_{21}, y_{22}, \dots, y_{2n_2}$. We call the underlying parameters μ_1 and μ_2, respectively, and further σ_1^2 and σ_2^2, respectively. The unbiased estimators are then for the expectations $\hat{\boldsymbol{\mu}}_1 = \bar{\boldsymbol{y}}_1$ and $\hat{\boldsymbol{\mu}}_2 = \bar{\boldsymbol{y}}_2$, respectively, and for the variances $\hat{\boldsymbol{\sigma}}_1^2 = \boldsymbol{s}_1^2$ and $\hat{\boldsymbol{\sigma}}_2^2 = \boldsymbol{s}_2^2$, respectively (according to Section 3.2).

4.3.1 A Confidence Interval for the Difference of Two Expectations – Equal Variances

If in the two-sample case the two variances are equal, usually from the two samples $y_{11}, y_{12}, \dots, y_{1n_1}$ and $y_{21}, y_{22}, \dots, y_{2n_2}$ a pooled estimator $\boldsymbol{s}^2 = \frac{(n_1-1)\cdot \boldsymbol{s}_1^2 + (n_2-1)\cdot \boldsymbol{s}_2^2}{n_1+n_2-2}$ of the common variance σ^2 is calculated.

The two-sided confidence interval is

$$\left[\bar{\boldsymbol{y}}_1 - \bar{\boldsymbol{y}}_2 - t\left(n_1 + n_2 - 2; 1 - \frac{\alpha}{2}\right) \boldsymbol{s}\sqrt{\frac{n_1 + n_2}{n_1 n_2}}; \bar{\boldsymbol{y}}_1 - \bar{\boldsymbol{y}}_2 + t\left(n_1 + n_2 - 2; 1 - \frac{\alpha}{2}\right) \boldsymbol{s}\sqrt{\frac{n_1 + n_2}{n_1 n_2}}\right]. \tag{4.16}$$

The lower $(1 - \alpha)$ confidence interval is given by

$$\left[\bar{\boldsymbol{y}}_1 - \bar{\boldsymbol{y}}_2 - t(n_1 + n_2 - 2; 1 - \alpha)\boldsymbol{s}\sqrt{\frac{n_1 + n_2}{n_1 n_2}}, \infty\right)$$

and the upper one by

$$\left(-\infty; \bar{\boldsymbol{y}}_1 - \bar{\boldsymbol{y}}_2 + t(n_1 + n_2 - 2; 1 - \alpha)\boldsymbol{s}\sqrt{\frac{n_1 + n_2}{n_1 n_2}}\right].$$

Problem 4.12 Calculate the confidence interval (4.16).

Solution

In R use:

```
> t.test(x, alternative= "two.sided", var.equal=TRUE,
     conf.level = 0.95)
```

Example

```
> x <- c(7.6, 13.2, 9.1, 10.6, 8.7, 10.6, 6.8, 9.9, 7.3,
        10.4, 13.3, 10.0, 9.5 )
> t.test(x, alternative= "two.sided", var.equal=TRUE,
        conf.level = 0.95)
        One Sample t-test
data:  x
t = 17.7289, df = 12, p-value = 5.67e-10
alternative hypothesis: true mean is not equal to 0
95 percent confidence interval:
  8.56863 10.96983
sample estimates:
mean of x
 9.769231
```

In the case of equal variances, it can be shown that optimal plans require equal sample sizes n_x and n_y. Thus $n_x = n_y = n$, and in the case where the half-expected width must be less than δ, we find n iteratively from

$$n = \left\lceil 2\sigma^2 \frac{t^2\left(2n-2; 1-\frac{\alpha}{2}\right)}{\delta^2(2n-2)} \frac{\Gamma^2\left(\frac{2n-1}{2}\right)}{\Gamma^2(n-1)} \right\rceil . \tag{4.17}$$

The reader may derive the sample size needed for one-sided intervals.

Problem 4.13 Calculate n using (4.17).

Solution
Use

```
> size.t.test(power=0.5,sig.level=,delta=,sd=, type =
      "two.sample")
```

Example
We calculate the sample size for $\alpha = 0.05$, $\delta = 0.5$, sd = 1.

```
> size.t.test(power=0.5,sig.level=0.05,delta=0.5,sd=1,type =
      "two.sample")
[1] 32
```

Problem 4.14 Derive the sample size formula for the construction of a one-sided confidence interval for the difference between two expectations for equal variances.

Solution

Analogously to (4.17) we obtain the formula

$$n = \left\lceil 2\sigma^2 \frac{t^2(2n-2; 1-\alpha)}{\delta^2(2n-2)} \frac{\Gamma^2\left(\frac{2n-1}{2}\right)}{\Gamma^2(n-1)} \right\rceil$$

and the corresponding R-command

```
> size.t.test(power=0.5,sig.level=,delta=,sd=,
    type="one.sample")
```

Example

We calculate the sample size for $\alpha = 0.05$, $\delta = 0.5$, sd = 1.

```
> size.t.test(power=0.5,sig.level=0.05,delta=0.5,sd=1,type =
    "one.sample")
[1] 18
```

We see that the sample size for a one-sided confidence interval with an analogous precision requirement as for the two-sided case is smaller than that for the two-sided case.

4.3.2 A Confidence Interval for the Difference of Two Expectations – Unequal Variances

If in the two-sample case, the two variances σ_1^2 and σ_2^2 are unequal, sample variances s_1^2 and s_2^2 from the two independent samples $y_{11}, y_{12}, \ldots, y_{1n_1}$ and $y_{21}, y_{22}, \ldots, y_{2n_2}$ are used.

The confidence interval

$$\left[\overline{y_1} - \overline{y_2} - t\left(f^*, 1-\frac{\alpha}{2}\right)\sqrt{\frac{s_1^2}{n_1} + \frac{s_2^2}{n_2}}, \overline{y_1} - \overline{y_2} + t\left(f^*, 1-\frac{\alpha}{2}\right)\sqrt{\frac{s_1^2}{n_1} + \frac{s_2^2}{n_2}}\right] \quad (4.18)$$

is an approximate $(1-\alpha)$ confidence interval (Welch 1947) with

$$f^* = \left\lceil \frac{\left(\frac{s_1^2}{n_1} + \frac{s_2^2}{n_2}\right)^2}{\frac{s_1^4}{(n_1-1){n_1}^2} + \frac{s_2^4}{(n_2-1){n_2}^2}} \right\rceil \text{ degrees of freedom.}$$

Example 4.2 Using R, determine a realised two-sided 0.95 confidence interval of μ for the random samples of normally distributed x- and y-data in Table 3.4. We assume that there are two independent random samples of the mice populations.

```
> x <- c(7.6, 13.2, 9.1, 10.6, 8.7, 10.6, 6.8, 9.9, 7.3,
         10.4, 13.3, 10.0, 9.5)
> y <- c(7.8, 11.1, 16.4, 13.7, 10.7, 12.3, 14.0, 11.9,
         8.8, 7.7, 8.9, 16.4, 10.2)
> t.test(x,y,alternative = "two.sided", mu = 0,
         var.equal = FALSE, conf.level=0.95)
```

```
        Welch Two Sample t-test
data:  x and y
t = -1.7849, df = 21.021, p-value = 0.08871
alternative hypothesis: true difference in means is not
  equal to 0
95 percent confidence interval:
 -3.8137583  0.2906814
sample estimates:
mean of x mean of y
 9.769231 11.530769
```

To determine the necessary sample sizes n_1 and $n_{2,}$ besides an upper bound for the half expected width δ, we need information about the two variances. Suppose that prior estimates $\boldsymbol{s}_1^2$ and $\boldsymbol{s}_2^2$ are available for the variances, which may possibly be unequal. For a two-sided confidence interval, we can calculate n_1 and n_2 approximately (by replacing the variances by their estimates) and iteratively from

$$n_1 = \left\lceil \frac{\sigma_1(\sigma_1 + \sigma_2)}{\delta^2} t^2 \left(f^*, 1 - \frac{\alpha}{2}\right) \right\rceil$$

and $n_2 = \left\lceil \dfrac{n_1 \sigma_2}{\sigma_1} \right\rceil$.

Problem 4.15 We would like to find a two-sided 99% confidence interval for the difference of the expectations of two normal distributions with unequal variances using independent samples from each population with power = 0.90 and variance ratio $\sigma_1{}^2/\sigma_2{}^2 = 4$. Given the minimum size of an experiment, we would like to find a two-sided 99% confidence interval for the difference of the expectations of two normal distributions with unequal variances using independent samples from each population and define the precision by $\delta = 0.4\sigma_x$. If we know that $\dfrac{\sigma_x^2}{\sigma_y^2} = 4$, we obtain $n_y = \left\lceil \dfrac{1}{2} n_x \right\rceil$.

Solution
Use first the OPDOE-command in R

```
> power.t.test (sd = , sig.level = , delta = , power = ,
    type = "two.sample", alternative = "two.sided")
```

This gives us the equal sample sizes due to the implementation of $\sigma_1 = \sigma_2$. We have assumed that $\sigma_1^2/\sigma_2^2 = 4$, hence $\sigma_1/\sigma_2 = 2$.

We may conjecture that n_1 is two times as large as σ_2 because $\sigma_1 = 2\sigma_2$. Hence we take $n_1 = 2n_2$.

Example
We define the precision by $\alpha = 0.01$, $\delta = 0.4\sigma$, and assume that $\dfrac{\sigma_1}{\sigma_2} = 2$.

```
> power.t.test(sd=1,sig.level=0.01,delta=0.4,power = 0.9,
  type = "two.sample", alternative = "two.sided")
```

```
    Two-sample t test power calculation

              n = 187.6586
          delta = 0.4
             sd = 1
      sig.level = 0.01
          power = 0.9
    alternative = two.sided

NOTE: n is number in *each* group
```

Hence $n_2 = 188$ and $n_1 = 2 * 188 = 376$.

4.3.3 A Confidence Interval for the Difference of Two Probabilities

Let us say that we are interested in a certain characteristic A from the elements of a population of size N. The number of elements in this population with characteristic A is $N(A)$. The population fraction with characteristic A is $\pi = N(A)/N$. We take a random sample of size n with replacement from this population and then the random variable $\boldsymbol{k}$ of elements with this characteristic A has a binomial distribution $B(n, \pi)$ with the parameters π and n.

The sample fraction $\boldsymbol{p} = \boldsymbol{k}/n$ is an unbiased estimator of π with expectation $E(\boldsymbol{p}) = \pi$ and variance $\text{var}(\boldsymbol{p}) = \pi(1 - \pi)/n$. If neither k nor $n - k$ is less than 5 and if the sample size n is large then the distribution of $\boldsymbol{k}$ can be approximated by a normal distribution with the mean $\mu = p$, the sample fraction, and variance $\sigma^2 = p(1 - p)/n$.

In practice the researcher is very often interested in the difference between the population fractions with characteristic A in the two populations I and II, namely π_1 and π_2. Suppose we have a random sample of size n_1 from population I and another independent sample of size n_2 from population II. Then the unbiased estimator of $\pi_1 - \pi_2$ is the difference of the sample fractions $\boldsymbol{p}_1 - \boldsymbol{p}_2$. If neither k_1 nor $n_1 - k_1$ and k_2 nor $n_2 - k_2$ is less than 5 and if the sample sizes n_1 and n_2 are large then the distribution of $\boldsymbol{k}$ can be approximated by a normal distribution with mean $\mu = p_1 - p_2$ and variance $\sigma^2 = p_1(1 - p_1)/n_1 + p_2(1 - p_2)/n_2$.

A confidence interval with confidence coefficient $1 - \alpha$ for $\pi_1 - \pi_2$ is then approximated with

$$\text{lower confidence limit} = p_1 - p_2 - Z_{1-\alpha/2}\sqrt{(p_1(1 - p_1)/n_1 + p_2(1 - p_2)/n_2)} \text{ and}$$

$$\text{upper confidence limit} = p_1 - p_2 + Z_{1-\alpha/2}\sqrt{(p_1(1 - p_1)/n_1 + p_2(1 - p_2)/n_2)}.$$

Example 4.3 Best et al. (1967) reported that from men aged 60–64 years at the beginning of the study, belonging to one of the two classes: (i) 'non-smoker' and (ii) 'those who reported that they smoked pipes only', the number of deaths during the succeeding six years was obtained.

The 2×2 contingency table of the observations is shown in Table 4.2.

Table 4.2 Confidence table of two kinds of smokers.

	Sample 1 (non-smokers)	Sample 2 (pipe smokers)
Dead	117	54
Alive	950	348
Total	$n_1 = 1067$	$n_2 = 402$

In R we can find the 95% confidence interval for the difference in the fraction of deceased as follows:

```
> CT2x2  <- cbind(c(117, 950), c(54, 348))
> CT2x2
     [,1] [,2]
[1,]  117   54
[2,]  950  348
> n1 <- sum(CT2x2[,1])
> n1
[1] 1067
> n2 <- sum(CT2x2[,2])
> n2
[1] 402
> prop.test(x=c(117,54), n=c(1067,402))
        2-sample test for equality of proportions with
        continuity correction

data:  c(117, 54) out of c(1067, 402)
X-squared = 1.4969, df = 1, p-value = 0.2212
alternative hypothesis: two.sided
95 percent confidence interval:
 -0.06463259  0.01528234
sample estimates:
   prop 1    prop 2
0.1096532 0.1343284
```

Hence the 95% confidence interval is: $-0.064\,632\,59 < \pi_1 - \pi_2 < 0.015\,282\,34$. The estimate of π_1 is $p_1 = 0.109\,653\,2$ and the estimate of π_2 is $p_2 = 0.134\,328\,4$. Calculation with the normal approximation in R by the formula above gives:

```
> p1 <- 0.1096532
> p2 <- 0.1343284
> diff <- p1- p2
> diff
[1] -0.0246752
> n1 <- 1067
> n2 <- 402
```

```
> z_0.975 = qnorm(0.975)
> z_0.975
[1] 1.959964
> width <- z_0.975*sqrt((p1*(1-p1)/n1) + (p2*(1-p2)/n2))
> width
[1] 0.03824509
> width <- z_0.975*sqrt((p1*(1-p1)/n1) + (p2*(1-p2)/n2))
> width
[1] 0.03824509
> CLlower <- diff - width
> CLlower
[1] -0.06292029
> CLupper <- diff + width
> CLupper
[1] 0.01356989
```

The difference in the results of the command `prop.test` is due to the use of Yates' continuity correction in the normal approximation, because the default command is `correct = TRUE`. Further defaults are `alternative = "two.sided"` and the default confidence level is `conf.level = 0.95`. The complete command is:

```
> prop.test(x=c(117,54), n=c(1067,402),alternative=
   "two.sided",conf.level=0.95,correct=TRUE)
        2-sample test for equality of proportions with
        continuity correction
data:  c(117, 54) out of c(1067, 402)
X-squared = 1.4969, df = 1, p-value = 0.2212
alternative hypothesis: two.sided
95 percent confidence interval:
 -0.06463259  0.01528234
sample estimates:
   prop 1    prop 2
0.1096532 0.1343284
```

If we do not want to use Yates' continuity correction the command is:

```
> prop.test(x=c(117,54), n=c(1067,402),alternative =
   "two.sided",
   conf.level = 0.95 , correct =FALSE)
        2-sample test for equality of proportions without
        continuity correction
data:  c(117, 54) out of c(1067, 402)
X-squared = 1.7285, df = 1, p-value = 0.1886
alternative hypothesis: two.sided
95 percent confidence interval:
 -0.06292021  0.01356996
sample estimates:
   prop 1    prop 2
0.1096532 0.1343284
```

Altman et al. (2002) recommend for the two sample unpaired case a different method for the confidence interval of the difference between two population proportions.

The recommended method can also be used for small data samples. The confidence interval calculation is method 10 of Newcombe (1998). Calculate for the first random sample of size n_1 with the sample fraction p_1 the lower limit l_1 and upper limit u_1 that define the $(1-\alpha)$ confidence interval for the first population proportion π_1. For the second sample of size n_2 calculate with the sample fraction p_2 the lower limit l_2 and upper limit u_2 that define the $(1-\alpha)$ confidence interval for the second population proportion π_2.

The $(1-\alpha)$ confidence interval for $\pi_1-\pi_2$ has lower limit $D-\sqrt{[(p_1-l_1)^2+(u_2-p_2)^2]}$ and upper limit $D+\sqrt{[(p_2-l_2)^2+(u_1-p_1)^2]}$ with $D=p_1-p_2$. Note that D is not generally at the midpoint of the confidence interval.

Example 4.4 Goodfield et al. (1992) reported adverse effects for dermatophyte onchomyosis (respiratory problems) for 5 patients in a random sample of 56 patients treated with terbinafine and for 0 patients in a random sample of 29 placebo treated patients.

In R we do the analysis as follows:

```
> binom.test(5, 56, p =0.5, alternative = "two.sided" ,
        conf.level = 0.95)
        Exact binomial test
data:  5 and 56
number of successes = 5, number of trials = 56, p-value =
     1.17e-10
alternative hypothesis: true probability of success is not
     equal to 0.5
95 percent confidence interval:
 0.02962984  0.19619344
sample estimates:
probability of success
            0.08928571
> binom.test(0, 29, p =0.5, alternative = "two.sided" ,
   conf.level = 0.95)
        Exact binomial test
data:  0 and 29
number of successes = 0, number of trials = 29, p-value =
   3.725e-09
alternative hypothesis: true probability of success is not
   equal to 0.5
95 percent confidence interval:
 0.0000000  0.1194449
sample estimates:
probability of success
                     0
```

We used the results of the `binom.test()` in the following commands:

```
> p1 <- 0.08928571
> l1 <- 0.02962984
> u1 <- 0.19619344
> p2 <-  0
> l2 <-  0.0000000
> u2 <-  0.1194449
> D <- p1-p2
> widthlower <-  sqrt((p1-l1)^2 + (u2-p2)^2)
> widthupper <- sqrt((p2-l2)^2 + (u1-p1)^2)
> CIlower <- D - widthlower
> CIupper <- D + widthupper
> CIlower
[1] -0.02970976
> CIupper
[1] 0.1961934
```

Hence the approximate (1 – 0.05) confidence interval is $-0.02970976 < \pi_1 - \pi_2 < 0.1961934$.

References

Altman, D.G., Machin, D., Bryant, T.N., and Gardner, M.J. (2002). *Statistics with Confidence; Confidence Intervals and Statistical Guidelines*, 2e. Bristol: British Medical Journal Books.

Best, E.W.R., Walker, C.B., Baker, P.M. et al. (1967). Summary of a Canadian Study on Smoking and Health. *Can. Med. Assoc. J.* 96 (15): 1104–1108.

Goodfield, M.J.D., Andrew, L., and Evans, E.G.V. (1992). Short-term treatment of dermatophyte onchomyotis with terbinafine. *BMJ* 304: 1151–1154.

Newcombe, R.G. (1998). Interval estimation for the difference between independent proportions: comparison of eleven methods. *Stat. Med.* 17: 873–890.

Pires, A.M. and Amado, C. (2008). Interval estimators for a binomial proportion: comparison of twenty methods. *RevStat Stat. J.* 6 (2): 165–197.

Rasch, D. and Schott, D. (2018). *Mathematical Statistics.* Oxford: Wiley.

Rasch, D. and Tiku, M.L. (eds. 1985) Robustness of statistical methods and nonparametric statistics. Proc. Conf. on Robustness of Statistical Methods and Nonparametric Statistics, Schwerin (DDR), May 29 June 2, 1983. Reidel Publ. Co. Dordrecht.

Welch, B.L. (1947). The generalization of students problem when several different population variances are involved. *Biometrika* 34: 28–35.

5

Analysis of Variance (ANOVA) – Fixed Effects Models

5.1   Introduction

In the analysis of variance, we assume that parameters of random variables depend on non-random variables, called factors. The values a factor can take we call factor levels or in short levels. We discuss cases where one, two or three factors have an influence on the observations.

An experimenter often has to find out in an experiment whether different values of several variables or of several factors have different results on the experimental material. If the effects of several factors have to be examined, the conventional method means to vary only one of these factors at once and to keep all other factors constant. To investigate the effect of p factors this way, p experiments have to be conducted. It can be that the results at the levels of a factor investigated depend on the constant levels of other factors, which means that interactions between the factors exist. The British statistician R. A. Fisher recommended experimental designs by varying the levels of all factors at the same time. For the statistical analysis of the experimental results of such designs (they are called factorial experiments), Fisher developed a statistical procedure: the analysis of variance. The first publication about this topic stemmed from Fisher and Mackenzie (1923), a paper about the analysis of field trials in Fisher's workplace at Rothamsted Experimental Station in Harpenden, UK. A good overview is given in Scheffé (1959) and in Rasch and Schott (2018).

The analysis of variance is based on the decomposition of the sum of squared deviations of the observations from the total mean of the experiment into components. Each of the components is assigned to a specific factor or to interactions of factors or to the experimental error. Further, a corresponding decomposition of the degrees of freedom belonging to sums of squared deviations is done. The analysis of variance is mainly used to estimate the effects of factor levels or to test statistical hypotheses (model I in this chapter), or to estimate components of variance that can be assigned to the different factors (model II – see Chapter 6).

The analysis of variance can be applied to several problems based on mathematical models called model I, model II and the mixed model, respectively. The problem leading to model I is as follows: all factor levels have been particularly selected and involved in the experiment because just these levels are of practical interest. The objective of the experiment is to find out whether the effects of the different levels (or factor level combinations) differ significantly or randomly from each other. The experimental question

Applied Statistics: Theory and Problem Solutions with R, First Edition.
Dieter Rasch, Rob Verdooren, and Jürgen Pilz.
© 2020 John Wiley & Sons Ltd. Published 2020 by John Wiley & Sons Ltd.

can be answered by a statistical test if particular assumptions are fulfilled. The statistical conclusion refers to (finite) factor levels specifically selected.

In this chapter, problems in a model I are discussed and these are the estimation of the effects and the interaction effects of the several factors and testing the significance of these effects.

We also show how to determine the optimal size of an experiment. For all these cases we assume that we have to plan an experiment with a type I risk $\alpha = 0.05$ and a power $1 - \beta = 0.95$.

5.1.1 Remarks about Program Packages

For the analysis, we can use the program package R as we have downloaded it. Those who like to analyse data by IBM-SPSS Statistics find programs in Rasch and Schott (2018) and who prefer SAS can find corresponding programs in Rasch et al. (2008). For experimental designs, we use in R the command

```
> install.packages("OPDOE")
```

and

```
> library(OPDOE)
```

`OPDOE` stands for 'optimal design of experiments' and was used in Chapter 3. The R syntax for calculating the sample size for analysis of variance (or for short ANOVA) can be found by `>size.anova`; a description of how to use `OPDOE` is found by

```
> help(size.anova)
```

Detailed instructions and examples are given in chapter 3 of Rasch et al. (2011).

5.2 Planning the Size of an Experiment

For planning the size of a balanced experiment, this means equal sample sizes for the effects, precision requirements are needed, as in Chapter 3. The following approach is valid for all sections of this chapter. Besides the precision requirement of the two risks α and β (or the power of the F-test $1 - \beta$) the non-centrality parameter λ of the non-central F-distribution with f_1 and f_2 degrees of freedom has to be given in advance. With the $(1 - \alpha)$- and the β-quantile of the non-central F-distribution with f_1 and f_2 degrees of freedom and the non-centrality parameter λ we have to solve the equation

$$F(f_1,f_2,0 \mid 1 - \alpha) = F(f_1,f_2,\lambda \mid \beta). \tag{5.1}$$

This equation plays an important role in all sections of this chapter. In addition to f_1, f_2, α, and β the difference δ between the largest and the smallest effect (main effect or in the following sections also interaction effect), to be tested against null, belongs to the precision requirement. We denote the solution λ in (5.1) by

$$\lambda = \lambda(\alpha,\ \beta,\ f_1,\ f_2).$$

Let E_{min}, E_{max} be the minimum and the maximum of q effects $E_1, E_2, \ldots, E_q$ of a fixed factor E or an interaction, respectively. Usually we standardise the precision requirement by the relative precision requirement $\delta = \frac{\lambda}{\sigma}$.

If $E_{max} - E_{min} \geq \delta$ then for the non-centrality parameter of the F-distribution (for even q) with $\overline{E} = \frac{1}{q}\sum_{i=1}^{q} E_i$ holds

$$\lambda = \sum_{i=1}^{q}(E_i - \overline{E})^2/\sigma^2 \leq \frac{\frac{q}{2}(E_{max} - \overline{E})^2 + \frac{q}{2}(E_{min} - \overline{E})^2}{\sigma^2}$$

$$\leq q(E_{max} - E_{min})^2/(2\sigma^2).$$

From this it follows

$$\lambda = \sum_{i=1}^{q}(E_i - \overline{E})^2/\sigma^2 \quad \leq q\delta^2/(2\sigma^2). \tag{5.2}$$

The minimal size of the experiment needed depends on λ accordingly to the exact position of all q effects. However, this is unknown when the experiment starts. We consider two extreme cases, the most favourable (resulting in the smallest minimal size n_{min}) and the least favourable (resulting in the largest minimal size n_{max}). The least favourable case leads to the smallest non-centrality parameter λ_{min} and by this to the so-called maximin size n_{max}. This occurs if the $q - 2$ non-extreme effects equal $\frac{E_{max}+E_{min}}{2}$. For $\overline{E} = 0$, $\sum_{i=1}^{q}(E_i - \overline{E})^2 = qE^2$ this is shown in the following scheme.

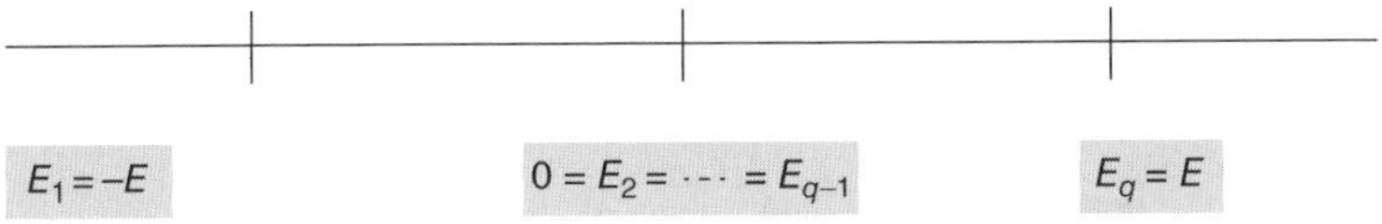

The most favourable case leads to the largest non-centrality parameter λ_{max} and by this to the so-called minimin size n_{min}. If $q = 2m$ (even) this is the case, if m of the E_i equal E_{min} and the m other E_i equal E_{max}. If $q = 2m + 1$ (odd) again m of the E_i should equal E_{min} and m other E_i should equal E_{max}, and the remaining effect should be equal to one of the two extremes E_{min} or E_{max}. For $\overline{E} = 0$, $\sum_{i=1}^{q}(E_i - \overline{E})^2 = qE^2$ this is shown in the following scheme for even q.

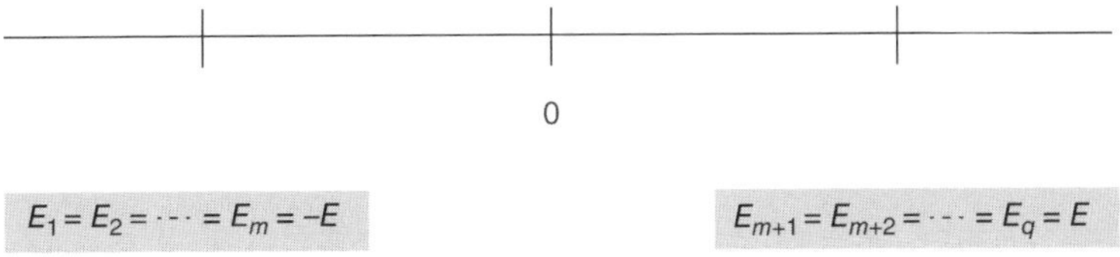

When we plan an experiment, we always obtain equal sub-class numbers. Therefore, we use models and ANOVA tables mainly for the equal subclass number case because we then have simpler formulae for the expected mean squares. In the analysis programs, unequal sub-class numbers are also possible.

In Section 5.3 we give a theoretical ANOVA – table (for random variables) with expected mean squares $E(\boldsymbol{MS})$. To find a proper $\boldsymbol{F}$-statistic for testing a null hypothesis corresponding to a fixed row in the table we proceed as follows. If the null hypothesis is correct the numerator and denominator of $\boldsymbol{F}$ have the same expectation. In general it is a ratio of two $\boldsymbol{MS}$ of a particular null hypothesis with the corresponding degrees of freedom. This ratio is centrally F-distributed if the numerator and the denominator in the case that the hypothesis is valid have the same expectation. This equality is, however, not sufficient if unequal subclass numbers occur, for instance it is not sufficient if the $\boldsymbol{MS}$ are not independent of each other. In this case, we obtain only a test statistic that is approximately F-distributed. Such cases occur in Chapter 6. We write $\tau = \delta/\sigma$.

5.3 One-Way Analysis of Variance

In this section, we investigate the effects of one factor.

From a populations or universes, $G_1, \dots, G_a$ random samples $\boldsymbol{Y}_1, \dots, \boldsymbol{Y}_a$ of size $n_1, \dots, n_a$, respectively are drawn independently of each other. We write $\boldsymbol{Y}_i = (\boldsymbol{y}_{i1}, \dots, \boldsymbol{y}_{in_i})^T$. The $\boldsymbol{y}_i$ are assumed to be distributed in the populations G_i as $N(\{\mu_i\}, \sigma^2 I_{n_i})$ with $\{\mu_i\} = (\mu_i, \cdots, \mu_i)^T$. Further we write $\mu_i = \mu + a_i \ (i = 1, \dots, a)$. Then we have one factor A with the factor levels A_i; $i = 1, \dots, a$ and write

$$\boldsymbol{y}_{ij} = \mu + a_i + \boldsymbol{e}_{ij} \quad (i = 1, \dots, a; j = 1, \dots, n_i). \tag{5.3}$$

We call μ the total mean and a_i the effect of the ith level of factor A. The total size of the experiment is $N = \sum_{i=1}^{a} n_i$.

In Table 5.1, we find the scheme of the observations of an experiment with a levels $A_1, \cdots, A_a$ of factor A and n_i observations for the ith level A_i of A. We use Equation (5.3) with the side conditions

$$E(\boldsymbol{e}_{ij}) = 0, \qquad \text{cov}(\boldsymbol{e}_{ij}, \boldsymbol{e}_{kl}) = \delta_{ik}\delta_{jl}\sigma^2.$$

For testing the hypothesis, $\boldsymbol{e}_{ij}$ and by this also $\boldsymbol{y}_{ij}$ is assumed to be normally distributed.

For testing hypotheses about $\mu + a_i$ further assumptions are not needed but to test hypotheses about a_i we need a so-called reparametrisation condition like $\sum_{i=1}^{a} n_i a_i = 0$ or $\sum_{i=1}^{a} a_i = 0$; both are equivalent if all $n_i = n$ and this we call the balanced case.

In this chapter, we use the point convention for writing sums. In the one-way case discussed in this section we have

$$\sum_{j=1}^{n_i} y_{ij} = Y_{i.}$$

Table 5.1 Observations y_{ij} of an experiment with a levels of a factor A.

	1	2	...	a
y_{ij}	y_{11}	y_{21}	$\cdots$	y_{a1}
	y_{12}	y_{22}	$\cdots$	y_{a2}
	$\vdots$	$\vdots$	$\vdots$	$\vdots$
	y_{1n_1}	y_{2n_2}	$\cdots$	y_{an_a}
n_i	n_1	n_2	$\cdots$	n_a
$Y_{i.}$	$Y_{1.}$	$Y_{2.}$	$\cdots$	$Y_{a.}$

and

$$\sum_{i=1}^{a}\sum_{j=1}^{n_i} y_{ij} = Y_{..}.$$

The arithmetic means are $\frac{1}{N}\sum_{i=1}^{a}\sum_{j=1}^{n} y_{ij} = \bar{y}_{..}$ and $\frac{1}{n_i}\sum_{j=1}^{n_i} y_{ij} = \bar{y}_{i\cdot}$

Estimators $\hat{\boldsymbol{a}}_i$ for a_i $(i = 1, \dots, a)$ and $\hat{\boldsymbol{\mu}}$ for μ in the model (5.3) are given by

$$\hat{\boldsymbol{\mu}}_1 = \frac{1}{a}\sum_{i=1}^{a} \bar{\boldsymbol{y}}_{i.}, \tag{5.4}$$

$$\hat{\boldsymbol{a}}_{1i} = \frac{a-1}{a}\bar{\boldsymbol{y}}_{i.} - \frac{1}{a}\sum_{j\neq i} \bar{\boldsymbol{y}}_j \tag{5.5}$$

if we assume $\sum_{i=1}^{a} a_i = 0$ and by

$$\hat{\boldsymbol{\mu}}_2 = \bar{\boldsymbol{y}}_{..}, \tag{5.6}$$

$$\hat{\boldsymbol{a}}_{2i} = \bar{\boldsymbol{y}}_{i.} - \bar{\boldsymbol{y}}_{..} \tag{5.7}$$

if we assume $\sum_{i=1}^{a} n_i a_i = 0$. For estimable functions, we drop the left subscripts and write the symbols as in (5.3).

The n_i are called sub-class numbers. Both estimators are identical in the balanced case if $n_i = n$ $(i = 1, \dots, a)$.

The reader may ask which reparametrisation condition he should use. There is no general answer. Besides the two forms above, many others are possible. However, fortunately many of the results derived below are independent of the side condition chosen. Often estimates of the a_i are less interesting than those for estimable functions of the parameters such as $\mu + a_i$ and $a_i - a_j$ and these estimable functions give the same answer for all side conditions.

The variance σ^2 in both cases is unbiasedly estimated by

$$\boldsymbol{s}^2 = \frac{\sum_{i=1}^{a}\sum_{j=1}^{n_i} \boldsymbol{y}_{ij}^2 - \frac{1}{N}\sum_{i=1}^{a} \boldsymbol{Y}_{i\cdot}^2}{N-a} = \boldsymbol{MS}_{\text{res}} \tag{5.8}$$

Table 5.2 gives the ANOVA table for model (5.3). In this table *SS* means sum of squares, *MS* means mean squares and *df* means degrees of freedom. We call this table a theoretical ANOVA table because we write the entries as random variables that are functions of the underlying random samples $\boldsymbol{Y}_1, \dots, \boldsymbol{Y}_a$. If we have observed data as realisations of the random samples then the column $E(\boldsymbol{MS})$ is dropped and nothing in the table is in bold print.

Table 5.3 is the empirical ANOVA table corresponding to Table 5.2.

5.3.1 Analysing Observations

Estimable functions of the model parameters are for instance $\mu + a_i (i = 1, \dots, a)$ or $a_i - a_j (i, j = 1, \dots, a; i \neq j)$ with the estimators (using (5.4)–(5.7))

$$\widehat{\mu + a_i} = \widehat{\mu_1 + a_{1i}} = \widehat{\mu_2 + a_{2i}} \tag{5.9}$$

Table 5.2 Theoretical ANOVA table: one-way classification, model I.

Source of variation	SS	df	MS	E(MS) for $n_i = n$
Main effect A	$\boldsymbol{SS}_A = \frac{1}{n_i}\sum_{i=1}^{a} \boldsymbol{Y}_{i.}^2 - \frac{\boldsymbol{Y}_{..}^2}{N}$	$a-1$	$\boldsymbol{MS}_A = \frac{\boldsymbol{SS}_A}{a-1}$	$\sigma^2 + \frac{n}{a-1}\sum a_i^2$
Residual	$\boldsymbol{SS}_{\text{res}} = \sum_{i=1}^{a}\sum_{j}^{n_i} \boldsymbol{y}_{ij}^2 - \frac{1}{n_i}\sum_{i=1}^{a} \boldsymbol{Y}_{i.}^2$	$N-a$	$\boldsymbol{MS}_{\text{res}} = \frac{\boldsymbol{SS}_{\text{res}}}{N-a}$	σ^2

Table 5.3 Empirical ANOVA table: one-way classification, model I .

Source of variation	SS	df	MS
Main effect A	$SS_A = \frac{1}{n_i}\sum_{i=1}^{a} Y_{i.}^2 - \frac{Y_{..}^2}{N}$	$a-1$	$MS_A = \frac{SS_A}{a-1}$
Residual	$SS_{\text{res}} = \sum_{i=1}^{a}\sum_{j}^{n_i} y_{ij}^2 - \frac{1}{n_i}\sum_{i=1}^{a} Y_{i.}^2$	$N-a$	$MS_{\text{res}} = \frac{SS_{\text{res}}}{N-a}$
Total	$SS_T = \sum_{i=1}^{a}\sum_{j}^{n_i} y_{ij}^2 - \frac{Y_{..}^2}{N}$	$N-1$	

and

$$\widehat{\boldsymbol{a}_i - \boldsymbol{a}_j} = \overline{\boldsymbol{y}}_{i.} - \overline{\boldsymbol{y}}_{j.} = \widehat{\boldsymbol{a}_{1i} - \boldsymbol{a}_{1j}} = \widehat{\boldsymbol{a}_{2i} - \boldsymbol{a}_{2j}} \tag{5.10}$$

respectively. They are independent of the special choice of the reparametrisation condition.

Besides point estimation, an objective of an experiment (model I) is to test the null hypothesis $H_0: a_i = a_j$ for all $i \neq j$ against the alternative that at least two of the effects a_i differ from each other. This null hypothesis corresponds to the assumption that the effects of the factor considered for all a levels are equal. The basis of the corresponding tests is the fact that the sum of squared deviations $\boldsymbol{SS}$ of $\boldsymbol{y}_{ij}$ from the total mean of the experiment $\overline{\boldsymbol{y}}$. can be broken down into independent components.

The total sum of squared deviations of the observations from the total mean of the experiment is

$$SS_T = \sum_{i=1}^{a}\sum_{j=1}^{n_i} (y_{ij} - \overline{y}_{..})^2 = \sum_{i=1}^{a}\sum_{j=1}^{n_i} (y_{ij} - \overline{y}_{i.})^2 + \sum_{i=1}^{a}\sum_{j=1}^{n_i} (\overline{y}_{i.} - \overline{y}_{..})^2.$$

The left-hand side is called $\boldsymbol{SS}$ total or for short $\boldsymbol{SS}_T$, the first component of the right-hand side is called $\boldsymbol{SS}$ within the treatments or levels of factor A (for short $\boldsymbol{SS}$ within $\boldsymbol{SS}_{\text{res}}$) and the last component of the right hand side $\boldsymbol{SS}$ between the treatments or levels of factor A ($\boldsymbol{SS}_A$), respectively.

We generally write

$$\boldsymbol{SS}_T = \sum_{ij} \boldsymbol{y}_{ij}^2 - \frac{1}{N}\boldsymbol{Y}_{..}^2, \tag{5.11}$$

$$\boldsymbol{SS}_{\text{res}} = \sum_{ij} y_{ij}^2 - \sum_i \frac{\boldsymbol{Y}_{i.}^2}{n_i}, \tag{5.12}$$

$$\boldsymbol{SS}_A = \sum_i \frac{\boldsymbol{Y}_{i.}{}^2}{n_i} - \frac{1}{N} Y_{..}^2. \tag{5.13}$$

It is known from Rasch and Schott (2018, theorem 5.4) that

$$\boldsymbol{F} = \frac{(N-a)\boldsymbol{SS}_A}{(a-1)\boldsymbol{SS}_{\text{res}}} \tag{5.14}$$

is distributed as $F(a-1, N-a, \lambda)$ with the non-centrality parameter

$$\lambda = \left[\sum_{\text{i}=1}^{\text{a}} n_i a_i^2 - \left(\sum_{\text{i}=1}^{\text{a}} n_i a_\text{i} \right)^2 / N \right] / \sigma^2.$$

If $H_0: \ a_1 = \ \dots \ = a_a$ is valid then we have $\lambda = 0$, and thus, $\boldsymbol{F}$ is $F(a-1, \ N-a)$ distributed. Therefore, the hypothesis $H_0: \ \ a_1 = \ \dots \ = a_a$ is tested by an F-test. The ratios $\boldsymbol{MS}_A = \frac{\boldsymbol{SS}_A}{a-1}$ and $\boldsymbol{MS}_{\text{res}} = \frac{\boldsymbol{SS}_{\text{res}}}{N-a}$ are called mean squares between treatments and within treatments or residual mean squares, respectively.

Example 5.1 We assume that the breeding value of three sires of a special cattle breed concerning milk fat in kilograms is tested via the milk fat performance of their daughters. In this case we have three levels of the factor sire, i.e. $a = 3$.

We assume that we model the milk fat performance of the daughters by a normally distributed random variable. Then we use the F-test in formula (5.14). We at first determine in the balanced case the numbers of daughters needed for such a performance test so that the hypothesis that the three sires have the same breeding value is erroneously rejected with a first type risk $\alpha = 0.05$. The power is at least $1 - \beta$ as long as $\delta \geq 2$, where δ is the difference between the largest and the smallest effect. The determination will be explained in Section 5.3.2.

The expectations of these $\boldsymbol{MS}$ are

$$E(\boldsymbol{MS}_A) = \sigma^2 + \frac{1}{a-1} \left[\sum_{i=1}^{a} n_i a_i^2 - \frac{1}{N} \left(\sum_{i=1}^{a} n_i a_i \right)^2 \right]$$

and

$$E(\boldsymbol{MS}_{\text{res}}) = \sigma^2.$$

Under the reparametrisation condition $\sum_{i=1}^{a} n_i a_i = 0$ we obtain

$$E(\boldsymbol{MS}_A) = \sigma^2 + \frac{1}{a-1} \sum_{i=1}^{a} n_i a_i^2.$$

Now the several steps in the simple ANOVA for model I can be summarised as follows.

We assumed that from systematically selected normally distributed populations with expectations $\mu + a_i$ and the same variance σ^2, representing the levels of a factor – also called treatments – independent random samples of size n_i have to be drawn. If possible, the size N of the experiment is determined in advance as small as possible so that a given precision requirement is fulfilled. That means we have to choose equal subclass numbers. However, even if an experiment is planned with equal subclass numbers, drop-outs may lead to unequal sub-class numbers.

For the N observations $\boldsymbol{y}_{ij}$ we assume model (5.3) with its side conditions. From the observations in Table 5.1 the column sums Y_{i}. and the number observations are initially calculated. The corresponding mean

$$\bar{y}_{i.} = \frac{\boldsymbol{Y}_{i.}}{n_i}$$

is the UMVUE (uniformly minimum variance unbiased estimator) under the assumed normal distribution, and for arbitrary distributions with finite second moments it is the BLUE (best linear unbiased estimator) of the $\mu + a_i$.

To test the null hypothesis $H_0: a_1 = \ldots = a_a$ we calculate the realisation

$$F = \frac{(N-a)SS_A}{(a-1)SS_{\text{res}}} \tag{5.15}$$

5.3.2 Determination of the Size of an Experiment

In the case of the one-way classification we determine the required experimental size for the most favourable as well as for the least favourable case, i.e. we are looking for the smallest n (for instance $n = 2q$) so that for $\lambda_{\max} = \lambda$ and for $\lambda_{\min} = -\lambda$, respectively, (5.2) is fulfilled.

The experimenter must select a size n in the interval $n_{\min} \leq n \leq n_{\max}$ but if he wants to be on the safe side, he must choose $n = n_{\max}$. The package OPDOE of R allows the determination of the minimal size for the most favourable and the least favourable case in dependence on α, β, δ, σ and the number a of the levels of the factor A. The corresponding algorithm stems from Lenth (1986) and Rasch et al. (1997). In any case one can show that the minimal experimental size is smallest for the balanced case if $n_1 = n_2 = \ldots = n_a = n$, which can be reached by planning the experiment.

Problem 5.1 Determine in a balanced design the sub-class number n in a one-way ANOVA for a precision determined by $\alpha = 0.05$, $\beta = 0.05$ and $\delta = 2\sigma$, and a test with (5.14). Unfortunately delta in the program below stands for $\tau = \delta/\sigma$.

Solution

The design function of the R-package OPDOE for the analysis of variance has for the one-way analysis of variance the form (unfortunately delta in the program below stands for $\tau = \delta/\sigma$).

```
> size.anova(model="a", a= ,alpha= ,beta= ,delta= ,case= )
```

Example

Determine $n_{\min}$ and $n_{\max}$ for $a = 3$, $\alpha = 0.05$, $\beta = 0.05$, and $\delta = 2\sigma$.

```
> size.anova(model="a", a=3, alpha=0.05, beta=0.05, delta=2,case="minimin")
n
7
> size.anova(model="a", a=3, alpha=0.05, beta=0.05, delta=2,case="maximin")

n
8
```

Table 5.4 Performances (milk fat in kg) y_{ij} of the daughters of three sires.

	Sire		
	B_1	B_2	B_3
y_{ij}	120	153	130
	155	144	138
	131	147	122
	130	139	131
	146	141	128
	138	150	135
	143	136	127
	151		131
$Y_{i.}$	1114	1010	1042
$\overline{y}_i$	139.25	144.2857	130.25

Table 5.5 ANOVA table for testing the hypothesis $H_0: a_1 = a_2 = a_3 = 0$ of Example 5.1.

Source of variation	SS	df	MS	F
Between sires	766.8	2	383.4	5.628
Within sires	1362.4	20	68.	
Corrected total	2129.22	22		

Now, one of the values $n_{\min} = 7$ or $n_{\max} = 8$ must be used. The experimental size for $a = 3$ is $N_{\min} = 3 \cdot n_{\min} = 21$ or $N_{\max} = 3 \cdot n_{\max} = 24$.

We like to be on the safe side and take $n = 8$. Unfortunately, one observation for sire 2 was lost so that we received unequal subclass numbers, as is seen in Table 5.4. Table 5.5 is the corresponding ANOVA table.

Problem 5.2 Calculate the entries in the ANOVA Table 5.3 and calculate estimates of (5.9) and (5.10).

Solution

Use the R-commands:

```
> mf1 <-  c(120, 155, 131, 130, 146, 138, 143, 151)
> mf2 <-  c(153, 144, 147, 139, 141, 150, 136)
> mf3 <-  c(130, 138, 122, 131, 128, 135, 127, 131)
> sire <- c(1,1,1,1,1,1,1,1,2,2,2,2,2,2,2,3,3,3,3,3,3,3,3)
> Pr5_2 <- data.frame(x=c(mf1,mf2,mf3),y=sire)
> meanB1 <- mean(Pr5_2$x[Pr5_2$y==1])
> meanB1
[1] 139.25
> meanB2 <- mean(Pr5_2$x[Pr5_2$y==2])
> meanB2
[1] 144.2857
```

```
> meanB3 <a5- mean(Pr5_2$x[Pr5_2$y==3])
> meanB3
[1] 130.25
> a1_a2 <- meanB1-meanB2
> a1_a2
[1] -5.035714
> a1_a3 <- meanB1-meanB3
> a1_a3
[1] 9
> a2_a3 <- meanB2-meanB3
> a2_a3
[1] 14.03571
```

Answers to (5.9) are: mean $B_1 = 139.25$, mean $B_2 = 144.2857$, mean $B_3 = 130.25$.
Answers to (5.10) are: $a_1 - a_2 = -5.035714$, $a_1 - a_3 = 9$, $a_2 - a_3 = 14.03571$.

Problem 5.3 Test the null hypothesis $H_0 : a_1 = a_2 = a_3 = 0$ with significance level $\alpha = 0.05$.

Solution

The data of Table 5.3 are already in an R-data file Pr5_2. Proceed with:

```
> Bull <- factor(Pr5_2$y)
> Bull
 [1] 1 1 1 1 1 1 1 2 2 2 2 2 2 2 3 3 3 3 3 3 3 3
Levels: 1 2 3
> MYaov <- aov(Pr5_2$x ~ Bull)
> MYaov
Call:
   aov(formula = Pr5_2$x ~ Bull)

Terms:
                     Bull Residuals
Sum of Squares   766.7888 1362.4286
Deg. of Freedom         2        20
Residual standard error: 8.253571
Estimated effects may be unbalanced
> summary(MYaov)
            Df Sum Sq Mean Sq F value Pr(>F)
Bull         2  766.8   383.4   5.628 0.0115 *
Residuals   20 1362.4    68.1
- - -
Signif. codes:0 "***" 0.001 "**" 0.01 "*" 0.05 "." 0.1 " " 1
```

The p-value Pr(>F) for the bulls is $0.0115 < 0.05$, hence $H_0 : a_1 = a_2 = a_3 = 0$ is rejected.
To calculate the SS for the corrected total:

```
> N <- length(Pr5_2[ ,1])
> N
```

```
[1] 23
> SSTot <- (N-1)*var(Pr5_2[ ,1])
> SSTot
[1] 2129.217
```

5.4 Two-Way Analysis of Variance

The two-way ANOVA is a procedure for experiments to investigate the effects of two factors. Let us investigate a varieties of wheat and b fertilisers in their effect on the yield (kilogram per hectare). The a varieties as well as the b fertilisers are assumed to be fixed (selected consciously), as always in this chapter with fixed effects. One of the factors is factor variety (factor A), and the other factor is fertiliser (factor B). In this and the next chapter the number of levels of a factor X is denoted by the same (small) letter x as the factor (a capital letter) itself. So, factor A has a, and factor B has b levels in the experiment. In experiments with two factors, the experimental material is classified in two directions. For this, we list the different possibilities:

(a) Observations occur in each level of factor A combined with each level of factor B. There are $a \cdot b$ combinations (classes) of factor levels. We say factor A is completely crossed with factor B or that we have a complete cross-classification.
   (a1) For each combination (class) of factor levels there exists one observation [($n_{ij} = 1$ with n_{ij} being the number of observations in class $(A_i. B_j)$.].
   (a2) For each combination (class) (i, j) the level i of factor A with the level j of factor B we have $n_{ij} \geqq 1$ observations, with at least one $n_{ij} > 1$. If all $n_{ij} = n$, we have a cross-classification with equal class numbers, called a balanced experimental design.

(b) At least one level of factor A occurs together with at least two levels of the factor B, and at least one level of factor B occurs together with at least two levels of the factor A, but we have no complete cross-classification. Then we say factor A is partially crossed with factor B, or we have an incomplete cross-classification.

(c) Each level of factor B occurs together with exactly one level of factor A. This is a nested classification of factor B within factor A. We also say that factor B is nested within factor A and write $B \prec A$.

Summarising the types of two-way classification we have:

(a) $n_{ij} = 1$ for all $(i, j) \rightarrow$ complete cross-classification with one observation per class
(b) $n_{ij} \geq 1$ for all $(i,j) \rightarrow$ complete cross-classification
(c) $n_{ij} = n \geq 1$ for all $(i, j) \rightarrow$ complete cross-classification with equal sub-class numbers
(d) At least one $n_{ij} = 0 \rightarrow$ incomplete cross-classification.

If $n_{kj} \neq 0$, then $n_{ij} = 0$ for $i \neq k$ (at least one $n_{ij} > 1$ and at least two $n_{ij} \neq 0$) $\rightarrow$ nested classification.

5.4.1 Cross-Classification ($A \times B$)

The observations y_{ijk} of a complete cross-classification are real numbers. In class (i,j) occur the observations y_{ijk}, $k = 1, \ldots, n_{ij}$. In block designs, we often have $n_{ij} = 1$ with single subclass numbers. If all $n_{ij} = n$ are equal, we have the case of equal subclass numbers, which we discuss now.

Without loss of generality, we represent the levels of factor A as the rows and the levels of factor B as the columns in the tables. When empty classes occur, i.e. some n_{ij} equal zero, we have an incomplete cross-classification; such a case occurs for incomplete block designs.

Let the random variables $\boldsymbol{y}_{ijk}$ in the class (i, j) be a random sample of a population associated with this class. The mean and variance of the population of such a class are called true mean and variance, respectively. The true mean of the class (i, j) is denoted by η_{ij}. Again we consider the case that the levels of the factors A and B are chosen consciously (model I).

We call

$$\mu = \overline{\eta}_{..} = \frac{\sum_{i=1}^{a}\sum_{j=1}^{b}\eta_{ij}}{ab}$$

the overall expectation of the experiment.

The difference $a_i = \overline{\eta}_{i.} - \mu$ is called the main effect of the ith level of factor A, the difference $b_j = \overline{\eta}_{.j} - \mu$ is called the main effect of the j-th level of factor B. The difference $a_{i|j} = \eta_{ij} - \overline{\eta}_{.j}$ is called the effect of the ith level of factor A under the condition that factor B occurs in the jth level. Analogously, $b_{j|i} = \eta_{ij} - \overline{\eta}_{i.}$ is called the effect of the jth level of factor B under the condition that factor A occurs in the ith level.

The distinction between the main effect and 'conditional effect' is important if the effects of the levels of one factor depend on the effect of the level of the other factor. In the analysis of variance, we then say that an interaction between the two factors exists. We define the effects of these interactions (and use them in place of the conditional results).

The interaction $(a, b)_{ij}$ between the ith level of factor A and the jth level of factor B in a two-way cross-classification is the difference between the conditional effect of the level A_i of factor A for a given level B_j of the factors B and the main effect of the level A_i of A.

Under the assumption above the random variable $\boldsymbol{y}_{ij}$ of the cross-classification varies randomly around the class mean in the form

$$\boldsymbol{y}_{ijk} = \eta_{ij} + \boldsymbol{e}_{ijk}.$$

We assume that the so-called error variables $\boldsymbol{e}_{ijk}$ are independent of each other $N(0, \sigma^2)$ distributed and write for a balanced design the model I equation:

$$\boldsymbol{y}_{ijk} = \mu + a_i + b_i + (a, b)_{ij} + \boldsymbol{e}_{ijk}, (i = 1, \dots, a; j = 1, \dots, b;\ \ k = 1, \dots, n) \tag{5.16}$$

with $(a, b)_{ij} = 0$ if $n = 0$. We assume the following side conditions:

$$\sum_{i=1}^{a} a_i = 0, \sum_{j=1}^{b} b_j = 0, \sum_{i=1}^{a} (a, b)_{ij} = 0,\ \ \forall j, \sum_{j=1}^{b} (a, b)_{ij} = 0,\ \ \forall i. \tag{5.17}$$

If in (5.14) all $(a, b)_{ij} = 0$ we call

$$\boldsymbol{y}_{ijk} = \mu + a_i + b_j + \boldsymbol{e}_{ijk}\ (i = 1, \dots, a; j = 1, \dots, b;\ \ k = 1, \dots n) \tag{5.18}$$

a model without interactions or an additive model, respectively.

5.4.1.1 Parameter Estimation

We can estimate the parameters in any model of ANOVA by the least squares method. We minimise in the case of model (5.16)

$$\sum_{i=1}^{a}\sum_{j=1}^{b}\sum_{k=1}^{n}(y_{ijk} - \mu - a_i - b_j - (a,b)_{ij})^2$$

under the side conditions (5.17) and receive (using the dot convention analogue to Section 5.3)

$$\hat{\mu} = \bar{y}_{...}, \hat{a}_i = \bar{y}_{i.} - \bar{y}_{...}, \hat{b}_j = \bar{y}_{.j.} - \bar{y}_{...}, \widehat{(a,b)}_{ij} = \bar{y}_{ij.} - \bar{y}_{i.} - \bar{y}_{.j.} + \bar{y}_{...}.$$

Models with Interactions We consider the model (5.15). Because $E(Y)$ is estimable we have

$$\eta_{ij} = \mu + a_i + b_j + (a,b)_{ij} \qquad \text{for all} \quad i,j \qquad \text{with} \qquad n_{ij} > 0$$

estimable. The BLUE of η_{ij} is

$$\hat{\boldsymbol{\eta}}_{ij} = \hat{\boldsymbol{\mu}} + \hat{\boldsymbol{a}}_i + \hat{\boldsymbol{b}}_j + \widehat{(\boldsymbol{a},\boldsymbol{b})}_{ij} = \bar{\boldsymbol{y}}_{ij}. \tag{5.19}$$

From (5.6) it follows

$$\operatorname{cov}(\hat{\boldsymbol{\eta}}_{ij}, \hat{\boldsymbol{\eta}}_{kl}) = \frac{\sigma^2}{n_{ij}}\delta_{ik}\delta_{jl}. \tag{5.20}$$

It is now easy to show that differences between a_i or between b_j are not estimable. All estimable functions of the components of (5.15) without further side conditions contain interaction effects $(a,b)_{ij}$. It follows from theorem 5.7 in Rasch and Schott (2018) that

$$L_A = a_i - a_k + \sum_{j=1}^{b} c_{ij}(b_j + (a,b)_{ij}) - \sum_{j=1}^{b} c_{kj}(b_j + (a,b)_{kj}) \quad \text{for} \quad i \neq k \tag{5.21}$$

or analogously

$$L_B = b_i - b_k + \sum_{j=1}^{a} d_{ji}(a_j + (a,b)_{ji}) - \sum_{j=1}^{a} d_{jk}(a_j + (a,b)_{jk}) \quad \text{for} \quad i \neq k$$

is estimable if $c_{rs} = 0$ for $n_{rs} = 0$ and $d_{rs} = 0$ for $n_{rs} = 0$ as well as

$$\sum_{j=1}^{b} c_{ij} = \sum_{j=1}^{b} c_{kj} = 1 \quad \left(\text{and} \ \sum_{j=1}^{a} d_{ji} = \sum_{j=1}^{a} d_{jk} = 1\right).$$

The BLUE of an estimable function of the form (5.21) is given by

$$\hat{\boldsymbol{L}}_A = \sum_{j=1}^{b} c_{ij}\bar{\boldsymbol{y}}_{ij.} - \sum_{j=1}^{b} c_{kj}\bar{\boldsymbol{y}}_{kj.} \tag{5.22}$$

with variance

$$\operatorname{var}(\hat{\boldsymbol{L}}_A) = \sigma^2 \sum_{j=1}^{b}\left(\frac{c_{ij}^2}{n_{ij}} + \frac{c_{kj}^2}{n_{kj}}\right). \tag{5.23}$$

We consider the following example.

Table 5.6 Results of testing pig fattening – fattening days (from 40 kg to 110 kg. For three test periods and two sexes) for the offspring of several boars.

		Sex	
		Male	Female
Test periods	1	91	
		84	
		86	99
	2	94	97
		92	89
		90	
		96	
	3	82	
		86	-

Example 5.6 From three test periods of testing pig fattening for male and female offspring of boars the number of fattening days an animal needed to grow from 40 kg to 110 kg have been recorded. The values are given in Table 5.6.

We obtain from (5.18)

$$\widehat{(a,b)}_{11} = \overline{y}_{11.} = 87 \quad \widehat{(a,b)}_{12} = \overline{y}_{12.} = 99$$

$$\widehat{(a,b)}_{21} = \overline{y}_{21.} = 93 \quad \widehat{(a,b)}_{22} = \overline{y}_{22.} = 93$$

$$\widehat{(a,b)}_{31} = \overline{y}_{31.} = 84.$$

The function $L_1 = b_1 - b_2 + (a,\ b)_{11} - (a,\ b)_{12}$ is estimable, because the condition of theorem 5.7 is fulfilled. The function $L_2 = b_1 - b_2 + (a,\ b)_{21} - (a,\ b)_{22}$ is also estimable. We get

$$\hat{L}_1 = \overline{\boldsymbol{y}}_{11.} - \overline{\boldsymbol{y}}_{12.} = -12, \quad \hat{L}_2 = \overline{\boldsymbol{y}}_{21.} - \overline{\boldsymbol{y}}_{22.} = 0.$$

Further $\text{var}(\hat{\boldsymbol{L}}_1) = \frac{4}{3}\sigma^2$ and $\text{var}(\hat{\boldsymbol{L}}_2) = \frac{3}{4}\sigma^2$.

Connected Incomplete Cross-Classifications In an incomplete cross-classification we have $(a,\ b)_{ij} = 0$ if $n_{ij} = 0$. Further we choose the factors A and B so that $a \geq b$.

An (incomplete) cross-classification is called connected if

$$W = ((a,b)_{ij}) \ (i,j = 1, \cdots, b-1)$$

is non-singular. If $|W| = 0$, then the cross-classification is disconnected.

Example 5.7 We consider a two-way cross-classification with $a = 5$, $b = 4$ and the subclass numbers

		Levels of B			
		B_1	B_2	B_3	B_4
	A_1	n	n	0	0
	A_2	n	n	0	0
Levels of A	A_3	n	n	0	0
	A_4	0	0	m	m
	A_5	0	0	m	m

Because $|W| = 0$ the design is disconnected. Here is $n_{.1} = n_{.2} = 3n$, $n_{.3} = n_{.4} = 2m$, $n_{1.} = n_{2.} = n_{3.} = 2n$, $n_{4.} = n_{5.} = 2m$, and the matrix W is given by

$$W = \begin{pmatrix} \frac{3}{2}n & -\frac{3}{2}n & 0 \\ -\frac{3}{2}n & \frac{3}{2}n & 0 \\ 0 & 0 & m \end{pmatrix}$$

The first row is (-1) times the second row so that W is singular. The term "disconnected cross-classification" can be illustrated by this example as follows. From the scheme of the sub-class numbers we see that the levels A_1, A_2, A_3, B_1, B_2 and A_4, A_5, B_3, B_4 form two separate cross classifications. If we add n further observations in $(A_2\ B_3)$, we obtain $n_{2.} = 3n$, $n_{.3} = 2m + n$ and W becomes with $|W| \neq 0$ connected,

$$W = \begin{pmatrix} \frac{5}{3}n & -\frac{4}{3}n & -\frac{n}{3} \\ -\frac{4}{3}n & \frac{5}{3n} & -\frac{n}{3} \\ -\frac{n}{3} & -\frac{n}{3} & m + \frac{2}{3}n \end{pmatrix}$$

With the knowledge of Testing Hypothesis of the following Section 5.4.1.2.1, on page 126 before the topic Models with interaction, we can easily see in a cross-classification of A with a levels and B with b levels in an additive model whether the scheme is disconnected. In R test the B effect in the model $y = A + B$ and/or if $df(B) < b - 1$; test the A effect in the model $y = B + A$, and/or if $df(A) < a - 1$ in the ANOVA table, then the scheme is disconnected.

5.4.1.2 Testing Hypotheses

In this section, testable hypotheses and tests of such hypotheses are considered.

Models without Interactions We start with model (5.18) and assume a connected cross-classification (W above non-singular).

If, as in (5.18), $n_{ij} = n$ (equal sub-class numbers), simplifications for the tests of hypotheses about the main effects result. We have the possibility further to construct an analysis of variance table, in which SS_A, SS_B, $SS_{\text{res}} = SS_R$ add to $SS_{\text{total}} = SS_{\text{T}}$.

If in model (5.18) $n \geq 1$ for all i and j, then the sum of squared deviations of the $\boldsymbol{y}_{ijk}$ from the total mean $\overline{\boldsymbol{y}}_{...}$ of the experiment

$$\boldsymbol{SS}_T = \sum_{i=1}^{a} \sum_{j=1}^{b} \sum_{k=1}^{n} (\overline{\boldsymbol{y}}_{ijk} - \overline{\boldsymbol{y}}\ldots)^2$$

can be written as

$$\boldsymbol{SS}_{\mathrm{T}} = \boldsymbol{SS}_A + \boldsymbol{SS}_B + \boldsymbol{SS}_{\mathrm{res}}$$

with

$$\boldsymbol{SS}_A = \frac{1}{bn}\sum_{i=1}^{a}\boldsymbol{Y}_{i..}^2 - \frac{1}{N}\boldsymbol{Y}_{...}^2, \quad \boldsymbol{SS}_B = \frac{1}{an}\sum_{j=1}^{b}\boldsymbol{Y}_{.j.}^2 - \frac{1}{N}\boldsymbol{Y}_{...}^2,$$

$$\boldsymbol{SS}_{\mathrm{res}} = \sum_{i=1}^{a}\sum_{j=1}^{b}\sum_{k=1}^{n}\boldsymbol{y}_{ijk}^2 - \frac{1}{bn}\sum_{i=1}^{a}\boldsymbol{Y}_{i..}^2 + \frac{1}{an}\sum_{j=1}^{b}\boldsymbol{Y}_{.j.}^2 + \frac{1}{N}\boldsymbol{Y}_{...}^2.$$

$\boldsymbol{SS}_A + \boldsymbol{SS}_B$ and $\boldsymbol{SS}_{\mathrm{res}}$ are independently distributed, and for normally distributed $\boldsymbol{y}_{ijk}$ we have $\frac{1}{\sigma^2}\boldsymbol{SS}_A$ distributed as $CS(a-1, \lambda_a)$, $\frac{1}{\sigma^2}\boldsymbol{SS}_B$ as $CS(b-1, \lambda_b)$ and $\frac{1}{\sigma^2}\boldsymbol{SS}_{\mathrm{res}}$ as $CS(N-a-b+1)$ with non-centrality parameters

$$\lambda_a = \frac{1}{\sigma^2}\sum_{i=1}^{a}(a_i - \bar{a}.)^2, \quad \lambda_b = \frac{1}{\sigma^2}\sum_{j=1}^{b}(b_j - \bar{b}.)^2.$$

Therefore, $\boldsymbol{F}_A = \frac{\boldsymbol{MS}_A}{\boldsymbol{MS}_{\mathrm{res}}}$ is non-centrally F-distributed with $a-1$ and $N-a-b+1$ degrees of freedom and non-centrality parameter $\lambda_a = \frac{1}{\sigma^2}\sum_{i=1}^{a}(a_i - \bar{a})^2$ and $\boldsymbol{F}_B = \frac{\boldsymbol{MS}_B}{\boldsymbol{MS}_{\mathrm{res}}}$ is non-centrally F-distributed with $b-1$ and $N-a-b+1$ degrees of freedom and non-centrality parameter $\lambda_b = \frac{1}{\sigma^2}\sum_{j=1}^{b}(b_j - \bar{b})^2$.

The realisations of these formulas are summarised in Table 5.7.

When the null hypothesis $H_{A0}: a_1 = a_2 = \cdots = a_a = 0$ is correct, then $\lambda_a = 0$. This null hypothesis can therefore be tested by $F_A = \frac{MS_A}{MS_{\mathrm{res}}}$. If $F_A > F(a-1, N-a-b+1, 1-\alpha)$ the null hypothesis is rejected with a first kind risk α. When the null hypothesis $H_{B0}: b_1 = b_2 = \cdots = b_b = 0$ is correct, then $\lambda_b = 0$. This null hypothesis can therefore be tested by $F_B = \frac{MS_B}{MS_{\mathrm{res}}}$. If $F_B > F(b-1, N-a-b+1, 1-\alpha)$ the null hypothesis is rejected with a first kind risk α.

Table 5.7 Empirical ANOVA table of a two-way cross-classification with equal subclass numbers ($n_{ij} = n$).

Source of Variation	SS	df	MS	F
Between the levels of A	$SS_A = \frac{1}{bn}\sum_{i=1}^{a} Y_{i..}^2 - \frac{1}{N}Y_{...}^2$	$a-1$	$\frac{SS_A}{a-1} = MS_A$	$\frac{MS_A}{MS_{\mathrm{res}}} = F_A$
Between the levels of B	$SS_B = \frac{1}{an}\sum_{j=1}^{b} Y_{.j.}^2 - \frac{1}{N}Y_{...}^2$	$b-1$	$\frac{SS_B}{b-1} = MS_B$	$\frac{MS_B}{MS_{\mathrm{res}}} = F_B$
Residual	$SS_{res} = SS_T - SS_A - SS_B$	$N-a-b+1$	$\frac{SS_{\mathrm{res}}}{N-a-b+1} = MS_{\mathrm{res}}$	

Problem 5.4 Determine the $(1-\alpha)$-quantile of the central F-distribution with df_1 and df_2 degrees of freedom.

Solution
Use the R command >qf(1 − α,df1,df2).

Example
We calculate the quantile of the central F-distribution for $1-\alpha = 0.95$, $df_1 = 10$ and $df_2 = 30$.

```
> qf(0.95,10,30)
[1] 2.16458
```

Problem 5.5 Determine the sample size for testing the hypothesis $H_{A0}: a_1 = a_2 = \cdots = a_a = 0$. Unfortunately delta in the program below stands for $\tau = \delta/\sigma$.

Solution
Use the OPDOE commands

```
> size.anova(model="axb", hypothesis="a", a=, b=, alpha=,beta=,
       delta=,cases="maximin")
```

or

```
> size.anova(model="axb", hypothesis="a", a=, b=, alpha=,beta=,
       delta=,cases="minimin")
```

Example
We choose $\alpha = 0.05$, $\beta = 0.2$, $a = 4$, $b = 2$, $\delta = 1$ and calculate the minimal and maximal subclass numbers n for the null hypothesis for A and B.

For testing the factor A we obtain

```
> size.anova(model="axb", hypothesis="a", a=4, b=2,
       alpha=0.05,beta=0.1, delta=1,cases="maximin")
 n
15
```

and

```
> size.anova(model="axb", hypothesis="a", a=4, b=2,
       alpha=0.05,beta=0.1, delta=1,cases="minimin")
n
8
```

For testing the factor B we exchange the entries for a and b and obtain

```
> size.anova(model="axb", hypothesis="a", a=2, b=4,
      alpha=0.05,beta=0.1, delta=1,cases="maximin")
n
6
```

Table 5.8 Observations (loss in per cent of dry mass, during storage of 300 days) of the experiment of Example 5.9 and results of first calculations.

		Forage crop	
		Green rye	Lucerne
Kind of storage	Glass in refrigerator	8.39	9.44
	Glass in barn	11.58	12.21
	Sack in refrigerator	5.42	5.56
	Sack in barn	9.53	10.39

and

```
> size.anova(model="axb", hypothesis="a", a=2, b=4,
        alpha=0.05,beta=0.1, delta=1,cases="minimin")
n
6
```

To test both hypotheses, the experimenter may use a subclass number between 8 and 15.

Example 5.9 Two forage crops (green rye and lucerne) are investigated concerning their loss of carotene during storage. Four storage possibilities (glass in a refrigerator, glass in a barn, sack in a refrigerator and sack in a barn) are chosen. The loss during storage is defined by the difference between the content of carotene at start and the content of carotene after storing for 300 days (in per cent of dry mass). The question is whether the kind of storage and/or of forage crop influences the loss during storage. We denote the kind of storage as factor A and the forage crop as factor B and arrange the observations (differences y_{ij}) in the form of Table 5.8. Because forage crops and kinds of storage have been selected consciously, we use for y_{ij} a model I and (5.18) as the model equation.

The analysis of variance assumes that the observations are realisations of random variables, which are, independently of each other, normally distributed with equal variances. Table 5.9 is the ANOVA table. As the F-tests show, only factor storage has a significant influence on the loss during storage; significant differences are found only between the kinds of storage, but not between the forage crops ($\alpha = 0.05$).

For the analysis of this design we can use in R the commands `>aov()` or `> lm()`.

Table 5.9 ANOVA Table of Example 5.9.

Source of variation	*SS*	*df*	*MS*	*F*
Between the storages	43.2261	3	14.4087	186.7
Between the forage crops	0.8978	1	0.8978	11.63
Residual	0.2315	3	0.0772	
Total	44.3554	7		

Use the command > `aov()` only for balanced two-way cross classifications.

The command >`lm()` gives more detailed information and can also be used for unbalanced two-way cross classifications.

```
> loss <- c(8.39, 11.58, 5.42, 9.53, 9.44, 12.21, 5.56, 10.39)
> storage <- c(1,2,3,4,1,2,3,4)
> crop <- c(1,1,1,1,2,2,2,2)
> Table_5_8 <- data.frame(cbind(loss,storage,crop))
> Table_5_8
   loss storage crop
1  8.39       1    1
2 11.58       2    1
3  5.42       3    1
4  9.53       4    1
5  9.44       1    2
6 12.21       2    2
7  5.56       3    2
8 10.39       4    2
> STORAGE <- factor(storage)
> CROP <- factor(crop)
> Anova1 <- aov(loss ~STORAGE + CROP, Table_5_8)
> Anova1
Call:
   aov(formula = loss ~ STORAGE + CROP, data = Table_5_8)
Terms:
                STORAGE    CROP Residuals
Sum of Squares  43.2261  0.8978    0.2315
Deg. of Freedom       3       1         3
Residual standard error: 0.2777889
Estimated effects may be unbalanced
> summary(Anova1)
            Df Sum Sq Mean Sq F value   Pr(>F)
STORAGE      3  43.23  14.409  186.72 0.000659 ***
CROP         1   0.90   0.898   11.63 0.042121 *
Residuals    3   0.23   0.077
- - -
Signif. codes: 0 "***" 0.001 "**" 0.01 "*" 0.05 "." 0.1 "" 1

> Anova2 <- lm(loss ~STORAGE + CROP, Table_5_8)
> Anova2
Call:
lm(formula = loss ~ STORAGE + CROP, data = Table_5_8)
Coefficients:
(Intercept)     STORAGE2     STORAGE3     STORAGE4        CROP2
      8.580        2.980       -3.425        1.045        0.670
> summary(Anova2)
Call:
lm(formula = loss ~ STORAGE + CROP, data = Table_5_8)
Residuals:
     1      2      3      4      5      6      7      8
-0.190  0.020  0.265 -0.095  0.190 -0.020 -0.265  0.095
```

```
Coefficients:
             Estimate Std. Error t value Pr(>|t|)
(Intercept)    8.5800     0.2196  39.069 3.69e-05 ***
STORAGE2       2.9800     0.2778  10.728  0.00173 **
STORAGE3      -3.4250     0.2778 -12.330  0.00115 **
STORAGE4       1.0450     0.2778   3.762  0.03285 *
CROP2          0.6700     0.1964   3.411  0.04212 *
- - -
Signif. codes: 0 "***" 0.001 "**" 0.01 "*" 0.05 "." 0.1 "" 1
Residual standard error: 0.2778 on 3 degrees of freedom
Multiple R-squared:  0.9948,    Adjusted R-squared:  0.9878
F-statistic:   143 on 4 and 3 DF,  p-value: 0.0009397
> anova(Anova2)
Analysis of Variance Table
Response: loss
          Df Sum Sq Mean Sq F value   Pr(>F)
STORAGE    3 43.226 14.4087 186.722 0.000659 ***
CROP       1  0.898  0.8978  11.635 0.042121 *
Residuals  3  0.231  0.0772
- - -
Signif. codes: 0 "***" 0.001 "**" 0.01 "*" 0.05 "." 0.1 "" 1
```

Example 5.10 We use the data of Table 5.6 with an additive model, which is an example of an unbalanced design. Hence, we must use the R command `>lm()`. For testing the hypothesis about the test periods effects we must use another formula in `>lm()` for the testing of the sex effects. We use for the tests a significance level $\alpha = 0.05$.

```
> days <- c(91,84,86,94,92,90,96,82,86,99,97,89)
> periods <- c(1,1,1,2,2,2,2,3,3,1,2,2)
> sex <- c(1,1,1,1,1,1,1,1,1,2,2,2)
> T_5_6 <- data.frame(cbind(days,periods,sex))
> T_5_6
   days periods sex
1    91       1   1
2    84       1   1
3    86       1   1
4    94       2   1
5    92       2   1
6    90       2   1
7    96       2   1
8    82       3   1
9    86       3   1
10   99       1   2
11   97       2   2
12   89       2   2
> # Prepare ANOVA for testing hypothesis Test periods effects
> Anova1 <- lm(days ~SEX + PERIODS, T_5_6)
> Anova1
Call:
lm(formula = days ~ SEX + PERIODS, data = T_5_6)
```

```
Coefficients:
(Intercept)         SEX2     PERIODS2     PERIODS3
      88.92         4.32         2.64        -4.92
> summary(Anova1)
Call:
lm(formula = days ~ SEX + PERIODS, data = T_5_6)
Residuals:
   Min     1Q Median     3Q    Max
 -6.88  -2.23   0.78   2.17   5.76
Coefficients:
            Estimate Std. Error t value Pr(>|t|)
(Intercept)   88.920      2.330  38.162 2.44e-10 ***
SEX2           4.320      3.051   1.416    0.194
PERIODS2       2.640      2.854   0.925    0.382
PERIODS3      -4.920      3.889  -1.265    0.241
- - -
Signif. codes: 0 "***" 0.001 "**" 0.01 "*" 0.05 "." 0.1 "" 1
Residual standard error: 4.403 on 8 degrees of freedom
Multiple R-squared:  0.5107,    Adjusted R-squared:  0.3272
F-statistic: 2.783 on 3 and 8 DF,  p-value: 0.1098
> anova(Anova1)
Analysis of Variance Table
Response: days
          Df Sum Sq Mean Sq F value  Pr(>F)
SEX        1  81.00   81.00  4.1774 0.07522 .
PERIODS    2  80.88   40.44  2.0856 0.18665
Residuals  8 155.12   19.39
- - -
> # Prepare ANOVA for testing Hypothesis about Sex effects
> Anova2 <- lm(days ~PERIODS + SEX, T_5_6)
> Anova2
Call:
lm(formula = days ~ PERIODS + SEX, data = T_5_6)
Coefficients:
(Intercept)     PERIODS2     PERIODS3          SEX2
      88.92         2.64        -4.92          4.32
> summary(Anova2)
Call:
lm(formula = days ~ PERIODS + SEX, data = T_5_6)
Residuals:
   Min     1Q Median     3Q    Max
 -6.88  -2.23   0.78   2.17   5.76
Coefficients:
            Estimate Std. Error t value Pr(>|t|)
(Intercept)   88.920      2.330  38.162 2.44e-10 ***
PERIODS2       2.640      2.854   0.925    0.382
PERIODS3      -4.920      3.889  -1.265    0.241
SEX2           4.320      3.051   1.416    0.194
- - -
Signif. codes: 0 "***" 0.001 "**" 0.01 "*" 0.05 "." 0.1 "" 1
Residual standard error: 4.403 on 8 degrees of freedom
```

```
Multiple R-squared:  0.5107,    Adjusted R-squared:  0.3272
F-statistic: 2.783 on 3 and 8 DF,  p-value: 0.1098
> anova(Anova2)
Analysis of Variance Table
Response: days
          Df Sum Sq Mean Sq F value  Pr(>F)
PERIODS    2 123.00   61.50  3.1717 0.09677 .
SEX        1  38.88   38.88  2.0052 0.19450
Residuals  8 155.12   19.39
- - -
Signif. codes: 0 "***" 0.001 "**" 0.01 "*" 0.05 "." 0.1 "" 1
```

From the ANOVA table of Anova1 we must use the last line of PERIODS for testing test period effects with F-value 2.0856 and p-value Pr(F>) = 0.18665, which is larger than $\alpha = 0.05$, hence the hypothesis of equal test period effects is not rejected.

From the ANOVA table of Anova2 we must use the last line of SEX for testing sex effects with F-value 2.00052 and p-value Pr(F>) = 0.19450, which is larger than $\alpha = 0.05$, hence the hypothesis of equal sex effects is not rejected.

Models with Interactions We consider now model (5.16) and assume a connected cross-classification.

The ANOVA table for this case is Table 5.10.

In the case of equal subclass numbers, we use the side conditions (5.17) and test the hypotheses

$$H_{A0}^{*}: \; a_1 = \ldots = a_a \; (= 0),$$

$$H_{B0}^{*}: \; b_1 = \ldots = b_b \; (= 0),$$

$$H_{AxB0}: \; (a,b)_{11} = \ldots = (a,b)_{ab} (= 0)$$

with the F-statistics

$$F_A = \frac{\left(\frac{1}{bn}\sum_{i=1}^{a} Y_{i.}^2 - \frac{1}{N} Y_{...}^2\right) ab(n-1)}{(a-1)SS_{\text{res}}}. \tag{5.24}$$

$$F_B = \frac{\left(\frac{1}{an}\sum_{j=1}^{b} Y_{.j.}^2 - \frac{1}{N} Y_{...}^2\right) ab(n-1)}{(b-1)SS_{\text{res}}}. \tag{5.25}$$

and

$$F_{AB} = \frac{\left(\frac{1}{n}\sum_{i=1}^{a}\sum_{j=1}^{b} Y_{ij.}^2 - \frac{1}{bn}\sum_{i=1}^{a} Y_{i..}^2 - \frac{1}{an}\sum_{j=1}^{b} Y_{.j.}^2 + \frac{1}{N} Y_{...}^2\right) ab(n-1)}{(a-1)(b-1)SS_{\text{res}}}, \tag{5.26}$$

SS_A, SS_B, SS_{AB} and SS_{res} are independently distributed, and for normally distributed y_{ijk} it is $\frac{1}{\sigma^2}SS_A$ as $CS(a-1, \lambda_a)$, $\frac{1}{\sigma^2}SS_B$ as $CS(b-1, \lambda_b)$, $\frac{1}{\sigma^2}SS_{AB}$ as $CS((a-1)(b-1), \lambda_{ab})$ and $\frac{1}{\sigma^2}SS_{\text{res}}$ as $CS(N-a-b+1)$ distributed with non-centrality parameters

Table 5.10 Analysis of variance table of a two-way cross-classification with equal sub-class numbers for model I with interactions under Condition (5.17).

Source of variation	SS	df	MS	E(MS)	F
Between rows (A)	$SS_A = \frac{1}{bn}\sum_i Y_{i..}^2 - \frac{1}{N}Y_{...}^2$	$a-1$	$\frac{SS_A}{a-1}$	$\sigma^2 + \frac{bn}{a-1}\sum_i a_i^2$	$\frac{ab(n-1)SS_A}{(a-1)SS_{\text{res}}}$
Between columns (B)	$SS_B = \frac{1}{an}\sum_j Y_{.j.}^2 - \frac{1}{N}Y_{...}^2$	$b-1$	$\frac{SS_B}{b-1}$	$\sigma^2 + \frac{an}{b-1}\sum_j b_j^2$	$\frac{ab(n-1)SS_B}{(b-1)SS_{\text{res}}}$
Interactions	$SS_{AB} = \frac{1}{n}\sum_{i,j} Y_{ij.}^2 - \frac{1}{bn}\sum_i Y_{i.}^2$ $= \frac{1}{an}\sum_i Y_{.j.}^2 + \frac{Y^2}{N}$	$(a-1)$ $\times(b-1)$	$\frac{SS_{AB}}{(a-1)(b-1)}$	$\sigma^2 + \frac{n\sum_{i,j}(a,b)_{ij}^2}{(a-1)(b-1)}$	$\frac{ab(n-1)SS_{AB}}{(a-1)(b-1)SS_{\text{res}}}$
Within classes (residual)	$SS_{\text{res}} = \sum_{i,j,k} y_{ijk}^2 - \frac{1}{n}\sum_{i,j} Y_{ij.}^2$	$ab(n-1)$	$\frac{SS_{\text{res}}}{ab(n-1)} = s^2$	σ^2	

$\lambda_a = \frac{1}{\sigma^2}\sum_{i=1}^{a}(a_i - \overline{a}.)^2$, $\lambda_b = \frac{1}{\sigma^2}\sum_{j=1}^{b}(b_j - \overline{b}.)^2$ and $\lambda_{ab} = \frac{1}{\sigma^2}\sum_{i=1}^{a}\sum_{j=1}^{b}[(a,b)_{ij} - \overline{(a,b)_{..}}]^2$, respectively.

Therefore $\boldsymbol{F}_A = \frac{\boldsymbol{MS}_A}{\boldsymbol{MS}_{\text{res}}}$ is non-centrally F-distributed with $a-1$ and $N-a-b+1$ degrees of freedom and non-centrality parameter $\lambda_a = \frac{1}{\sigma^2}\sum_{i=1}^{a}(a_i - \overline{a})^2$, $\boldsymbol{F}_B = \frac{\boldsymbol{MS}_B}{\boldsymbol{MS}_{\text{res}}}$ is non-centrally F-distributed with $b-1$ and $N-a-b+1$ degrees of freedom and non-centrality parameter $\lambda_b = \frac{1}{\sigma^2}\sum_{j=1}^{b}(b_j - \overline{b})^2$ and $\boldsymbol{F}_{AB} = \frac{\boldsymbol{MS}_{AB}}{\boldsymbol{MS}_{\text{res}}}$ is non-centrally F-distributed with $(a-1)(b-1)$ and $N-a-b+1$ degrees of freedom and non-centrality parameter

$$\lambda_{ab} = \frac{1}{\sigma^2}\sum_{i=1}^{a}\sum_{j=1}^{b}[(a,b)_{ij} - \overline{(a,b)_{..}}]^2.$$

When the null hypotheses above are correct, then the corresponding non-centrality parameter is zero. The corresponding null hypothesis can therefore be tested by $F_A = \frac{MS_A}{MS_{\text{res}}}$, $F_B = \frac{MS_B}{MS_{\text{res}}}$ and $F_{AB} = \frac{MSA_B}{MS_{\text{res}}}$, respectively. If $F_A > F(a-1,\ N-a-b+1,\ 1-\alpha)$ the null hypothesis H_{A0}^* is rejected with a first kind risk α. If $F_B > F(b-1,\ N-a-b+1,\ 1-\alpha)$ the null hypothesis H_{B0}^* is rejected with a first kind risk α. Finally, if $F_{AB} > F((a-1)(b-1), N-a-b+1, 1-\alpha)$ the null hypothesis $H_{A\times B0}$ is rejected with a first kind risk α.

Example 5.12 We consider again a similar experiment as in Example 5.9 and the loss data in storages in glass and sack with four observations per sub-class as shown in Table 5.11. Table 5.12 is the ANOVA table. Due to the F-test, H_{0A} has to be rejected but H_{0B} and H_{0AB} not.

Problem 5.6 Calculate the entries of Table 5.11 and give the commands for Table 5.12.

Solution

Make a data frame and use the **R** command >`lm()` to get the ANOVA table.

Because we have a balanced two-way cross-classification we do not need to use a different formula in > `lm()` for the test of the main effects.

Table 5.11 Observations of the carotene storage experiment of Example 5.12.

		Forage crop	
		Green rye	Lucerne
Kind of storage	Glass	8.39	9.44
		7.68	10.12
		9.46	8.79
		8.12	8.89
	Sack	5.42	5.56
		6.21	4.78
		4.98	6.18
		6.04	5.91

Table 5.12 ANOVA table for the carotene storage experiment of Example 5.12.

Source of variation	*SS*	*df*	*MS*	*F*
Between the kind of storage	41.6347	1	41.6347	101.70
Between the forage crops	0.7098	1	0.7098	1.73
Interactions	0.9073	1	0.9073	2.22
Within classes (residual)	4.9128	12	0.4094	
Total	48.1646	15		

Example

```
> loss <- c(8.39,7.68,9.46,8.12,5.42,6.21,4.98,6.04,9.44,10.12,
  8.79,8.89,5.56,4.78,6.18,5.91)
> storage <- c(1,1,1,1,2,2,2,2,1,1,1,1,2,2,2,2)
> crop <- c(1,1,1,1,1,1,1,1,2,2,2,2,2,2,2,2)
> T_5_11 <- data.frame(cbind(loss,storage,crop))
> T_5_11
loss storage crop
1 8.39 1 1
2 7.68 1 1
3 9.46 1 1
4 8.12 1 1
5 5.42 2 1
6 6.21 2 1
7 4.98 2 1
8 6.04 2 1
9 9.44 1 2
10 10.12 1 2
11 8.79 1 2
12 8.89 1 2
13 5.56 2 2
14 4.78 2 2
15 6.18 2 2
16 5.91 2 2
> STORAGE <- factor(storage)
> CROP <- factor(crop)
> Anova <- lm(loss ~ STORAGE + CROP + STORAGE*CROP, T_5_11)
> Anova
Call:
lm(formula = loss ~ STORAGE + CROP + STORAGE * CROP, data = T_5_11)

Coefficients:
(Intercept) STORAGE2 CROP2 STORAGE2:CROP2
8.4125 -2.7500 0.8975 -0.9525

> summary(Anova)
Call:
lm(formula = loss ~ STORAGE + CROP + STORAGE * CROP, data = T_5_11)
Residuals:
```

```
Min 1Q Median 3Q Max
-0.8275 -0.4450 -0.0350 0.4200 1.0475
Coefficients:
                 Estimate Std. Error        t value Pr(>|t|)
(Intercept)          v8.4125 0.3199          26.295 5.60e-12 ***
STORAGE2            -2.7500 0.4524          -6.078 5.52e-05 ***
CROP2                v0.8975 0.4524           1.984 0.0706 .
STORAGE2:CROP2      -0.9525 0.6398          -1.489 0.1624
- - -
Signif. codes: 0 "***" 0.001 "**" 0.01 "*" 0.05 "." 0.1 "" 1
Residual standard error: 0.6398 on 12° of freedom
Multiple R-squared: 0.898, Adjusted R-squared: 0.8725
F-statistic: 35.22 on 3 and 12 DF, p-value: 3.155e-06

> anova(Anova)
Analysis of Variance Table

Response: loss
               Df  Sum Sq  Mean Sq    F value   Pr(>F)
STORAGE        1   41.635  41.63510   1.6965    3.27e-07 ***
CROP           1   0.710   0.710      1.7338    0.2125
STORAGE:CROP   1   0.907   0.907      2.2161    0.1624
Residuals      12  4.913   0.409
Signif. codes: 0 "***" 0.001 "**" 0.01 "*" 0.05 "." 0.1 "" 1
```

Problem 5.7 Calculate the sample size for testing $H_{AA0}: (a, b)_{11} = (a, b)_{12} = \cdots = (a, b)_{ab} = 0$. Unfortunately delta in the program below stands for $\tau = \delta/\sigma$.

Solution

Use the OPDOE command

```
>size.anova(model = "axb", hypothesis = "axb", a=, b=, alpha=, beta=,
 delta=, cases=)
```

Example

How many replications in the four subclasses are needed to test the hypothesis

$$H_{0AB} : (a, b)_{11} = \ldots = (a, b)_{\text{ab}} (= 0)$$

with the precision requirements $\alpha = 0.05$, $\beta = 0.1$ and $\delta = 2\sigma$?

```
> size.anova(model="axb", hypothesis="axb", a=2, b=2,
 alpha=0.05,beta=0.1, delta=2, cases="minimin")
n
4

> size.anova(model="axb", hypothesis="axb", a=2, b=2,
 alpha=0.05,beta=0.1, delta=2, cases="maximin")
n
6
```

Therefore, the sub-class number has to be between 4 and 6.

5.4.2 Nested Classification (*A*>*B*)

A nested classification is a classification with super- and sub-ordinated factors, where the levels of a sub-ordinated or nested factor are considered as further subdivision of the levels of the super-ordinated factor. Each level of the nested factor occurs in just one level of the super-ordinated factor. An example is the subdivision of the United States into states (super-ordinated factor A) and counties (nested factor B).

As for the cross-classification we assume that the random variables $\boldsymbol{y}_{ijk}$ vary randomly from the expectations η_{ij}, i.e.

$$\boldsymbol{y}_{ijk} = \eta_{ij} + \boldsymbol{e}_{ijk} \quad (i = 1, \dots, a; j = 1, \dots, b_i; k = 1, \dots, n_{ij}),$$

and that $\boldsymbol{e}_{ijk}$ are, independently of each other, $N(0, \sigma^2)$-distributed. With

$$\mu = \overline{\eta}.. = \frac{\sum_{i=1}^{a}\sum_{j=1}^{b_i} \eta_{ij} n_{ij}}{N}$$

the total mean of the experiment is defined.

In nested classification, interactions cannot occur.

The difference $a_i = \overline{\eta}_{i.} - \mu$ is called the effect of the ith level of factor A, the difference $b_{ij} = \eta_{ij} - \eta_{i}.$ is the effect of the jth level of B within the ith level of A.

The model equation for $\boldsymbol{y}_{ijk}$ in the balanced case then reads

$$\boldsymbol{y}_{ijk} = \mu + a_i + b_{ij} + \boldsymbol{e}_{ijk}, i = 1, \dots, a; j = 1, \dots b, k = 1, \dots n \tag{5.27}$$

(interactions do not exist).

Usual side conditions are

$$\sum_{i=1}^{a} a_i = 0, \qquad \sum_{j=1}^{b} b_{ij} = 0 \text{ (for all } i).$$

or

$$\sum_{i=1}^{a} N_{i.} a_i = 0, \quad \sum_{j=1}^{b} n_{ij} b_{ij} = 0 \text{ (for all } i). \tag{5.28}$$

Minimising

$$\sum_{i=1}^{a}\sum_{j=1}^{b}\sum_{k=1}^{n_{ij}} (y_{ijk} - \mu - a_i - b_{ij})^2,$$

under the side conditions (5.28), we obtain the BLUE

$$\hat{\boldsymbol{\mu}} = \overline{\boldsymbol{y}}_{...}, \quad \hat{\boldsymbol{a}}_i = \overline{\boldsymbol{y}}_{i..} - \overline{\boldsymbol{y}}_{...}, \quad \hat{\boldsymbol{b}}_{ij} = \overline{\boldsymbol{y}}_{ij.} - \overline{\boldsymbol{y}}_{i..}. \tag{5.29}$$

The total sum of squares is again split into components

$$\begin{aligned} \boldsymbol{SS}_T &= \sum_{i,j,k} (\boldsymbol{y}_{ijk} - \overline{\boldsymbol{y}}_{...})^2 \\ &= \sum_{i,j,k} (\overline{\boldsymbol{y}}_{i..} - \overline{\boldsymbol{y}}_{...})^2 + \sum_{i,j,k} (\overline{\boldsymbol{y}}_{ij.} - \overline{\boldsymbol{y}}_{i..})^2 + \sum_{i,j,k} (\boldsymbol{y}_{ijk} - \overline{\boldsymbol{y}}_{ij.})^2 \end{aligned}$$

or

$$SS_{\mathrm{T}} = SS_A + SS_{B\ \mathrm{in}\ A} + SS_{\mathrm{res}},$$

where SS_A is the SS between the A levels, $SS_{B\ \mathrm{in}\ A}$ is the SS between the B levels within the A levels and SS_{res} is the SS within the classes (B levels).

The SS are written in the form

$$SS_T = \sum_{i,j,k} y_{ijk}^2 - \frac{Y_{...}^2}{N}, \quad SS_A = \sum_i \frac{Y_{i..}^2}{N_{i.}} - \frac{Y_{...}^2}{N},$$

$$SS_{B\ \mathrm{in}\ A} = \sum_{i,j} \frac{Y_{ij.}^2}{n_{ij.}} - \sum_i \frac{Y_{i..}^2}{N_{i.}}, \quad SS_{\mathrm{res}} = \sum_{i,j,k} y_{ijk}^2 - \sum_{i,j} \frac{Y_{ij.}^2}{n_{ij}}.$$

Here and in the sequel, we assume the side conditions (5.28).

The expectations of the MS are given in Table 5.13.

MS_A, $MS_{B\ \mathrm{in}\ A}$ and MS_{res} in Table 5.13 are independently of each other distributed as $CS(a-1, \lambda_a)$, $CS(B.-a, \lambda_{b\ \mathrm{in}\ a})$ and $CS(N-B.)$, respectively, where

$$\lambda_a = \frac{1}{\sigma^2}\sum_{i=1}^{a}(a_i - \bar{a}.)^2, \quad \lambda_{b\ \mathrm{in}\ a} = \frac{1}{\sigma^2}\sum_{i=1}^{a}\sum_{j=1}^{b}(b_{ij} - \bar{b}_{i.})^2.$$

Therefore

$$F_A = \frac{MS_A}{MS_{\mathrm{res}}} \text{ is distributed as } F(a-1, N-B_., \lambda_a)$$

and

$$F_B = \frac{MS_B}{MS_{\mathrm{res}}} \text{ is distributed as } F(B_. - a, N-B_., \lambda_{b\ \mathrm{in}\ a}).$$

Using this the null hypothesis H_{A0}: $a_1 = \ldots = a_a$, can be tested by $F_A = \frac{MS_A}{MS_{\mathrm{res}}}$, which, under H_{A0} is distributed as $F(a-1, N-B_.)$.

The null hypothesis H_{B0}: $b_{11} = \ldots = b_{ab}$, can be tested by $F_B = \frac{MS_{B\ \mathrm{in}\ A}}{MS_{\mathrm{res}}}$, which, under H_{B0}, is distributed as $F(B_. - a, N-B_.)$.

Table 5.13 Theoretical ANOVA table of the two-way nested classification for model I.

Source of variation	*SS*	*df*	*MS*	*E(MS)*
Between A levels	$SS_A = \sum_i \frac{Y_{i..}^2}{N_{i.}} - \frac{Y_{...}^2}{N}$	$a-1$	$MS_A = MS_A = \frac{SS_A}{a-1}$	$\sigma^2 + \frac{1}{a-1}\sum_i N_{i.}a_i^2$
Between B levels within A levels	$SS_{B\ \mathrm{in}\ A} = \sum_{i,j} \frac{Y_{ij.}^2}{n_{ij}} - \sum_i \frac{Y_{i..}^2}{N_{i.}}$	$B.-a$	$MS_{B\ \mathrm{in}\ A} = \frac{SS_{B\ \mathrm{in}\ A}}{B.-a}$	$\sigma^2 + \frac{1}{B.-a}\sum_{i,j} n_{ij}b_{ij}^2$
Within B levels (residual)	$SS_{\mathrm{res}} = \sum_{i,j,k} y_{ijk}^2 - \sum_{i,j} \frac{Y_{ij.}^2}{n_{ij}}$	$N-B.$	$MS_{\mathrm{res}} = \frac{SS_{\mathrm{res}}}{N-B.}$	σ^2

Table 5.14 Observations of the example.

	Levels of *A*											
	A_1				A_2				A_3			
	Levels of *B*											
Observation	B_{11}	B_{12}	B_{13}	B_{14}	B_{21}	B_{22}	B_{23}	B_{24}	B_{31}	B_{32}	B_{33}	B_{34}
1	30	0	7	28	24	14	20	20	14	14	18	−25
2	−19	20	5	15	16	11	18	−12	−18	8	16	13
3	−31	32	3	20	−18	27	8	0	33	−19	7	36
4	−14	11	−5	20	11	11	32	−5	−9	6	−4	18
5	−14	13	8	−48	−10	8	−25	44	−7	−38	21	−7

Problem 5.8 Calculate the ANOVA table with the realised sum of squares of Table 5.14.

Solution

We use in R the command `>lm()`.

Example

We consider an example of the balanced case with $a = 3$ levels of factor A, $b = 4$ levels of factor B within each level of factor A. The data are shown in Table 5.14.

```
> y1 <- c(30,-19,-31,-14,-14,0,20,32,11,13,7,5,3,-5,8,28,15,20,20,-48)
> y2 <- c(24,16,-18,11,-10,14,11,27,11,8,20,18,8,32,-25,20,-12,0,-5,44)
> y3 <- c(14,-18,33,-9,-7,14,8,-19,6,-38,18,16,7,-4,21,-25,13,36,18,-7)
> a1 <- rep(1,20)
> a2 <- rep(2,20)
> a3 <- rep(3,20)
> b1 <- c(11,11,11,11,11,12,12,12,12,12,13,13,13,13,13,14,14,14,14,14)
> b2 <- c(21,21,21,21,21,22,22,22,22,22,23,23,23,23,23,24,24,24,24,24)
> b3 <- c(31,31,31,31,31,32,32,32,32,32,33,33,33,33,33,34,34,34,34,34)
> y <- c(y1,y2,y3)
> a <- c(a1,a2,a3)
> b <- c(b1,b2,b3)
> T_5_14 <- data.frame(cbind(y,a,b))
> A <- factor(a)
> B <- factor(b)
> Anova <- lm(y~A + A/B, T_5_14)
> anova(Anova)
Analysis of Variance Table
Response: y
          Df  Sum Sq Mean Sq F value Pr(>F)
A          2   441.2  220.62  0.5759 0.5661
A:B        9  2656.9  295.21  0.7706 0.6437
Residuals 48 18388.8  383.10
```

We now show how, for the nested classification, the minimal experimental size can be determined. We choose for testing the effects of A.

Problem 5.9 Determine the subclass number for fixed precision to test $H_{A0}: a_1 = a_2 = \cdots = a_a = 0$ and $H_{B0}: b_1 = b_2 = \cdots = b_b = 0$. Unfortunately delta in the program below stands for $\tau = \delta/\sigma$.

Solution

Choose first for testing H_{A0}: $a_1 = \ldots = a_a$, the OPDOE command

```
> size.anova(model="a>b",hypothesis="a",a=,b=,alpha=,beta=,
      delta=,cases="minimin")
```

or

```
case ="maximin".
```

Choose for testing H_{B0}: $b_{11} = \ldots = b_{ab}$, the OPDOE command

```
> size.anova(model="a>b",hypothesis="b",a=,b=,alpha=,beta=,
    delta=,case="minimin")
```

or

```
case = "maximin".
```

Example

We use $\alpha = 0.05$, $a = 5$, $b = 8$, $\beta = 0.05$, and $\delta = 1$.

```
> size.anova(model="a>b",hypothesis="a",a=5,b=8,alpha=0.05,beta=0.05,
    delta=1,case="minimin")
n
3
> size.anova(model="a>b",hypothesis="a",a=5,b=8,alpha=0.05,
     beta=0.05,delta=1,case="maximin")
n
5
```

We have to choose between three and five observations per level of factor B. For testing the effects of the factor B we use $\alpha = 0.01$, $a = 5$, $b = 8$, $\beta = 0.1$, and $\delta = 1$.

```
> size.anova(model="a>b",hypothesis="b",a=5,b=8,alpha=0.01,
      beta=0.1,delta=1,case="minimin")
n
5
> size.anova(model="a>b",hypothesis="b",a=5,b=8,alpha=0.01,
      beta=0.1,delta=1,case="maximin")
 n
84 .
```

5.5 Three-Way Classification

The principle underlying the two-way ANOVA (two-way classification) is also useful if more than two factors occur in an experiment. In this section, we only give a short overview of the cases with three factors without proving all statements because the principles of proof are similar to those in the case with two factors.

We consider the case with three factors because it often occurs in applications, which can be handled with a justifiable number of pages, and because besides the cross-classification and the nested classification a mixed classification occurs. At this point, we make some remarks about the numerical analysis of experiments using ANOVA. Certainly, a general computer program for arbitrary classifications and numbers of factors with unequal class numbers can be elaborated. However, such a program, even with modern computers, is not easy to apply because the data matrices easily obtain several tens of thousands of rows. Therefore, we give for some special cases of the three-way analysis of variance numerical solutions for which easy-to-use programs are written using R.

Problems with more than three factors are described in Hartung et al. (1997) and in Rasch et al. (2008).

5.5.1 Complete Cross-Classification ($A \times B \times C$)

We assume that the observations of an experiment are influenced by three factors A, B, and C with a, b, and c levels $A_1, \dots, A_a$, $B_1, \dots, B_b$, and $C_1, \dots, C_c$, respectively. For each possible combination (A_i, B_j, C_k) let $n \geq 1$ observations y_{ijkl} $(l = 1, \cdots, n)$ be present. If the subclass numbers n_{ijkl} are not all equal or if some of them are zero but the classification is connected, we must use in R different models in `>lm()`. For the testing of interaction $A \times B$, in the formula $A \times B$ must be placed before $A \times B \times C$; for the testing of interaction $A \times C$, in the formula $A \times C$ must be placed before $A \times B \times C$; for the testing of interaction $B \times C$ in the formula $B \times C$ must be placed before $A \times B \times C$. An example of the analysis of an unbalanced three-way cross-classification with R is described at the end of this section.

Each combination (A_i, B_j, C_k) $(i = 1, \cdots, a; j = 1, \cdots, b; \quad k = 1, \cdots, c)$ of factor levels is called a class and is characterised by (i, j, k). The expectation in the population associated with the class (i, j, k) is $\eta_{ijk.}$

We define

$$\overline{\eta}_{i..} = \frac{\sum_{j,k} \eta_{ijk}}{bc} \text{ the expectation of the } i\text{th level of factor } A$$

$$\overline{\eta}_{.j.} = \frac{\sum_{i,k} \eta_{ijk}}{ac} \text{ the expectation the } j\text{th level of factor } B$$

and

$$\overline{\eta}_{..k} = \frac{\sum_{i,j} \eta_{ijk}}{ab} \text{ the expectation of the } k\text{th level of factor } C.$$

The overall expectation is

$$\mu = \overline{\eta}_{...} = \frac{\sum_{i,j,k} \eta_{ijk}}{abc}.$$

The main effects of the factors A, B, and C we define by

$$a_i = \overline{\eta}_{i..} - \mu, \qquad b_j = \overline{\eta}_{.j.} - \mu, \qquad c_k = \overline{\eta}_{..k} - \mu.$$

Assuming that the experiment is performed at a particular level C_k of the factor C we have a two-way classification with the factors A and B, and the conditional interactions between the levels of the factors A and B for fixed k are given by

$$\eta_{ijk} - \overline{\eta}_{i.k} - \overline{\eta}_{.jk} + \overline{\eta}_{..k}.$$

The interactions $(a, b)_{ij}$ between the ith A level and the jth B level are the means over all C levels, i.e. $(a, b)_{ij}$ is defined as

$$(a, b)_{ij} = \overline{\eta}_{ij.} - \overline{\eta}_{i..} - \overline{\eta}_{.j.} + \mu.$$

The interactions between A levels and C levels $(a, c)_{ik}$ and between B levels and C levels $(b, c)_{jk}$ are defined by

$$(a, c)_{ik} = \overline{\eta}_{i.k} - \overline{\eta}_{i..} - \overline{\eta}_{..k} + \mu$$

and

$$(b, c)_{jk} = \overline{\eta}_{.jk} - \overline{\eta}_{.j.} - \overline{\eta}_{..k} + \mu,$$

respectively.

The difference between the conditional interactions between the levels of two of the three factors for the given level of the third factor and the (unconditional) interaction of these two factors depends only on the indices of the levels of the factors, and not on the factor for which the interaction of two factors is calculated. We call it the second order interaction $(a, b, c)_{ijk}$ (between the levels of three factors). Without loss of generality we write

$$(a, b, c)_{ijk} = \overline{\eta}_{ijk} - \overline{\eta}_{ij.} - \overline{\eta}_{i.k} - \overline{\eta}_{.jk} + \overline{\eta}_{i..} + \overline{\eta}_{.j..} + \overline{\eta}_{..k} - \mu.$$

The interactions between the levels of two factors are called first-order interactions. From the definition of the main effect and the interactions we write for η_{ijk}

$$\eta_{ijk} = \mu + a_i + b_j + c_k + (a, b)_{ij} + (a, c)_{ik} + (b, c)_{jk} + (a, b, c)_{ijk}.$$

Under the definitions above, the side conditions for all values of the indices not occurring in the summation at any time are

$$\sum_i a_i = \sum_j b_j = \sum_k c_k = \sum_i (a, b)_{ij} = \sum_j (a, b)_{ij} = \sum_i (a, c)_{ik} = \sum_k (a, c)_{ik} =$$
$$= \sum_j (b, c)_{jk} = \sum_k (b, c)_{jk} = \sum_i (a, b, c)_{ijk} = \sum_j (a, b, c)_{ijk} = \sum_k (a, b, c)_{ijk} = 0.$$

The n observations $\boldsymbol{y}_{ijkl}$ in each class are assumed to be independent of each other and $N(0, \sigma^2)$-distributed. The variable (called error term) $\boldsymbol{e}_{ijkl}$ is the difference between $\boldsymbol{y}_{ijkl}$ and the expectation η_{ijk} of the class, i.e. we have

$$\boldsymbol{y}_{ijkl} = \eta_{ijk} + \boldsymbol{e}_{ijkl}$$

or

$$\boldsymbol{y}_{ijkl} = \mu + a_i + b_j + c_k + (a, b)_{ij} + (a, c)_{ik} + (b, c)_{jk} + (a, b, c)_{ijk} + \boldsymbol{e}_{ijkl}. \tag{5.30}$$

By the least squares method we obtain the following estimators:

$$\overline{\boldsymbol{y}}_{....} = \frac{1}{abcn} \sum_{i,j,k,l} \boldsymbol{y}_{ijkl} \text{ for } \mu$$

as well as

$$\hat{\boldsymbol{a}}_i = \overline{\boldsymbol{y}}_{i...} - \overline{\boldsymbol{y}}_{....}$$

$$\hat{\boldsymbol{b}}_j = \overline{\boldsymbol{y}}_{.j..} - \overline{\boldsymbol{y}}_{....}$$

$$\hat{\boldsymbol{c}}_k = \overline{\boldsymbol{y}}_{..k.} - \overline{\boldsymbol{y}}_{...}\overline{\boldsymbol{y}}_{....};$$

$$\widehat{(\boldsymbol{a},\boldsymbol{b})}_{ij} = \overline{\boldsymbol{y}}_{ij..} - \overline{\boldsymbol{y}}_{i...} - \overline{\boldsymbol{y}}_{.j..} + \overline{\boldsymbol{y}}_{....}$$

$$\widehat{(\boldsymbol{a},\boldsymbol{c})}_{ik} = \overline{\boldsymbol{y}}_{i.k.} - \overline{\boldsymbol{y}}_{i...} - \overline{\boldsymbol{y}}_{.k..} + \overline{\boldsymbol{y}}_{....}$$

$$\widehat{(\boldsymbol{b},\boldsymbol{c})}_{jk} = \overline{\boldsymbol{y}}_{.jk.} - \overline{\boldsymbol{y}}_{.j..} - \overline{\boldsymbol{y}}_{..k.} + \overline{\boldsymbol{y}}_{....}$$

$$\widehat{(\boldsymbol{a},\boldsymbol{b},\boldsymbol{c})}_{ijk} = \overline{\boldsymbol{y}}_{ijk.} - \overline{\boldsymbol{y}}_{ij..} - \overline{\boldsymbol{y}}_{\text{i.k.}} - \overline{\boldsymbol{y}}_{\text{.jk.}} + \overline{\boldsymbol{y}}_{\text{i...}} + \overline{\boldsymbol{y}}_{\text{.j.}} + \overline{\boldsymbol{y}}_{\text{..k.}} - \overline{\boldsymbol{y}}_{....}.$$

If any of the interaction effects in (5.30) are zero the corresponding estimator above is dropped the others remain unchanged. The following model equations lead to different SS_{res}, as shown in Table 5.15.

$$\boldsymbol{y}_{ijkl} = \mu + a_i + b_j + c_k + (a,b)_{ij} + (a,c)_{ik} + (b,c)_{jk} + \boldsymbol{e}_{ijkl}. \tag{5.31}$$

We may split the overall sum of squares $\sum_{i,j,k,l} (\boldsymbol{y}_{ijkl} - \overline{y}_{....})^2$ into eight components: three corresponding with the main effects, three with the first order interactions, one with the second order interaction, and one with the error term or the residual. The corresponding SS are shown in the ANOVA table (Table 5.15). In this table N is, again, the total number of observations, $N = abcn$.

The following hypotheses can be tested (H_{0x} is one of the hypotheses $H_{0A}, \dots, H_{0ABC}$; SS_x is the corresponding $\boldsymbol{SS}$).

$$H_{0A} : a_i = 0 \quad (\text{for all } i),$$
$$H_{0B} : b_j = 0 \quad (\text{for all } j),$$
$$H_{0C} : c_k = 0 \quad (\text{for all } k),$$

$$H_{AB0} : (a,b)_{ij} = 0 \qquad (\text{for all } i,j),$$
$$H_{AC0} : (a,c)_{ik} = 0 \qquad (\text{for all } i,k),$$
$$H_{BC0} : (b,c)_{jk} = 0 \qquad (\text{for all } j,k),$$
$$H_{ABC0} : (a,b,c)_{ijk} = 0 \qquad (\text{for all } i,j,k, \ \textit{if } n > 1).$$

Under the hypothesis H_{x0}, $\frac{1}{\sigma^2}\boldsymbol{SS}_x$ and $\frac{1}{\sigma^2}\boldsymbol{SS}_{\text{res}}$ are independent of each other, with the df given in the ANOVA table, centrally χ^2-distributed. Therefore, the test statistics given in column $\boldsymbol{F}$ of the ANOVA table are, with the corresponding degrees of freedom, centrally F-distributed. For $n = 1$ all hypotheses except H_{ABC0} can be tested under the assumption $(a, b, c)_{ijk} = 0$ for all i, j, k because then $\frac{1}{\sigma^2}\boldsymbol{SS}_{ABC} = \frac{1}{\sigma^2}\boldsymbol{SS}_{\text{res}}$ and $\frac{1}{\sigma^2}\boldsymbol{SS}_x$, under $H_{x0}(x = A, B, C, \text{ etc.})$, are independent of each other centrally χ^2--distributed.

Table 5.15 ANOVA table of a three-way cross-classification with equal subclass numbers (model I).

Source of variation	SS	df
Between A levels	$SS_A = \frac{1}{bcn}\sum_i Y_{i\ldots}^2 - \frac{1}{N}Y_{\ldots\ldots}^2$	$a-1$
Between B levels	$SS_B = \frac{1}{acn}\sum_j Y_{\cdot j\cdot\cdot}^2 - \frac{1}{N}Y_{\ldots\ldots}^2$	$b-1$
Between C levels	$SS_C = \frac{1}{abn}\sum_k Y_{\cdot\cdot k\cdot}^2 - \frac{1}{N}Y_{\ldots\ldots}^2$	$c-1$
Interaction $A \times B$	$SS_{AB} = \frac{1}{cn}\sum_{i,j} Y_{ij\cdot\cdot}^2 - \frac{1}{bcn}\sum_i Y_{i\ldots}^2 - \frac{1}{acn}\sum_j Y_{\cdot j\cdot\cdot}^2 + \frac{Y_{\ldots\ldots}^2}{N}$	$(a-1)(b-1)$
Interaction $A \times C$	$SS_{AC} = \frac{1}{bn}\sum_{i,k} Y_{i\cdot k\cdot}^2 - \frac{1}{bcn}\sum_i Y_{i\ldots}^2 - \frac{1}{abn}\sum_k Y_{\cdot\cdot k\cdot}^2 + \frac{Y_{\ldots\ldots}^2}{N}$	$(a-1)(c-1)$
Interaction $B \times C$	$SS_{BC} = \frac{1}{an}\sum_{j,k} Y_{\cdot jk\cdot}^2 - \frac{1}{acn}\sum_i Y_{\cdot j\cdot\cdot}^2 - \frac{1}{abn}\sum_k Y_{\cdot\cdot k\cdot}^2 + \frac{Y_{\ldots\ldots}^2}{N}$	$(b-1)(c-1)$
Interaction $A \times B \times C$	$SS_{\mathrm{ABC}} = \sum_{ijkl} y_{ijkl}^2 - (\sum_{ijkl} yijkl)^2/N - SS_A - SS_B - SS_C - SS_{AB} - SS_{AC} - SS_{BC} - SS_{res}$	$(a-1)(b-1)(c-1)$
Within the classes (residual), (5.30)	$SS_{\mathrm{res}} = \sum_{i,j,k,l} y_{ijkl}^2 - \frac{1}{n}\sum_{i,j,k} Y_{ijk\cdot}^2$	$abc(n-1)$
Within the classes (residual), (5.31)	$SS_{\mathrm{res}} = \sum_{i,j,k,l} y_{ijkl}^2 - \frac{1}{bn}\sum_{i,k} Y_{i\cdot k\cdot}^2 - \frac{1}{cn}\sum_{i,j} Y_{ij\cdot\cdot}^2 - \frac{1}{an}\sum_{j,k} Y_{\cdot jk\cdot}^2 + \frac{1}{acn}\sum_i Y_{\cdot j\cdot\cdot}^2 + \frac{1}{abn}\sum_k Y_{\cdot\cdot k\cdot}^2 + \frac{1}{bcn}\sum_i Y_{i\ldots}^2 - \frac{1}{N}Y_{\ldots\ldots}^2$	$(a-1)(b-1)(c-1) + abc(n-1)$

MS in (5.30)	E(MS) in (5.30)	F in (5.30)
$MS_A = \frac{SS_A}{a-1}$	$\sigma^2 + \frac{bcn}{a-1}\sum a_i^2$	$\frac{abc(n-1)}{a-1}\frac{SS_A}{SS_{\mathrm{res}}}$
$MS_B = \frac{SS_B}{b-1}$	$\sigma^2 + \frac{acn}{b-1}\sum b_j^2$	$\frac{abc(n-1)}{b-1}\frac{SS_B}{SS_{\mathrm{res}}}$
$MS_C = \frac{SS_C}{c-1}$	$\sigma^2 + \frac{abn}{c-1}\sum c_k^2$	$\frac{abc(n-1)}{c-1}\frac{SS_C}{SS_{\mathrm{res}}}$
$MS_{AB} = \frac{SS_{AB}}{(a-1)(b-1)}$	$\sigma^2 + \frac{cn}{(a-1)(b-1)}\sum (a,b)^2$	$\frac{abc(n-1)}{(a-1)(b-1)}\frac{SS_{AB}}{SS_{\mathrm{res}}}$
$MS_{AC} = \frac{SS_{AC}}{(a-1)(c-1)}$	$\sigma^2 + \frac{bn}{(a-1)(c-1)}\sum (a,c)^2$	$\frac{abc(n-1)}{(a-1)(c-1)}\frac{SS_{AC}}{SS_{\mathrm{res}}}$
$MS_{BC} = \frac{SS_{BC}}{(b-1)(c-1)}$	$\sigma^2 + \frac{an}{(b-1)(c-1)}\sum (b,c)^2$	$\frac{abc(n-1)}{(b-1)(c-1)}\frac{SS_{BC}}{SS_{\mathrm{res}}}$
$MS_{ABC} = \frac{SS_{ABC}}{(a-1)(b-1)(c-1)}$	$\sigma^2 + \frac{n}{(a-1)(b-1)(c-1)}\sum (a,b,c)^2$	$\frac{abc(n-1)}{(a-1)(b-1)(c-1)}\frac{SS_{ABC}}{SS_{\mathrm{res}}}$
$MS_{\mathrm{res}} = s^2 = \frac{SS_{\mathrm{res}}}{abc(n-1)}$	σ^2	

The test statistic F_x is given by

$$F_x = \frac{(a-1)(b-1)(c-1)}{df_x} \cdot \frac{SS_x}{SS_{\text{res}}}.$$

If the null hypotheses are not true, we have non-central distributions with non-centrality parameters analogous to those in Section 5.4.

Example 5.16 We consider a three-way analysis of variance with class numbers $n = 2$; as factors we use the forage group (A), the kind of storage (B – barn or refrigerator) and the packaging material (C – glass or sack) (Table 5.16). We have $a = b = c = 2$ and $n = 2$. Table 5.17 is the ANOVA table of the examples; the F-tests are done under the assumption that all second order interactions vanish using the SS_{res} defined above. Only between the kinds of storage significant differences ($\alpha = 0.05$) can be found, i.e. only the hypothesis H_A is rejected.

Problem 5.10 Calculate the ANOVA table of a three-way cross-classification for model (5.30).

Solution

Make a data-frame of the observations and use the command `>lm()`.

Example

We use the data of Table 5.16 and calculate the entries of Table 5.17.

```
> y <- c(8.39,8.69,5.42,6.13,11.58,10.56,9.53,8.78,
   9.44,10.1,5.56,4.97,12.21,11.87,10.39,9.96)
> storage <- c(rep(1,4),rep(2,4),rep(1,4),rep(2,4))
> material <-  c(rep(1,2),rep(2,2),rep(1,2),rep(2,2),rep(1,2),
    rep(2,2),rep(1,2),rep(2,2))
> crop <- c(rep(1,8),rep(2,8))
> T_5_16 <- data.frame(cbind(y, storage, material, crop))
> STOR <- factor(storage)
> MAT <- factor(material)
> CROP <- factor(crop)
> Anova <- lm(y~ STOR*MAT*CROP,
                  T_5_16)
> Anova
```

Table 5.16 Three-way classification with factors kind of storage, packaging material and forage crop.

		Forage crop	
Kind of storage	**Packaging material**	**Green rye**	**Lucerne**
Refrigerator	Glass	8.39	9.44
		8.69	10.1
	Sack	5.42	5.56
		6.13	4.97
Barn	Glass	11.58	12.21
		10.56	11.87
	Sack	9.53	10.39
		8.78	9.96

```
Call:
lm(formula = y ~ STOR + MAT + CROP + STOR * MAT + STOR * CROP +  MAT *
CROP + STOR * MAT * CROP)
Coefficients:
     (Intercept)          STOR2                 MAT2
           CROP2    STOR2:MAT2      STOR2:CROP2    MAT2:CROP2
STOR2:MAT2:CROP2
           8.540          2.530             -2.765         1.230
0.850          -0.260            -1.740
           1.790
> summary(Anova)
Call:
lm(formula = y ~ STOR + MAT + CROP + STOR * MAT + STOR * CROP
+    MAT * CROP + STOR * MAT * CROP)
Residuals:
    Min      1Q  Median      3Q     Max
-0.5100 -0.3038  0.0000  0.3038  0.5100
Coefficients:
            Estimate Std. Error t value Pr(>|t|)(Intercept)
8.5400  0.3202  26.670  4.2e-09 ***
STOR2          2.5300     0.4529   5.587 0.000518 ***
MAT2          -2.7650     0.4529  -6.106 0.000288 ***
CROP2          1.2300     0.4529   2.716 0.026407 * STOR2:MAT2
0.8500 0.6404 1.327 0.221058 STOR2:CROP2  -0.2600   0.6404 -
0.406 0.695401 MAT2:CROP2 -1.7400  0.6404  -2.717 0.026374 *
STOR2:MAT2:CROP2   1.7900     0.9057   1.976 0.083517 .
- - -
Signif. codes: 0 "***" 0.001 "**" 0.01 "*" 0.05 "." 0.1 "" 1
Residual standard error: 0.4529 on 8 degrees of freedom
Multiple R-squared:  0.9799,    Adjusted R-squared:  0.9624
F-statistic: 55.84 on 7 and 8 DF,  p-value: 3.645e-06
> anova(Anova)
Analysis of Variance Table
Response: y
              Df Sum Sq Mean Sq  F value    Pr(>F)
STOR           1 42.837  42.837 208.8847 5.138e-07 ***
MAT            1 30.526  30.526 148.8510 1.889e-06 ***
CROP           1  1.836   1.836   8.9529  0.017277 *
STOR:MAT       1  3.045   3.045  14.8483  0.004854 **
STOR:CROP      1  0.403   0.403   1.9662  0.198439
MAT:CROP       1  0.714   0.714   3.4818  0.099021 .
STOR:MAT:CROP  1  0.801   0.801   3.9060  0.083517 .
Residuals      8  1.641   0.205
- - -
Signif. codes: 0 "***" 0.001 "**" 0.01 "*" 0.05 "." 0.1 "" 1
> N <- length(y)
> N
[1] 16
> SST <- (N-1)*var(y)
> SST
[1] 81.80258
> df <- N-1
 df
[1] 15
```

If the subclass numbers n_{ijkl} are not all equal or if some of them are zero but the classification is connected, we must change the formula for testing the two-factor interactions. In the following example, we give an unbalanced three-way cross-classification.

Table 5.17 ANOVA Table for data of Table 5.16.

Source of variation	*SS*	*df*	*MS*	*F*
Between kind of storage	42.837	1	42.837	208.8847
Between forage crops	1.836	1	1.836	8.9529
Between packaging material	30.526	1	30.526	148.8510
Interaction kind of storage × packaging material	3.045	1	3.045	14.8483
Interaction kind of storage × forage crops	0.403	1	0.403	1.9662
Interaction forage crops × packaging material	0.714	1	0.714	3.4818
Interaction storage × forage × material	0.801	1	0.801	3.9060
Residual	1.641	8	0.205	
Total	81.80258	15		

Example 5.17 From Kuehl (1994) we change an exercise to an unbalanced one. The California brown shrimp spawn at sea and the hatched eggs undergo larval transformation while being transported towards the shore. By the time they transform to post-larval stage they enter estuaries, where they grow rapidly into sub-adults and migrate back offshore as they approach sexual maturity. The shrimp encounter wide temperature and salinity variations in their life cycle because of their migrations during the cycle. Thus, a knowledge of how temperature and salinity affect their growth and survival is of great importance to understanding their life history and ecology. From the standpoint of mariculture, another important factor is stocking density in the culture tanks that affects intraspecific competition. The investigators wanted to know how water temperature, water salinity, and density of shrimp populations influenced the growth rate of shrimp raised in aquaria and whether the factor acted independently on the shrimp populations. A factorial arrangement was used with three factors: T (temperature: 25 °C, 35 °C); S (salinity of the water: 10%, 25%, 40%); and D (density of shrimp in the aquarium: 80 shrimp/40 l, 160 shrimp/40 l). The levels were those considered most likely to exhibit an effect if the factor was influential on shrimp growth. The experiment design consisted of three replicate aquaria for each of the 12 treatment combinations of the $2 \times 3 \times 2$ factorial. Each of the 12 treatment combinations was randomly assigned to three aquaria for a completely randomised design. The 36 aquaria were stocked with post-larval shrimp at the beginning of the test. The weight gain of the shrimp in four weeks for each of the 36 aquaria is shown in Table 5.18 on a per-shrimp basis. From the balanced experiment we have discarded at random three aquaria results to give the analysis of an unbalanced $2 \times 3 \times 2$ factorial. The missing data are indicated by an asterisk.

In **R** the missing data must be indicated by NA (= not available).

```
> y1 <- c(86,52,73,544,NA, 482,390,290,397)
> y2 <- c(53,73,86,393,398,NA,249,265,243)
> y3 <- c(439,436,349,249,245,330,NA,277,205)
```

Table 5.18 Water temperature (*T*), water salinity (*S*), and density of shrimp populations (*D*) and the weight gain (mg) of shrimp.

T	*D*	*S*	Weight gain (mg)
1 (25 °C)	1 (80)	1 (10%)	86, 52, 73
		2 (25%)	544, *, 482
		3 (40%)	390, 290, 397
	2 (160)	1	53, 73, 86
		2	393, 398, *
		3	249, 265, 243
2 (35 °C)	1	1	439, 436, 349
		2	249, 245, 330
		3	*, 277, 205
	2	1	324, 305, 364
		2	352, 267, 316
		3	188, 223, 281

```
> y4 <- c(324,305,364,352,267,316,188,23,281)
> t1 <- c(rep(1,18))
> t2 <- c(rep(2,18))
> d1 <- c(rep(1,9), rep(2,9))
> d2 <- c(rep(1,9), rep(2,9))
> s1 <- c(rep(1,3),rep(2,3),rep(3,3))
> s2 <- c(rep(1,3),rep(2,3),rep(3,3))
> s3 <- c(rep(1,3),rep(2,3),rep(3,3))
> s4 <- c(rep(1,3),rep(2,3),rep(3,3))
> y <- c(y1,y2,y3,y4)
> t <- c(t1,t2)
> d <- c(d1,d2)
> s <- c(s1,s2,s3,s4)
> table <- data.frame(cbind(y,t,d,s))
> T <- factor(t)
> S <- factor(s)
> D <- factor(d)
> anova1 <- lm (y ~ T + D + S + T*S + T*D + S*D + T*S*D , table)
> anova(anova1)
Analysis of Variance Table
Response: y
          Df Sum Sq Mean Sq F value    Pr(>F)
T          1  11012   11012  3.6988  0.068115 .
D          1  27697   27697  9.3031  0.006082 **
S          2 101840   50920 17.1037 3.911e-05 ***
T:S        2 354666  177333 59.5651 2.212e-09 ***
T:D        1   2226    2226  0.7478  0.396945
D:S        2   6300    3150  1.0581  0.364910
T:D:S      2  17923    8961  3.0100  0.070891 .
Residuals 21  62520    2977
- - -
```

```
Signif. codes: 0 "***" 0.001 "**" 0.01 "*" 0.05 "." 0.1 "" 1
> anova2 <- lm (y ~ T + D + S + T*S + S*D + T*D + T*S*D)
> anova(anova2)
Analysis of Variance Table
Response: y
          Df Sum Sq Mean Sq F value    Pr(>F)
T          1  11012   11012  3.6988  0.068115 .
D          1  27697   27697  9.3031  0.006082 **
S          2 101840   50920 17.1037 3.911e-05 ***
T:S        2 354666  177333 59.5651 2.212e-09 ***
D:S        2   7055    3527  1.1848  0.325428
T:D        1   1472    1472  0.4943  0.489729
T:D:S      2  17923    8961  3.0100  0.070891 .
Residuals 21  62520    2977
- - -
Signif. codes:  0 "***" 0.001 "**" 0.01 "*" 0.05 "." 0.1 " " 1
> anova3 <- lm ( y ~ T + D + S + T*D + S*D + T*S + T*S*D)
> anova(anova3)
Analysis of Variance Table
Response: y
          Df Sum Sq Mean Sq F value    Pr(>F)
T          1  11012   11012  3.6988  0.068115 .
D          1  27697   27697  9.3031  0.006082 **
S          2 101840   50920 17.1037 3.911e-05 ***
T:D        1   1212    1212  0.4071  0.530337
D:S        2   9700    4850  1.6291  0.219931
T:S        2 352281  176140 59.1644 2.349e-09 ***
T:D:S      2  17923    8961  3.0100  0.070891 .
Residuals 21  62520    2977
- - -
```

Note that in the three ANOVA tables the *SS(T:D:S)* remains the same, and the test of this interaction *TDS* is correct.

The test for the interaction *DS* is given in anova1, the test for interaction *TD* is given in anova2 and the test of interaction *TS* is given in anova3.

This is due to the fact that the program `lm()` gives a hierarchical analysis of the variance table; each factor is corrected for his predecessors. In a balanced design the order of the interactions is not important, but for unbalanced design you must give the correct order to find a test for the second-order interactions.

Problem 5.11 Calculate the minimal subclass number to test the hypotheses of the main effect and interactions in a three-way cross-classification under model (5.30). Unfortunately delta in the program below stands for $\tau = \delta/\sigma$.

Solution

Use in the `OPDOE` program for testing `hypothesis = "a", "b", "c", "ab", "ac", "bc"` and `cases = "minimin"` or `cases = "maximin"`.

Example 5.18 We try to plan an experiment with three cross-classified factors A with $a = 3$ levels, B with $b = 4$ levels and C with $c = 5$ levels and assume model (5.31). We further put $\alpha = 0.01$, $\beta = 0.1$, and $\delta = 0.5$. We demonstrate the program for a hypothesis of the main effects of A and the interaction effects $A \times B$.

```
>size.anova(model = "axbxc",hypothesis = "a",a = 3,b = 4,c = 5,
 alpha = 0.01,beta = 0.1,delta = 0.5, case = "minimin")
n
5
>size.anova(model = "axbxc",hypothesis = "a",a = 3,b = 4,c = 5,alpha = 0.01,
 beta = 0.1,delta = 0.5,case = "maximin")
n
8
>size.anova(model = "axbxc",hypothesis = "axb",a = 3,b = 4, c = 5,
alpha = 0.01,beta = 0.1,delta = 0.5,case = "minimin")
n
7
>size.anova(model = "axbxc",hypothesis = "axb",a = 3,b = 4, c = 5,
 alpha = 0.01,beta = 0.1,delta = 0.5, case = "maximin")
 n
38
```

5.5.2 Nested Classification ($C \prec B \prec A$)

We speak about a three-way nested classification if factor C is sub-ordinated to factor B and factor B is sub-ordinated to factor A, i.e. if $C \prec B \prec A$. We assume that the random variable y_{ijkl} varies randomly with expected value $\eta_{ijk}(i = 1, \dots, a; j = 1, \dots, b_i;\ k = 1, \dots, c_{ij})$, i.e. we assume

$$\boldsymbol{y}_{ijkl} = \eta_{ijk} + \boldsymbol{e}_{ijkl} \qquad (l = 1, \dots, n_{ijk}),$$

where $\boldsymbol{e}_{ijkl}$, independent of each other, are $N(0, \sigma^2)$-distributed. Using

$$\mu = \eta_{\dots} = \frac{\sum_{i=1}^{a}\sum_{j=1}^{b_i}\sum_{k=1}^{c_{ij}} \eta_{ijk} n_{ijk}}{N}$$

we define the total mean of the experiment as $N = \sum_{i=1}^{a}\sum_{j=1}^{b}\sum_{k=1}^{c} n_{ijk}$.

The difference $a_i = \overline{\eta}_{i..} - \mu$ is called the effect of the ith level of A, the difference $b_{ij} = \overline{\eta}_{ij.} - \overline{\eta}_{i..}$ is called the effect of the jth level of B within the ith level of A and the difference $c_{ijk} = \eta_{ijk} - \overline{\eta}_{ij.}$ is called the effect of the kth level of C within the jth level of B and the ith level of A.

Then we model the observations using

$$\boldsymbol{y}_{ijkl} = \mu + a_i + b_{ij} + c_{ijk} + \boldsymbol{e}_{ijkl}. \tag{5.32}$$

There exist no interactions. We consider (5.32) with $N_{ij.} = \sum_k n_{ijk}; N_{i..} = \sum_{jk} n_{ijk}$ under the side conditions

$$\sum_{i=1}^{a} N_{i..} a_i = \sum_{j=1}^{b_i} N_{ij.} b_{ij} = \sum_{k=1}^{c_{ij}} n_{ijk} c_{ijk} = 0.$$

Minimising

$$\sum_{l=1}^{n_{ijk}}\sum_{i=1}^{a}\sum_{j=1}^{b_i}\sum_{k=1}^{c_{ij}}(y_{ijkl}-\mu-a_i-b_{ij}-c_{ijk})^2 \tag{5.33}$$

under the side conditions above, leads to the BLUE of the parameters as follows

$$\hat{\mu}=\bar{y}_{....},\quad \hat{a}_{ii}=\bar{y}_{i...}-\bar{y}_{....},\quad \hat{b}_{ij}=\bar{y}_{ij..}-\bar{y}_{i...},\quad \hat{c}_{ijk}=\bar{y}_{ijk.}-\bar{y}_{ij..}.$$

In a three-way nested classification we have

$$SS_T = SS_A + SS_{B\text{ in }A} + SS_{C\text{ in }B\text{ (and }A)} + SS_{\text{res}}$$

with $N=\sum_{i=1}^{a}\sum_{j=1}^{b}\sum_{k=1}^{c} n_{ijk}, N_{ij.}=\sum_k n_{ijk}; N_{i..}=\sum_{jk} n_{ijk}$ and

$$SS_T=\sum_{i,j,k,l} y_{ijkl}^2-\frac{Y_{....}^2}{N},\qquad SS_A=\sum_i\frac{Y_{i...}^2}{N_{i..}}-\frac{Y_{....}^2}{N},$$

$$SS_{B\ in\ A}=\sum_{i,j}\frac{Y_{ij..}^2}{N_{ij.}}-\sum_i\frac{Y_{i...}^2}{N_{i..}},\qquad SS_{C\ in\ B}=\sum_{i,j,k}\frac{Y_{ijk.}^2}{n_{ijk}}-\sum_{i,j}\frac{Y_{ij..}^2}{N_{ij.}},$$

$$SS_{\text{res}}=\sum_{i,j,k,l} y_{ijkl}^2-\sum_{i,j,k}\frac{Y_{ijk.}^2}{n_{ijk}}.$$

The variables $\frac{1}{\sigma^2}SS_A$ up to $\frac{1}{\sigma^2}SS_{C\text{ in }B}$ are, with $B_{..}=\sum_{ij} b_{ij}, C_{...}=\sum_{ijk} c_{ijk}$, pairwise independently $CS(a-1,\lambda_a)$, $CS(B.-a,\lambda_b)$, $CS(C..-B.,\lambda_c)$, respectively, and $\frac{1}{\sigma^2}SS_{\text{res}}$ is $CS(N-C..)$-distributed. The non-centrality parameters λ_a, λ_b and λ_c vanish under the null hypotheses $H_{A0}: a_i=0\ (i=1,\ldots.a)$, $H_{B0}: b_{ij}=0\ (i=1,\ldots.a;\ j=1,\ldots,b_i)$, $H_{C0}: c_{ijk}=0\ (i=1,\ldots.a;\ j=1,\ldots,b_i;\ k=1,\ldots,c_{ij})$, so that the usual F statistics can be used. Table 5.19 shows the SS and MS for calculating the F-statistics. If H_{A0} is valid F_A is $F(a-1,\ N-C..)$-distributed. If H_{B0} is valid then F_B is $F(B-a,\ N-C..)$-distributed, and if H_{C0} is valid then F_c is $F(C..-B.,\ N-C..)$-distributed.

Example 5.19 We consider a three-way nested classification with $a=2,\ b=2,\ c=3,\ n=4,\ N=48$ and the data in Table 5.20.

Problem 5.12 Calculate the ANOVA table for the data of Table 5.20.

Solution

After making the data-frame of the observations use > `lm()`

Example

```
> y1 <- c(93,89,97 105 109 107,94 106 102 101,99,98,89 102 104,97,81,83,85,
 91,87,91,82,85)
> y2 <- c(97,93,95,91,88,92,94,82,80,84,83,81,83,88,87,86,82,89,
 93,81,81,82,80,85)
> a1 <- rep(1, 24)
> a2 <- rep(2, 24)
```

Table 5.19 ANOVA table of a three-way nested classification for model I.

Source of Variation	SS	df	MS	E(MS)	F
Between A	$\sum_i \frac{Y_{i\cdots}^2}{N_{i\cdot\cdot}} - \frac{Y_{\cdots\cdot}^2}{N}$	$a-1$	$\frac{SS}{a-1}$	$\sigma^2 + \frac{1}{a-1}\sum_{i=1}^{n} N_{i.}a_i^2$	$\frac{MS_A}{MS_{\text{res}}} = F_A$
Between B in A	$\sum_{i,j} \frac{Y_{ij\cdot\cdot}^2}{N_{ij\cdot}} - \sum_i \frac{Y_{\cdots\cdot}^2}{N_{i\cdot\cdot}}$	$B. - a$	$\frac{SS_{B \text{ in } A}}{B. - a}$	$\sigma^2 + \frac{1}{B. - a}\sum_{i,j} N_{ij.}b_{ij}^2$	$\frac{MS_{B \text{ in } A}}{MS_{\text{res}}} = F_B$
Between C in B and A	$\sum_{i,j,k} \frac{Y_{ijk.}^2}{n_{ijk}} - \sum_{i,j} \frac{Y_{ij\cdot\cdot}^2}{N_{ij\cdot}}$	$C.. - B.$	$\frac{SS_{C \text{ in } B}}{C.. - B.}$	$\sigma^2 + \frac{1}{C.. - B.}\sum_{i,j,k} n_{ijk}c_{ijk}^2$	$\frac{MS_{C \text{ in } B}}{MS_{\text{res}}} = F_C$
Residual	$\sum_{i,j,k,l} Y_{ijkl}^2 - \sum_{i,j,k} \frac{Y_{ijk.}^2}{n_{ijk}}$	$N - C..$	$\frac{SS_{\text{res}}}{N - C..}$	σ^2	
Total	$\sum_{i,j,k,l} Y_{ijkl}^2 - \frac{Y_{\cdots\cdot}^2}{N}$	$N-1$			

Table 5.20 Observations of a three-way nested classification.

Factor A	A_1						A_2					
B	B_{11}			B_{12}			B_{21}			B_{22}		
C	C_{111}	C_{112}	C_{113}	C_{121}	C_{122}	C_{123}	C_{211}	C_{212}	C_{213}	C_{221}	C_{222}	C_{223}
1	93	109	102	89	81	87	97	88	80	83	82	81
2	89	107	101	102	83	91	93	92	84	88	89	82
3	97	94	99	104	85	82	95	94	83	87	93	80
4	105	106	98	97	91	85	91	82	81	86	81	85

```
> b11_12 <- c(rep(11, 12), rep(12, 12))
> b21_22 <- c(rep(21, 12), rep(22, 12))
> c1 <- c(rep(111,4), rep(112,4), rep(113,4), rep(121,4),
 rep(122,4), rep(123,4))
> c2 <- c(rep(211,4), rep(212,4), rep(213,4), rep(221,4),
 rep(222,4), rep(223,4))
> y <- c(y1,y2)
> a <- c(a1,a2)
> b <- c(b11_12, b21_22)
> c <- c(c1,c2)
> T_5_19 <- data.frame(cbind[y,a,b,c])
> A <- factor(a)
> B <- factor(b)
> C <- factor(c)
> Anova <- lm(y ~ A + A/B + A/B/C, T_5_19)
> anova(Anova)
Analysis of Variance Table
Response: y
                    Df    Sum Sq  Mean Sq  F value    Pr(>F)
```

```
A            1    833.33   833.33    39.3959    2.972e-07 ***
A : B        2    707.42   353.71    16.7216    7.313e-06 ***
A : B : C    8    875.67   109.46    5.1747     0.000243 ***
Residuals   36    761.50   21.15
```

Now we show how to calculate the minimal subclass numbers for the three F-tests in the nested classification using **R**.

Problem 5.13 Determine the minimal subclass numbers for the three tests of the main effects.

Unfortunately delta in the program below stands for $\tau = \delta/\sigma$.

Solution

Use in `OPDOE` the command `>size.anova(model = "a>b>c", hypothesis=,a=,b=,c=,alpha=,beta=,delta=,cases=)` with `case="minimin"` or `case="maximin"`.

Example

We try to plan an experiment for three-way nested classification factors A with $a = 3$ levels, B in A with $b = 3$ levels and C in B in A with $c = 4$ levels and assume model (5.32). We further put $\alpha = 0.01$, $\beta = 0.1$ and $\delta = 0.5$.

```
> size.anova(model="a>b>c",hypothesis="a",a=3,b=3,c=4,
       lpha=0.01,beta=0.1,delta=0.5,case="minimin")
n
8
> size.anova(model="a>b>c",hypothesis="a",a=3,b=3,c=4,
      alpha=0.01,beta=0.1,delta=0.5,case="maximin")
 n
12
> size.anova(model="a>b>c",hypothesis="b",a=3,b=3,c=4,
      alpha=0.01,beta=0.1,delta=0.5,case="minimin")
 n
1
> size.anova(model="a>b>c",hypothesis="c",a=3,b=3,c=4,
      alpha=0.01,beta=0.1,delta=0.5,case="minimin")
 n
18
> size.anova(model="a>b>c",hypothesis="c",a=3,b=3,c=4,
      alpha=0.01,beta=0.1,delta=0.5,case="maximin")
  n
302
```

As we can see, the minimin sizes and the maximin sizes differ more for the nested factors.

5.5.3 Mixed Classifications

In experiments with three or more factors besides a cross-classification or a nested classification, we often find a further type of classification, the so-called mixed (partially

Table 5.21 Observations of a mixed classification type (A≻B) × C with $a = 2$, $b = 3$, $c = 2$, $n = 2$.

	A_1			A_2		
	B_{11}	B_{12}	B_{13}	B_{21}	B_{22}	B_{23}
C_1	288	355	329	310	303	299
	295	369	343	282	321	328
C_2	278	336	320	288	302	289
	272	342	315	287	297	284

nested) classifications. In the three-way ANOVA, two mixed classifications occur (Rasch 1971). We consider the case that the birth weight of piglets is observed in a three-way classification with factors boar, sow within boar and gender of the piglet. The latter is cross-classified with the nested factors boar and sow.

5.5.3.1 Cross-Classification between Two Factors where One of Them Is Sub-Ordinated to a Third Factor ((*B* ≺ *A*)xC)

If in a balanced experiment a factor B is sub-ordinated to a factor A and both are cross-classified with a factor C then the corresponding model equation is given by

$$\begin{cases} y_{ijkl} = \mu + a_i + b_{ij} + c_k + (a, c)_{ik} + (b, c)_{jk(i)} + e_{ijkl} \\ (i = 1, \dots, a;\ \ j = 1, \dots,\ b;\quad k = 1, \dots,\ c; l = 1, \dots, n), \end{cases} \tag{5.34}$$

In (5.34) μ is the general experimental mean, a_i is the effect of the ith level of factor A; b_{ij} is the effect of the jth level of factor B within the ith level of factor A; c_k is the effect of the kth level of factor C. Further $(a, c)_{ik}$ and $(b, c)_{jk(i)}$ are the corresponding interaction effects and $\boldsymbol{e}_{ijkl}$ are the random error terms.

As usual, the error terms $\boldsymbol{e}_{ijkl}$ are independently distributed with expectation zero and the same variance σ^2; for testing and confidence estimation normality is assumed in addition.

Model (5.34) is considered under the side conditions for all indices not occurring in the summation

$$\sum_{i=1}^{a} a_i = \sum_{j=1}^{b} b_{ij} = \sum_{k=1}^{c} c_k = \sum_{i=1}^{a} (a, c)_{ik} = \sum_{k=1}^{c} (a, c)_{ik} = \sum_{j=1}^{b} (b, c)_{jk} = \sum_{k=1}^{c} (b, c)_{jk(i)} = 0 \tag{5.35}$$

and

$$E(\boldsymbol{e}_{ijkl}) = 0, \quad E(\boldsymbol{e}_{ijkl}\boldsymbol{e}_{i'j'k'l'}) = \delta_{ii'}\delta_{jj'}\delta_{kk'}\delta_{ll'}\sigma^2, \quad \sigma^2 = \mathrm{var}(\boldsymbol{e}_{ijkl}) \tag{5.36}$$

(for all *i,j, k, l*).

The observations

$$y_{ijkl} \quad (i = 1, \dots, a; j = 1, \dots,\ b; k = 1, \dots,\ c; l = 1, \dots, n)$$

are allocated as shown in Table 5.21 (we restrict ourselves to the so-called balanced case where the number of B levels is equal for all A levels and the subclass numbers are equal).

Example 5.20 In Table 5.21 the arrangement of observations in a mixed classification of type (A≻B) × C is shown. How to analyse such data is demonstrated in Problem 5.14. As an example, we consider testing pig fattening for male and female (factor *C*) offspring of sows (factor *B*) nested in boars (factor *A*). The observed character is the number of fattening days an animal needed to grow up from 40 kg to 110 kg.

For the sum of squared deviations of the random variables

$$y_{ijkl} \quad (i = 1, \ldots, a; j = 1, \ldots, b; k = 1, \ldots, c; l = 1, \ldots, n)$$

from their arithmetic mean

$$\boldsymbol{SS}_T = \sum_{i,j,k,l} (\boldsymbol{y}_{ijkl} - \bar{\boldsymbol{y}}_{\ldots.})^2 = \sum_{i,j,k,l} \boldsymbol{y}_{ijkl}^2 - \frac{\boldsymbol{Y}_{\ldots.}^2}{N}, \quad (N = abcn)$$

we have

$$\boldsymbol{SS}_T = \boldsymbol{SS}_A + \boldsymbol{SS}_{B\,\mathrm{in}\,A} + \boldsymbol{SS}_C + \boldsymbol{SS}_{A\times C} + \boldsymbol{SS}_{B\times C\,\mathrm{in}\,A} + \boldsymbol{SS}_{\mathrm{res}},$$

where

$$\boldsymbol{SS}_A = \frac{1}{bcn} \sum_{i=1}^{a} \boldsymbol{Y}_{i\ldots}^2 - \frac{\boldsymbol{Y}_{\ldots}^2}{N}$$

is the ***SS*** between the levels of *A*,

$$\boldsymbol{SS}_{B\text{ in }A} = \frac{1}{cn} \sum_{i=1}^{a} \sum_{j=1}^{b} \boldsymbol{Y}_{ij\cdot\cdot}^2 - \frac{1}{bcn} \sum_{i=1}^{a} \boldsymbol{Y}_{i\ldots}^2$$

is the ***SS*** between the levels of *B* within the levels of *A*,

$$\boldsymbol{SS}_C = \frac{1}{abn} \sum_{k=1}^{c} \boldsymbol{Y}_{\cdot\cdot k\cdot}^2 - \frac{\boldsymbol{Y}_{\ldots.}^2}{N}$$

is the ***SS*** between the levels of *C*,

$$\boldsymbol{SS}_{A\times C} = \frac{1}{bn} \sum_{i=1}^{a} \sum_{k=1}^{c} \boldsymbol{Y}_{i\cdot k\cdot}^2 - \frac{1}{bcn} \sum_{i=1}^{a} \boldsymbol{Y}_{i\ldots}^2 - \frac{1}{abn} \sum_{k=1}^{c} \boldsymbol{Y}_{\cdot\cdot k\cdot}^2 + \frac{\boldsymbol{Y}_{\ldots.}^2}{N}$$

is the ***SS*** for the interactions *A* × *C*,

$$\boldsymbol{SS}_{B\times C\text{ in }A} = \frac{1}{n} \sum_{i=1}^{a} \sum_{j=1}^{b} \sum_{k=1}^{c} \boldsymbol{Y}_{ijk\cdot}^2 - \frac{1}{cn} \sum_{i=1}^{a} \sum_{j=1}^{b} \boldsymbol{Y}_{ij\cdot\cdot}^2 - \frac{1}{bn} \sum_{i=1}^{a} \sum_{k=1}^{c} \boldsymbol{Y}_{i\cdot k\cdot}^2 + \frac{1}{bcn} \sum_{i=1}^{a} \boldsymbol{Y}_{i\ldots}^2$$

is the ***SS*** for the interactions *B* × *C* within the levels of *A* and

$$\boldsymbol{SS}_{\mathrm{res}} = \sum_{i,j,k,l} \boldsymbol{y}_{ijk}^2 - \frac{1}{n} \sum_{i=1}^{a} \sum_{j=1}^{b} \sum_{k=1}^{c} \boldsymbol{Y}_{ijk\cdot}^2$$

the ***SS*** within the classes. The $N - 1$ degrees of freedom of SS_T corresponding to the components of SS_T can be split into six components. These components are given in Table 5.22. In the third column of Table 5.22 we find the ***MS*** obtained from the ***SS*** by division with the degrees of freedom.

Table 5.22 ANOVA table for a balanced three-way mixed classification $(B \prec A)xC$ $(n > 1)$.

Source of variation	SS
Between the levels of A	$SS_A = \frac{1}{bcn}\sum_{i=1}^{a} Y_{i\cdots}^2 - \frac{1}{N}Y_{\cdots\cdot}^2$
Between the levels of B within the levels of A	$SS_{B\text{ in}1} = \frac{1}{cn}\sum_{i=1}^{a}\sum_{j=1}^{b} Y_{ij\cdot\cdot}^2 - \frac{1}{bcn}\sum_{i=1}^{a} Y_{i\cdots}^2$
Between the levels of C	$SS_c = \frac{1}{abn}\sum_{k=1}^{a} Y_{\cdot\cdot k\cdot}^2 - \frac{1}{N}Y_{\cdots\cdot}^2$
Interaction $A \times C$	$SS_{A\times C} = \frac{1}{bn}\sum_{i=1}^{a}\sum_{k=1}^{a} Y_{i\times k\times}^2 - \frac{1}{bcn}\sum_{i=1}^{a} Y_{iL}^2 - \frac{1}{abn}\sum_{k=1}^{e} Y_{\times\times k\times}^2 + \frac{1}{N}Y_{\cdots\cdot}^2$
Interaction $B \times C$ within the levels of A	$SS_{B\times C\text{ in }A} = \frac{1}{n}\sum_{i=1}^{a}\sum_{j=1}^{b}\sum_{k=1}^{e} Y_{ijk\cdot}^2 - \frac{1}{cn}\sum_{i=1}^{a}\sum_{j=1}^{b} Y_{ij\cdot\cdot}^2 - \frac{1}{bn}\sum_{i=1}^{a}\sum_{k=1}^{e} Y_{i\cdot k\cdot}^2 - \frac{1}{bcn}\sum_{i=1}^{a} Y_{i\cdots}^2$
Residual	$SS_{\text{res}} = \sum_{i=1}^{a}\sum_{j=1}^{b}\sum_{k=1}^{c}\sum_{l=1}^{n} Y_{ijkl}^2 - \frac{1}{n}\sum_{i=1}^{a}\sum_{j=1}^{b}\sum_{k=1}^{c} Y_{ijk\cdot}^2$

df	MS	E(MS)
$a-1$	$MS_A = \frac{SS_A}{a-1}$	$\sigma^2 + \frac{bcn}{a-1}\sum_{i=1}^{a} a_i^2$
$a(b-1)$	$MS_{B\text{ in }A} = \frac{SS_{B\text{ in }A}}{a(b-1)}$	$\sigma^2 + \frac{cn}{a(b-1)}\sum_{i=1}^{a}\sum_{j=1}^{b} b_{ij}^2$
$c-1$	$MS_C = \frac{SS_c}{c-1}$	$\sigma^2 + \frac{abn}{c-1}\sum_{k=1}^{c} c_k^2$
$(a-1)(c-1)$	$MS_{A\times C} = \frac{SS_{A\times C}}{(a-1)(c-1)}$	$\sigma^2 + \frac{bn}{(a-1)(c-1)}\sum_{i=1}^{a}\sum_{k=1}^{c}(a,c)_{ik}^2$
$a(b-1)(c-1)$	$MS_{B\times C\text{ in }A} = \frac{SS_{B\times C\text{ in }A}}{a(b-1)(c-1)}$	$\sigma^2 + \frac{n}{a(b-1)(c-1)}\sum_{i=1}^{a}\sum_{j=1}^{b}\sum_{k=1}^{c}(b,c)_{jk(i)}^2$
$N-abc$	$MS_{\text{res}} = \frac{SS_{\text{res}}}{abc(n-1)}$	σ^2

The hypothesis $H_{0A}: a_i = 0$ can be tested by help of the statistic $F_A = \frac{MS_A}{MS_{\text{res}}}$ which, under H_{0A}, is F-distributed with $a-1$ and $N-abc$ degrees of freedom. In Table 5.22 we see that in our model, we test the hypothesis over all effects $(a_i, b_{ij}, \dots, (a,b,c)_{ijk})$ by using the ratios of the corresponding MS and MS_{res} as test statistic.

Problem 5.14 Calculate the empirical ANOVA table and perform all possible F-tests for Example 5.20.

Solution

Make in R the data-frame for the observations of Table 5.21 and use >`lm()`

Example

```
> y1 <- c(288,295,278,272,355,369,336,342,329,343,320,315)
> y2 <- c(310,282,288,287,303,321,302,297,299,328,289,284)
> a <- c(rep(1,12), rep(2,12))
> b1 <- c(rep(11,4), rep(12,4), rep(13,4))
> b2 <- c(rep(21,4), rep(22,4), rep(23,4))
> c1 <- c(rep(1,2),rep(2,2),rep(1,2),rep(2,2),rep(1,2),rep(2,2))
> c2 <- c(rep(1,2),rep(2,2),rep(1,2),rep(2,2),rep(1,2),rep(2,2))
> y <- c(y1,y2)
> b <- c(b1,b2)
> c <- c(c1,c2)
> T_5_20 <- data.frame(cbind(y,a,b,c))
> A <- factor(a)
> B <- factor(b)
> C <- factor(c)
> Anova <- lm ( y ~A + A/B + C + A*C + (A/B)*C )
> summary(Anova)
Call:
lm(formula = y ~ A + A/B + C + A * C + (A/B) * C)
Residuals:
    Min      1Q  Median      3Q     Max
-14.500  -3.125   0.000   3.125  14.500
Coefficients: (12 not defined because of singularities)
            Estimate Std. Error t value Pr(>|t|)
(Intercept)  291.500      7.272  40.088 3.75e-14 ***
A2            22.000     10.283   2.139 0.053653 .
C2           -16.500     10.283  -1.605 0.134580
A1:B12        70.500     10.283   6.856 1.76e-05 ***
A2:B12            NA         NA      NA       NA
A1:B13        44.500     10.283   4.327 0.000983 ***
A2:B13            NA         NA      NA       NA
A1:B21            NA         NA      NA       NA
A2:B21       -17.500     10.283  -1.702 0.114542
A1:B22            NA         NA      NA       NA
A2:B22        -1.500     10.283  -0.146 0.886450
A1:B23            NA         NA      NA       NA
A2:B23            NA         NA      NA       NA
A2:C2        -10.500     14.543  -0.722 0.484130
A1:B12:C2     -6.500     14.543  -0.447 0.662872
A2:B12:C2         NA         NA      NA       NA
A1:B13:C2     -2.000     14.543  -0.138 0.892898
A2:B13:C2         NA         NA      NA       NA
A1:B21:C2         NA         NA      NA       NA
A2:B21:C2     18.500     14.543   1.272 0.227441
A1:B22:C2         NA         NA      NA       NA
A2:B22:C2     14.500     14.543   0.997 0.338426
A1:B23:C2         NA         NA      NA       NA
```

```
A2:B23:C2          NA          NA       NA         NA
- - -
Signif. codes: 0 "***" 0.001 "**" 0.01 "*" 0.05 "." 0.1 "" 1
Residual standard error: 10.28 on 12 degrees of freedom
Multiple R-squared:  0.9193,    Adjusted R-squared:  0.8453
F-statistic: 12.42 on 11 and 12 DF,  p-value: 6.34e-05
> anova(Anova)
Analysis of Variance Table
Response: y
          Df Sum Sq Mean Sq F value    Pr(>F)
A          1 2646.0 2646.00 25.0213 0.0003082 ***
C          1 1872.7 1872.67 17.7084 0.0012142 **
A:B        4 9701.3 2425.33 22.9346 1.511e-05 ***
A:C        1   16.7   16.67  0.1576 0.6983418
A:B:C      4  211.7   52.92  0.5004 0.7362183
Residuals 12 1269.0  105.75
- - -
Signif. codes: 0 "***" 0.001 "**" 0.01 "*" 0.05 "." 0.1 "" 1
```

How to determine the sample sizes is shown in Problem 5.15.

Problem 5.15 Determine the minimin and maximin sample sizes for testing $H_{A0}: a_i = 0$ (for all i).

Unfortunately delta in the program below stands for $\tau = \delta/\sigma$.

Solution

Use the `OPDOE` package `>size.anova(model = "[a>b]xc", hypothesis = "a",a=, b=, c=, alpha=, beta = 2, delta = 5, case = "minimin")`

or `case = "maximin"`.

Example

Here and in the following problems we use $a = 5$, $b = 4$, $c = 2$, $\alpha = 0.05$, $\beta = 0.2$, $\delta = 0.5$.

```
> size.anova(model="(a> b)xc", hypothesis="a",a=5, b=4, c=2,
        alpha=0.05, beta=0.2, delta=0.5, case="minimin")
n
5
> size.anova(model="(a> b)xc", hypothesis="a",a=5, b=4, c=2,
       alpha=0.05, beta=0.2, delta=0.5, case="maximin")
 n
13
```

Problem 5.16 Determine the minimin and maximin sample sizes for testing $H_{C0}: c_k = 0$ (for all k).

Unfortunately delta in the program below stands for $\tau = \delta/\sigma$.

Solution

Use the `OPDOE` package `>size.anova(model = "[a>b]xc", hypothesis= "c",a=, b=, c=, alpha=, beta=, delta=, case = "minimin")`

or `case = "maximin"`.

Example

```
> size.anova(model="(a>b)xc", hypothesis="c",a=5, b=4, c=2,
        alpha=0.05, beta=0.2, delta=0.5, case="minimin")
n
4
> size.anova(model="(a>b)xc", hypothesis="c",a=5, b=4, c=2,
        alpha=0.05, beta=0.2, delta=0.5, case="maximin")
n
4
```

Problem 5.17 Determine the minimin and maximin sample sizes for testing $H_{A\times C0}:(ac)_{ik}=0$ (for all i and k).

Unfortunately delta in the program below stands for $\tau = \delta/\sigma$.

Solution

Use the OPDOE package

```
> size.anova(model="(a>b)xc", hypothesis="axc",a=, b=, c=,
    alpha=, beta=, delta=, case="minimin")
```

or `case = "maximin"`.

Example

```
> size.anova(model="(a>b)xc", hypothesis="axc",a=5, b=4,
        c=2, alpha=0.05, beta=0.2, delta=0.5, case="minimin")
n
8
> size.anova(model="(axb)>c", hypothesis="axb",a=5, b=4,
        c=2, alpha=0.05, beta=0.2, delta=0.5, case="maximin")
 n
70
```

5.5.3.2 Cross-Classification of Two Factors, in which a Third Factor is Nested ($C \prec (A \times B)$)

If two cross-classified factors ($A \times B$) are super-ordered to a third factor (C) we have another mixed classification. The model equation for the random observations in a balanced design is given by

$$\boldsymbol{y}_{ijkl} = \mu + a_i + b_j + c_{ijk} + (a,b)_{ij} + \boldsymbol{e}_{ijkl},$$
$$(i = 1, \dots, a; j = 1, \dots, b; k = 1, \dots c; l = 1, \dots, n) \tag{5.37}$$

This is again the situation of model I; the error terms $\boldsymbol{e}_{ijkl}$ again fulfil condition (5.36).

We assume that for all values of the indices not occurring in the summation, we have the side conditions

$$\sum_{i=1}^{a} a_i = \sum_{j=1}^{b} b_j = \sum_{k=1}^{c} c_{ijk} = \sum_{i=1}^{a} (a,b)_{ij} = \sum_{j=1}^{b} (a,b)_{ij} = 0. \tag{5.38}$$

The total sum of squared deviations can be split into components

$$SS_T = SS_A + SS_B + SS_{C \text{ in } AB} + SS_{A\times B} + SS_{\text{res}}$$

with

$$SS_A = \sum_{i=1}^{a} \frac{Y_{i\cdots}^2}{bcn} - \frac{Y_{\cdots\cdot}^2}{N},$$

the SS between the A levels,

$$SS_B = \sum_{j=1}^{a} \frac{Y_{\cdot j\cdot\cdot}^2}{acn} - \frac{Y_{\cdots\cdot}^2}{N},$$

the SS between the B levels,

$$SS_{C \text{ in } AB} = \sum_{i=1}^{a}\sum_{j=1}^{b}\sum_{k=1}^{c} \frac{Y_{ijk\cdot}^2}{n} - \sum_{i=1}^{a}\sum_{j=1}^{b} \frac{Y_{ij\cdot\cdot}^2}{cn},$$

the SS between the C levels within the $A \times B$ combinations,

$$SS_{A\times B} = \sum_{i=1}^{a}\sum_{j=1}^{b} \frac{Y_{ij\cdot\cdot}^2}{cn} - \sum_{i=1}^{a} \frac{Y_{i\cdots}^2}{bcn} - \sum_{j=1}^{b} \frac{Y_{\cdot j\cdot\cdot}^2}{acn} + \frac{Y_{\cdots\cdot}^2}{N},$$

the SS for the interactions between factor A and factor B, and

$$SS_{\text{res}} = \sum_{i=1}^{a}\sum_{j=1}^{b}\sum_{k=1}^{c}\sum_{l=1}^{n} y_{ijkl}^2 - \sum_{i=1}^{a}\sum_{j=1}^{b}\sum_{k=1}^{c} \frac{Y_{ijk\cdot}^2}{n}.$$

The expectations of the MS in this model are shown in Table 5.23 and the hypotheses

$$H_{A0}:\ a_i = 0,\quad H_{B0}:\ b_j = 0,\quad H_{C0}:\ c_{ijk} = 0, H_{AB0}:\ (a,b)_{ij} = 0,$$

where the zero values are assumed to hold for all indices used in the hypotheses, can be tested by using the corresponding F-statistic as the ratios of MS_A, MS_B, MS_C, and $MS_{A\times B}$, respectively, (as numerator) and MS_{res} (as denominator).

Example 5.21 As an example, consider the mast performance of offspring of sires of beef cattle (factor C) of different genotypes (factor A) in several years (factor B). If each sire occurs in just one (genotype × year) combination then the structure of Table 5.24 is given.

Problem 5.18 Calculate the ANOVA table and the F-tests for Example 5.21.

Solution

Make a data-frame of the data of Table 5.24 and use in R > `lm()`.

Example

```
> y1 <- c(58,60,65,62,55,57,68,65)
> y2 <- c(62,61,71,73,63,62,68,72)
> y3 <- c(58,61,59,70,59,63,65,67)
> a1 <- c(rep(1,2),rep(2,2),rep(1,2),rep(2,2),rep(1,2),rep(2,2))
```

Table 5.23 ANOVA table and expectations of the MS for model I of a balanced three-way ANOVA *A* with *B* cross-classified, *C* nested in the *A* x *B* combinations.

Source of variation	SS
Between *A* levels	$SS_A = \sum_{i=1}^{a} \frac{Y_{i\cdots}^2}{bcn} - \frac{Y_{\cdots\cdot}^2}{N}$
Between *B* levels	$SS_B = \sum_{j=1}^{a} \frac{Y_{\cdot j\cdot\cdot}^2}{acn} - \frac{Y_{\cdots\cdot}^2}{N}$
Between *C* levels in $A \times B$ combinations	$SS_{C\text{ in }AB} = \sum_{i=1}^{a}\sum_{j=1}^{b}\sum_{k=1}^{c} \frac{Y_{ijk\cdot}^2}{n} - \sum_{i=1}^{a}\sum_{j=1}^{b} \frac{Y_{ij\cdot\cdot}^2}{cn}$
Interaction $A \times B$	$SS_{A\times B} = \sum_{i=1}^{a}\sum_{j=1}^{b} \frac{Y_{ij\cdot\cdot}^2}{cn} - \sum_{i=1}^{a} \frac{Y_{i\cdots}^2}{bcn} - \sum_{j=1}^{b} \frac{Y_{\cdot j\cdot\cdot}^2}{acn} + \frac{Y_{\cdots\cdot}^2}{N}$
Residual	$SS_{\text{res}} = \sum_{i=1}^{a}\sum_{j=1}^{b}\sum_{k=1}^{c}\sum_{l=1}^{n} y_{ijkl}^2 - \sum_{i=1}^{a}\sum_{j=1}^{b}\sum_{k=1}^{c} \frac{Y_{ijk\cdot}^2}{n}$

df	*MS*	*E*(*MS*) under (5.38)
$a-1$	$MS_A = \frac{SS_A}{a-1}$	$\sigma^2 + \frac{bcn}{a-1}\sum_{i=1}^{a} a_i^2$
$b-1$	$MS_B = \frac{SS_B}{b-1}$	$\sigma^2 + \frac{acn}{b-1}\sum_{j=1}^{b} b_j^2$
$ab(c-1)$	$MS_{C\text{ in }AB} = \frac{SS_{C\text{ in }AB}}{ab(c-1)}$	$\sigma^2 + \frac{n}{ab(c-1)}\sum_{i=1}^{a}\sum_{j=1}^{b}\sum_{k=1}^{c} c_{ijk}^2$
$(a-1)(b-1)$	$MS_{A\times B} = \frac{SS_{A\times B}}{(a-1)(b-1)}$	$\sigma^2 + \frac{cn}{(a-1)(b-1)}\sum_{i=1}^{a}\sum_{j=1}^{b} (a,b)_{ij}^2$
$N-abc$	$MS_{res} = \frac{SS_{res}}{N-abc}$	σ^2

```
> a2 <- c(rep(1,2),rep(2,2),rep(1,2),rep(2,2),rep(1,2),rep(2,2))
> b <- c(rep(1,8), rep(2,8),rep(3,8))
> c1 <-c(rep(111,4),rep(112,4))
> c2 <-c(rep(121,4),rep(122,4))
> c3 <-c(rep(131,4),rep(132,4))
> y <- c(y1,y2,y3)
```

Table 5.24 Observations of a mixed classification type (A×B)>C with $a = 2$, $b = 3$, $c = 2$, $n = 2$.

	B_1		B_2		B_3	
	C_{111}	C_{112}	C_{121}	C_{122}	C_{131}	C_{132}
A_1	58	55	62	63	58	59
	60	57	61	62	61	63
A_2	65	68	71	68	59	65
	62	65	73	72	70	67

```
> a <- c(a1,a2)
> c <- c(c1,c2,c3)
> A <- factor(a)
> B <- factor(b)
> C <- factor(c)
> Anova <- lm ( y ~A + B + (A*B)/C + A*B )
> anova(Anova)
Analysis of Variance Table
Response: y
          Df  Sum Sq Mean Sq F value    Pr(>F)
A          1 308.167 308.167 37.3535 5.242e-05 ***
B          2 117.000  58.500  7.0909   0.00927 **
A:B        2  16.333   8.167  0.9899   0.40002
A:B:C      6  27.500   4.583  0.5556   0.75746
Residuals 12  99.000   8.250
- - -
Signif. codes: 0 "***" 0.001 "**" 0.01 "*" 0.05 "." 0.1 "" 1
```

Now let us determine sample sizes.

Problem 5.19 Determine the minimin and maximin sample sizes for testing $H_{A0}: a_i = 0$ (for all i).

Unfortunately delta in the program below stands for $\tau = \delta / \sigma$.

Solution

Use the `OPDOE` package command `> size.anova(model = "(axb) > c", hypothesis = "a",a=, b=, c=, alpha=, beta=, delta=, case = "minimin")`

or `case = "maximin"`.

Example

```
> size.anova(model="(axb)>c", hypothesis="a",a=6, b=5, c=4,
      alpha=0.05, beta=0.1, delta=0.5, case="minimin")
n
3
> size.anova(model="(axb)>c", hypothesis="a",a=6, b=5, c=4,
       alpha=0.05, beta=0.1, delta=0.5, case="maximin")
n
7
```

Problem 5.20 Determine the minimin and maximin sample sizes for testing $H_{A \times C0}: (ac)_{ik} = 0$ (for all i and k).

Unfortunately delta in the program below stands for $\tau = \delta / \sigma$.

Solution

Use the `OPDOE` package with command

```
> size.anova(model="(axb)>c", hypothesis="axb",a=, b=, c=,
      alpha=, beta=, delta=, case="minimin")
```

or `case = "maximin"`.

Example

```
> size.anova(model="(a>b)xc", hypothesis="axc",a=5, b=4,
                c=2,
      alpha=0.05, beta=0.2, delta=0.5, case="minimin")
n
5
> size.anova(model="(axb)>c", hypothesis="axb",a=5, b=4,
                c=2,
      alpha=0.05, beta=0.2, delta=0.5, case="maximin")
 n
70
```

References

Fisher, R.A. and Mackenzie, W.A. (1923). Studies in crop variation. II. The manurial response of different potato varieties. *Journal of Agricultural Sciences* 13: 311–320.

Hartung, J., Elpelt, B., and Voet, B. (1997). *Modellkatalog Varianzanalyse.* München: Oldenburg Verlag.

Kuehl, R.O. (1994). *Statistical Principles of Research Design and Analysis.* Belmont, California: Duxbury Press.

Lenth, R.V. (1986). Computing non-central Beta probabilities. *Appl. Statistics* 36: 241–243.

Rasch, D. (1971). Mixed classification the three-way analysis of variance. *Biom. Z.* 13: 1–20.

Rasch, D. and Schott, D. (2018). *Mathematical Statistics.* Oxford: Wiley.

Rasch, D., Wang, M., and Herrendörfer, G. (1997). Determination of the size of an experiment for the F-test in the analysis of variance. Model I. In: *Advances in Statistical Software 6. The 9th Conference on the Scientific Use of Statistical Software.* Heidelberg: Springer.

Rasch, D., Herrendörfer, G., Bock, J., Victor, N., and Guiard, V. Hrsg. (2008). *Verfahrensbibliothek Versuchsplanung und - auswertung,* 2. verbesserte Auflage in einem Band mit CD. R. Oldenbourg Verlag München Wien.

Rasch, D., Pilz, J., Verdooren, R., and Gebhardt, A. (2011). *Optimal Experimental Design with R.* Boca Raton: Chapman and Hall.

Scheffé, H. (1959). *The Analysis of Variance.* New York, Hoboken: Wiley.

6

Analysis of Variance – Models with Random Effects

6.1 Introduction

Whereas in Chapter 5 the general structure of analysis of variance models are introduced and investigated for the case that all effects are fixed real numbers (ANOVA model I), we now consider the same models but assume that all factor levels have randomly been drawn from a universe of factor levels. We call this the ANOVA model II. Therefore, effects (except the overall mean μ) are random variables and not parameters, which have to be estimated. Instead of estimating the random effects, we estimate and test the variances of these effects – called variance components. The terms main effect and the interaction effect are defined as in Chapter 5 (but these effects are now random variables).

6.2 One-Way Classification

We characterise methods of variance component estimation for the simplest case, the one-way ANOVA, and demonstrate most of them by some data set. For this, we assume that a sample of a levels of a random factor $\boldsymbol{A}$ has been drawn from the universe of factor levels, which is assumed to be large. In order to be not too abstract let us assume that the levels are sires. From the ith sire a random sample of n_i daughters is drawn and their milk yield y_{ij} recorded. This case is called balanced if for each of the sires the same number n of daughters has been selected. If the n_i are not all equal to n we called it an unbalanced design.

We consider the model

$$\boldsymbol{y}_{ij} = \mu + \boldsymbol{a}_i + \boldsymbol{e}_{ij} \qquad i = 1, \dots, a; j = 1, \dots, n_{\mathrm{i}} \tag{6.1}$$

The $\boldsymbol{a}_i$ are the main effects of the levels A_i. They are random variables. The $\boldsymbol{e}_{ij}$ are the errors and also random. The constant μ is the overall mean. Model (6.1) is completed by the assumptions

$$E(\boldsymbol{a}_i) = 0, var(\boldsymbol{a}_i) = \sigma_a^2, E(\boldsymbol{e}_{ij}) = 0, var(\boldsymbol{e}_{ijj}) = \sigma^2 \quad \text{for all } i \text{ and } j; \tag{6.2}$$

all random components on the right-hand side of (6.1) are independent.

The variances σ_a^2 and σ^2 are called variance components. The total number of observations is always denoted by N, in the balanced case we have $N = an$. From (6.2) it follows

$$var(\boldsymbol{y}_{ij}) = \sigma_a^2 + \sigma^2. \tag{6.3}$$

Applied Statistics: Theory and Problem Solutions with R, First Edition.
Dieter Rasch, Rob Verdooren, and Jürgen Pilz.

Table 6.1 Expected mean squares of the one-way ANOVA model II.

Source of variation	df	*MS*	*E*(*MS*)Unbalanced case	*E*(*MS*)Balanced case
Factor *A*	$a-1$	$MS_A = \frac{SS_A}{a-1}$	$\sigma^2 + \frac{1}{a-1}\left[N - \frac{\sum_{i=1}^{a} n_i^2}{N}\right]\sigma_a^2$	$\sigma^2 + n\sigma_a^2$
Residual	$a(n-1)$	$MS_{\text{res}} = \frac{SS_{\text{res}}}{N-1}$	σ^2	σ^2

Let us assume that all the random variables in (6.1) are normally distributed even if this is not needed for all the estimation methods. Then it follows that $\boldsymbol{a}_i$ and $\boldsymbol{e}_{ij}$ are independent of each other $N(0;\, \sigma_a^{\,2})$- and $N(0;\, \sigma^2)$-distributed respectively. The $\boldsymbol{y}_{ij}$ are not independent of each other $N(\mu;\, \sigma^2 + \sigma_a^{\,2})$-distributed. The dependence exists between variables within the same factor level (class) because

$$cov(\boldsymbol{y}_{ij}, \boldsymbol{y}_{ik}) = cov(\mu + \boldsymbol{a}_i + \boldsymbol{e}_{ij}, \mu + \boldsymbol{a}_i + \boldsymbol{e}_{ik}) = cov(\boldsymbol{a}_i, \boldsymbol{a}_i) = var(\boldsymbol{a}_i) = \sigma_a^2 \text{ if } j \neq k.$$

We call $cov(\boldsymbol{y}_{ij}, \boldsymbol{y}_{ik}),\ i = 1, \dots, a;\ j = 1, \dots, n$ the covariance within classes.

A standardised measure of this dependence is the intra-class correlation coefficient

$$\rho_{\text{IC}} = \frac{\sigma_a^{\,2}}{\sigma_a^{\,2} + \sigma^2} \tag{6.4}$$

The ANOVA table is that of Table 5.2 but the expected mean squares (*MS*) differ from those in Table 5.2 and are given in Table 6.1.

6.2.1 Estimation of the Variance Components

For the one-way classification, we describe several methods of estimation and compare them with each other. The analysis of variance method is the simplest one and stems from the originator of the analysis of variance, R. A. Fisher. In Henderson's fundamental paper from 1953, it was mentioned as method I. An estimator is defined as a random mapping from the sample space into the parameter space. The parameter space of variances and variance components is the positive real line. When realisations of such a mapping can be outside of the parameter space, we should not call them estimators. Nevertheless, this term is in use in the estimation of variance components following the citation "A good man with his groping intuitions! Still knows the path that is true and fit." Or its German original in Goethe's Faust, Prolog. (Der Herr): "Ein guter Mensch, in seinem dunklen Drange, ist sich des rechten Weges wohl bewußt."

We use below the notation $\boldsymbol{Y} = (\boldsymbol{y}_{11}, \dots, \boldsymbol{y}_{a,n_a})$ for the vector of all observations. The variance of $\boldsymbol{Y}$ is a matrix

$$V = \text{var}(\boldsymbol{Y}) = \begin{pmatrix} \sigma_a^2 + \sigma^2 & \cdots & \sigma_a^2 \\ \vdots & \ddots & \vdots \\ \sigma_a^2 & \cdots & \sigma_a^2 + \sigma^2 \end{pmatrix} \tag{6.5}$$

6.2.1.1 ANOVA Method

The ANOVA method does not need the normal assumption; it follows in all classifications the same algorithm.

Algorithm for estimating variance components by the analysis of variance method:

1. Obtain from the column *E*(***MS***) in any of the ANOVA tables (here and in the further sections) the formula for each variance component
2. Replace *E*(***MS***) by ***MS*** and σ^2 in the equations by the corresponding $\boldsymbol{s}^2$ to obtain the estimators or by s^2 to receive the estimates of the corresponding variance components.

Because in the equations differences of *MS* occur it may happen that, we receive negative estimators and estimates of positive parameters of the variance components. However, the ANOVA method for estimating the variance components have for normally distributed variables ***Y*** a positive probability that negative estimators, see Verdooren (1982). Nevertheless, the estimators are unbiased and for normally distributed variables ***Y*** in the balanced case they have minimum variance (best quadratic unbiased estimators), but negative estimates are impermissible, see Verdooren (1980).

We now demonstrate that algorithm in our one-way case.

1. $\sigma^2 = E(\boldsymbol{MS}_{\text{res}}),\ \sigma_a^2 = \dfrac{E(\boldsymbol{MS}_A) - E(\boldsymbol{MS}_{\text{res}})}{\frac{1}{a-1}\left[N - \dfrac{\sum_{i=1}^a n_i^2}{N}\right]},$

2. $\boldsymbol{s}_a^2 = \dfrac{(a-1)(\boldsymbol{MS}_A - \boldsymbol{MS}_{\text{res}})}{\left[N - \dfrac{\sum_{i=1}^a n_i^2}{N}\right]},\ \boldsymbol{s}^2 = \boldsymbol{MS}_{\text{res}};\ s_a^2 = \dfrac{(a-1)(MS_A - MS_{\text{res}})}{\left[N - \dfrac{\sum_{i=1}^a n_i^2}{N}\right]},\ s^2 = MS_{\text{res}}.$

The estimate s_a^2 is negative if $MS_{\text{res}} > MS_A$.

Problem 6.1 Estimate the variance components using the ANOVA method.

Solution

In the base package of R the ANOVA table is given with the command `aov()` but the expected mean squares, *E*(***MS***), are not given. Using the *E(**MS**)* to estimate the variance components directly we must use the R package `VCA` (variance component analysis).

Example

We use the milk fat performances (in kilograms) of the daughters of 10 sires randomly selected from a cattle population (including the bold entries) shown in Table 6.2 (with bold entries only used when equal sub-class numbers are necessary). First, we give the balanced example with 12 observations per sire (or bull).

```
> #  usual ANOVA Table:
> b1 <- c(120,155,131,130,140,140,142,146,130,152,115,146)
> b2 <- c(152,144,147,103,131,102,102,150,159,132,102,160)
> b3 <- c(130,138,123,135,138,152,159,128,137,144,154,131)
> b4 <- c(149,107,143,133,139,102,103,110,103,138,124,117)
> b5 <- c(110,142,124,109,154,135,118,116,150,148,138,115)
> b6 <- c(157,107,146,133,104,119,107,138,147,152,124,142)
> b7 <- c(119,158,140,108,138,154,156,145,150,124,100,140)
> b8 <- c(150,135,150,125,104,150,140,103,132,128,122,154)
> b9 <- c(144,112,123,121,132,144,132,129,103,140,106,152)
```

Table 6.2 Milk fat performances y_{ij} of daughters of ten sires.

	Sire (bull)									
	B_1	B_2	B_3	B_4	B_5	B_6	B_7	B_8	B_9	B_{10}
	120	152	130	149	110	157	119	150	144	159
	155	144	138	107	142	107	158	135	112	105
	131	147	123	143	124	146	140	150	123	103
	130	103	135	133	109	133	108	125	121	105
	140	131	138	139	154	104	138	104	132	144
	140	102	152	102	135	119	154	150	144	129
	142	102	159	103	118	107	156	140	132	119
	146	150	128	110	116	138	145	103	129	100
	130	159	137	103	150	147	150	132	103	115
	152	132	144	138	148	152	124	128	140	146
	115	102	154	**124**	138	124	100	122	106	108
	146	160	**131**	**117**	115	142	**140**	154	152	119
n_i	12	12	11 12	10 12	12	12	11 12	12	12	12

```
> b10 <- c(159,105,103,105,144,129,119,100,115,146,108,119)
> bull <- rep(1:10, each = 12)
> y <- c(b1,b2,b3,b4,b5,b6,b7,b8,b9,b10)
> Table_6_2b <- data.frame(bull, y)
> # the b is used for balanced
> BULL <- factor(bull)
> Anova1b <- aov(y ~ BULL)
> Anova1b
Call:
   aov(formula = y ~ BULL)
Terms:
 7                       BULL Residuals
Sum of Squares    3814.63   33547.33
Deg. of Freedom         9        110
Residual standard error: 17.46356
Estimated effects may be unbalanced
```

Now ANOVA Table using the *E(MS)* column to estimate the variance components.
Use for this the package VCA.

```
> library(VCA)
> Anova2b <- anovaVCA(y ~ bull,Table_6_2b, VarVC.method = c("scm"))
> Anova2b
Result Variance Component Analysis:
-----------------------------------

  Name  DF        SS           MS         VC         %Total     SD
1 total 116.76977                         314.88179  100        17.744909
2 bull  9         3814.633333  423.848148 9.906033   3.145953   3.147385
3 error 110       33547.333333 304.975758 304.975758 96.854047  17.463555
```

Table 6.3 ANOVA table of model II with *E*(**MS**) of the example of Problem 6.1.

Source of variation	*df*	*SS*	*MS*	*E(MS)*
Bulls	9	3 814.63	423.85	$\sigma^2 + 12\sigma^2{}_{\text{Bulls}}$
Residual	110	33 547.33	304.98	σ^2

```
  CV[%]
1 13.547456
2 2.40289
3 13.332654
Mean: 130.9833 (N = 120)
Experimental Design: balanced  |  Method: ANOVA
```

#"scm" = Searle, S.R, Casella, G., McCulloch, C.E. (1992), Variance Components, Wiley New York

Remark

In this balanced one-way classification the ANOVA table of model II with *E*(***MS***) is given in Table 6.3.

Hence the VC = variance components estimates are $s^2 = 304.98$ and $s^2_{\text{Bulls}} = (423.85 - 304.98)/12 = 9.906$.

Now we use the unbalanced example, hence the bold data are deleted in Table 6.2.

```
> b3[12] <- NA
> b3
 [1] 130 138 123 135 138 152 159 128 137 144 154  NA
> b4[11] <- NA
> b4[12] <- NA
> b4
 [1] 149 107 143 133 139 102 103 110 103 138  NA  NA
> b7[12] <- NA
> b7
 [1] 119 158 140 108 138 154 156 145 150 124 100  NA
>  y <- c(b1,b2,b3,b4,b5,b6,b7,b8,b9,b10)
>  bull <- rep(1:10, each = 12)
> Table_6_2u <- data.frame(bull, y)
> # the u is used for unbalanced
> Anova1u <- aov(y ~ BULL)
> Anova1u
Call:
   aov(formula = y ~ BULL)
Terms:
                      BULL Residuals
Sum of Squares     3609.11  33426.03
Deg. of Freedom          9       106
Residual standard error: 17.75781
```

```
Estimated effects may be unbalanced
4 observations deleted due to missingness
```

Now the ANOVA table using the *E(MS)* column to estimate the variance components.

```
> Anova2u <- anovaVCA(y ~ bull,Table_6_2u,VarVC.method = c("scm"))
There are 4 missing values for the response variable (obs: 36, 47, 48, 84)!
> Anova2u
Result Variance Component Analysis:
-----------------------------------

  Name  DF         SS           MS         VC         %Total    SD
1 total 113.684098                         322.728113 100       17.964635
2 bull  9          3609.106113  401.01179  7.38819    2.289292  2.718123
3 error 106        33426.031818 315.339923 315.339923 97.710708 17.757813
  CV[%]
1 13.704443
2 2.073538
3 13.546668
Mean: 131.0862 (N = 116, 4 observations removed due to missing data)
Experimental Design: unbalanced  |  Method: ANOVA
```

Remark

In this unbalanced one-way classification ANOVA table model II with *E(MS)* is given in Table 6.4.

The coefficient of σ^2_{Bulls} in *E(MS)* is $(1/(10-1)^*[116-(7^*12^2+2^*11^2+10^2)/116]=$ 11.5958.

Hence VC = variance components estimates are $s^2 = 315.340$ and $s^2_{\text{Bulls}} = (401.012 - 315.340)/11.5958 = 7.38819$.

6.2.1.2 Maximum Likelihood Method

Now we use the assumption that $\boldsymbol{y}_{ij}$ in (6.1) is normally distributed with variance from (6.3). Further we assume equal sub-class numbers, i.e. $N = an$ because otherwise the description becomes too complicated. Those interested in the general case may read Sarhai and Ojeda (2004, 2005). Harville (1977) gives a good background of the maximum likelihood approaches.

The density function of the vector of all observations $\boldsymbol{Y}=(\boldsymbol{y}_{11},\ldots,\boldsymbol{y}_{a,n_a})$ is (with $\oplus$ *for direct sum*)

$$f(Y|\mu,\sigma^2,\sigma_a^2)=\frac{1}{(2\pi)^{\frac{N}{2}}|V|^{\frac{1}{2}}}e^{\left[\frac{1}{2}(Y-\mu 1_N)^T V^{-1}(Y-\mu 1_N)\right]}$$
$$=\frac{e^{\frac{1}{2\sigma^2}(Y-\mu 1_N)^T(Y-\mu 1_N)+\frac{\sigma_a^2}{2\sigma^2(\sigma^2+n\sigma_a^2)}(Y-\mu 1_N)^T\bigoplus_{i=1}^{a}1_{n,n}(Y-\mu 1_N)}}{(2\pi)^{\frac{N}{2}}(\sigma^2)^{\frac{a}{2}(n-1)}(\sigma^2+n\sigma_a^2)^{\frac{n}{2}}} \quad (6.6)$$

Table 6.4 ANOVA table of model II of the unbalanced one-way classification with EMS.

Source of variation	*df*	*SS*	*MS*	*E(MS)*
Bulls	9	3 609.11	401.012	$\sigma^2 + 11.5958\,\sigma^2_{\text{Bulls}}$
Residual	106	33 426.03	315.340	σ^2

and this becomes

$$f(Y \mid \mu, \sigma^2, \sigma_a^2) = \frac{\exp\left[-\frac{1}{2}\left(\frac{SS_{\text{res}}}{\sigma^2} + \frac{SS_{\text{A}}}{\sigma^2+n\sigma_a^2} + \frac{an(\bar{y}_{..}-\mu)^2}{\sigma^2+n\sigma_a^2}\right)\right]}{(2\pi)^{\frac{N}{2}}(\sigma^2)^{\frac{a}{2}(n-1)}(\sigma^2+n\sigma_a^2)^{\frac{a}{2}}} = L \tag{6.7}$$

with SS_{res} and SS_A from Table 5.2.

We obtain the maximum-likelihood – estimates $\widetilde{\sigma}^2$, $\widetilde{\sigma}_{a,}^2$ and $\widetilde{\mu}$ by zeroing the derivatives of ln L with respect to the three unknown parameters and obtain

$$0 = \frac{-an}{\widetilde{\sigma}^2 + n\widetilde{\sigma}_a^2}(\bar{y}_{..} - \widetilde{\mu})$$

$$0 = -\frac{a(n-1)}{2\widetilde{\sigma}^2} - \frac{a}{2(\widetilde{\sigma}^2 + n\widetilde{\sigma}_a^2)} + \frac{SS_{\text{res}}}{2\widetilde{\sigma}^4} + \frac{SS_A}{2(\widetilde{\sigma}^2 + n\widetilde{\sigma}_a^2)^2}$$

$$0 = -\frac{na}{2(\widetilde{\sigma}^2 + n\widetilde{\sigma}_a^2)} + \frac{nSS_A}{2(\widetilde{\sigma}^2 + n\widetilde{\sigma}_a^2)^2}.$$

From the first equation (after transition to random variables) it follows for the estimators

$$\widetilde{\mu} = \bar{y}_{..}\ \hat{\mu} = \bar{y}_{..}$$

and from the two other equations

$$a(\widetilde{\boldsymbol{\sigma}}^2 + n\widetilde{\boldsymbol{\sigma}}_a^2) = \boldsymbol{SS}_A a(\widetilde{\boldsymbol{\sigma}}^2 + n\widetilde{\boldsymbol{\sigma}}_a^2) = \boldsymbol{SS}_A$$

and

$$\widetilde{\boldsymbol{\sigma}}^2 = \frac{\boldsymbol{SS}_{\text{res}}}{a(n-1)} = \boldsymbol{s}^2 = \boldsymbol{MS}_{\text{res}} \tag{6.8}$$

hence

$$\widetilde{\boldsymbol{\sigma}}_a^2 = \frac{1}{n}\left[\frac{\boldsymbol{SS}_{\text{A}}}{a} - \boldsymbol{MS}_{\text{res}}\right] = \frac{1}{n}\left[\left(1-\frac{1}{a}\right)\boldsymbol{MS}_{\text{A}} - \boldsymbol{MS}_{\text{res}}\right]. \tag{6.9}$$

Because the matrix of the second derivatives is negative definite, we really reach maxima.

Note that for a random sample of size n from a normal variable with distribution $N(\mu, \sigma^2)$ the maximum likelihood ML estimate of μ is the sample mean $\bar{y}$ and the ML estimate for σ^2 is $[(n-1)/n]s^2$, where s^2 is the sample variance.

Problem 6.2 Estimate the variance components using the ML method.

Solution

In the base package of R the ML estimates of the variance components are not possible. We use therefore the R package `lme4`.

Example

We use the data of Table 6.2 in Problem 6.1 (including the bold entries). We use from there the data frame `Table_6_2b`.

For this balanced one-way classification, the ML estimates are as follows.

```
> # We use the package lme4.
> library(lme4)
```

```
> MLbalanced <- lmer (y ~1+(1|bull),data=Table_6_2b,REML = FALSE)
> summary(MLbalanced)
Linear mixed model fit by maximum likelihood  ['lmerMod']
Formula: y ~ 1 + (1 | bull)
   Data: Table_6_2b
     AIC      BIC   logLik deviance df.resid
  1035.2   1043.6   -514.6   1029.2      117
Scaled residuals:
    Min      1Q  Median      3Q     Max
-1.8318 -0.7580  0.1010  0.8049  1.7189
Random effects:
 Groups   Name        Variance Std.Dev.
 bull     (Intercept)   6.374   2.525
 Residual             304.976  17.464
Number of obs: 120, groups:  bull, 10
Fixed effects:
            Estimate Std. Error t value
(Intercept)  130.983      1.783   73.47
```

We use now the unbalanced data of Table 6.2 by deleting the bold data. We use from there the data frame `Table_6_2u`.

```
> library(lme4)
> MLunbalanced <- lmer (y ~1+(1|bull),data=Table_6_2u,REML = FALSE)
> summary(MLunbalanced)
Linear mixed model fit by maximum likelihood  ['lmerMod']
Formula: y ~ 1 + (1 | bull)
   Data: Table_6_2u
     AIC      BIC   logLik deviance df.resid
  1004.0   1012.3   -499.0    998.0      113
Scaled residuals:
    Min      1Q  Median      3Q     Max
-1.7746 -0.8416  0.1290  0.8101  1.6326
Random effects:
 Groups   Name        Variance Std.Dev.
 bull     (Intercept)   3.248   1.802
 Residual             316.029  17.777
Number of obs: 116, groups:  bull, 10
Fixed effects:
            Estimate Std. Error t value
(Intercept)  131.084      1.746   75.06
```

6.2.1.3 *REML* – Estimation

Anderson and Bancroft (1952, p. 320) introduced a restricted maximum likelihood (REML) method. There are extensions by Thompson (1962) and a generalisation by Patterson and Thompson (1971). This method uses a translation invariant restricted likelihood function depending on the variance components to be estimated only and not on the fixed effect μ. This restricted likelihood function is a function of the sufficient statistics for the variance components. The latter is then derived with respect to the variance components under the restriction that the solutions are non-negative.

The method REML can be found in Searle et al. (1992). The method means that the likelihood function of TY is maximised in place of the likelihood function of $\boldsymbol{Y}$. T is a $(N - a - 1) \times N$ matrix, whose rows are $N - a - 1$ linear independent rows of $I_N - X(X^TX)^-X^T$ with X from (6.1) in Rasch and Schott (2018).

The (natural) logarithm of the likelihood function of TY is with V in (6.5)

$$\ln L = \frac{1}{2}(N - a - 1)\ln(2\pi) - \frac{1}{2}(N - a - 1)\ln\sigma^2 - \frac{1}{2}\ln\left(\left|\frac{\sigma_a^2}{\sigma^2}TVT^T\right|\right)$$
$$- \frac{1}{2\sigma^2 y^T T^T \frac{\sigma_a^2}{\sigma^2} TVT^T Ty}\frac{\sigma_a^2}{\sigma^2}TVT^T.$$

Now we differentiate this function with respect to σ^2 and $\frac{\sigma_a^2}{\sigma^2}$ and zeroing these derivatives. The arising equation we solve iteratively and gain the estimators.

$$\boldsymbol{s}^2 = \min\left[\boldsymbol{MS}_{\text{res}}, \frac{\boldsymbol{SS}_{\text{res}} + \boldsymbol{SS}_A}{an - 1}\right], \tag{6.10}$$

$$\boldsymbol{s}_a^2 = \frac{1}{n}\max\left\{[\boldsymbol{MS}_A - \boldsymbol{MS}_{\text{res}}]; 0\right\}. \tag{6.11}$$

Because the matrix of second derivatives is negative definite, we find the maxima.

Note that for a random sample of size n from a normal variable with distribution $N(\mu, \sigma^2)$ the REML estimate of μ is the sample mean $\bar{y}$ and the REML estimate for σ^2 is s^2, where s^2 is the sample variance.

This REML method is increasingly in use in applications – especially in animal breeding; even for not normally distributed variables. Another method MINQUE (minimum norm quadratic unbiased estimator) to estimate variance components, which is not based on normally variables, needs an idea of the starting value of the variance components. Using an iterative procedure by inserting the outcomes of the previous MINQUE procedure in the next MINQUE procedure gives the same result as the REML procedure! Hence, even for not normally distributed model effects we can use the REML.

Furthermore, for balanced designs with ANOVA estimators when the estimates are not negative, the REML estimates give the same answers. If the ANOVA estimate is negative because MS_A is smaller than MS_{res}, this is an inadmissible estimate. However, the REML estimate gives in such cases the correct answer zero. Hence, in practice the REML estimators are preferred for estimating the variance components. The REML method should always be used if the data are not balanced.

Besides methods based on the frequency approach generally used in this book there are Bayesian methods of variance component estimation. In this approach, we assume that the parameters of the distribution, and by this especially the variance components, are random variables with some prior knowledge about their distribution. This prior knowledge is sometimes given by an a priori distribution, sometimes by data from an earlier experiment. This prior distribution is combined with the likelihood of the sample resulting in a posterior distribution. The estimator is used in such a way that it minimises the so-called Bayes risk. See Tiao and Tan (1965), Federer (1968), Klotz et al. (1969), Gelman et al. (1995).

Problem 6.3 Estimate the variance components using the REML method.

Solution
We use the R package `lme4`.

Example
For this balanced one-way classification we use for REML estimates the data frame Table 6.2 (including the bold entries).

```
> library(lme4)
> REMLbalanced <- lmer (y ~1+(1|bull),data=Table_6_2b,REML = TRUE)
> summary(REMLbalanced)
Linear mixed model fit by REML ['lmerMod']
Formula: y ~ 1 + (1 | bull)
   Data: Table_6_2b
REML criterion at convergence: 1026.2
Scaled residuals:
    Min      1Q  Median      3Q     Max
-1.8547 -0.7579  0.1034  0.8338  1.7646
Random effects:
 Groups   Name        Variance Std.Dev.
 bull     (Intercept)   9.906   3.147
 Residual             304.976  17.464
Number of obs: 120, groups:  bull, 10
Fixed effects:
            Estimate Std. Error t value
(Intercept)  130.983      1.879   69.69
```

The estimates of the variance component using the REML method are therefore

$$s_{\text{bull}}^2 = 9.9, s_{\text{res}}^2 = 304.98.$$

Note: in the balanced case, the REML estimates are equal to the ANOVA estimates because the ANOVA estimate for σ_A^2 is positive.

We use the unbalanced data of Table 6.2 (excluding the bold entries) of Problem 6.1. For this unbalanced one-way classification we use for REML estimates the data frame `Table_6_2u`.

We use the package `lme4`.

```
> library(lme4)
> REMLunbalanced <- lmer (y ~1+(1|bull),data=Table_6_2u,REML = TRUE)
> summary(REMLunbalanced)
Linear mixed model fit by REML ['lmerMod']
Formula: y ~ 1 + (1 | bull)
   Data: Table_6_2u

REML criterion at convergence: 995
Scaled residuals:
    Min      1Q  Median      3Q     Max
-1.7979 -0.8001  0.1156  0.8381  1.6872
Random effects:
 Groups   Name        Variance Std.Dev.
 bull     (Intercept)   6.802   2.608
```

```
 Residual                315.901   17.774
Number of obs: 116, groups:  bull, 10
Fixed effects:
            Estimate Std. Error t value
(Intercept)  131.083      1.845   71.03
```

The estimates of the variance component using the REML method are therefore

$$s^2_{\text{bull}} = 6.80, s^2_{\text{res}} = 315.90.$$

Note: in the unbalanced case the REML estimates are not equal to the ANOVA estimates.

6.2.2 Tests of Hypotheses and Confidence Intervals

For the balanced one-way random model to construct the confidence intervals for $\sigma^2{}_a$ and σ^2 and to tests hypotheses about these variance components we need besides (6.2) a further side condition in the model equation (6.1) about the distribution of $\boldsymbol{y}_{ij}$. We assume now that the $\boldsymbol{y}_{ij}$ are not independent of each other $N(\mu, \sigma_a^2 + \sigma^2)$-distributed. Then for the distribution of $\boldsymbol{MS}_A$ and $\boldsymbol{MS}_{\text{res}}$ we know from theorem 6.5 in (Rasch and Schott (2018)) that for the special case of equal sub-class numbers (balanced case) the quadratic forms $\boldsymbol{q}_1 = \frac{\boldsymbol{SS}_{\text{res}}}{\sigma^2}$ and $\boldsymbol{q}_2 = \frac{\boldsymbol{SS}_A}{\sigma^2+n\sigma_a^2}$ are independent of each other CS$[a(n-1)]$- and CS$[a-1]$- distributed, respectively. From this, it follows that

$$\boldsymbol{F} = \frac{\boldsymbol{SS}_{\text{A}}}{\boldsymbol{SS}_{\text{res}}} \frac{a(n-1)}{a-1} \tag{6.12}$$

is $[(\sigma^2 + n\,\sigma^2{}_{\text{A}})/\sigma^2]\,F[a-1, a(n-1)]$-distributed. Under the null hypothesis $H_0 : \sigma_a^2 = 0$, $\boldsymbol{F}$ is distributed as $F[a-1,\ a(n-1)]$. If we have for the p-value $\Pr(F[a-1, a(n-1)] > \boldsymbol{F}) \leq \alpha$, then $H_0 : \sigma_a^2 = 0$ is rejected at significance level α.

For an unbalanced design the distribution of $\boldsymbol{SS}_A$ is a linear combination of $(a-1)$ independent CS[1] variables, where the coefficients are functions of σ^2 and $\sigma^2{}_{\text{A}}$. Hence $\boldsymbol{F}$ is not distributed as $[((\sigma^2 + n\,\sigma^2{}_{\text{A}})/\sigma^2)]\boldsymbol{F}[a-1, a(n-1)]$, but under the null hypothesis $H_0 : \sigma_a^2 = 0$, $\boldsymbol{F}$ is distributed as $F[a-1, a(n-1)]$.

Problem 6.4 Test the null hypothesis $H_0 : \sigma_a^2 = 0$ for the balanced case with significance level $\alpha = 0.05$.

Solution

From an ANOVA table made by R, for example with `aov()`, we can calculate (6.12) in R and find the p-value by R using the command `1 - pf( )`.

Example

For the balanced data of Table 6.2 of Problem 6.1 we have already found the ANOVA table with `aov()`. This is the ANOVA table for a fixed model, but the data provided are the same when we want to use the bulls as a random effect.

```
> Anova1b <- aov(y ~ BULL)
> Anova1b
Call:
   aov(formula = y ~ BULL)
```

```
Terms:
                          BULL Residuals
Sum of Squares    3814.63   33547.33
Deg. of Freedom         9        110
Residual standard error: 17.46356
Estimated effects may be unbalanced
#  Insert the R commands:
> F <- (3814.63/33547.33)*(110/9)
> F
[1] 1.389775
> p_value <- 1 - pf(F,9,110)
> p_value
[1] 0.2013481
```

Conclusion

The F-test gives the p-value $= 0.2013 > 0.05$ hence $H_0 : \sigma_a^2 = 0$ is not rejected.

The random model can also be found with `aov()`.

```
> randombullb <- aov(y ~ Error(BULL), data = Table_6_2b)
> randombullb
Call:
aov(formula = y ~ Error(BULL), data = Table_6_2b)
Grand Mean: 130.9833
Stratum 1: BULL
Terms:
                    Residuals
Sum of Squares    3814.633
Deg. of Freedom          9
Residual standard error: 20.58757
Stratum 2: Within
Terms:
                    Residuals
Sum of Squares    33547.33
Deg. of Freedom        110
Residual standard error: 17.46356
```

For the calculation of the p-value belonging to the F-test, see above.

Problem 6.5 Test the null hypothesis $H_0 : \sigma_a^2 = 0$ for the unbalanced case with significance level $\alpha = 0.05$.

Solution

From an ANOVA table made using R, for example with `aov()`, we can calculate (6.12) in R and find the p-value using R with the command `1 - pf()`.

Example

For the unbalanced data of Table 6.2 of Problem 6.1 we have already found the ANOVA table with `aov( )`.

```
> Anova1u <- aov(y ~ BULL)
> Anova1u
Call:
   aov(formula = y ~ BULL)
Terms:
                        BULL Residuals
Sum of Squares   3609.11   33426.03
Deg. of Freedom        9        106
Residual standard error: 17.75781
Estimated effects may be unbalanced
4 observations deleted due to missingness
#  Insert the R commands:
> F <- (3609.11/33426.03)*(106/9)
> F
[1] 1.271682
> p_value <- 1 - pf(F,9,106)
> p_value <- 1 - pf(F,9,106)
> p_value
[1] 0.2608504
```

Conclusion

The F-test gives the p-value = 0.2609 > 0.05 hence $H_0 : \sigma_a^2 = 0$ is not rejected.

When we want to use the random model the ANOVA can also be found also with aov().

```
> randombullu <- aov(y ~ Error(BULL), data = Table_6_2u)
> randombullu
Call:
aov(formula = y ~ Error(BULL), data = Table_6_2u)
Grand Mean: 131.0862
Stratum 1: BULL
Terms:
                Residuals
Sum of Squares   3609.106
Deg. of Freedom         9
Residual standard error: 20.02528
Stratum 2: Within
Terms:
                Residuals
Sum of Squares   33426.03
Deg. of Freedom       106
Residual standard error: 17.75781
```

For the calculation of the p-value belonging to the F-test, see above.

In the R package lme4 the test of the null hypothesis $H_0 : \sigma_a^2 = 0$ with $\alpha = 0.05$ does not proceed with the F-test but with the likelihood ratio (LR) test. Usually the LR test of models of fit are done with the maximum likelihood (ML) estimates. However, for random models like model II the LR test is appropriate even if the models are fit by the REML estimates. However, for the so-called mixed models with fixed and random effects as described in Chapter 7, the REML fit that differs in its fixed effects is inappropriate; in that case we must use the fit using ML estimates. However, for the one-way model II we must use the LR with the ML because the model under H_0 would only be fitted by the command `lm()`, which uses the ML.

We fit with the R package `lme4` first the largest model for the balanced data of Problem 6.1 with the data-frame `Table_6_2b` and with ML.

```
> library(lme4)
> lmer1 <-   lmer(y ~ 1 + (1| bull), data = Table_6_2b,REML = FALSE)
```

Then we fit the model belonging to the null hypothesis $H_0 : \sigma_a^2 = 0$ with `lm( )` from the R base package.

```
> lmer2 <-   lm(y ~ 1, data = Table_6_2b)
> # The likelihood-ratio test is then done by
> anova(lmer1, lmer2, refit = FALSE)
Data: Table_6_2b
Models:
lmer2: y ~ 1
lmer1: y ~ 1 + (1 | bull)
       Df    AIC    BIC  logLik deviance  Chisq Chi Df Pr(>Chisq)
lmer2a  2 1033.5 1039.0 -514.73   1029.5
lmer1a  3 1035.2 1043.6 -514.61   1029.2 0.2443      1     0.6211
```

Note: we call the LR test statistic Chisq and it is given as 0.2443;

$$\text{Chisq} = -2^*(-514.73) + 2^*(-514.61) = 0.24 \text{ from the column logLik.}$$

$$\Pr(\text{Cs}\,[1] \geq 0.2443) = 0.6211 > 0.05.$$

Hence the null hypothesis $H_0 : \sigma_a^2 = 0$ is not rejected if we use the significance level $\alpha = 0.05$.

The exact F-test gives the p-value = 0.2609 and the LR test the p-value 0.6211.

The LR test does not work well for small data sets.

Based on the results above we can construct confidence intervals for the variance component σ^2. Because $\boldsymbol{q}_1$ is CS$[a(n-1)]$-distributed,

$$\left[\frac{\boldsymbol{SS}_{\text{res}}}{\text{CS}\left[a(n-1) \mid 1-\frac{\alpha}{2}\right]}, \frac{\boldsymbol{SS}_{\text{res}}}{\text{CS}\left[a(n-1) \mid \frac{\alpha}{2}\right]}\right] \tag{6.13}$$

is a $(1-\alpha)$ confidence interval for σ^2 if $n = n_1 = \cdots = n_a$ and further that

$$\left[\frac{\boldsymbol{MS}_{\text{A}} - \boldsymbol{MS}_{\text{res}}F(a-1, N-a, 1-\alpha/2)}{\boldsymbol{MS}_{\text{A}} + (n-1)\boldsymbol{MS}_{\text{res}}F(a-1, N-a, 1-\alpha/2)}, \frac{\boldsymbol{MS}_{\text{A}} - \boldsymbol{MS}_{\text{res}}F(a-1, N-a, \alpha/2)}{\boldsymbol{MS}_{\text{A}} + (n-1)\boldsymbol{MS}_{\text{res}}F(a-1, N-a, \alpha/2)}\right] \tag{6.14}$$

is a $(1-\alpha)$ confidence interval for the intra-class correlation coefficient (ICC) $\frac{\sigma_a^2}{\sigma^2+\sigma_a^2}$ if $n = n_1 = \ldots = n_a$.

An exact $(1-\alpha)$ confidence interval for σ_a^2/σ^2 for balanced data is with $F = MS_A/MS_{\text{res}}$:

$$[(F/F(a-1, N-a, 1-\alpha/2) - 1)/n,\ \ (F/F(a-1, N-a, \alpha/2) - 1)/n]$$

An approximate $(1-\alpha)$ confidence interval for σ_a^2 for balanced data is given by Williams (1962)

$$[(\text{SS}_A\,(1 - F(a-1, N-a, 1-\alpha/2)/F)/(n\text{CS}\,[a(n-1) \mid 1-\alpha/2],$$

$$(\text{SS}_A\,(1 - F(a-1, N-a, \alpha/2)/F)]/(n\text{CS}\,[a(n-1) \mid \alpha/2].$$

An approximate $(1-\alpha)$ confidence interval for σ_a^2 in the case of unequal sub-class numbers is obtained by Seely and Lee (1994).

Problem 6.6 Construct a $(1-\alpha)$ confidence interval for σ_a^2 and a $(1-\alpha)$ confidence interval for the ICC $\frac{\sigma_a^2}{\sigma^2+\sigma_a^2}$ with $\alpha = 0.05$.

Solution

From the ANOVA table of a balanced design with model II we obtain SS_A= SSA and SS_{res}= SSRes with the corresponding degrees of freedom. With R commands, we solve (6.13) and (6.14).

Example

For the balanced data of Table 6.2 of Problem 6.1 we have found the ANOVA table with `aov( )`.

```
> Anova1b <- aov(y ~ BULL)
> Anova1b
Call:
   aov(formula = y ~ BULL)
Terms:
                    BULL Residuals
Sum of Squares   3814.63  33547.33
Deg. of Freedom        9       110
Residual standard error: 17.46356
Estimated effects may be unbalanced
> #  Insert the R commands:
> SSA <- 3814.63
> SSRes <- 33547.33
> dfA <- 9
> dfRes <- 110
> MSA <- SSA/dfA
> MSRes <- SSRes/dfRes
> chiu <- qchisq(0.975, dfRes)
> chil <- qchisq(0.025, dfRes)
> LowerCI1 <- SSRes/chiu
> UpperCI1 <- SSRes/chil
> LowerCI1   # lower CI Limit for variance
[1] 238.0652
```

```
> UpperCI1    # upper CI Limit for variance
[1] 404.8331
> MSRes       # estimate variance
[1] 304.9757
> Fu <- qf(0.975, dfA ,dfRes)
> Fl <- qf(0.025, dfA, dfRes)
> LowerCI2 <- (MSA - MSRes*Fu)/(MSA + (12-1)*MSRes*Fu)
> UpperCI2 <- (MSA - MSRes*Fl)/(MSA + (12-1)*MSRes*Fl)
> LowerCI2 #lower CI Limit intra-class correlation coeffi-
cient
[1] -0.03246049
> UpperCI2 #upper CI Limit intra-class correlation coeffi-
cient
[1] 0.2366987
> varA <- (MSA - MSRes)/12
> intraCC <- varA/(varA + MSRes)
> intraCC   # estimate intra-class correlation coefficient
[1] 0.03145944
```

Remark

The intra-class correlation coefficient 0.0315 is positive, hence `LowerCI2` must be 0.

6.2.3 Expectation and Variances of the ANOVA Estimators

Because the estimators obtained using the ANOVA method are unbiased, we get

$$E(\boldsymbol{s}_a^2) = \sigma_a^2$$

and

$$E(\boldsymbol{s}^2) = \sigma^2.$$

The variances and covariance of the ANOVA estimators of the variance components in the balanced case for normally distributed y_{ij} are:

$$\text{var}(\boldsymbol{s}^2) = \frac{2\sigma^4}{a(n-1)}, \tag{6.15}$$

$$\text{var}(\boldsymbol{s}_a^2) = \frac{2}{n^2}\left[\frac{(\sigma^2 + n\sigma_a^2)^2}{a-1} + \frac{\sigma^4}{a(n-1)}\right] \tag{6.16}$$

$$\text{cov}(\boldsymbol{s}^2, \boldsymbol{s}_a^2) = \frac{-2\sigma^4}{na(n-1)}. \tag{6.17}$$

Estimators for the variances and covariance in (6.15)–(6.17) can be obtained by replacing the quantities σ^2 and σ_a^2 occurring in these formulae by their estimators $\boldsymbol{s}^2$ and $\boldsymbol{s}_a^2$ respectively. These estimators of the variances and covariance components are biased. We get

$$\text{var}(\boldsymbol{s}^2) = \frac{2\boldsymbol{s}^4}{a(n-1)+2} \tag{6.18}$$

$$\widehat{\text{var}(s_a{}^2)} = \frac{2}{n^2}\left[\frac{s^2+s_a^2}{a+1} - \frac{s^2}{a(n-1)+2}\right] \tag{6.19}$$

and

$$\widehat{\text{cov}(s^2, s_a^2)} = \frac{-2s^4}{n[a(n-1)+2]}. \tag{6.20}$$

Problem 6.7 Estimate the variance of the ANOVA estimators of the variance components σ^2 and σ_a^2.

Solution

From an ANOVA table for a balanced design we find s^2 and s_a^2 with the df_A and df_{res}.

Using R we insert these into (6.18)–(6.20).

Example

From the balanced data of Table 6.2 of Problem 6.1 we find for s^2 and $s^2{}_A$ using df_A and df_{res}.

```
> Anova2b <- anovaVCA(y ~ bull,Table_6_2b, VarVC.method = c("scm"))
> Anova2b
Result Variance Component Analysis:
-----------------------------------

  Name  DF        SS           MS         VC         %Total    SD
1 total 116.76977                         314.88179  100       17.744909
2 bull  9         3814.633333  423.848148 9.906033   3.145953  3.147385
3 error 110       33547.333333 304.975758 304.975758 96.854047 17.463555
  CV[%]
1 13.547456
2 2.40289
3 13.332654
Mean: 130.9833 (N = 120)
Experimental Design: balanced  |  Method: ANOVA
> s2rest <- 304.975758
> s2A <- 9.906033
> n <- 12
> a <- 10
> var_s2rest <- (2*s2rest*s2rest)/(a*(n-1) + 2)
> var_s2rest  # Estimate of the variance of s2rest
[1] 1660.897
> var_s2A <- (2/(n*n))*((s2rest + s2A)/(a+1) - (s2rest/(a*(n-1)+2)))
> var_s2A  # Estimate of variance of s2A
[1] 0.3597586
```

Hence we obtain the estimates $s^2 = 304.98$ and $s_a^2 = 9.91$.

Deriving the formulae for $\text{var}(s_a^2)$ and $\text{cov}(s_a^2.s^2)$ for unequal n_i is cumbersome. The derivation can be found in Hammersley (1949) and by another method in Hartley (1967). Townsend (1968, appendix IV) gives a derivation for the case of simple unbalanced designs.

6.3 Two-Way Classification

Here and in Section 6.4, we consider mainly the estimators of the variance components with the analysis of variance method and the REML method.

6.3.1 Two-Way Cross Classification

In the two-way cross-classification our model is

$$\boldsymbol{y}_{ijk} = \mu + \boldsymbol{a}_i + \boldsymbol{b}_j + (\boldsymbol{a},\boldsymbol{b})_{ij} + \boldsymbol{e}_{ijk}$$
$$(i = 1, \dots , a; j = 1, \dots , b; k = 1, \dots , n_{ij}) \tag{6.21}$$

with side conditions that $\boldsymbol{a}_i$, $\boldsymbol{b}_j$, $(\boldsymbol{a},\boldsymbol{b})_{ij}$, and $\boldsymbol{e}_{ijk}$ are uncorrelated and:

$$E(\boldsymbol{a}_i) = E(\boldsymbol{b}_j) = E((\boldsymbol{a},\boldsymbol{b})_{ij}) = E(\boldsymbol{a}_i, \boldsymbol{b}_j) = E(\boldsymbol{a}_i(\boldsymbol{a},\boldsymbol{b})_{ij}) = E(\boldsymbol{b}_j(\boldsymbol{a},\boldsymbol{b})_{ij}) = 0$$
$$E(\boldsymbol{e}_{ijk}) = E(\boldsymbol{a}_i\boldsymbol{e}_{ijk}) = E(\boldsymbol{b}_j\boldsymbol{e}_{ijk}) = E((\boldsymbol{a},\boldsymbol{b})_{ij}\boldsymbol{e}_{ijk}) = 0 \quad \text{for all} \quad i, j, k$$
$$var(\boldsymbol{a}_i) = \sigma_a^2 \quad \text{for all} \quad i, \quad var(\boldsymbol{b}_j) = \sigma_b^2 \quad \text{for all} \quad j$$
$$var((\boldsymbol{a},\boldsymbol{b})_{ij}) = \sigma_{ab}^2 \quad \text{for all} \quad i, j, \quad var(\boldsymbol{e}_{ijk}) = \sigma^2 \quad \text{for all} \quad i, j, k.$$

For testing and confidence intervals we additionally assume that $\boldsymbol{y}_{ijk}$ is normally distributed.

In a balanced two-way cross-classification ($n_{ij} = n$ for all i, j) and normally distributed $\boldsymbol{y}_{ijk}$ the sum of squares in Table 5.7 used as a theoretical table with random variables are stochastically independent, and we have

$$\frac{\boldsymbol{SS}_A}{bn\sigma_a^2 + n\sigma_{ab}^2 + \sigma^2} \quad \text{is} \quad \text{CS}(a-1)$$
$$\frac{\boldsymbol{SS}_B}{an\sigma_b^2 + n\sigma_{ab}^2 + \sigma^2} \quad \text{is} \quad \text{CS}(b-1)$$
$$\frac{\boldsymbol{SS}_{AB}}{n\sigma_{ab}^2 + \sigma^2} \quad \text{is} \quad \text{CS}[(a-1)(b-1)]$$

distributed.

To test the hypotheses:

$$H_{A0} : \sigma_a^2 = 0, \quad H_{B0} : \sigma_b^2 = 0, \quad H_{AB0} : \sigma_{ab}^2 = 0,$$

we use the following facts.

The statistic

$$\boldsymbol{F}_A = \frac{\boldsymbol{SS}_A}{\boldsymbol{SS}_{AB}}(b-1)$$

is the $\frac{bn\sigma_a^2+n\sigma_{ab}^2+\sigma^2}{n\sigma_{ab}^2+\sigma^2}$-fold of a random variable distributed as $F[a-1, (a-1)(b-1)]$. If H_{A0} is true $\boldsymbol{F}_A$ is $F[a-1, (a-1)(b-1)]$-distributed.

The statistic

$$\boldsymbol{F}_B = \frac{\boldsymbol{SS}_B}{\boldsymbol{SS}_{AB}}(a-1)$$

is the $\frac{an\sigma_b^2+n\sigma_{ab}^2+\sigma^2}{n\sigma_{ab}^2+\sigma^2}$-fold of a random variable distributed as $F[b-1, (a-1)(b-1)]$. If H_{B0} is true $\boldsymbol{F}_B$ is $F[b-1, (a-1)(b-1)]$-distributed.

The statistic

$$\boldsymbol{F}_{AB} = \frac{\boldsymbol{SS}_{AB}}{\boldsymbol{SS}_{\text{res}}} \cdot \frac{ab(n-1)}{(a-1)(b-1)}$$

is the $\frac{n\sigma_{ab}^2+\sigma^2}{\sigma^2}$-fold of a random variable distributed as $F[(a-1)(b-1), ab(n-1)]$. If H_{AB0} is true, $\boldsymbol{F}_{AB}$ is $F[(a-1)(b-1), ab(n-1)]$-distributed.

The hypotheses H_{A0}, H_{B0}, and H_{AB0} are tested by the statistics $\boldsymbol{F}_A$, $\boldsymbol{F}_B$, and $\boldsymbol{F}_{AB}$ respectively. If the observed F-value is larger than the $(1-a)$-quantile of the central F-distribution with the corresponding degrees of freedom we may conjecture that the corresponding variance component is positive and not zero.

Problem 6.8 Test for the balanced case the hypotheses:

$$H_{A0} : \sigma_a^2 = 0, \quad H_{B0} : \sigma_b^2 = 0, \quad H_{AB0} : \sigma_{ab}^2 = 0$$

with significance level $\alpha = 0.05$ for each hypothesis.

Solution

In R we will use the package `base` to make first the ANOVA table for fixed effects. Then we do the test for the hypotheses according to the random model by making the correct F-statistics. Then we use the package `lme4` to test the hypotheses for model II of the two-way classification.

In a factory four operators (A) are chosen at random, and four machines (B) are chosen at random. Each operator must make at a machine a certain product according to a certain specification. The deviations of the specification y of three products chosen at random are given in Table 6.5.

```
> y1 <- c(8.4,8.2,7.1,-1.0,1.2,-0.9,0.5,-1.1,0.4,-0.2,0.8,0.7)
> y2 <- c(7.9,7.3,6.8,0.6,0.3,-1.4,1.0,2.6,0.8,1.1,0.2,2.3)
> y3 <- c(6.6,5.8,3.4,0.0,0.7,-1.0,-0.1,-0.1,0.1,1.2,0.6,0.3)
> y4 <- c(4.6,3.6,4.5,0.3,1.0,0.9,-0.6,0.7,0.0,-1.0,1.6,1.0)
> y <-c(y1,y2,y3,y4)
> a <- rep(1:4, each = 12)
> b0 <- c(1,1,1,2,2,2,3,3,3,4,4,4)
> b <-c(b0,b0,b0,b0)
> problem7 <- data.frame(cbind(a,b,y))
> A <- factor(a)
> B <- factor(b)
    > fixedanova <- aov( y ~ A + B + A*B, problem7)
    > summary(fixedanova)
                    Df Sum Sq Mean Sq F value  Pr(>F)
    A                3   9.17    3.06   3.722 0.02108 *
    B                3 306.24  102.08 124.299 < 2e-16 ***
    A:B              9  25.46    2.83   3.445 0.00454 **
    Residuals       32  26.28    0.82
    —
Signif. codes:  0 '***' 0.001 '**' 0.01 '*' 0.05 '.' 0.1 ' ' 1
```

Table 6.5 Deviations of the specification *y* of three at random chosen products from four operators (*A*) and four machines (*B*).

y	Prod	A_1	A_2	A_3	A_4
B_1	1	8.4	7.9	6.6	4.6
	2	8.2	7.3	5.8	3.6
	3	7.1	6.8	3.4	4.5
B_2	1	−1.0	0.6	0.0	0.3
	2	1.2	0.3	0.7	1.0
	3	−0.9	−1.4	−1.0	0.9
B_3	1	0.5	1.0	−0.1	−0.6
	2	−1.1	2.6	−0.1	0.7
	3	0.4	0.8	0.1	0.0
B_4	1	−0.2	1.1	1.2	−1.0
	2	0.8	0.2	0.6	1.6
	3	0.7	2.3	0.3	1.0

The test for $H_{AB0} : \sigma^2_{ab} = 0$ is correctly given with F-value 3.445 = MS_{AB}/MS_{res} = 2.83/0.82.

Conclusion

The p-value $AB = \Pr(F(9, 32) > 3.445) = 0.00454 < 0.05$ hence $H_{AB0} : \sigma^2_{ab} = 0$ is rejected.

The test for H_0: $\sigma^2_a = 0$ is not correctly given in the fixed ANOVA table. The F-statistic is MS_A/MS_{AB} = 3.06/2.83 = 1.08 and the p-value $A = \Pr(F(3,9) > 1.08)$ must be calculated in R.

```
> FA <- 3.06/2.83
> FA
[1] 1.081272
> p_valueA <- 1-pf(FA,3,9)
> p_valueA
[1] 0.40517
```

Conclusion

Because we have a p-value $A = 0.40527 > 0.05$, H_0: $\sigma^2_a = 0$ is not rejected.

The test for H_0: $\sigma^2{}_b = 0$ is not correctly given in the fixed ANOVA table. The F-statistic is

$F_B = \frac{MS_B}{MS_{AB}}$ = 102.08/2.83 = 36.07 and the p-value $B = \Pr(F(3.9) > 36.07)$ must be calculated in R.

```
> FB <- 102.08/2.83
> FB
[1] 36.07067
> p_valueB <- 1-pf(FB, 3,9)
```

```
> p_valueB
[1] 2.412119e-05
```

Conclusion

Because we have a p-value $B = 0.000024 < 0.05$, the H_0: $\sigma_b^2 = 0$ is rejected.

With the R package lme4 we get using the REML:

```
> library(lme4)
> model1 <- lmer( y ~1 + (1|a) + (1|b)+(1|a:b),data= problem7)
> model2 <- lmer( y ~1 + (1|a) + (1|b)   , data= problem7)
> model3 <- lmer( y ~1   + (1|b) + (1|a:b), data= problem7)
> model4 <- lmer( y ~1 + (1|a)   + (1|a:b), data= problem7)
> # We use the Likelihood-Ratio test with the REML estimates.
> # test H0: Variance component of A*B is zero
> anova(model1, model2, refit = FALSE)
  Data: problem7
  Models:
  model2: y ~ 1 + (1 | a) + (1 | b)
  model1: y ~ 1 + (1 | a) + (1 | b) + (1 | a:b)
         Df    AIC    BIC  logLik deviance  Chisq Chi Df Pr(>Chisq)
  model2  4 172.02 179.51 -82.011   164.02
  model1  5 167.54 176.90 -78.769   157.54 6.4826      1    0.01089 *
  ---
Signif. codes:  0 '***' 0.001 '**' 0.01 '*' 0.05 '.' 0.1 ' ' 1
```

Conclusion

Because the p-value Pr(>Chisq) $= 0.01089 < 0.05$, the hypothesis H_{AB0} : $\sigma_{ab}^2 = 0$ is rejected.

```
> # test H0: Variance component of A is zero
> Data: problem7
  Models:
  model3: y ~ 1 + (1 | b) + (1 | a:b)
  model1: y ~ 1 + (1 | a) + (1 | b) + (1 | a:b)
         Df    AIC    BIC  logLik deviance  Chisq Chi Df Pr(>Chisq)
  model3  4 165.55 173.03 -78.773   157.55
 model1  5 167.54 176.90 -78.769   157.54 0.0068      1     0.9341
```

Conclusion

Because the p-value Pr(>Chisq) $= 0.9341$ is larger than 0.05, the hypothesis H_{A0} : $\sigma_a^2 = 0$ is not rejected.

```
> # test H0: Variance component of B is zero
> anova(model1, model4, refit = FALSE)
  Data: problem7
  Models:
  model4: y ~ 1 + (1 | a) + (1 | a:b)
  model1: y ~ 1 + (1 | a) + (1 | b) + (1 | a:b)
         Df    AIC    BIC  logLik deviance  Chisq Chi Df Pr(>Chisq)
  model4  4 185.80 193.29 -88.901   177.80
  model1  5 167.54 176.90 -78.769   157.54 20.263      1   6.75e-06 ***
  ---
Signif. codes:  0 '***' 0.001 '**' 0.01 '*' 0.05 '.' 0.1 ' ' 1
```

Conclusion

Because the p-value Pr(>Chisq) = 0.00000675 is smaller than 0.05, the hypothesis H_{B0} : $\sigma_b^2 = 0$ is rejected.

With the package lme4 we get using the ML:

```
> library(lme4)
> model1 <- lmer( y ~1 + (1|a) + (1|b)+(1|a:b),data= problem7)
> model2 <- lmer( y ~1 + (1|a) + (1|b)   , data= problem7)
> model3 <- lmer( y ~1  + (1|b) + (1|a:b), data= problem7)
> model4 <- lmer( y ~1 + (1|a)  + (1|a:b), data= problem7)
> # We use the Likelihood-Ratio test with the ML estimates
> # test H0: Variance component of A*B is zero
> anova(model1, model2, refit = TRUE)
    refitting model(s) with ML (instead of REML)
    Data: problem7
    Models:
    model2: y ~ 1 + (1 | a) + (1 | b)
    model1: y ~ 1 + (1 | a) + (1 | b) + (1 | a:b)
           Df    AIC    BIC  logLik deviance  Chisq Chi Df Pr(>Chisq)
    model2  4 174.49 181.97 -83.243   166.49
    model1  5 169.98 179.34 -79.992   159.98 6.5028      1    0.01077 *
    ---
    Signif. codes:  0 '***' 0.001 '**' 0.01 '*' 0.05 '.' 0.1 ' ' 1
```

Conclusion

Because the p-value Pr(>Chisq) = 0.01077 is smaller than 0.05, the hypothesis H_{AB0} : $\sigma_{ab}^2 = 0$ is rejected.

```
> # test H0: Variance component of A is zero
> anova(model1, model3, refit = TRUE)
    refitting model(s) with ML (instead of REML)
    Data: problem7
    Models:
    model3: y ~ 1 + (1 | b) + (1 | a:b)
    model1: y ~ 1 + (1 | a) + (1 | b) + (1 | a:b)
             Df    AIC    BIC  logLik deviance  Chisq Chi Df Pr(>Chisq)
    model3   4    167.99 175.47 -79.994   159.99
model1   5    169.98 179.34 -79.992   159.98 0.0042     1 0.9483
```

Conclusion

Because the p-value Pr(>Chisq) = 0.9483 is larger than 0.05, the hypothesis H_{A0} : $\sigma_a^2 = 0$ is not rejected.

```
> # test H0: Variance component of B is zero
> anova(model1, model4, refit = TRUE)
    refitting model(s) with ML (instead of REML)
    Data: problem7
    Models:
    model4: y ~ 1 + (1 | a) + (1 | a:b)
    model1: y ~ 1 + (1 | a) + (1 | b) + (1 | a:b)
           Df    AIC    BIC  logLik deviance  Chisq Chi Df Pr(>Chisq)
    model4  4 186.86 194.34 -89.430   178.86
```

```
model1  5 169.98 179.34 -79.992    159.98 18.876       1  1.395e-05 ***
---
Signif. codes:  0 '***' 0.001 '**' 0.01 '*' 0.05 '.' 0.1 ' ' 1
```

Conclusion

Because the p-value Pr(>Chisq) = 0.00001395 is smaller than 0.05, the hypothesis H_{B0} : $\sigma_b^2 = 0$ is rejected.

In Table 5.7 we have to replace, for the random model with the ANOVA method, the expected mean squares by

$$\left.\begin{aligned} E(\boldsymbol{MS}_A) &= bn\sigma_a^2 + n\sigma_{ab}^2 + \sigma^2 \\ E(\boldsymbol{MS}_B) &= an\sigma_b^2 + n\sigma_{ab}^2 + \sigma^2 \\ E(\boldsymbol{MS}_{AB}) &= n\sigma_{ab}^2 + \sigma^2 \\ E(\boldsymbol{MS}_{\text{res}}) &= \sigma^2 \end{aligned}\right\}. \tag{6.22}$$

Now we can directly find the ANOVA estimators for the variance components.

Problem 6.9 Derive the estimators of the variance components with the ANOVA method using the REML.

Solution

The algorithm of the analysis of variance method shown in Section 6.2 provides estimators of the variance components of the balanced case

$$\left.\begin{aligned} \boldsymbol{s}^2 &= \boldsymbol{MS}_{\text{res}}, & \boldsymbol{s}_{ab}^2 &= \tfrac{1}{n}(\boldsymbol{MS}_{AB} - \boldsymbol{MS}_{\text{res}}) \\ \boldsymbol{s}_b^2 &= \tfrac{1}{an}(\boldsymbol{MS}_B - \boldsymbol{MS}_{AB}) & \boldsymbol{s}_a^2 &= \tfrac{1}{bn}(\boldsymbol{MS}_A - \boldsymbol{MS}_{AB}) \end{aligned}\right\}. \tag{6.23}$$

In R we will use the packages `VCA` and `lme4` to estimate the variance components of model II for the two-way classification.

We use the data of Problem 6.7 where we already have in R the data-frame `problem7`.

```
> library(VCA)
> anova8 <- anovaVCA(y~a + b + a:b,problem7,VarVC.method = c("scm"))
> anova8
    Result Variance Component Analysis:
    -----------------------------------

  Name  DF       SS         MS         VC       %Total    SD       CV[%]
1 total 3.956536                       9.780486 100       3.127377 171.167719
2 a     3        9.170625   3.056875   0.018981 0.194075  0.137773 7.540614
3 b     3        306.242292 102.080764 8.270972 84.566065 2.87593  157.405508
4 a:b   9        25.461875  2.829097   0.669282 6.843038  0.818097 44.776109
5 error 32       26.28      0.82125    0.82125  8.396822  0.906228 49.599733

  Mean: 1.827083 (N = 48)
Experimental Design: balanced  |  Method: ANOVA
```

We found the following estimates of the variance components

$$s_a^2 = 0.019; s_b^2 = 8.271; s_{ab}^2 = 0.669; s^2 = 0.821.$$

Now we estimate the variance components using the REML method. Because the data of Problem 6.7 are balanced, the REML estimates are the same as the ANOVA estimates, because the ANOVA estimates of the variance components are positive.

```
> library(lme4)
> y.lmer <- lmer( y ~1 + (1|a) + (1|b) + (1|a:b), data= problem7)
> summary(y.lmer)
    Linear mixed model fit by REML ['lmerMod']
    Formula: y ~ 1 + (1 | a) + (1 | b) + (1 | a:b)
       Data: problem7
    REML criterion at convergence: 157.5
    Scaled residuals:
         Min       1Q   Median       3Q      Max
    -2.30598 -0.41655  0.01979  0.51536  1.58012
    Random effects:
     Groups   Name        Variance Std.Dev.
     a:b      (Intercept) 0.66928  0.8181
     b        (Intercept) 8.27097  2.8759
     a        (Intercept) 0.01898  0.1378
     Residual             0.82125  0.9062
    Number of obs: 48, groups:  a:b, 16; b, 4; a, 4
    Fixed effects:
                Estimate Std. Error t value
(Intercept)    1.827      1.460   1.251
```

6.3.2 Two-Way Nested Classification

The two-way nested classification is a special case of the incomplete two-way cross-classification, it is maximally unconnected. The formulae for the estimators of the variance components become very simple. We use the notation of Section 5.3.2, but now the a_i and b_j in (5.33) are random variables. The model equation (5.33) then becomes

$$\boldsymbol{y}_{ijk} = \mu + \boldsymbol{a}_i + \boldsymbol{b}_{j(i)} + \boldsymbol{e}_{k(ij)}, i = 1, \ldots, a; j = 1, \ldots, b; k = 1, \ldots, n_{i.j}, \tag{6.24}$$

with the conditions of uncorrelated $\boldsymbol{a}_i$, $\boldsymbol{b}_{ij}$, and $\boldsymbol{e}_{ijk}$ and

$$E(\boldsymbol{a}_i) = E(\boldsymbol{b}_{ij}) = \text{cov}(\boldsymbol{a}_i, \boldsymbol{e}_{ijk}) = \text{cov}(\boldsymbol{b}_{ij}, \boldsymbol{e}_{ijk}) = 0$$

$$var(\boldsymbol{a}_i) = \sigma_a^2, var(\boldsymbol{b}_{ij}) = \sigma_b^2, var(\boldsymbol{e}_{ijk}) = \sigma^2, \text{for all } i, j, k.$$

Further, we assume for tests that the random variables $\boldsymbol{a}_i$ are $N(0, \sigma_a^2)$, $\boldsymbol{b}_{ij}$ are $N(0, \sigma_b^2)$, and $\boldsymbol{e}_{ijk}$ are $N(0, \sigma^2)$.

We use here and in other sections the columns source of variation, sum of squares and degrees of freedom of the corresponding ANOVA tables in Chapter 5 because these columns are the same for models with fixed and random effects. The column expected mean squares and F-test statistics must be derived for models with random effects anew. We now replace in Table 5.13 the column of $E(\boldsymbol{MS})$ for the random model (see Table 6.6).

Table 6.6 The column *E*(*MS*) of the two-way nested classification for model II.

Source of variation	E(MS)
Between A levels	$\sigma^2 + \lambda_2\sigma_b^2 + \lambda_3\sigma_a^2$
Between B levels within A levels	$\sigma^2 + \lambda_1\sigma_b^2$
Within B levels (residual)	σ^2

In this table the positive coefficients λ_i are defined by

$$\lambda_1 = \frac{1}{B_{.} - a}\left(N - \sum_{i=1}^{a} \frac{\sum_{j=1}^{b} n_{ij}^2}{N_{i.}}\right)$$

$$\lambda_2 = \frac{1}{a-1}\sum_{i=1}^{a}\sum_{j=1}^{b} n_{ij}^2\left(\frac{1}{N_{i.}} - \frac{1}{N}\right)$$

$$\lambda_3 = \frac{1}{a-1}\left(N - \frac{1}{N}\sum_{i=1}^{a} N_{i.}^2\right). \tag{6.25}$$

Problem 6.10 Derive the ANOVA estimates for all variance components; give also the REML estimates.

Solution

From the analysis of variance method, we obtain the estimators of the variance components by

$$\left.\begin{aligned} s^2 &= MS_{\text{res}} \\ s_b &= \frac{1}{\lambda_1}(MS_{B \text{ in } A} - MS_{\text{res}}) \\ s_a^2 &= \frac{1}{\lambda_3}\left(MS_A - \frac{\lambda_2}{\lambda_1}MS_{B \text{ in } A} - \left(1 - \frac{\lambda_2}{\lambda_1}\right)MS_{\text{res}}\right) \end{aligned}\right\} \tag{6.26}$$

with

$$\lambda_1' = (B_{.} - a)\lambda_1, \lambda_2' = (a-1)\lambda_2, \lambda_3' = (a-1)\lambda_3$$

and

$$\lambda_4 = N + \frac{1}{N}\sum_{i=1}^{a} N_{i.}^2, \lambda_5 = \lambda_1'^2\left[(\lambda_4 - N)\lambda_4 - \frac{2}{N}\sum_{i=1}^{a} N_{i.}^3\right].$$

The formulae for the variances of the variance components are given as formulae (6.60) and (6.61) in Rasch and Schott (2018).

Example

For testing the performance of boars, the performance of the offspring of boars under unique feeding, fattening and slaughter are measured. From the results of testing, two boars b_1, b_2 were randomly selected. For each boar, we observed the offspring of several sows. As well as from B_1 and also from B_2 three observations y (number the fattening

Table 6.7 Data of the example for Problem 6.9.

Number of fattening days y_{ijk}.	Boars	B_1			B_2		
	Sows	S_{11}	S_{12}	S_{13}	S_{21}	S_{22}	S_{23}
Offspring	y_{ijk}	93	107	109	89	87	81
		89	99	107	102	91	83
		97		94	104	82	85
		105		106	97		91
	n_{ij}	4	2	4	4	3	4
	$n_{i.}$		10			11	

days from 40 kg up to 110 kg) are available. The variance components for boars and sows (within boars) and within sows must be estimated.

Table 6.7 shows the observations y_{ijk}. In this case we have $a = 2$, $b_1 = 3$, $b_2 = 3$.

Using the R package VCA we find the ANOVA estimates.

```
> y1 <- c(93,89,97,105,107,99,109,107,94,106)
> y2 <- c(89,102,104,97,87,91,82,81,83,85,91)
> y <- c(y1,y2)
> b1 <- rep(1,10)
> b2 <- rep(2, 11)
> b <- c(b1,b2)
> s1 <- rep(1,4)
> s2 <- rep(2,2)
> s3 <- rep(3,4)
> s5 <- rep(4,4)
> s4 <- rep(4,4)
> s5 <- rep(5,3)
> s6 <- rep(6,4)
> s <- c(s1,s2,s3,s4,s5,s6)
> problem9 <- data.frame(b,s,y)
> library(VCA)
> anova9 <- anovaVCA(y~ b+ b:s,problem9,VarVC.method =c("scm"))
> anova9
Result Variance Component Analysis:
-----------------------------------

  Name  DF        SS         MS         VC        %Total    SD       CV[%]
1 total 3.520746                        105.29654 100       10.26141 10.785266
2 b     1         568.535065 568.535065 40.933538 38.874532 6.397932 6.724553
3 b:s   4         531.369697 132.842424 28.318558 26.894101 5.321518 5.593187
4 error 15        540.666667 36.044444  36.044444 34.231366 6.003703 6.310198
Mean: 95.14286 (N = 21)
Experimental Design: unbalanced  |  Method: ANOVA
```

Now we will use the REML estimates.

```
> library(lme4)
> lmer9 <- lmer( y ~1 + (1|b) + (1|b:s), data= problem9)
> summary(lmer9)
Linear mixed model fit by REML ['lmerMod']
Formula: y ~ 1 + (1 | b) + (1 | b:s)
   Data: problem9
REML criterion at convergence: 139.2
```

```
Scaled residuals:
     Min       1Q   Median       3Q      Max
-1.48912 -0.65918  0.00967  0.74445  1.34738
Random effects:
 Groups   Name        Variance Std.Dev.
 b:s      (Intercept) 26.18    5.117
 b        (Intercept) 46.90    6.848
 Residual             35.77    5.980
Number of obs: 21, groups:  b:s, 6; b, 2
Fixed effects:
            Estimate Std. Error t value
(Intercept)    95.39       5.44   17.54
```

Note: the REML estimates are different from the ANOVA estimates due to the unbalanced data.

Problem 6.11 Test the hypotheses $H_{A0} : \sigma_a^2 = 0$ and $H_{B0} : \sigma_b^2 = 0$ with $\alpha = 0.05$ for each hypothesis.

Solution

In R we will use the package `lme4` and fit several models to apply the ML test. Because the data are unbalanced, we will not use first the ANOVA table for fixed effects, because there are no exact F-tests for the hypotheses.

Example

We use the data of Table 6.7 of Problem 6.9. In R we have already made the data-frame `problem9`.

```
> library(lme4)
> lmer1 <- lmer( y ~1 + (1|b) + (1|b:s), data= problem9)
> lmer2 <- lmer( y ~1 + (1|b) , data= problem9)
> lmer3 <- lmer( y ~1  + (1|b:s), data= problem9)
> # test H0: Variance component of B is zero
> anova(lmer1, lmer2, refit = TRUE)
refitting model(s) with ML (instead of REML)
Data: problem9
Models:
lmer2: y ~ 1 + (1 | b)
lmer1: y ~ 1 + (1 | b) + (1 | b:s)
      Df    AIC    BIC  logLik deviance  Chisq Chi Df Pr(>Chisq)
lmer2  3 153.52 156.66 -73.761   147.52
lmer1  4 152.01 156.18 -72.003   144.01 3.5162      1    0.06077 .
—
Signif. codes:  0 '***' 0.001 '**' 0.01 '*' 0.05 '.' 0.1 ' ' 1
> # test H0: Variance component of A is zero
> anova(lmer1, lmer3, refit = TRUE)
refitting model(s) with ML (instead of REML)
Data: problem9
Models:
lmer3: y ~ 1 + (1 | b:s)
```

```
lmer1: y ~ 1 + (1 | b) + (1 | b:s)
      Df    AIC    BIC  logLik deviance  Chisq Chi Df Pr(>Chisq)
lmer3  3 150.60 153.73 -72.300   144.60
lmer1  4 152.01 156.18 -72.003   144.01 0.5931      1     0.4412
```

Conclusion

Because the p-values Pr(>Chisq) for both tests are larger than 0.05, the hypotheses H_{A0} : $\sigma_a^2 = 0$ and H_{B0} : $\sigma_b^2 = 0$ are not rejected.

6.4 Three-Way Classification

We have four three-way classifications and proceed in all sections in the same way. At first, we complete the corresponding ANOVA table of Chapter 5 using the column $E(\boldsymbol{MS})$ for the model with random effects. Then we use the ANOVA method only and use the algorithm of the analysis of variance method in Section 6.4.1 to obtain estimators of the variance components in the balanced case.

6.4.1 Three-Way Cross-Classification with Equal Sub-Class Numbers

We start with the model equation for the balanced model

$$\boldsymbol{y}_{ijkl} = \mu + \boldsymbol{a}_i + \boldsymbol{b}_i + \boldsymbol{c}_k + (\boldsymbol{a}, \boldsymbol{b})_{ij} + (\boldsymbol{a}, \boldsymbol{c})_{ik} + (\boldsymbol{b}, \boldsymbol{c})_{jk} + (\boldsymbol{a}, \boldsymbol{b}, \boldsymbol{c})_{i.jk} + \boldsymbol{e}_{ijkl}, i = 1, \dots, a;$$
$$j = 1, \dots, b; k = 1, \dots, c, l = 1, \dots, n \tag{6.27}$$

with the side conditions that the expectations of all random variables of the right-hand side of (6.27) are equal to zero and all covariances between different random variables of the right-hand side of (6.27) vanish: $var(\boldsymbol{a}_{\mathrm{i}}) = \sigma_{\mathrm{a}}^2$, $var(\boldsymbol{b}_{\mathrm{j}}) = \sigma_{\mathrm{b}}^2$, $var(\boldsymbol{c}_{\mathrm{k}}) = \sigma_{\mathrm{c}}^2$, $var(\boldsymbol{a},\boldsymbol{b})_{\mathrm{ij}} = \sigma_{\mathrm{ab}}^2$, $var(\boldsymbol{a},\boldsymbol{c})_{\mathrm{ik}} = \sigma_{\mathrm{ac}}^2$, $var(\boldsymbol{b},\boldsymbol{c})_{\mathrm{jk}} = \sigma_{\mathrm{bc}}^2$, $var(\boldsymbol{a},\boldsymbol{b},\boldsymbol{c})_{\mathrm{ijk}} = \sigma_{\mathrm{abc}}^2$. Further we assume for tests that the $\boldsymbol{y}_{ijkl}$ are normally distributed.

Table 6.8 shows the $E(\boldsymbol{MS})$ for this case as the new addendum of Table 5.15.

Table 6.8 The column $E(MS)$ as supplement for model II to the analysis of variance Table 5.15.

Source of variation	$E(MS)$
Between A levels	$\sigma^2 + n\sigma_{abc}^2 + cn\sigma_{ab}^2 + bn\sigma_{ac}^2 + bcn\sigma_a^2$
Between B levels	$\sigma^2 + n\sigma_{abc}^2 + cn\sigma_{ab}^2 + an\sigma_{bc}^2 + acn\sigma_b^2$
Between C levels	$\sigma^2 + n\sigma_{abc}^2 + an\sigma_{bc}^2 + bn\sigma_{ac}^2 + abn\sigma_c^2$
Interaction $A \times B$	$\sigma^2 + n\sigma_{abc}^2 + cn\sigma_{ab}^2$
Interaction $A \times C$	$\sigma^2 + n\sigma_{abc}^2 + bn\sigma_{ac}^2$
Interaction $B \times C$	$\sigma^2 + n\sigma_{abc}^2 + an\sigma_{bc}^2$
Interaction $A \times B \times C$	$\sigma^2 + n\sigma_{abc}^2$
Within the sub-classes (residual)	σ^2

Problem 6.12 Use the analysis of variance method to obtain the estimators for the variance components by solving

$$MS_A = s^2 + ns^2_{abc} + cns^2_{ab} + bns^2_{ac} + bcns^2_a$$
$$MS_B = s^2 + ns^2_{abc} + cns^2_{ab} + ans^2_{bc} + acns^2_b$$
$$MS_C = s^2 + ns^2_{abc} + ans^2_{bc} + bns^2_{ac} + abns^2_c$$
$$MS_{AB} = s^2 + ns^2_{abc} + cns^2_{ab}$$
$$MS_{AC} = s^2 + ns^2_{abc} + bns^2_{ac}$$
$$MS_{BC} = s^2 + ns^2_{abc} + ans^2_{bc}$$
$$MS_{ABC} = s^2 + ns^2_{abc}$$
$$MS_{\text{res}} = s^2.$$

Solution
In R we use the package VCA for the ANOVA estimates and package lme4 for the REML estimates.

Example
In a factory, from the population of operators we take a random sample A of three operators; from the population of machines we take a random sample B of three machines; from the production material we have a population of batches, where we take a random sample of C of three batches. From the product produced by operator A_i on machine B_j from the batch C_k we determine a characteristic with value y_{ijk}. One is interested in the sources of variation in y_{ijk}. The data in Table 6.9 are from Kuehl, page 212 (1994).

In R we use for the ANOVA estimates of the variance components the package VCA and for the REML estimates the package lme4 .

```
> y1 <- c(0.60,0.48,0.98,0.93,1.37,1.50)
> y2 <- c(1.69,2.01,2.21,2.48,3.31,2.84)
> y3 <- c(3.47,3.30,5.68,5.11,5.74,5.38)
> y4 <- c(0.05,0.12,0.15,0.26,0.72,0.51)
> y5 <- c(0.11,0.09,0.23,0.35,0.78,1.11)
> y6 <- c(0.06,0.19,0.40,0.75,2.10,1.18)
> y7 <- c(0.07,0.06,0.07,0.21,0.40,0.57)
> y8 <- c(0.08,0.14,0.23,0.35,0.72,0.88)
```

Table 6.9 (Kuehl 1994) Observations of products produced by operator A_i on machine B_j from the batch C_k.

y	A_1			A_2			A_3		
	B_{11}	B_{12}	B_{13}	B_{21}	B_{22}	B_{23}	B_{31}	B_{32}	B_{33}
C_1	0.60	1.69	3.47	0.05	0.11	0.06	0.07	0.08	0.22
	0.48	2.01	3.30	0.12	0.09	0.19	0.06	0.14	0.17
C_2	0.98	2.21	5.68	0.15	0.23	0.40	0.07	0.23	0.43
	0.93	2.48	5.11	0.26	0.35	0.75	0.21	0.35	0.35
C_3	1.37	3.31	5.74	0.72	0.78	2.10	0.40	0.72	1.95
	1.50	2.84	5.38	0.51	1.11	1.18	0.57	0.88	2.87

```
> y9 <- c(0.22,0.17,0.43,0.35,1.95,2.87)
> y <- c(y1,y2,y3,y4,y5,y6,y7,y8,y9)
> a <- rep(1:3, each = 18)
> b1 <- rep(1:3, each =6)
> b <- c(b1,b1,b1)
> c1 <- rep(1:3, each = 2)
> c <- c(c1,c1,c1,c1,c1,c1,c1,c1,c1)
> problem11 <- data.frame(a,b,c,y)
> library(VCA)
> anova11<-anovaVCA(y~a+b+c+a:b+a:c+b:c+a:b:c,problem11,
      VarVC.method=   c("scm"))
> anova11
  Result Variance Component Analysis:
  -----------------------------------

    Name  DF        SS        MS        VC        %Total    SD        CV[%]
  1 total 5.818679                      3.099903  100       1.760654  139.796089
  2 a     2         58.134344 29.067172 1.322089  42.649381 1.149821  91.295924
  3 b     2         26.2828   13.1414   0.41299   13.32267  0.642643  51.025898
  4 c     2         12.460844 6.230422  0.29758   9.599644  0.545509  43.313431
  5 a:b   4         20.618956 5.154739  0.824737  26.60524  0.90815   72.107197
  6 a:c   4         1.284578  0.321144  0.019138  0.617358  0.138338  10.984077
  7 b:c   4         3.036656  0.759164  0.092141  2.972375  0.303547  24.101653
  8 a:b:c 8         1.650556  0.206319  0.07509   2.422343  0.274026  21.757693
  9 error 27        1.51575   0.056139  0.056139  1.810989  0.236936  18.812776

  Mean: 1.259444 (N = 54)
Experimental Design: balanced  |  Method: ANOVA

> library(lme4)
> lmer11 <- lmer(y~1+(1|a)+(1|b)+(1|c)+(1|a:b)+(1|a:c)+
      (1|b:c)+(1|a:b:c) , data= problem11)
> summary(lmer11)
Linear mixed model fit by REML ['lmerMod']
Formula: y ~ 1 + (1 | a) + (1 | b) + (1 | c) + (1 | a:b) + (1 | a:c) +
    +(1 | b:c) + (1 | a:b:c)
   Data: problem11
REML criterion at convergence: 80.5
Scaled residuals:
     Min       1Q   Median       3Q      Max
-1.96769 -0.42854 -0.02282  0.27936  2.50366
Random effects:
 Groups   Name        Variance Std.Dev.
 a:b:c    (Intercept) 0.07509  0.2740
 b:c      (Intercept) 0.09214  0.3035
 a:c      (Intercept) 0.01914  0.1383
 a:b      (Intercept) 0.82474  0.9082
 c        (Intercept) 0.29758  0.5455
 b        (Intercept) 0.41299  0.6426
 a        (Intercept) 1.32209  1.1498
 Residual             0.05614  0.2369
Number of obs: 54, groups:  a:b:c, 27; b:c, 9; a:c, 9; a:b, 9; c, 3; b, 3; a, 3
Fixed effects:
            Estimate Std. Error t value
(Intercept)   1.2594     0.8862   1.421
```

If model equation (6.27), including its conditions about expectations and covariances of the components of $\boldsymbol{y}_{ijkl}$, is valid and $\boldsymbol{y}_{ijkl}$ is multivariate normally distributed with the marginal distributions

$$N(\mu, \sigma_a^2 + \sigma_b^2 + \sigma_c^2 + \sigma_{ab}^2 + \sigma_{ac}^2 + \sigma_{bc}^2 + \sigma_{abc}^2 + \sigma^2)$$

Table 6.10 Test statistics for testing hypotheses and distributions of these test statistics.

Test statistic	H_0	Distribution of the test statistic under H_0
$\boldsymbol{F}_{AB} = \frac{\boldsymbol{MS}_{AB}}{\boldsymbol{MS}_{ABC}}$	$\sigma^2_{ab} = 0$	$F[(a-1)(b-1),\ (a-1)(b-1)(c-1)]$
$\boldsymbol{F}_{AC} = \frac{\boldsymbol{MS}_{AC}}{\boldsymbol{MS}_{ABC}}$	$\sigma^2_{ac} = 0$	$F[(a-1)(c-1),\ (a-1)(b-1)(c-1)]$
$\boldsymbol{F}_{BC} = \frac{\boldsymbol{MS}_{BC}}{\boldsymbol{MS}_{ABC}}$	$\sigma^2_{bc} = 0$	$F[(b-1)(c-1),\ (a-1)(b-1)(c-1)]$
$\boldsymbol{F}_{ABC} = \frac{\boldsymbol{MS}_{ABC}}{\boldsymbol{MS}_{res}}$	$\sigma^2_{abc} = 0$	$F[(a-1)(b-1)(c-1),\ N-abc]$

then the $\frac{SS_X}{E(MS_X)}$ are CS(df_X)-distributed ($X = A, B, C, AB, AC, BC, ABC$) with $\boldsymbol{SS}_X$, and df_X from Table 5.15 and $E(\boldsymbol{MS}_X)$ from Table 6.8.

Problem 6.13 Derive the F-tests with significance level $\alpha = 0.05$ for testing each of the null hypotheses $H_{AB0} : \sigma^2_{ab} = 0; H_{AC0} : \sigma^2_{ac} = 0; H_{BC0} : \sigma^2_{bc} = 0; H_{ABC0} : \sigma^2_{abc} = 0$.

Solution

The formulae can be found in Table 6.10.

Example

We have already made the data-frame `problem11`. We first make the ANOVA table for fixed effects.

```
> A <- factor(a)
> B <- factor(b)
> C <- factor(c)
> fixedanova <- aov( y ~ A+B+C+A:B+A:C+B:C+A:B:C, problem11)
> summary(fixedanova)
            Df Sum Sq Mean Sq F value    Pr(>F)
A            2  58.13  29.067 517.772  < 2e-16 ***
B            2  26.28  13.141 234.087  < 2e-16 ***
C            2  12.46   6.230 110.982 9.45e-14 ***
A:B          4  20.62   5.155  91.821 2.59e-15 ***
A:C          4   1.28   0.321   5.721  0.00181 **
B:C          4   3.04   0.759  13.523 3.57e-06 ***
A:B:C        8   1.65   0.206   3.675  0.00510 **
Residuals   27   1.52   0.056
—
Signif. codes:  0 '***' 0.001 '**' 0.01 '*' 0.05 '.' 0.1 ' ' 1
> # test for H0 variance component AB is 0
> FAB <- 5.155/0.206
> FAB
[1] 25.02427
> p_valueAB <- 1-pf(FAB,4,8)
> p_valueAB
```

```
[1] 0.0001411135
> # test for H0 variance component AC is 0
> FAC <- 0.321/0.206
> FAC
[1] 1.558252
> p_valueAC <- 1-pf(FAC,4,8)
> p_valueAC
[1] 0.274647
> # test for H0 variance component BC is 0
> FBC <- 0.759/0.206
> FBC
[1] 3.684466
> p_valueBC <- 1-pf(FBC,4,8)
> p_valueBC
[1] 0.05505248
> # test for H0 variance component ABC is 0
> FABC <- 0.206/0.056
> FABC
[1] 3.678571
> p_valueABC <-1-pf(FABC,8,27)
> p_valueABC
[1] 0.005070746
> # Only the p_valueABC is correct found in the fixed ANOVA table.
```

Conclusion

The hypotheses $H_{AC0} : \sigma^2_{ac} = 0, H_{BC0} : \sigma^2_{bc} = 0$ are not rejected.

Problem 6.14 Derive the approximate F-tests with significance level $\alpha = 0.05$ for testing each of the null hypotheses $H_{A0} : \sigma^2_a = 0; H_{B0} : \sigma^2_b = 0; H_{C0} : \sigma^2_c = 0$.

Solution

For testing the null hypothesis $H_{A0} : \sigma^2_a = 0; H_{B0} : \sigma^2_b = 0; H_{C0} : \sigma^2_c = 0$ we need a result from Satterthwaite (1946) because $E(\boldsymbol{MS}_X)$ (X = A, B, C) under the null hypothesis is not equal to any $E(\boldsymbol{MS})$ in other rows of the ANOVA table. Therefore, we use instead linear combinations of the $E(\boldsymbol{MS})$ so that $E(\boldsymbol{MS}_X)$ under H_{0X} equals the $E(\boldsymbol{MS})$ of this linear combination. We find

$$E(\boldsymbol{MS}_A) = bcn\sigma^2_a + E(\boldsymbol{MS}_{AB} + \boldsymbol{MS}_{AC} - \boldsymbol{MS}_{ABC})$$
$$E(\boldsymbol{MS}_B) = acn\sigma^2_b + E(\boldsymbol{MS}_{AB} + \boldsymbol{MS}_{BC} - \boldsymbol{MS}_{ABC})$$
$$E(\boldsymbol{MS}_C) = abn\sigma^2_c + E(\boldsymbol{MS}_{AC} + \boldsymbol{MS}_{BC} - \boldsymbol{MS}_{ABC}).$$

We use this to construct test statistics for the null hypotheses $H_{A0} : \sigma^2_a = 0; H_{B0} : \sigma^2_b = 0; H_{C0} : \sigma^2_c = 0$, which are approximately F-distributed.

These test statistics for approximate F-tests are:

$$\boldsymbol{F}_A = \frac{\boldsymbol{MS}_A}{\boldsymbol{MS}_{AB} + \boldsymbol{MS}_{AC} - \boldsymbol{MS}_{ABC}}$$

and this is, under H_{A0}, approximately $F(a-1, r_a)$-distributed;

$$\boldsymbol{F}_B = \frac{\boldsymbol{MS}_B}{\boldsymbol{MS}_{AB} + \boldsymbol{MS}_{BC} - \boldsymbol{MS}_{ABC}}$$

and this is under H_{B0}, approximately $F(b-1, r_b)$-distributed and

$$\boldsymbol{F}_C = \frac{\boldsymbol{MS}_C}{\boldsymbol{MS}_{BC} + \boldsymbol{MS}_{AC} - \boldsymbol{MS}_{ABC}}$$

and this is, under H_{C0}, approximately $F(c-1, r_c)$-distributed.

The approximate F-test has, respectively, denominator degrees of freedom r_a, r_b, r_c, which are determined by the Satterthwaite's method. Let v_i be the degrees of freedom of MS_i then the degree of freedom r of $(MS_1 + MS_2 - MS_3)$ is approximated by:

$$r = (MS_1 + MS_2 - MS_3)^2/[(MS_1)^2/v_1 + (MS_2)^2/v_2 + (-MS_3)^2/v_3].$$

See also Gaylor and Hopper (1969) about estimating the df of linear combinations of mean squares by Satterthwaite's formula.

Example

In the output of Problem 6.14 we have already found the ANOVA table for the fixed effects model.

```
> # test for H0 variance component A is 0
> denominatorA <- 5.155 + 0.321 - 0.206
> FA <- 29.067/denominatorA
> FA
[1] 5.51556
> rA <- (denominatorA^2)/(5.155^2/4 + 0.321^2/4 +0.206^2/8)
> rA
[1] 4.161002
> p_valueA <- 1-pf(FA, 2, rA)
> p_valueA
[1] 0.06759019
> # test for H0 variance component B is 0
> denominatorB <- 5.155 + 0.759 - 0.206
> FB <- 13.141/denominatorB
> FB
[1] 2.302207
> rB <- (denominatorB^2)/(5.155^2/4 + 0.759^2/4 + 0.206^2/8)
> rB
[1] 4.796419
> p_valueB <- 1-pf(FB, 2, rB)
> p_valueB
[1] 0.1991247
> # test for H0 variance component C is 0
> denominatorC <- 0.759 + 0.321 - 0.206
> FC <- 6.230/denominatorC
> FC
[1] 7.128146
> rC <- (denominatorC^2)/(0.759^2/4 + 0.321^2/4 + 0.206^2/8)
> rC
```

```
[1] 4.362887
> p_valueC <- 1-pf(FC, 2, rC)
> p_valueC
[1] 0.0421972
```

Only the hypothesis for $H_{C0} : \sigma_c^2 = 0$ is rejected.

Remark

To do the LR test we proceed with the R package `lme4` and start with the largest model with all the random variables, like we have done in Problem 6.12. Then we can test a certain variance component by fitting the model without this variance component. We demonstrate this for the test of H_0: $\sigma^2_{\text{abc}} = 0$.

```
> library(lme4)
> lmer121 <- lmer(y~1+(1|a)+(1|b)+(1|c)+(1|a:b)+(1|a:c)+
     (1|b:c)+(1|a:b:c) , data= problem11)
> lmer122 <- lmer(y~1+(1|a)+(1|b)+(1|c)+(1|a:b)+(1|a:c)+
     (1|b:c) , data= problem11)
> # Test H0 the variance component of ABC is 0
> anova(lmer121, lmer122, refit = TRUE)
   refitting model(s) with ML (instead of REML)
   Data: problem11
   Models:
   lmer122: y ~ 1 + (1 | a) + (1 | b) + (1 | c) + (1 | a:b) + (1 | a:c) +
   lmer122:     (1 | b:c)
   lmer121: y ~ 1 + (1 | a) + (1 | b) + (1 | c) + (1 | a:b) + (1 | a:c) +
   lmer121:     (1 | b:c) + (1 | a:b:c)
           Df     AIC    BIC  logLik deviance  Chisq Chi Df Pr(>Chisq)
   lmer122  8 104.218 120.13 -44.109   88.218
   lmer121  9  99.934 117.83 -40.967   81.934 6.2837      1    0.01219 *
   —
Signif. codes:  0 '***' 0.001 '**' 0.01 '*' 0.05 '.' 0.1 ' ' 1
```

Conclusion

Because the p-value Pr(>Chisq) = 0.01219 is smaller than 0.05, the hypothesis H_0: $\sigma^2_{\text{abc}} = 0$ is rejected.

6.4.2 Three-Way Nested Classification

For the three-way nested classification $C \prec B \prec A$ we assume the following model equation

$$\boldsymbol{y}_{ijkl} = \mu + \boldsymbol{a}_i + \boldsymbol{b}_{j(i)} + \boldsymbol{c}_{k(ij)} + \boldsymbol{e}_{l(ijk)},\ i = 1, \dots, a;\ j = 1, \dots, b_i; k = 1, \dots, c_{ij}, l = 1, \dots, n_{ijk}. \tag{6.28}$$

The conditions are: all random variables of the right-hand side of (6.28) have expectation 0 and are pairwise uncorrelated and var($\boldsymbol{a}_i$) = σ_a^2 for all i, var($\boldsymbol{b}_{ij}$) = σ_b^2 for all $i.j$, var($\boldsymbol{c}_{ijk}$) = σ_c^2 for all i, j, k and var($\boldsymbol{e}_{ijkl}$) = σ^2 for all i, j, k, l.

We find the ***SS***, ***df***, and ***MS*** of the three-way nested analysis of variance in Table 5.18. The $E(\boldsymbol{MS})$ for the random model can be found in Table 6.11.

Table 6.11 Expectations of the *MS* of a three-way nested classification for model II.

Source of variation	*E*(*MS*)
Between the *A* levels	$\sigma^2 + \sigma_a^2 \frac{N-\frac{D}{N}}{a-1} + \sigma_b^2 \frac{\lambda_3-\frac{E}{N}}{a-1} + \sigma_c^2 \frac{\lambda_2-\frac{F}{N}}{a-1}$
Between the *B* levels within the *A* levels	$\sigma^2 + \sigma_b^2 \frac{N-\lambda_3}{B-a} + \sigma_c^2 \frac{\lambda_1-\lambda_2}{B-a}$
Between the *C* levels within the *B* and *A* levels	$\sigma^2 + \sigma_c^2 \frac{N-\lambda_1}{C-B}$
Residual	σ^2

In Table 6.11 we use the abbreviations:

$$D = \sum_{i=1}^{a} N_{i..}^2,\quad E_i = \sum_{j=1}^{b} N_{ij.}^2, E = \sum_{i=1}^{a} E_i$$

$$F_{ij} = \sum_{k=1}^{c} n_{ijk}^2, F_i = \sum_{j=1}^{b} F_{ij}, F = \sum_{i=1}^{a} F_i,$$

$$\lambda_1 = \sum_{ij} \frac{F_i}{N_{ij.}}, \lambda_2 = \sum_i \frac{F_i}{N_{i..}}, \lambda_3 = \sum_i \frac{E_i}{N_{i..}}$$

Problem 6.15 Derive by the analysis of variance method the ANOVA estimates of the variance components in Table 6.11 and also the REML estimates.

Solution

$$s^2 = MS_{\text{res}}$$

$$s_c^2 = \frac{C_{..} - B_{.}}{N - \lambda_1}(MS_{C \text{ in } B} - MS_{\text{res}})$$

$$s_b^2 = \frac{B_{.} - a}{N - \lambda_3}\left(MS_{B \text{ in } A} - MS_{\text{res}} - \frac{\lambda_1 - \lambda_2}{B_{.} - a} s_c^2\right)$$

$$s_a^2 = \frac{a - 1}{n - \frac{D}{N}}\left(MS_A - MS_{\text{res}} - \frac{\lambda_2 - \frac{F}{N}}{a - 1} s_c^2 - \frac{\lambda_3 - \frac{F}{N}}{a - 1} s_b^2\right).$$

For the ANOVA estimates we use the R package `VCA` and for the REML estimates we use the R package `lme4`.

Example

In an experiment we have the random factor A with two classes; within the levels A_i we have the nested random factor B with two classes and within levels B_j we have the nested random factor C with respectively two and three classes. The observations y_{ijkl} are shown in Table 6.12.

For the data of such an experiment we first make in R the data-frame `problem13`.

```
> y <-  c(1, 2, 4, 3, 3.5, 4, 3.5, 4.5, 5, 7, 8, 6, 7, 8.5, 9, 10)
> a <-  rep(1:2, each = 8)
```

Table 6.12 Observations y_{ijkl} of a three-way nested classification model II.

			y	
A_1	B_{11}	C_{111}	1	2
		C_{112}	4	
	B_{12}	C_{121}	3	3.5
		C_{122}	4	3.5
		C_{123}	4.5	
A_2	B_{21}	C_{211}	5	7
		C_{212}	8	
	B_{22}	C_{221}	6	7
		C_{222}	8.5	9
		C_{223}	10	

```
> b1 <-  rep(1:2, times = c(3,5))
> b <- c(b1,b1)
> c <- c(1, 1, 2, 1, 1, 2, 2, 3, 1, 1, 2, 1, 1, 2, 2, 3)
> problem13 <- data.frame(a,b,c,y)
> library(VCA)
> anova13 <- anovaVCA(y~ a+ a:b+b:c,problem13,VarVC.method =c("scm"))
> anova13
Result Variance Component Analysis:
-----------------------------------

  Name  DF       SS       MS       VC        %Total    SD       CV[%]
1 total 1.470778                   11.649835 100       3.413186 63.501126
2 a     1        76.5625  76.5625  9.41335   80.802432 3.068118 57.081272
3 a:b   2        7.354167 3.677083 0.145241  1.246718  0.381104 7.090314
4 b:c   3        15.0875  5.029167 1.452819  12.470722 1.205329 22.42473
5 error 9        5.745833 0.638426 0.638426  5.480128  0.799016 14.865406

Mean: 5.375 (N = 16)
Experimental Design: unbalanced  |  Method: ANOVA
> library(lme4)
>lmer13 <- lmer( y ~1+(1|a)+(1|a:b)+(1|b:c), data= problem13)
> summary(lmer13)
Linear mixed model fit by REML ['lmerMod']
Formula: y ~ 1 + (1 | a) + (1 | a:b) + (1 | b:c)
   Data: problem13
REML criterion at convergence: 50.9
Scaled residuals:
     Min       1Q   Median       3Q      Max
-1.45734 -0.49905  0.07056  0.47646  1.22502
Random effects:
 Groups   Name        Variance Std.Dev.
 b:c      (Intercept) 1.6184   1.272
 a:b      (Intercept) 0.0000   0.000
 a        (Intercept) 9.4985   3.082
 Residual             0.5746   0.758
Number of obs: 16, groups:  b:c, 5; a:b, 4; a, 2
Fixed effects:
```

```
              Estimate Std. Error t value
(Intercept)      5.594      2.261   2.474
```

Hence the estimates of the variance components are $s_a^2 = 9.50; s^2_{b \text{ in } a} = 0; s^2_{c \text{ in } ab} = 1.62$ and $s^2 = 0.57$.

6.4.3 Three-Way Mixed Classifications

We consider the mixed classifications in Sections 5.5.3.1 and 5.5.3.2 but assume now random effects.

6.4.3.1 Cross-Classification Between Two Factors Where One of Them is Sub-Ordinated to a Third Factor ((*B* ≺ *A*)×*C*)

If in a balanced experiment a factor *B* is sub-ordinated to a factor *A* and both are cross-classified with a factor *C* then the corresponding model equation is given by

$$\boldsymbol{y}_{ijkl} = \mu + \boldsymbol{a}_i + \boldsymbol{b}_{j(i)} + \boldsymbol{c}_k + (\boldsymbol{a},\boldsymbol{c})_{ik} + (\boldsymbol{b},\boldsymbol{c})_{jk(i)} + \boldsymbol{e}_{ijkl}; i = 1, \dots, a; j = 1, \dots, b;$$
$$k = 1, \dots, c; l = 1, \dots, n \tag{6.29}$$

where μ is the general experimental mean, $\boldsymbol{a}_i$ is the random effect of the ith level of factor *A* with $E(\boldsymbol{a}_i) = 0$, $var(\boldsymbol{a}_i) = \sigma_a^2$. Further, $\boldsymbol{b}_{ij}$ is the random effect of the jth level of factor *B* within the ith level of factor *A*, with $E(\boldsymbol{b}_{ij}) = 0$, $var(\boldsymbol{b}_{ij}) = \sigma_b^2$ and $\boldsymbol{c}_k$ is the random effect of the *k*th level of factor *C*, with $E(\boldsymbol{c}_k) = 0$, $var(\boldsymbol{c}_k) = \sigma_c^2$. Further, $(\boldsymbol{a},\boldsymbol{c})_{ik}$ and$(\boldsymbol{b},\boldsymbol{c})_{jk(i)}$ are the corresponding random interaction effects with $E((\boldsymbol{a},\boldsymbol{c})_{ik}) = 0$, $var((\boldsymbol{a},\boldsymbol{c})_{ik}) = \sigma_{ac}^2$ and $E((\boldsymbol{b},\boldsymbol{c})_{jk(i)}) = 0$, $var\ ((\boldsymbol{b},\boldsymbol{c})_{jk(i)}) = \sigma_{bc}^2$. $\boldsymbol{e}_{ijkl}$ is the random error term with $E(\boldsymbol{e}_{ijkl}) = 0$, $var\ (\boldsymbol{e}_{ijkl}) = \sigma^2$. All the right-hand side random effects are uncorrelated with each other.

The ANOVA table with df and expected mean squares $E(\boldsymbol{MS})$ is as follows, $(N = abcn)$.

Problem 6.16 Derive by the analysis of variance method the ANOVA estimates of the variance components in Table 6.13 and also the REML estimates.

Solution

For the ANOVA estimates of the variance components, we use the R package `VCA`. For the REML estimates, we use the R package `lme4`.

Table 6.13 ANOVA table with df and expected mean squares *E*(*MS*) of model (6.29).

Source of variation	df	E(MS)
Between the levels of *A*	$a - 1$	$bcn\,\sigma_a^2 + cn\,\sigma_{b(a)}^2 + bn\,\sigma_{ac}^2 + n\,\sigma_{bc(a)}^2 + \sigma^2$
Between the levels of *B* within the levels of *A*	$a(b - 1)$	$cn\,\sigma_{b(a)}^2 + n\,\sigma_{bc(a)}^2 + \sigma^2$
Between the levels of *C*	$c - 1$	$abn\,\sigma_c^2 + bn\,\sigma_{ac}^2 + n\,\sigma_{bc(a)}^2 + \sigma^2$
Interaction *A* × *C*	$(a - 1)(c - 1)$	$bn\,\sigma_{ac}^2 + n\,\sigma_{bc(a)}^2 + \sigma^2$
Interaction *B* × *C* within the levels of *A*	$a(b - 1)(c - 1)$	$n\,\sigma_{bc(a)}^2 + \sigma^2$
Residual	$N - abc$	σ^2
Corrected total	$N - 1$	

Example

The factor A is random with three classes, the factor B within the levels of A is random with two levels, the factor C is random with three classes. The data are y and given in Table 6.14.

```
> y1 <-  c(141.2, 142.6, 135.7, 136.8, 163.2, 163.3)
> y2 <-  c(51.2, 51.4, 143.0, 143.3, 181.4, 180.3)
> y3 <-  c(189.8, 190.3, 132.4, 130.3, 173.6, 173.9)
> y4 <-  c(191.5, 193.0, 134.4, 130.0, 174.9, 175.6)
> y5 <-  c(141.9, 142.7, 137.4, 135.2, 166.6, 165.5)
> y6 <-  c(145.5, 144.7, 141.1, 139.1, 175.0, 172.0)
>  a  <-  rep(1 : 3, each = 12)
>  b  <-  rep(c(11,12,21,22,31,32) , each = 6)
> c1 <- rep(1:3, each = 2)
> y  <- c(y1, y2, y3, y4, y5, y6)
> c  <- c(c1, c1, c1, c1, c1,c1)
> problem14 <- data.frame(a, b, c, y)
> library(VCA)
> anova14 <- anovaVCA(y~ a + a/b + c + a:c + b:c ,problem14,      VarVC.method =c("scm"))
> anova14
Result Variance Component Analysis:
-----------------------------------
  Name  DF        SS            MS            VC           %Total    SD        CV[%]
1 total 11.119385                             1193.339549 100       34.544747 22.903438
2 a     2         5290.877222  2645.438611  10.75066     0.900889  3.27882   2.173883
3 a:b   3         1529.105     509.701667   0*           0*        0*        0*
4 c     2         8467.617222  4233.808611  86.670243    7.262832  9.309685  6.172394
5 a:c   4         12775.062778 3193.765694  501.682257   42.040194 22.398265 14.850225
6 b:c   6         7122.22      1187.036667  592.800278   49.675742 24.34749  16.142577
7 error 18        25.85        1.436111     1.436111     0.120344  1.198379  0.794534
Mean: 150.8278 (N = 36)
Experimental Design: balanced  |  Method: ANOVA | * VC set to 0 | adapted MS used for total DF
>library(lme4)
> lmer14 <- lmer(y~1+(1|a)+(1|a:b)+(1|c)+(1|a:c)+(1|b:c),problem14)
> summary(lmer14)
Linear mixed model fit by REML ['lmerMod']
Formula: y ~ 1 + (1 | a) + (1 | a:b) + (1 | c) + (1 | a:c) + (1 | b:c)
   Data: problem14

REML criterion at convergence: 236
Scaled residuals:
     Min       1Q    Median       3Q      Max
-1.84123 -0.44611 -0.02025  0.45375  1.83040
Random effects:
 Groups   Name        Variance  Std.Dev.
 b:c      (Intercept) 4.799e+02 2.191e+01
 a:c      (Intercept) 5.124e+02 2.264e+01
 a:b      (Intercept) 1.329e-15 3.646e-08
```

Table 6.14 Data in a three-way mixed classification (($B \prec A$)xC) model II.

y	A_1		A_2		A_3	
	B_{11}	B_{12}	B_{21}	B_{22}	B_{31}	B_{32}
C_1	141.2	91.2	189.8	191.5	141.9	145.5
	142.6	91.4	190.3	193.0	142.7	144.7
C_2	135.7	143.0	132.4	134.4	137.4	141.1
	136.8	143.3	130.3	130.0	135.2	139.1
C_3	163.2	181.4	173.6	174.9	166.6	175.0
	163.3	180.3	173.9	175.6	165.5	172.0

```
 c          (Intercept) 1.019e+02 1.009e+01
 a          (Intercept) 8.109e-13 9.005e-07
 Residual               1.436e+00 1.198e+00
Number of obs: 36, groups:  b:c, 18; a:c, 9; a:b, 6; c, 3; a, 3
Fixed effects:
            Estimate Std. Error t value
(Intercept)   150.83      10.84   13.91
```

Hence the estimates of the variance components are $s_a^2 = 0; s_{b \text{ in } a}^2 = 0; s_c^2 = 101.9;$ $s_{ab}^2 = 0; s_{ac}^2 = 512.4; s^2{}_{bc} = 479.9$ and $s^2 = 1.44$.

6.4.3.2 Cross-Classification of Two Factors in Which a Third Factor is Nested ($C \prec (A \times B)$)

The model equation for this type is

$$y_{ijkl} = \mu + \boldsymbol{a}_i + \boldsymbol{b}_j + \boldsymbol{c}_{k(ij)} + (\boldsymbol{a},\boldsymbol{b})_{ij} + \boldsymbol{e}_{ijkl}; i = 1, \dots, a; j = 1, \dots, b; k = 1, \dots, c; l = 1, \dots, n \tag{6.30}$$

where μ is the general experimental mean, $\boldsymbol{a}_i$ is the effect of the ith level of factor A with $E(\mathbf{a}_i) = 0$, $var(\boldsymbol{a}_i) = \sigma_a^2$. Further, $\boldsymbol{b}_j$ is the effect of the jth level of factor B with $E(\boldsymbol{b}_j) = 0$, $var\ (\boldsymbol{b}_j) = \sigma_b^2$; $\boldsymbol{c}_{ijk}$ is the effect of the kth level of factor C within the combinations of $A \times B$, with $E(\boldsymbol{c}_{ijk}) = 0$, $var\ (\boldsymbol{c}_{ijk}) = \sigma_{c(ab)}^2$. Further $(\boldsymbol{a},\boldsymbol{b})_{ij}$ is the corresponding random interaction effect with $E((\boldsymbol{a},\boldsymbol{b})_{ij}) = 0$, $var\ ((\boldsymbol{a},\boldsymbol{b})_{ij}) = \sigma_{ab}^2$ and $\boldsymbol{e}_{ijkl}$ is the random error term with $E(\boldsymbol{e}_{ijkl}) = 0$, $var\ (\boldsymbol{e}_{ijkl}) = \sigma^2$. The right-hand side random effects are uncorrelated with each other.

The ANOVA table with df and expected mean squares $E(\boldsymbol{MS})$ is as follows, ($N = abcn$).

Problem 6.17 Derive by the analysis of variance method the ANOVA estimates of the variance components in Table 6.15 and also the REML estimates.

Solution

For the ANOVA estimates of the variance components we use the R package VCA. For the REML estimates we use the R package `lme4`.

Example

The factor A is random with three classes, the factor B is random with two classes, the factor C within the levels of $A \times B$ is random with two levels. The data are y and comes from Kuehl, page 254 (1994), see Table 6.16.

Table 6.15 ANOVA table with df and expected mean squares *E(MS)* of model (6.30).

Source of variation	*df*	*E(MS)*
Between A levels	$a - 1$	$bcn\sigma_a^2 + n\sigma_{c(ab)}^2 + cn_{ab}^2 + \sigma^2$
Between B levels	$b - 1$	$acn\sigma_b^2 + n\sigma_{c(ab)}^2 + cn_{ab}^2 + \sigma^2$
Between C levels in $A \times B$ combinations	$ab(c - 1)$	$cn\ \sigma_{ab}^2 + \sigma^2$
Interaction $A \times B$	$(a - 1)(b - 1)$	$n\sigma_{c(ab)}^2 + cn\ \sigma_{ab}^2 + \sigma^2$
Residual	$N - abc$	σ^2

Table 6.16 Data in a three-way mixed classification $C \prec (AxB)$ model II.

y	A_1		A_2		A_3	
	C_{111}	C_{112}	C_{121}	C_{122}	C_{131}	C_{132}
B_1	3.833	3.819	3.756	3.882	3.720	3.729
	3.866	3.853	3.757	3.871	3.720	3.768
	C_{211}	C_{212}	C_{221}	C_{222}	C_{231}	C_{232}
B_2	3.932	3.884	3.832	3.917	3.776	3.833
	3.943	3.888	3.829	3.915	3.777	3.827

```
> y1 <- c( 3.833, 3.866, 3.932, 3.943)
> y2  <- c( 3.819, 3.853, 3.884, 3.888)
> y3 <-  c( 3.756, 3.757, 3.832, 3.829)
> y4 <-  c(  3.882, 3.871, 3.917, 3.915)
> y5 <-  c( 3.720, 3.720, 3.776, 3.777)
> y6 <-  c( 3.729, 3.768, 3.833, 3.827)
> y <- c(y1,y2,y3,y4,y5,y6)
> a <- rep(1 : 3, each = 8)
> b1 <- rep(1 : 2, each = 2)
> b <- c(b1,b1,b1,b1,b1,b1)
> c<- rep(1 : 12, each = 2)
> problem15 <- data.frame(a,b,c,y)
> library(VCA)
> anova15 <- anovaVCA(y~ a+b+(a:b)/c,problem15,
         VarVC.method =c("scm"))
> anova15
Result Variance Component Analysis:
-----------------------------------------------

  Name  DF        SS       MS       VC       %Total     SD       CV[%]
1 total 4.399143                    0.007625 100        0.087323 2.279801
2 a      2        0.049641 0.024821 0.00309  40.523749 0.055588 1.451282
3 b      1        0.025285 0.025285 0.002099 27.523462 0.045812 1.196048
4 a:b    2        2e-04    1e-04    0*       0*         0*       0*
5 a:b:c 6         0.028219 0.004703 0.002267 29.726104 0.04761  1.242985
6 error 12        0.002038 0.00017  0.00017  2.226685  0.01303  0.340194
Mean: 3.830292 (N = 24)
Experimental Design: balanced  |  Method: ANOVA | * VC set to 0 | adapted MS used for total DF.
> library(lme4)
>lmer15 <- lmer(y~1+(1|a)+(1|b)+(1|a:b)+(1|(a:b):c),problem15)
> summary(lmer15)
Linear mixed model fit by REML ['lmerMod']
Formula: y ~ 1 + (1 | a) + (1 | b) + (1 | a:b) + (1 | (a:b):c)
   Data: problem15
REML criterion at convergence: -91.9
Scaled residuals:
     Min       1Q  Median      3Q     Max
-1.5008 -0.2124 -0.1088  0.2623  1.4922
Random effects:
 Groups   Name        Variance  Std.Dev.
 (a:b):c  (Intercept) 0.0016913 0.04113
 a:b      (Intercept) 0.0000000 0.00000
 a        (Intercept) 0.0026585 0.05156
 b        (Intercept) 0.0018111 0.04256
 Residual             0.0001698 0.01303
Number of obs: 24, groups:  (a:b):c, 12; a:b, 6; a, 3; b, 2
Fixed effects:
            Estimate Std. Error t value
(Intercept)  3.83029    0.04404   86.97.
```

Hence the estimates of the variance components are $s_a^2 = 0.0027; s_b^2 = 00018; s_{c \text{ in } a}^2 = 0.0017; s_{ab}^2 = 0$; and $s^2 = 0.0002$.

Finally, a general remark on the examples for estimating variance components. Usually, a huge number of the corresponding factor levels is needed and often available. For demonstrating the R-programs, we used smaller data sets to show how to proceed.

References

Anderson, R.L. and Bancroft, T.A. (1952). *Statistical Theory in Research*. New York: McGraw-Hill.

Federer, W.T. (1968). Non-negative estimators for components of variance. *Appl. Stat.* 17: 171–174.

Gaylor, D.W. and Hopper, F.N. (1969). Estimating the degree of freedom for linear combinations of mean squares by Satterthwaite's formula. *Technometrics* 11: 691–706.

Gelman, A., Carlin, J.B., Stern, H.S., and Rubin, D.B. (1995). *Bayesian Data Analysis*. New York: Chapman and Hall.

Hammersley, J.M. (1949). The unbiased estimate and standard error of the interclass variance. *Metron* 15: 189–204.

Hartley, H.O. (1967). Expectations, variances and covariances of ANOVA mean squares by "synthesis". *Biometrics* 23: 105–114.

Harville, D.A. (1977). Maximum-likelihood approaches to variance component estimation and to related problems. *J. Am. Stat. Assoc.* 72: 320–340.

Henderson, C.R. (1953). Estimation of variance and covariance components. *Biometrics* 9: 226–252.

Klotz, J.H., Milton, R.C., and Zacks, S. (1969). Mean square efficiency of estimators of variance components. *J. Am. Stat. Assoc.* 64: 1383–1402.

Kuehl, R.O. (1994). *Statistical Principles of Research Design and Analysis*. Belmont, California: Duxbury Press.

Patterson, H.D. and Thompson, R. (1971). Recovery of inter-block information when block sizes are unequal. *Biometrika* 58: 545–554.

Rasch, D. and Schott, D. (2018). *Mathematical Statistics*. Oxford: Wiley.

Sarhai, H. and Ojeda, M.M. (2004). *Analysis of Variance for Random Models, Balanced Data*. Basel/Berlin: Birkhäuser, Boston.

Sarhai, H. and Ojeda, M.M. (2005). *Analysis of Variance for Random Models, Unbalanced Data*. Basel/Berlin: Birkhäuser, Boston.

Satterthwaite, F.E. (1946). An approximate distribution of estimates of variance components. *Biom. Bull.* 2: 110–114.

Searle, S.R., Casella, G., and McCulloch, C.E. (1992). *Variance Components*. New York/Chichester/Brisbane/Toronto/Singapore: Wiley.

Seely, J.F. and Lee, Y. (1994). A note on Satterthwaite confidence interval for a variance. *Commun. Stat.* 23: 859–869.

Thompson, W.A. Jr. (1962). Negative estimates of variance components. *Ann. Math. Stat.* 33: 273–289.

Tiao, G.C. and Tan, W.Y. (1965). Bayesian analysis of random effects models in the analysis of variance I: posterior distribution of variance components. *Biometrika* 52: 35–53.

Townsend, E.C. (1968) *Unbiased estimators of variance components in simple unbalanced designs*. PhD thesis, Cornell Univ., Ithaca, USA.
Verdooren, L.R. (1980). On estimation of variance components. *Stat. Neerl.* 34: 83–106.
Verdooren, L.R. (1982). How large is the probability for the estimate of variance components to be negative? *Biom. J.* 24: 339–360.
Williams, J.S. (1962). A confidence interval for variance components. *Biometrika* 49: 278–281.

7

Analysis of Variance – Mixed Models

7.1 Introduction

In mixed models, as well fixed effects as in Chapter 5, random effects as in Chapter 6 also occur, i.e. mixed models are models where in the model equation at least one, but not all, effects are random variables.

In mixed models, we discuss problems of variance component estimation and of estimating and testing fixed effects.

Therefore, we need expected mean squares from the tables of Chapter 5 if the effect defining a row is fixed and the tables of Chapter 6 if the effect defining a row is random.

An interaction is random if at least one of the factors involved is random. In discussing the different classifications, we use the same procedure as in Chapters 5 and 6. Of course, in a one-way classification a mixed model is impossible. We therefore start with the two-way classification.

7.2 Two-Way Classification

We discuss here the cross-classification where we consider the factor A to be fixed without loss of generality. If the factor B is fixed, we rename both factors. In the nested classification, two-mixed models occur when the super-ordinate factor or the nested factor is random. We write random factors with their elements in bold.

7.2.1 Balanced Two-Way Cross-Classification

We consider two cross-classified factors A (fixed) and $\boldsymbol{B}$ (random) and their interactions $A\boldsymbol{B}$.

The model equation of the balanced case is

$$\boldsymbol{y}_{ijk} = \mu + a_i + \boldsymbol{b}_j + (\boldsymbol{a}, \boldsymbol{b})_{ij} + \boldsymbol{e}_{ijk}, i = 1, \dots, a;\ j = 1, \dots, b, k = 1, \dots, n \tag{7.1}$$

with side conditions

$$\sum_{i=1}^{a} a_{\mathrm{i}} = 0,$$

$$\mathrm{var}(\boldsymbol{b}_j) = \sigma_b^2 \quad \text{for all } j, \quad \mathrm{cov}(\boldsymbol{b}_j, \boldsymbol{b}_k) = 0, \quad \text{for all } j, k \text{ with } j \neq k,$$

Applied Statistics: Theory and Problem Solutions with R, First Edition.
Dieter Rasch, Rob Verdooren, and Jürgen Pilz.

$$var((\boldsymbol{a},\boldsymbol{b})_{ij}) = \sigma^2_{ab} \quad \text{for all } i,j, \tag{7.2}$$

$$\text{cov}(\boldsymbol{b}_j, (\boldsymbol{a},\boldsymbol{b})_{ij'}) = \text{cov}(\boldsymbol{b}_{j'}, \boldsymbol{e}_{ijk}) = \text{cov}((\boldsymbol{a},\boldsymbol{b})_{i'j'}, \boldsymbol{e}_{ijk}) = 0$$

and

$$\text{cov}((\boldsymbol{a},\boldsymbol{b})_{ij}, (\boldsymbol{a},\boldsymbol{b})_{ij'}) = 0 \;\; (j \neq j')\,(\text{case I}) \tag{7.3}$$

or

$$\sum_{i=1}^{a} (\boldsymbol{a},\boldsymbol{b})_{ij} = 0 \quad \text{for all } j\,(\text{case II) with var}((\boldsymbol{a},\boldsymbol{b})_{\text{ij}}) = \sigma^{*2}_{ab}. \tag{7.4}$$

If we use (7.4) the term $\overline{a} = \frac{1}{a}\sum_{i=1}^{a} a_i$ vanishes, and $(\boldsymbol{a},\boldsymbol{b})_{ij}$ and $(\boldsymbol{a},\boldsymbol{b})_{i'j}$ $(i \neq i'; j = 1, \dots, b)$ are correlated; the covariance is $\text{cov}((\boldsymbol{a},\boldsymbol{b})_{ij}, (\boldsymbol{a},\boldsymbol{b})_{i'j}) = \sigma_{ab}$ for all j and $i \neq i'$.

Because $\text{var}\left(\sum_{i=1}^{a} (\boldsymbol{a},\boldsymbol{b})_{ij}\right) = \text{var}(0) = 0 = \sum_{i=1}^{a} \sigma^{*2}_{ab} + \sum_{i=1}^{a} \sum_{\substack{i'=1 \\ i\neq i'}}^{a} \sigma_{ab} = a\sigma^{*2}_{ab} + a(a-1)\sigma_{ab}$,
the covariance between the $(\boldsymbol{a},\boldsymbol{b})_{ij}$ and $(\boldsymbol{a},\boldsymbol{b})_{i'j}(i \neq i'; j = 1, \dots, b)$ is $\sigma_{ab} = -\frac{1}{a-1}\sigma^{*2}_{ab}$.

In case II we define $\text{var}(\boldsymbol{bj}) = \sigma^{*2}_b$ for all j.

Searle (1971) and Searle et al. (1992) clearly recorded the relations between the two cases. He showed that σ^2_b in case I and σ^{*2}_b in case II changed in their meaning.

Problem 7.1 Derive the analysis of variance (ANOVA) estimators of all variance components using case I in the fourth column of Table 7.1.

Solution

Use the algorithm for ANOVA estimators from Chapter 6 and get

$$\left.\begin{aligned} s^2 &= MS_{\text{res}}, \\ s^2_{ab} &= \frac{1}{n}(MS_{AB} - MS_{\text{res}}), \\ s^2_b &= \frac{1}{an}(MS_B - MS_{AB}). \end{aligned}\right\} \tag{7.5}$$

Example 7.1 In a variety trial, five barley varieties were tested, in a completely randomised design with two plots per variety, at six randomly chosen locations in a certain arable region. The yield per plot y in kilograms is given in Table 7.2. Here is $a = 5$ and $n = 2$.

We use first the R package VCA (variance component analysis) to get the ANOVA estimates for the variance components.

```
> y1 <- c(119.8,98.9,86.9,77.7,98.9,86.4,92.3,103.1,81.0,80.7,
       146.6,140.4)
> y2 <- c(124.8,115.7,96.0,94.1,104.1,100.3,99.6,129.6,98.3,84.2,
       145.7,148.1)
> y3 <- c(121.4,111.9,77.1,66.7,89.0,79.9,97.3,105.1,105.4,92.3,
        142.0,145.5)
> y4 <- c(140.8,125.5,101.8,91.8,89.3,61.9,131.3,139.9,109.7,97.2,
         191.5,187.7)
```

Table 7.1 Expectations of the *MS* in Table 5.10 for a Mixed model (Levels of *A* fixed) for two side conditions.

Source of Variation	***df***	***MS***	***E*(*MS*) Case I**	***E*(*MS*) Case II**
Between rows (*A*)	$a-1$	$\frac{SS_A}{a-1}$	$\frac{bn}{a-1}\sum_{i=1}^{a}(a_i-\overline{a})^2+n\sigma_{ab}^2+\sigma^2$	$\frac{bn}{a-1}\sum_{i=1}^{a}a_i^2+\frac{na}{a-1}\sigma_{ab}^{*2}+\sigma^2$
Between columns (***B***)	$b-1$	$\frac{SS_B}{b-1}$	$an\sigma_b^2+n\sigma_{ab}^2+\sigma^2$	$an\sigma_b^{*2}+\sigma^2$
Interactions	$(a-1)(b-1)$	$\frac{SS_{AB}}{(a-1)(b-1)}$	$n\sigma_{ab}^2+\sigma^2$	$\frac{na}{a-1}\sigma_{ab}^{*2}+\sigma^2$
Within classes (residual)	$ab(n-1)$	$\frac{SS_{res}}{ab(n-1)}=s^2$	σ^2	σ^2

Table 7.2 Yield per plot *y* in kilograms of a variety trial.

y	Variety				
	1	2	3	4	5
Location					
1	119.8	124.8	121.4	140.8	124.0
	98.9	115.7	111.9	125.5	106.2
2	86.9	96.0	77.1	101.8	78.9
	77.7	94.1	66.7	91.8	67.4
3	98.9	104.1	89.0	89.3	69.1
	86.4	100.3	79.9	61.9	76.7
4	92.3	99.6	97.3	131.3	108.4
	103.1	129.6	105.1	139.9	116.5
5	81.0	98.3	105.4	109.7	119.7
	80.7	84.2	92.3	97.2	100.4
6	146.6	145.7	142.0	191.5	150.7
	140.4	148.1	145.5	187.7	142.2

```
> y5 <- c(124.0,106.2,78.9,67.4,69.1,76.7,108.4,116.5,119.7,100.4,
        150.7,142.2)
> y <- c(y1, y2, y3, y4, y5)
> variety<- rep(1:5, each = 12)
> loc1 <- rep(1:6 , each = 2)
> location <- c(loc1,loc1,loc1,loc1,loc1)
> V <- factor(variety)
> L <- factor(location)
> problem7.1 <- data.frame( V, L, y)
> library(VCA)
> MI <- anovaMM(y ~ V + (L) + (V:L), VarVC.method ="scm",problem7.1)
> MI
ANOVA-Type Estimation of Mixed Model:
- - - - - - - - - - - - - - - - - - - -
        [Fixed Effects]
       int          V1          V2          V3          V4          V5
105.016667   -3.958333   6.691667   -2.216667   17.350000   0.000000
        [Variance Components]
  Name  DF    SS          MS          VC          %Total      SD          CV[%]
1 total 7.2808                        839.052767  100         28.966408   26.675023
2 L     5     34656.704   6931.3408   666.885492  79.480757   25.824126   23.781311
3 V:L   20    5249.717667 262.485883  90.318608   10.764354   9.50361     8.751828
4 error 30    2455.46     81.848667   81.848667   9.754889    9.047025    8.331361
Mean: 108.59 (N = 60)
Experimental Design: balanced  |  Method: ANOVA
```

Now we estimate the restricted maximum likelihood (REML*)* estimates for the variance components.

```
> library(lme4)
> MIreml <- lmer( y ~ V + (1|L) + (1|V:L), problem7.1)
> summary(MIreml)
Linear mixed model fit by REML ['lmerMod']
```

```
Formula: y ~ V + (1 | L) + (1 | V:L)
   Data: problem7.1
REML criterion at convergence: 456.3
Scaled residuals:
     Min       1Q   Median       3Q      Max
-2.36249 -0.51610  0.08722  0.58540  1.63431
Random effects:
 Groups   Name        Variance Std.Dev.
 V:L      (Intercept)  90.32    9.504
 L        (Intercept) 666.89   25.824
 Residual              81.85    9.047
Number of obs: 60, groups:  V:L, 30; L, 6
Fixed effects:
            Estimate Std. Error t value
(Intercept)  101.058     11.533   8.762
V2            10.650      6.614   1.610
V3             1.742      6.614   0.263
V4            21.308      6.614   3.222
V5             3.958      6.614   0.598
Correlation of Fixed Effects:
   (Intr) V2     V3     V4
V2 -0.287
V3 -0.287  0.500
V4 -0.287  0.500  0.500
V5 -0.287  0.500  0.500  0.500
```

Because the design is balanced and the ANOVA estimates for the variance components are positive, the ANOVA estimates are the same as the REML estimates.

Note that the intercept 101.058 is the mean of variety *V1*.

The estimate of the fixed effect *V2* is the difference of the means of variety *V2* and variety *V1*: 111.7083–101.0583 = 10.650. Analogously estimating *V3* = 1.742 is the difference between the means of variety *V3* and *V1*: 102.8–101.0583, etc.

Problem 7.2 Derive the ANOVA estimators of all variance components using case II in the last column of Table 7.1.

Solution

Use the algorithm for ANOVA estimators from Chapter 6 and obtain

$$\left.\begin{aligned} s^2 &= MS_{\text{res}}, \\ s_{ab}^{*2} &= \frac{a-1}{na}(MS_{AB} - MS_{\text{res}}), \\ s_b^{*2} &= \frac{1}{an}(MS_B - M_{\text{res}}). \end{aligned}\right\} \tag{7.6}$$

Example

We use the dataset of Problem 7.1 with the data-frame `problem7.1`.

From the output of the VCA package we found from the ANOVA table the mean squares MSRes = 81.848667, MSAB = 262.485883, MSB = 6931.3408. Further, $a = 5$ and $n = 2$.

```
> MSRes <- 81.848667
> MSAB <- 262.405883
> MSB <- 6931.3408
> s2 <- MSRes
> s2
[1] 81.84867
> a <- 5
> n <- 2
> s2ab <- ((a-1)/(n*a))*(MSAB - MSRes)
> s2ab
[1] 72.22289
> s2b <- (1/(a*n))*(MSB-MSRes)
> s2b
[1] 684.9492
```

We found the following estimates of the variance components

$$s_b^2 = 684.94;\ s_{ab}^2 = 72.22;\ s^2 = 81.85.$$

The null hypotheses that can be tested are:

H_{01}: 'all a_i are equal'
H_{02}: '$\sigma_b^2 = 0$'
H_{03}: '$\sigma_{ab}^2 = 0$.'

If H_{01} is true, then $E(\boldsymbol{MS}_A)$ equals $E(\boldsymbol{MS}_{AB})$, for case I and case II, and we test H_{01} using

$$\boldsymbol{F}_A = \frac{\boldsymbol{MS}_A}{\boldsymbol{MS}_{AB}}, \tag{7.7}$$

which, under H_{01}, has an F-distribution with $(a-1)$ and $(a-1)(b-1)$ degrees of freedom.

Problem 7.3 Test H_{01}: 'all a_i are equal' against H_{A1}: 'not all a_i are equal' with significance level $\alpha = 0.05$.

Example
We use the dataset of Problem 7.1 with the data-frame `problem7.1`.
To get the MSA we use the ANOVA table from the base package with `aov()`.

```
> Anovafixed <- aov( y ~ V *L , data = problem7.1)
> summary(Anovafixed)
            Df Sum Sq Mean Sq F value   Pr(>F)
V            4   3630     908  11.089 1.22e-05 ***
L            5  34657    6931  84.685  < 2e-16 ***
V:L         20   5250     262   3.207  0.00196 **
Residuals   30   2455      82
```

```
---
Signif. codes:  0 '***' 0.001 '**' 0.01 '*' 0.05 '.' 0.1 ' ' 1
> MSA <- 908
> dfA <- 4
> MSAB <- 262
> dfAB <- 20
> FA <- MSA/MSAB
> FA
[1] 3.465649
> p_value <- 1- pf(FA, dfA, dfAB)
> p_value
[1] 0.02631101
```

Because the p-value $= 0.026 < 0.05$, H_{01}: 'all a_i are zero' is, with significance level $\alpha = 0.05$, rejected.

Problem 7.4 Show the F-statistics and their degrees of freedom to test H_{02}: '$\sigma_b^2 = 0$' and H_{03}: '$\sigma_{ab}^2 = 0$' each with significance level $\alpha = 0.05$.

Solution

If H_{02} is true, then for case I $E(\boldsymbol{MS}_B)$ equals $E(\boldsymbol{MS}_{AB})$, and we test H_{02} using

$$\boldsymbol{F}_{BI} = \frac{\boldsymbol{MS}_B}{\boldsymbol{MS}_{AB}} \tag{7.8}$$

which under H_{02} has an F-distribution with $(b-1)$ and $(a-1)(b-1)$ degrees of freedom.

If H_{02} is true, then for case II $E(\boldsymbol{MS}_B)$ equals $E(\boldsymbol{MS}_{\text{res}})$, and we test H_{02} using

$$\boldsymbol{F}_{BII} = \boldsymbol{MS}_B/\boldsymbol{MS}_{\text{res}}.$$

If H_{03} is true, then for case I and case II $E(\boldsymbol{MS}_{AB})$ equals $E(\boldsymbol{MS}_{\text{res}})$ and we test H_{03} using

$$\boldsymbol{F}_{AB} = \frac{\boldsymbol{MS}_{AB}}{\boldsymbol{MS}_{\text{res}}} \tag{7.9}$$

which under H_{02} has an F-distribution with $(a-1)(b-1)$ and $ab(n-1)$ degrees of freedom.

Example

We use the same dataset of Problem 7.1 with the data-frame `problem7.1`. To get the MSA we use the fixed ANOVA table of Problem 7.3 .

For case I we get:

```
> MSB <- 6931
> dfB <- 5
> MSAB <- 262
> dfAB <- 20
> FBI <- MSB/MSAB
> FBI
[1] 26.4542
```

```
> p_value <- 1-pf(FBI, dfB, dfAB)
> p_value
[1] 3.582138e-08
```

Because the p-value $= 3.58\ \text{e} - 08 < 0.05$, we reject for case I the hypothesis H_{02}: '$\sigma_b^2 = 0$' with significance level $\alpha = 0.05$.

For case II we get:

```
> MSB <- 6931
> dfB <- 5
> MSRes <- 82
> dfRes <- 30
> FBII <- MSB/MSRes
> FBII
[1] 84.52439
> p_value <- 1-pf(FBII, dfB, dfRes)
> p_value
[1] 1.110223e-16
```

Because the p-value $= 1.11\ \text{e} - 16 < 0.05$, we reject for case II the hypothesis H_{02}: '$\sigma_b^2 = 0$' with significance level $\alpha = 0.05$.

For the test of H_{03}: '$\sigma_{ab}^2 = 0$' with significance level $\alpha = 0.05$ we use the fixed ANOVA table of Problem 7.3.

```
> MSAB <- 262
> dfAB <- 20
> MSRes <- 82
> dfRes <- 30
> FAB <- MSAB/MSRes
> p_value <- 1 - pf(FAB, dfAB , dfRes)
> p_value
[1] 0.002014215
```

Because the p-value $= 0.02 < 0.05$, we reject H_{03}: '$\sigma_{ab}^2 = 0$' with significance level $\alpha = 0.05$.

Problem 7.5 Estimate the variance components in an unbalanced two-way mixed model for Example 7.2.

Example 7.2 In a variety trial, five barley varieties were tested, in a completely randomised design with two plots per variety, at six randomly chosen locations in a certain arable region. Due to heavy rain and stormy weather, some plots were destroyed. The yield per plot y in kilograms is given in Table 7.3. These are indicated with a * .

Solution and Example

In unbalanced two-way mixed models, we can estimate the variance components with the package VCA, but the ANOVA estimates are different from the REML estimates. The REML estimates are preferred.

Table 7.3 Yield per plot *y* in kilograms of a variety-location, two-way classification.

y	Variety				
	1	2	3	4	5
Location					
1	119.8	124.8	121.4	140.8	124.0
	98.9	*	111.9	125.5	106.2
2	86.9	96.0	77.1	101.8	78.9
	77.7	94.1	66.7	*	67.4
3	98.9	104.1	89.0	89.3	69.1
	86.4	100.3	79.9	61.9	76.7
4	92.3	99.6	*	131.3	108.4
	103.1	129.6	105.1	139.9	116.5
5	81.0	98.3	105.4	109.7	119.7
	80.7	84.2	92.3	97.2	100.4
6	146.6	145.7	142.0	191.5	*
	140.4	148.1	145.5	187.7	142.2

```
> y1 <- c(119.8,98.9,86.9,77.7,98.9,86.4,92.3,103.1,81.0,80.7,146.6,140.4)
> y2 <- c(124.8, NA,96.0,94.1,104.1,100.3,99.6,129.6,98.3,84.2,145.7,148.1)
> y3 <- c(121.4,111.9,77.1,66.7,89.0,79.9, NA,105.1,105.4,92.3,142.0,145.5)
> y4 <- c(140.8,125.5,101.8,
        NA,89.3,61.9,131.3,139.9,109.7,97.2,191.5,187.7)
> y5 <- c(124.0,106.2,78.9,67.4,69.1,76.7,108.4,116.5,119.7,100.4,
        NA,142.2)
> y <- c(y1, y2, y3, y4, y5)
> variety<- rep(1:5, each = 12)
> loc1 <- rep(1:6 , each = 2)
> location <- c(loc1,loc1,loc1,loc1,loc1)
> V <- factor(variety)
> L <- factor(location)
> problem7.2 <- data.frame( V, L, y)
> library(VCA)
> M7.2VCA <- anovaMM( y ~V + (L) + (V:L), VarVC.method = "scm", problem7.2)
There are 4 missing values for the response variable (obs: 14, 31, 40, 59)!
> M7.2VCA
ANOVA-Type Estimation of Mixed Model:
- - - - - - - - - - - - - - - - - - - - -
        [Fixed Effects]

       int          V1          V2          V3          V4          V5
104.510508   -3.452175    7.872415   -0.956779   18.614055    0.000000
        [Variance Components]

  Name  DF       SS            MS            VC         %Total     SD
1 total 7.456441                             832.675268 100        28.856113
2 L     5        31494.535185 6298.907037 650.405479 78.11034   25.503048
3 V:L   20       5223.503982  261.175199  93.904019  11.277388 9.690409
4 error 26       2297.51      88.365769   88.365769  10.612273 9.400307
  CV[%]
1 26.666155
```

```
2 23.567562
3 8.954981
4 8.686896
Mean: 108.2125 (N = 56, 4 observations removed due to missing data)
Experimental Design: unbalanced  |  Method: ANOVA
> library(lme4)
> M7.2 <- lmer( y ~ V + (1|L) + (1|V:L), problem7.2)
> summary(M7.2)
Linear mixed model fit by REML ['lmerMod']
Formula: y ~ V + (1 | L) + (1 | V:L)
   Data: problem7.2
REML criterion at convergence: 427.3
Scaled residuals:
     Min       1Q   Median       3Q      Max
-2.33220 -0.41156  0.09677  0.60403  1.62288
Random effects:
 Groups   Name        Variance Std.Dev.
 V:L      (Intercept)  92.88    9.637
 L        (Intercept) 651.93   25.533
 Residual              86.21    9.285
Number of obs: 56, groups:  V:L, 30; L, 6
Fixed effects:
            Estimate Std. Error t value
(Intercept)  101.058     11.459   8.819
V2            11.326      6.806   1.664
V3             2.495      6.806   0.367
V4            22.066      6.806   3.242
V5             3.452      6.806   0.507
Correlation of Fixed Effects:
   (Intr) V2     V3     V4
V2 -0.291
V3 -0.291  0.489
V4 -0.291  0.489  0.489
V5 -0.291  0.489  0.489  0.489
```

For the test of the fixed effect V we use the R package `lmerTest`. The default is Satterthwaite's method to calculate the df of the denominator in the F-test.

```
> library(lmerTest)
> M7.2test <- lmer( y ~ V + (1|L) + (1|V:L), problem7.2)
> TestV <- anova(M7.2test)
> TestV
Type III Analysis of Variance Table with Satterthwaite's method
  Sum Sq Mean Sq NumDF DenDF F value  Pr(>F)
V 1189.9  297.46     4 20.82  3.4504 0.02586 *
- - -
Signif. codes:  0 '***' 0.001 '**' 0.01 '*' 0.05 '.' 0.1 ' ' 1
```

A more detailed test result can be found with the step function of `lmerTest`.

```
> M7.2s <-step(M7.2test, reduce.fixed = FALSE, reduce.random = FALSE)
> M7.2s
Backward reduced random-effect table:
           Eliminated npar  logLik    AIC     LRT Df Pr(>Chisq)
<none>                   8 -213.63 443.27
(1 | L)             0    7 -227.36 468.73 27.4600  1  1.604e-07 ***
(1 | V:L)           0    7 -217.54 449.07  7.8074  1   0.005203 **
- - -
```

```
Signif. codes:  0 '***' 0.001 '**' 0.01 '*' 0.05 '.' 0.1 ' ' 1
Backward reduced fixed-effect table:
Degrees of freedom method: Satterthwaite
  Eliminated Sum Sq Mean Sq NumDF DenDF F value  Pr(>F)
V          0 1189.9  297.46     4 20.82  3.4504 0.02586 *
---
Signif. codes:  0 '***' 0.001 '**' 0.01 '*' 0.05 '.' 0.1 ' ' 1
Model found:
y ~ V + (1 | L) + (1 | V:L)
```

For the test of the fixed effect *V*, we can also use the Kenward–Roger's method to calculate the df of the denominator in the *F*-test with the R package `lmerTest`.

```
> TestVKR <- anova(M7.2test, ddf = "Kenward-Roger")
> TestVKR
Type III Analysis of Variance Table with Kenward-Roger's method
  Sum Sq Mean Sq NumDF  DenDF F value  Pr(>F)
V 1187.7  296.92     4 19.821   3.444 0.02712 *
---
Signif. codes:  0 '***' 0.001 '**' 0.01 '*' 0.05 '.' 0.1 ' ' 1
> M7.2sKR <-step(M7.2test, ddf = "Kenward-Roger",
    reduce.fixed = FALSE, reduce.random = FALSE)
> M7.2sKR
Backward reduced random-effect table:
          Eliminated npar  logLik    AIC     LRT Df Pr(>Chisq)
<none>                 8 -213.63 443.27
(1 | L)            0   7 -227.36 468.73 27.4600  1  1.604e-07 ***
(1 | V:L)          0   7 -217.54 449.07  7.8074  1   0.005203 **
---
Signif. codes:  0 '***' 0.001 '**' 0.01 '*' 0.05 '.' 0.1 ' ' 1
Backward reduced fixed-effect table:
Degrees of freedom method: Kenward-Roger
  Eliminated Sum Sq Mean Sq NumDF  DenDF F value  Pr(>F)
V          0 1187.7  296.92     4 19.821   3.444 0.02712 *
---
Signif. codes:  0 '***' 0.001 '**' 0.01 '*' 0.05 '.' 0.1 ' ' 1
Model found:
y ~ V + (1 | L) + (1 | V:L)
```

We consider now a mixed model with $n = 1$. In such a mixed model we have no residual because *df* (residual) = 0. The EMS of the interaction *AB* in case I is now $\sigma^2_{ab} + \sigma^2$. The EMS of the random factor *B* in case I is now $a\sigma^2_b + \sigma^2_{ab} + \sigma^2$ and the EMS of the fixed factor *A* is now $(b/(a-1))\sum_{i=1}^{a}(a_i - \overline{a})^2 + \sigma^2_{ab} + \sigma^2$. If we denote $\sigma^{*2} = \sigma^2_{ab} + \sigma^2$ then the new residual is the interaction of *AB* and we have a mixed model without interaction, with the fixed factor *A* with df(*A*) $= a - 1$, the random factor *B* with $df(B) = b - 1$ and the residual with $df(\text{res}) = (a - 1)(b - 1)$. The fixed model ANOVA table is the same for the mixed model ANOVA table if $n = 1$.

Example 7.3 For an experiment we randomly select 12 farms from a similar arable region. The varieties are the levels of a fixed factor *A* of varieties and the 12 farms are levels of a random factor B. In each farm the varieties are laid down in a completely randomised design. Both A and B are cross-classified. The yield y in decitonnes per hectare was measured. The results are given in Table 7.4.

The analysis in R proceeds as follows:

```
> y1 <- c(32,28,30,44,43,48,42,42,39,44,40,42)
> y2 <- c(48,52,47,55,53,57,64,64,64,59,58,57)
> y3 <- c(25,25,34,28,26,33,40,42,47,34,27,32)
> y4 <- c(33,38,44,39,38,37,53,41,47,54,50,46)
> y5 <- c(48,27,38,21,30,36,38,29,23,33,36,36)
> y6 <- c(29,27,31,31,33,26,27,33,32,31,30,35)
> y <- c(y1,y2,y3,y4,y5,y6)
> a <- rep(1 : 6 , each = 12)
> b1 <- rep(1:12)
> b <- c(b1, b1, b1, b1, b1, b1)
> A <- factor(a)
> B <- factor(b)
> Table7.2 <- data.frame(A, B, y)
> ANOVAfixed <- aov(y ~A + B, data = Table7.2)
> summary(ANOVAfixed)
            Df Sum Sq Mean Sq F value   Pr(>F)
A            5   5696  1139.2  34.372 9.27e-16 ***
B           11    732    66.5   2.007   0.0452 *
Residuals   55   1823    33.1
- - -
Signif. codes:  0 '***' 0.001 '**' 0.01 '*' 0.05 '.' 0.1 ' ' 1

> library(VCA)
> M7.2 <- anovaMM(y ~ A + (B) , VarVC.method ="scm",Table7.2)
> M7.2
ANOVA-Type Estimation of Mixed Model:
- - - - - - - - - - - - - - - - - - - -
        [Fixed Effects]
      int        A1        A2        A3        A4        A5        A6
30.416667  9.083333 26.083333  2.333333 12.916667  2.500000  0.000000
        [Variance Components]
```

Table 7.4 Yields of 6 varieties tested on 12 randomly chosen farms.

	A: varieties					
***B*: farms**	**1**	**2**	**3**	**4**	**5**	**6**
1	32	48	25	33	48	29
2	28	52	25	38	27	27
3	30	47	34	44	38	31
4	44	55	28	39	21	31
5	43	53	26	38	30	33
6	48	57	33	37	36	26
7	42	64	40	53	38	27
8	42	64	42	41	29	33
9	39	64	47	47	23	32
10	44	59	34	54	33	31
11	40	58	27	50	36	30
12	42	57	32	46	36	35

```
  Name  DF        SS          MS        VC        %Total     SD       CV[%]
1 total 59.819803                       38.708333 100        6.221602 15.856827
2 B       11      731.819444  66.52904  5.564141  14.374531 2.358843 6.011919
3 error 55        1822.930556 33.144192 33.144192 85.625469 5.757099 14.672961
Mean: 39.23611 (N = 72)
Experimental Design: balanced  |  Method: ANOVA
```

We obtained $\boldsymbol{F}_A = \frac{MS_A}{MS_{AB}} = 1139.247/33.144 = 34.372$, and therefore we found significant differences between the varieties, because `Pr(>F) = 9.27e-16<0.05`. For $\boldsymbol{F}_B$ we obtained $\boldsymbol{F}_B = 2.007$ and this means that the variance component for the farms has `Pr(>F) = 0.0452<0.05`, hence it is, with $\alpha = 0.05$, significantly larger than zero.

To find the minimum size of the experiment that will satisfy given precision requirements, we must remember that only the degrees of freedom of the corresponding F-statistic influence the power of the test and by this the size of the experiment. To test the hypothesis H_{01} that the fixed factor has no influence on the observations, we have $(a-1)$ and $(a-1)(b-1)$ degrees of freedom of numerator and denominator, respectively. Thus the sub-class number n does not influence the size needed and therefore is chosen as small as possible. If we know that there are no interactions we choose $n = 1$, but if interactions may occur we choose $n = 2$. Because the number of levels a of the factor under test is fixed, we can only choose b, the size of the sample of B levels to fulfil precision requirements.

Analogously to the sample size determination in Chapters 5 and 6 we use the R `OPDOE` (optimal design of experiments) package.

Problem 7.6 Determine the minimin and maximin number of levels of the random factor to test H_{01}: 'all a_i are zero' if no interactions are expected for given values of α, β, δ/σ, a, and n.

Solution

We put $n = 1$ and use the R `OPDOE` program, where we choose for "cases" either "minimin" or "maximin". Please remember that delta in the R program OPDOE corresponds to δ/σ.

```
> size_b.two_way_cross.mixed_model_a_fixed_a(alpha, beta, δ/σ,a,1, "cases",).
```

Example

We want to test the null hypothesis that six wheat varieties do not differ in their yields with the precision requirements $\alpha = 0.05$, $\beta = 0.1$, $\sigma = 1$, $\delta = 1.6$ and $n = 1$.

```
> size_b.two_way_cross.mixed_model_a_fixed_a(0.05,0.1,1.6,6,1, "minimin")
[1] 6
```

for the minimal number of B levels and

```
> size_b.two_way_cross.mixed_model_a_fixed_a(0.05,0.1,1.6,6,1, "maximin")
[1] 15
```

for the maximal number of B levels.

Problem 7.7 Determine the minimin and maximin number of levels of the random factor to test H_{01}: 'all a_i are equal' if interactions are expected.

Solution

We put $n = 2$ and use the R program OPDOE

```
> size_b.two_way_cross.mixed_model_a_fixed_a(0.05,0.1,1.6,6,2, "minimin")
```

and

```
> size_b.two_way_cross.mixed_model_a_fixed_a(0.05,0.1,1.6,6,2, "maximin").
```

Example

We want to test the null hypothesis that six wheat varieties do not differ in their yields with the precision requirements $\alpha = 0.05$, $\beta = 0.1$, $\sigma = 1$, $\delta = 1.6$ and $n = 2$.

```
> size_b.two_way_cross.mixed_model_a_fixed_a(0.05,0.1,1.6,6,2, "minimin")
[1] 4
```

for the minimal number of B levels and

```
> size_b.two_way_cross.mixed_model_a_fixed_a(0.05,0.1,1.6,6,2, "maximin")
[1] 8
```

for the maximal number of B levels.

The choice $b = 12$ in Example 7.3 seems to be acceptable.

7.2.2 Two-Way Nested Classification

In the two-way nested classification, we have two model equations depending on which factor is random.

We use the model equation if the factor B is random, the nested factor A is fixed

$$y_{ijk} = \mu + a_i + \boldsymbol{b}_{j(i)} + \boldsymbol{e}_{k(i,j)},$$
$$(i = 1, \dots, a; j = 1, \dots, b; \ k = 1, \dots, n). \tag{7.10}$$

The side conditions are

$$\text{var}(\boldsymbol{b}_{j(i)}) = \sigma_b^2 \quad \text{for all } j, \quad \text{cov}(\boldsymbol{b}_{j(i)}, \boldsymbol{b}_{k(i)}) = 0, \quad \text{for all } j,k \text{ with } j \neq k,$$
$$\sum_{i=1}^{a} a_i = 0, \tag{7.11}$$
$$\text{cov}(\boldsymbol{b}_{j(i)}, \boldsymbol{e}_{ijk}) = 0$$

Let the levels of A be randomly selected from the level population and the levels of B fixed, the model equation is then

$$y_{ijk} = \mu + \boldsymbol{a}_i + b_{j(i)} + \boldsymbol{e}_{k(i.j)}$$
$$(i = 1, \dots, a; j = 1, \dots, b; k = 1, \dots, n) \tag{7.12}$$

with corresponding side conditions.

Expectations of all random variables are zero, $\text{var}(\boldsymbol{a}_i) = \sigma_a^2$ for all i; $\text{var}(\boldsymbol{e}_{k(i,j)}) = \sigma^2$ for all i,j,k; all covariances between different random variables on the right-hand side of (7.12) are zero.

Table 7.5 *E(MS)* for balanced nested mixed models.

Source of variation	*A* fixed *B* random	*A* random *B* fixed
Between *A* levels	$\sigma^2 + n\,\sigma^2_{bina} + \frac{bn}{a-1}\sum_{i=1}^{a} a_i^2$	$\sigma^2 + bn\,\sigma^2_a$
Between *B* levels within *A*	$\sigma^2 + n\,\sigma^2_{bina}$	$\sigma^2 + \frac{bn}{a(b-1)}\sum_{i,j} b_{ij}^2$
Residual	σ^2	σ^2

The columns ***SS***, *df* and ***MS*** in the corresponding ANOVA table are model independent and given in Table 5.13. The expectations of the ***MS*** for both models are in Table 7.5

Example 7.4 Example 17.10 from Ott and Longnecker (2001). Researchers conducted an experiment to determine the content uniformity of film-coated tablets produced for a cardiovascular drug used to lower blood pressure. They obtained a random sample of three batches from each of two blending sites; within each batch, they assayed a random sample of five tablets to determine content uniformity y. The data are shown in Table 7.6.

Problem 7.8 Estimate the variance components if the nested factor ***B*** is random by the ANOVA method.

Solution

From the second column of Table 7.5 we obtain by the algorithm in Chapter 5

$$\hat{\sigma}^2 + n\hat{\sigma}^2_{b\text{ in }a} = \boldsymbol{MS}_{B\text{ in }A},$$

$$\hat{\sigma}^2 = \boldsymbol{MS}_{\text{res}}$$

and from this we obtain

$$\hat{\sigma}^2 = \boldsymbol{MS}_{\text{res}} \tag{7.13}$$

Table 7.6 Data of an experiment to determine the content uniformity of film-coated tablets produced for a cardiovascular drug.

Site	1			2		
Batches within site	**1**	**2**	**3**	**1**	**2**	**3**
Tablets within each batch	5.03	4.64	5.10	5.05	5.46	4.90
	5.10	4.73	5.15	4.96	5.15	4.95
	5.25	4.82	5.20	5.12	5.18	4.86
	4.98	4.95	5.08	5.12	5.18	4.86
	5.05	5.06	5.14	5.05	5.11	5.07

and

$$\widehat{\sigma}^2_{b \text{ in } a} = \frac{1}{n}(\boldsymbol{MS}_{B \text{ in } A} - \boldsymbol{MS}_{\text{res}}).$$

Example

Use the data of Example 7.4 and create the data-frame `example7.4`.

Fixed factor A is sites. Random factor $\boldsymbol{B}$ is batches within sites with $n = 5$.

We can use the following command in the R base package

```
> nestedM <- aov( y ~A + Error(A/B), data = example7.4)
```

to get the mean squares and do in R (7.13):

```
> y1 <- c(5.03, 5.10, 5.25, 4.98, 5.05)
> y2 <- c(4.64, 4.73, 4.82, 4.95, 5.06)
> y3 <- c(5.10, 5.15, 5.20, 5.08, 5.14)
> y4 <- c(5.05, 4.96, 5.12, 5.12, 5.05)
> y5 <- c(5.46, 5.15, 5.18, 5.18, 5.11)
> y6 <- c(4.90, 4.95, 4.86, 4.86, 5.07)
> y <- c(y1, y2, y3, y4, y5, y6)
> a <- rep (1 :2, each = 15)
> b1 <- rep(1:3, each = 5)
> b <- c(b1, b1)
> A <- factor(a)
> B <- factor(b)
> example7.4 <- data.frame(A, B, y)
> nestedM <- aov( y ~A + Error(A/B), data = example7.4)
> nestedM
Call:
aov(formula = y ~ A + Error(A/B), data = example7.4)
Grand Mean: 5.043333
Stratum 1: A
Terms:
                         A
Sum of Squares  0.01825333
Deg. of Freedom          1
Estimated effects are balanced
Stratum 2: A:B
Terms:
                Residuals
Sum of Squares  0.4540133
Deg. of Freedom         4
Residual standard error: 0.3369026
Stratum 3: Within
Terms:
                Residuals
Sum of Squares     0.2902
Deg. of Freedom        24
Residual standard error: 0.1099621
```

```
> MSres <- 0.2902/24
> s2 <- MSres
> s2
[1] 0.01209167
>  MSBinA <- 0.4540133/4
>  s2BinA <-  (MSBinA-MSres)/5
>  s2BinA
[1] 0.02028233
```

With the R package VCA we can get directly the ANOVA estimates of the variance components.

```
> M7.4 <- anovaMM( y ~ A/(B), VarVC.method= "scm", example7.4)
> M7.4


ANOVA-Type Estimation of Mixed Model:
- - - - - - - - - - - - - - - - - - - -

        [Fixed Effects]

      int        A1        A2
 5.068000 -0.049333  0.000000


        [Variance Components]

  Name  DF       SS       MS       VC       %Total    SD       CV[%]
1 total 7.896362                   0.032374 100       0.179928 3.567636
2 A:B   4        0.454013 0.113503 0.020282 62.650069 0.142416 2.823848
3 error 24       0.2902   0.012092 0.012092 37.349931 0.109962 2.180346

Mean: 5.043333 (N = 30)

Experimental Design: balanced  |  Method: ANOVA
```

With the R package lme4 we get directly the REML estimates of the variance components.

```
> library(lme4)
> model7.4 <- lmer( y ~ A + (1|A/B) , data = example7.4, REML = TRUE)
> model7.4
Linear mixed model fit by REML ['lmerMod']
Formula: y ~ A + (1 | A/B)
   Data: example7.4
REML criterion at convergence: -29.7928
Random effects:
 Groups   Name        Std.Dev.
 B:A      (Intercept) 0.14242
 A        (Intercept) 0.06261
 Residual             0.10996
Number of obs: 30, groups:  B:A, 6; A, 2
Fixed Effects:
(Intercept)          A2
    5.01867     0.04933
convergence code 0; 1 optimizer warnings; 0 lme4 warnings
> as.data.frame(summary(fit)$varcor)
```

```
       grp        var1 var2       vcov      sdcor
1      A:B (Intercept) <NA> 0.02028233  0.1424161
2 Residual        <NA> <NA> 0.01209167  0.1099621
> M7.4$aov.tab
             DF        SS          MS         VC    %Total        SD    CV[%]
total  7.896362        NA          NA 0.03237400 100.00000 0.1799278 3.567636
A:B    4.000000 0.4540133  0.11350333 0.02028233  62.65007 0.1424161 2.823848
```

Problem 7.9 How can we test the null hypothesis that the effects of all the levels of factor A are equal, H_{A0}: 'all a_i are equal' against H_{AA}: 'not all a_i are equal', with significance level $\alpha = 0.05$?

Solution

The null hypothesis H_{A0}: 'all a_i are equal' is tested using the test statistic:

$$F_A = \frac{MS_A}{MS_{B \text{ in } A}} \tag{7.14}$$

which, under H_0, has an F-distribution with $(a-1)$ and $a(b-1)$ degrees of freedom.

Example

From Problem 7.8 we use the ANOVA table from `nestedM` .

```
> MSA <- 0.0182533 / 1
> MSBinA <- 0.4540133 / 4
> FA <- MSA / MSBinA
> FA
[1] 0.1608173
> p_value <- 1-pf(FA , 1, 4)
> p_value
[1] 0.7089036
```

Because the p_value $= 0.7089036 > 0.05$ we cannot reject H_{A0} .

Problem 7.10 How can we test the null hypothesis $H_{B0} : \sigma^2_{b \text{ in } a} = 0$ against $H_{BA}: \sigma^2_{b \text{ in } a} > 0$ with significance level $\alpha = 0.05$?

Solution

The null hypothesis $H_{0B} : \sigma^2_{b \text{ in } a} = 0$ is tested using the test statistic:

$$F_{B \text{ in } A} = \frac{MS_{B \text{ in } A}}{MS_{\text{res}}} \tag{7.15}$$

which under H_0 has an F-distribution with $a(b-1)$ and $ab(n-1)$ degrees of freedom.

Example

From Problem 7.8 and Problem 7.9 we use the R output .

```
> MSBinA <-0.4540133 / 4
> MSres <- 0.2902 / 24
> FBinA <- MSBinA / MSres
```

```
> FBinA
[1] 9.386905
> p_value <- 1 - pf(FBinA , 4, 24 )
> p_value
[1] 0.0001028394
```

Because the $p\_value = 0.0001028394 < 0.05$ H_{0B} is rejected.

Problem 7.11 Find the minimum size of the experiment which for a given number a of levels of the factor A will satisfy the precision requirements given by α; β, δ, σ.

Solution
Because in (7.14) no n occurs, we set $n = 1$ and use the R OPDOE command

```
> size_b.two_way_nested.b_random_a_fixed_a(α,β,δ/σ,a, "minimin")
```

Example
We choose a = 6, $\alpha = 0.05$; $\beta = 0.1$ $\delta = 1.2\ \sigma$ with $\sigma = 1$ and get

```
> size_b.two_way_nested.b_random_a_fixed_a(0.05,0.1,1.2,6, "minimin")
[1] 9
```

We consider now the model (7.12) if the factor A is random and the nested factor B is fixed.

Problem 7.12 Estimate the variance components σ^2 and σ_a^2 by the analysis of variance method.

Solution
From the third column of Table 7.5 we obtain by the algorithm in Chapter 6 that the variance components σ^2 and σ_a^2 can be estimated due to the balancedness by the analysis of variance method as

$$s^2 = MS_{\text{res}} \tag{7.16}$$

$$s^2{}_a = (MS_{\text{A}} - MS_{\text{res}})/(bn)$$

Example
As an example we use the data from Example 7.5.

Example 7.5 In an experiment, three batches were drawn at random from a production process. Three samples from each batch were drawn at random and for the concentration of a certain element y were analysed with two methods. Hence, we have a random factor A of batches and a fixed factor B of methods within batches. The data are shown in Table 7.7.

```
> y1 <- c( 1, 3, 4, 3, 6, 4)
> y2 <- c(7, 9, 8, 11, 9, 10)
> y3 <- c(4, 6, 4, 7, 5, 9)
```

Table 7.7 Data from Example 7.5 with a random factor ***A*** of batches and a fixed factor *B* of methods within batches.

		y
A_1	B_{11}	1 3 4
	B_{12}	3 6 4
A_2	B_{21}	7 9 8
	B_{22}	11 9 10
A_3	B_{31}	4 6 4
	B_{32}	7 5 9

```
> y <- c(y1, y2, y3)
> a <- rep (1 : 3 , each = 6)
> b1 <- rep(1:2 , each = 3)
> b  <-  c(b1, b1, b1)
> A <- factor(a)
> B <- factor(b)
> example7.5 <- data.frame ( A , B, y)
```

The R package VCA cannot give the estimates of the variance components for this model with random ***A*** and fixed *B* within ***A***. Hence we first use in the R base package aov() to get the ANOVA table. Here we have $b = 2$ and $n = 3$.

```
> ANOVAfixed <- aov( y ~ A + A/B , example7.5)
> ANOVAfixed
Call:
   aov(formula = y ~ A + A/B, data = example7.5a)
Terms:
                        A      A:B Residuals
Sum of Squares   91.44444 18.33333  24.00000
Deg. of Freedom         2        3        12
Residual standard error: 1.414214
Estimated effects may be unbalanced
> MSA <- 91.44444/2
> MSres <- 24/12
> s2 <- MSres
> s2
[1] 2
> b <- 2
> n <- 3
> s2a <- (MSA-MSres)/(b*n)
> s2a
[1] 7.287037
> library(lme4)
> M7.9REML <- lmer( y ~ (1|A) + A/B , example7.5)
> M7.9REML
Linear mixed model fit by REML ['lmerMod']
Formula: y ~ (1 | A) + A/B
   Data: example7.5a
REML criterion at convergence: 48.964
Random effects:
```

```
 Groups   Name        Std.Dev.
 A        (Intercept) 2.910
 Residual             1.414
Number of obs: 18, groups:  A, 3
Fixed Effects:
(Intercept)           A2            A3          A1:B2         A2:B2         A3:B2
      2.667         5.333         2.000         1.667         2.000         2.333
> as.data.frame(summary(M7.9REMLa)$varcor)
       grp        var1 var2     vcov    sdcor
1        A (Intercept) <NA> 8.467612 2.909916
2 Residual        <NA> <NA> 2.000000 1.414214
```

Hence, the REML estimate 8.467612 of the variance component for A is different from the ANOVA estimate 7.287037. This is due to the fact that the convergence criteria is met but the final Hessian matrix is not positive definite.

Problem 7.13 Test the null hypothesis H_{B0}: 'the effects of all the levels of factor B are equal' against H_{B0}: 'the effects of all the levels of factor B are not equal' with significance level $\alpha = 0.05$.

Solution
The null hypothesis H_{B0} is tested using the test statistic:

$$\boldsymbol{F}_B = \frac{\boldsymbol{MS}_B}{\boldsymbol{MS}_{\text{res}}}, \tag{7.17}$$

which under H_B has an F-distribution with $a(b-1)$ and $ab(n-1)$ degrees of freedom.

Example
Using the output of R in Problem 7.12 with the data of Example 7.5 we get:

```
> MSB <- 34.16667   /3
> MSB
[1] 11.38889
> MSres <- 1.944444
> FB <- MSB / MSres
> p_value <- 1-pf(FB,3,12)
> p_value
[1] 0.01056774
```

Because the p-value $= 0.01056774 < 0.05$ H_{0B} is rejected.

Problem 7.14 Test the null hypothesis $H_{A0} : \sigma_A^2 = 0$ against H_{AA}: $\sigma_A^2 > 0$ with significance level $\alpha = 0.05$.

Solution
In Table 7.5 we see that under $H_{A0} : \sigma_A^2 = 0$ the $E(\boldsymbol{MS})$ between the A levels and the residuals are identical and therefore we can test H_{0A} with the test statistic

$$\boldsymbol{F}_A = \frac{\boldsymbol{MS}_A}{\boldsymbol{MS}_{\text{res}}} \tag{7.18}$$

which under H_{A0} has an F-distribution with $a-1$ and $ab(n-1)$ degrees of freedom.

Example

Using the output of R in Problem 7.13 with the data of Example 7.5 we get

```
> MSA <- 97/2
> MSres <- 1.944444
> FA <- MSA / MSres
> FA
> p_value <- 1-pf(FA,2,12)
> p_value
[1] 5.315505e-05
```

Because the p-value $= 0.0000532 < 0.05$ H_{0A} is rejected.

To find for testing $H_{0A} : \sigma_A^2 = 0$ the minimum size of the experiment that will satisfy given precision requirements, we cannot, as above, put $n = 1$ because the sub-class number occurs in the degrees of freedom in the denominator of (7.17).

Problem 7.15 Show how we can determine the minimin and the maximin sample size to test the null hypothesis $H_{A0} : \sigma_A^2 = 0$.

Solution

Use the R OPDOE-command in

```
> size_n.two_way_nested.a_random_b_fixed_b(α,β,δ/σ,a,b, "cases")
```

and "cases" must later be inserted for "minimin" or "maximin", to search for the value of a that needs the lowest values.

Example

We choose for $b = 10$ the precision $\alpha = 0.05$, $\delta = 0.1$, and $\delta/\sigma = 0.9$, and start with $a = 2$.

```
> size_n.two_way_nested.a_random_b_fixed_b(0.05,0.1,0.9,2,10, "minimin")
[1] 7
```

and

```
> size_n.two_way_nested.a_random_b_fixed_b(0.05,0.1,0.9,2,10, "maximin")
[1] 63
```

If we increase a using $a = 5$ this leads to smaller minimin sizes but higher maximin sizes.

```
> size_n.two_way_nested.a_random_b_fixed_b(0.05,0.1,0.9,5,10, "minimin")
[1] 5
```

and

```
> size_n.two_way_nested.a_random_b_fixed_b(0.05,0.1,0.9,5,10, "maximin")
[1] 89
```

Turning now to $a = 20$ gives

```
> size_n.two_way_nested.a_random_b_fixed_b(0.05,0.1,0.9,20,10, "minimin")
[1] 3
```

and

```
> size_n.two_ay_nested.a_random_b_fixed_b(0.05,0.1,0.9,20,10, "maximin")
[1] 158
```

From this we can recommend using $a = 2$, because this gives the smallest maximin size whereas the minimin size increases irrelevantly with decreasing a.

In mixed models, it may happen that variance components of the random effects as well as the fixed effects have to be estimated. If the distribution of $\boldsymbol{Y}$ is normal, we can use the maximum likelihood method. In the unbalanced case (see Rasch and Schott (2018)) the problems are difficult to solve.

We differentiate the likelihood function with respect to the fixed effects and the variance components. The derivatives are set at zero, and we get simultaneous equations that we solve by iteration. Hartley and Rao (1967) give the formulae and proposals for numerical solutions of the simultaneous equations. The numerical solution is elaborate.

7.3 Three-Way Layout

We are interested here in models with at least one fixed factor and for determining the minimum size of the experiment; we are interested only in hypotheses about fixed factors.

For three-way analysis of variance, we have the following types of classification:

- Cross-classification. Observations are possible for all combinations of the levels of the three factors A, B and C (symbol: A × B × C).
- Nested classification. The levels of C are nested within B, and those of B are nested within A (symbol $A \succ B \succ C$).
- Mixed classification. We have two cases:
  - Case 1. A and B are cross-classified, and C is nested within the classes (i,j) (symbol $(A \times B) \succ C$).
  - Case 2. B is nested in A, and C is cross-classified with all $A \times B$ combinations (symbol$(A \succ B) \times C$).

For the examples, we use six levels of a factor $A : a = 6$, $\alpha = 0.05$; $\delta = 0.1$, and $\delta = \sigma$. If possible, we fix the number of fixed factors B and C by $b = 5$ and $c = 4$ respectively.

7.3.1 Three-Way Analysis of Variance – Cross-Classification $A \times B \times C$

The observations y_{ijkl} in the combinations of factor levels are as follows (model I is used for all factors fixed in Chapter 5 and model II for all factors random in Chapter 6):

- Model III. The levels of A and B are fixed; the levels of C are randomly selected $A \times B \times \boldsymbol{C}$.
- Model IV. The levels of A are fixed; the levels of B and C are randomly selected $A \times \boldsymbol{B} \times \boldsymbol{C}$

The missing model II is one with three random factors and will not be discussed here (this is without loss of generality because we can rename the factors so that the first one(s) is (are) fixed).

For model III the model equation is given by:

$$\boldsymbol{y}_{ijkl} = \mu + a_i + b_j + \boldsymbol{c}_k + (a,b)_{ij} + (\boldsymbol{a},\boldsymbol{c})_{ik} + (\boldsymbol{b},\boldsymbol{c})_{jk} + (\boldsymbol{a},\boldsymbol{b},\boldsymbol{c})_{ijk} + \boldsymbol{e}_{ijkl},$$
$$i = 1, \dots, a; j = 1, \dots, b; k = 1, \dots, c; l = 1, \dots, n_{ijk}.$$

The model becomes complete under the conditions that the random variables on the right-hand side of the equation are uncorrelated and have, with the same suffixes, equal variances, and all fixed effects besides μ over j and k sum up to zero.

For model IV the model equation is given by:

$$\boldsymbol{y}_{ijkl} = \mu + a_i + \boldsymbol{b}_j + c_k + (\boldsymbol{a}, \boldsymbol{b})_{ij} + (\boldsymbol{a}, \boldsymbol{c})_{ik} + (\boldsymbol{b}, \boldsymbol{c})_{jk} + (\boldsymbol{a}, \boldsymbol{b}, \boldsymbol{c})_{ijk} + \boldsymbol{e}_{ijkl},$$
$$i = 1, \dots, a; j = 1, \dots, b; k = 1, \dots, c; l = 1, \dots, n_{ijk}.$$

The model becomes complete under the conditions that the random variables on the right-hand side of the equation are uncorrelated and have with the same suffixes equal variances and that the c_k add up to zero. We consider in the sequel mainly the balanced case with a_i and $n_{ijk} = n$.

The analysis of variance table for both models is Table 7.8.

In Table 7.9 the expected mean squares for model III and model IV are given.

To find the appropriate F-test for testing the hypothesis

$H_{A0}: a_i = 0, \forall\, i$ against H_{AA} : at least one $a_i \neq 0$ for model III or IV, and
$H_{B0}: b_j = 0, \forall\, j$ against H_{BA} : at least one $b_j \neq 0$ for model III

we will demonstrate the algorithm for these models step by step; in the next sections we present only the results.

- Step 1. Define the null hypothesis that all the a_i are zero.
- Step 2. Choose the appropriate model (III or IV).
  Model III
- Step 3. Find the $E(\boldsymbol{MS})$ column in the ANOVA table that corresponds to the model. Table 7.9, second column.
- Step 4. In the table, find the row for the factor that appears in the null hypothesis. Main effect A.
- Step 5. Change the $E(\boldsymbol{MS})$ in this row to what it would be if the null hypothesis were true.

$\sigma^2 + bn\sigma^2_{ac} + \frac{bcn}{a-1} \sum_i a_i^2$ becomes $\sigma^2 + bn\sigma^2_{ac}$ if the hypothesis is true.

- Step 6. Search in the table (in the same column) for the row that now has the same $E(\boldsymbol{MS})$ as you found in the 5th step. Interaction $A \times C$
- Step 7. The F-value is now the value of the $\boldsymbol{MS}$ of the row found in the 4th step divided by the value of the MS of the row found in the 6th step.

$$\boldsymbol{F} = \frac{\boldsymbol{MS}_A}{\boldsymbol{MS}_{A\times B}} \tag{7.19}$$

which under H_0 has an F-distribution with $f_1 = a - 1$ and $f_2 = ab(n-1)$ degrees of freedom.

Problem 7.16 Use the algorithm above to find the test statistic for testing $H_{A0}: a_i = 0, \forall\, i$ against H_{AA} : at least one $a_i \neq 0$ for model III.

Table 7.8 ANOVA table – three-way ANOVA – cross-classification, balanced case.

Source of variation	*SS*	*df*	*MS*
Main effect A	$SS_A = \frac{1}{bcn}\sum_i Y_{i...}^2 - \frac{1}{N}Y_{....}^2$	$a-1$	$\frac{SS_A}{a-1}$
Main effect B	$SS_B = \frac{1}{acn}\sum_j Y_{.j..}^2 - \frac{1}{N}Y_{....}^2$	$b-1$	$\frac{SS_B}{b-1}$
Main effect C	$SS_C = \frac{1}{abn}\sum_k Y_{..k.}^2 - \frac{1}{N}Y_{....}^2$	$c-1$	$\frac{SS_C}{c-1}$
Interaction $A \times B$	$SS_{AB} = \frac{1}{cn}\sum_{i,j} Y_{ij..}^2 - \frac{1}{bcn}\sum_i Y_{i...}^2 - \frac{1}{acn}\sum_j Y_{.j..}^2 + \frac{1}{N}Y_{....}^2$	(a − 1)(b − 1)	$\frac{SS_{AB}}{(a-1)(b-1)}$
Interaction $A \times C$	$SS_{AC} = \frac{1}{bn}\sum_{i,k} Y_{i.k.}^2 - \frac{1}{bcn}\sum_i Y_{i...}^2 - \frac{1}{abn}\sum_k Y_{..k.}^2 + \frac{1}{N}Y_{....}^2$	$(a-1)(c-1)$	$\frac{SS_{AC}}{(a-1)(c-1)}$
Interaction $B \times C$	$SS_{BC} = \frac{1}{an}\sum_{j,k} Y_{.jk.}^2 - \frac{1}{acn}\sum_j Y_{.j..}^2 - \frac{1}{abn}\sum_k Y_{..k.}^2 + \frac{1}{N}Y_{....}^2$	$(b-1)(c-1)$	$\frac{SS_{BC}}{(b-1)(c-1)}$
Interaction $A \times B \times C$	$SS_{ABC} = SS_T - SS_A - SS_B - SS_C - SS_{AB} - SS_{AC} - SS_{BC} - SS_R$	$(a-1)(b-1)\,(c-1)$	$\frac{SS_{ABC}}{(a-1)(b-1)(c-1)}$
Residual	$SS_R = \sum_{i,j,k,l} y_{ijkl}^2 - \frac{1}{n}\sum_{i,j,k} Y_{ijk.}^2$	$abc(n-1)$	$\frac{SS_R}{abc(n-1)}$
Total	$SS_T = \sum_{i,j,k,l} y_{ijkl}^2 - \frac{1}{N}Y_{....}^2$	$N-1$	

Table 7.9 Expected mean squares for the three-way cross-classification – balanced case.

Source of variation	Mixed model A, B fixed, C random (model III)	Mixed model A fixed, B,C random (model IV)
Main effect A	$\sigma^2 + bn\sigma^2_{ac} + \frac{bcn}{a-1}\sum_i a_i^2$	$\sigma^2 + n\sigma^2_{abc} + bn\sigma^2_{ac} + cn\sigma^2_{ab} + \frac{bcn}{a-1}\sum_i a_i^2$
Main effect B	$\sigma^2 + an\sigma^2_{bc} + \frac{acn}{b-1}\sum_j b_j^2$	$\sigma^2 + an\sigma^2_{bc} + acn\sigma^2_b$
Main effect C	$\sigma^2 + abn\sigma^2_c$	$\sigma^2 + an\sigma^2_{bc} + abn\sigma^2_c$
Interaction $A \times B$	$\sigma^2 + n\sigma^2_{abc} + \frac{cn}{(a-1)(b-1)}\sum_{i,j}(a,b)^2_{ij}$	$\sigma^2 + n\sigma^2_{abc} + cn\sigma^2_{ab}$
Interaction $A \times C$	$\sigma^2 + bn\sigma^2_{ac}$	$\sigma^2 + n\sigma^2_{abc} + bn\sigma^2_{ac}$
Interaction $B \times C$	$\sigma^2 + an\sigma^2_{bc}$	$\sigma^2 + an\sigma^2_{bc}$
Interaction $A \times B \times C$	$\sigma^2 + n\sigma^2_{abc}$	$\sigma^2 + n\sigma^2_{abc}$
Residual	σ^2	σ^2

Solution

$$F = \frac{MS_A}{MS_{A\times C}}. \tag{7.20}$$

Problem 7.17 Use the algorithm above to find the test statistic for testing $H_{B0}: b_j = 0, \forall j$ against H_{BA} : at least one $b_j \neq 0$ for model III.

Solution

$$F = \frac{MS_B}{MS_{B\times C}}. \tag{7.21}$$

Problem 7.18 Use the algorithm above to find the test statistic for testing $H_{AB0}: (a,b)_{ij} = 0, \forall i,j$ against H_{ABA} : at least one $(a,b)_{ij} \neq 0$ for model III.

Solution

$$F = \frac{MS_{A\times B}}{MS_{A\times B\times C}}. \tag{7.22}$$

To test $H_{A0}: a_i = 0, \forall i$ against H_{AA} : at least one $a_i \neq 0$ for model IV our algorithm gives no solution and this means that for this hypothesis no exact F-test exists.

We can only derive a test statistic for an approximate F-test as described in lemma 6.3 in Rasch and Schott (2018).

For this, we use step 8 (in place of step 6):

Search in the table (in the same column) rows for which a linear combination of the $E(\boldsymbol{MS})$ gives the same $E(\boldsymbol{MS})$ as found in the 5th step.

E(MS) of interaction $A \times B +$ interaction $A \times C -$ interaction $A \times B \times C$.

The expectation of the linear combination $\boldsymbol{MS}_{A\times B} + \boldsymbol{MS}_{A\times C} - \boldsymbol{MS}_{A\times B\times C}$ is $\sigma^2 + n\sigma^2_{abc} + bn\sigma^2_{ac} + cn\sigma^2_{ab}$. An approximate F-test is based on the ratio

$$F = \frac{MS_A}{MS_{A\times B} + MS_{A\times C} - MS_{A\times B\times C}}.$$

However, the degrees of freedom depend upon the variance components of the random effects in the model or its estimates. We have, under $H_{A0}: a_i = 0$, approximately $a - 1$ and

$$f = \frac{(MS_{AB} + MS_{AC} - MS_{ABC})^2}{\frac{MS^2_{AB}}{(a-1)(b-1)} + \frac{MS^2_{AC}}{(a-1)(c-1)} + \frac{MS^2_{ABC}}{(a-1)(b-1)(c-1)}}$$

degrees of freedom of the corresponding approximate F-test with the Satterthwaite procedure.

To determine the size of the sub-classes we use for model III the R-package OPDOE. For model IV we refer to approximate sample sizes given in Rasch et al. (2012).

Because in the case of three-way classifications there are too many situations where we like to demonstrate the R program for determining the size of the experiment, we do not use examples from our consultation practice, but only artificial examples. The risk of the first kind is fixed at $\alpha = 0.05$; the risk of the second kind if $\delta = c\sigma$ is fixed at $\beta = 0.1$, $a = 5$ and $b = 5$.

Problem 7.19 The minimin and maximin number of levels of the random factor $\boldsymbol{C}$ to test $H_{A0}: a_i = 0, \forall\, i$ against H_{AA}: at least one $a_i \neq 0$ has to be calculated for model III of the three-way analysis of variance – cross-classification.

Solution

We use the R OPDOE command

size_c.three_way_cross.model_3_a(α,β,δ/σ,a,b,n,"cases") using $n = 2$ and "cases" ="minimin" or "maximin". The sub-class number is arbitrary because the degrees of freedom of (7.19) do not depend on n.

Example

We use $\delta/\sigma = 0.5$ and obtain

```
>  size_c.three_way_cross.model_3_a(0.05,0.1,0.5,6,5,2, "minimin")
[1] 6
```

and

```
> size_c.three_way_cross.model_3_a(0.05,0.1,0.5,6,5,2, "maximin")
[1] 15
```

This means that we select between $c = 6$ and $c = 15$ levels of the random factor C and the total size of the experiment is then $N = abcn = 60c$.

Problem 7.20 Calculate the minimin and maximin number of levels of the random factor C to test $H_{B0}: b_j = 0, \forall\, j$ against H_{BA}: at least one $b_j \neq 0$ for model III of the three-way analysis of variance – cross-classification.

Solution
We interchange A and B and use the command `size_c.three_way_cross.model_3`; the sub-class number is arbitrary because the degrees of freedom of (7.20) do not depend on n.

```
> size_c.three_way_cross.model_3_a(α,β,δ/σ,b,a,n,"cases")
```

using $n = 2$ and `"cases"` = `"minimin"` or `"maximin"`.

Example
We use $\delta/\sigma = 0.5$ and obtain

```
> size_c.three_way_cross.model_3_a(0.05,0.1,0.5,5,6,2, "minimin")
[1] 6
> size_c.three_way_cross.model_3_a(0.05,0.1,0.5,5,6,2, "maximin")
[1] 12
```

This means we need between 6 and 12 levels of the factor C.

Problem 7.21 The minimin and maximin number of levels of the random factor $\boldsymbol{C}$ to test $H_{AB0}: (a,b)_{ij} = 0, \forall\, i, j$ against H_{ABA} : at least one $(a,b)_{ij} \neq 0$ has to be calculated for model III of the three-way analysis of variance – cross-classification.

Solution
We use the R `OPDOE` -command

```
> size_c.three_way_cross.model_3_axb(α,β,δ/σ,b,a,n,"cases")
```

the sub-class number is arbitrary because the degrees of freedom of (7.22) do not depend on n.

Example
We take $\delta/\sigma = 1$ and obtain

```
> size_c.three_way_cross.model_3_axb(0.05,0.1,1,5,6,2, "minimin")
[1] 3

> size_c.three_way_cross.model_3_axb(0.05,0.1,1,5,6,2, "maximin")
[1] 27
```

Thus, we need between 3 and 27 levels of the factor C.

Problem 7.22 Estimate the variance components for model III and model IV of the three-way analysis of variance – cross-classification.

Solution
For model III we use the second column of Table 7.9. We obtain the following equations:

$$MS_A = \sigma^2 + bn\sigma^2_{ac} + \frac{bcn}{a-1}\sum_i a_i^2$$
$$MS_B = \sigma^2 + an\sigma^2_{bc} + \frac{acn}{b-1}\sum_j b_j^2$$

$$MS_C = \sigma^2 + abn\sigma_c^2$$

$$MS_{AB} = \sigma^2 + n\sigma_{abc}^2 + \frac{cn}{(a-1)(b-1)}\sum_{i,j}(a,b)_{ij}^2$$

$$MS_{AC} = \sigma^2 + bn\sigma_{ac}^2$$

$$MS_{BC} = \sigma^2 + an\sigma_{bc}^2$$

$$MS_{ABC} = \sigma^2 + n\sigma_{abc}^2$$

$$MS_{\text{res}} = \sigma^2.$$

From this, we obtain by the ANOVA method the estimators

$$s_c^2 = \frac{1}{abn}(MS_C - MS_{\text{res}}),$$

$$s_{ac}^2 = \frac{1}{bn}(MS_{AC} - MS_{\text{res}}),$$

$$s_{bc}^2 = \frac{1}{an}(MS_{BC} - MS_{\text{res}}),$$

$$s_{abc}^2 = \frac{1}{n}(MS_{ABC} - MS_{\text{res}}),$$

$$s^2 = MS_{\text{res}}.$$

For model IV, we use the third column of Table 7.9. We obtain the following equations:

$$MS_A = \sigma^2 + n\sigma_{abc}^2 + bn\sigma_{ac}^2 + cn\sigma_{ab}^2 + \frac{bcn}{a-1}\sum_i a_i^2,$$

$$MS_B = \sigma^2 + an\sigma_{bc}^2 + acn\sigma_b^2,$$

$$MS_C = \sigma^2 + an\sigma_{bc}^2 + abn\sigma_c^2.$$

$$MS_{AB} = \sigma^2 + n\sigma_{abc}^2 + cn\sigma_{ab}^2,$$

$$MS_{AC} = \sigma^2 + n\sigma_{abc}^2 + bn\sigma_{ac}^2,$$

$$MS_{BC} = \sigma^2 + an\sigma_{bc}^2,$$

$$MS_{ABC} = \sigma^2 + n\sigma_{abc}^2,$$

$$MS_{\text{res}} = \sigma^2.$$

From this, we obtain by the ANOVA method the estimators

$$s_b^2 = \frac{1}{acn}(MS_B - MS_{BC}),$$

$$s_c^2 = \frac{1}{abn}(MS_C - MS_{BC}),$$

$$s_{ab}^2 = \frac{1}{cn}(MS_{AB} - MS_{ABC}),$$

$$s_{ac}^2 = \frac{1}{bn}(MS_{AC} - MS_{ABC}),$$

$$s_{bc}^2 = \frac{1}{an}(MS_{BC} - MS_{\text{res}}),$$

$$s_{abc}^2 = \frac{1}{n}(MS_{ABC} - MS_{\text{res}}),$$

$$s^2 = MS_{\text{res}}.$$

7.3.2 Three-Way Analysis of Variance – Nested Classification $A \succ B \succ C$

For the three-way nested classification with at least one fixed factor, we have the following seven models

- *Model III*. Factor A random, the other fixed
- *Model IV*. Factor B random, the other fixed
- *Model V*. Factor C random, the other fixed
- *Model VI*. Factor A fixed, the other random
- *Model VII*. Factor B fixed, the other random
- *Model VIII*. Factor C fixed, the other random.

For all models, the ANOVA table is the same and given as Table 7.10

7.3.2.1 Three-Way Analysis of Variance – Nested Classification – Model III – Balanced Case

The model equation is:

$$\boldsymbol{y}_{ijkl} = \mu + \boldsymbol{a}_i + b_{j(i)} + c_{k(ij)} + \boldsymbol{e}_{ijkl}\ i = 1, \ldots, a, j = 1, \ldots b, k = 1, \ldots, c, l = 1, \ldots, n.$$

The model becomes complete under the conditions that the random variables on the right-hand side of the equation are uncorrelated and have with the same suffixes equal variances, and all fixed effects besides μ over j and k sum up to zero.

Table 7.10 ANOVA table of the three-way nested classification – unbalanced case.

Source of variation	*SS*	*df*	*MS*
Between A	$SSA = \sum_i \frac{Y_{i\cdots}^2}{N_{i\cdot\cdot}} - \frac{Y_{\cdots\cdot}^2}{N}$	$a - 1$	$\frac{SSA}{a-1}$
Between B in A	$SS_{\text{B in}A} = \sum_{i,j} \frac{Y_{ij\cdot\cdot}^2}{N_{ij\cdot}} - \sum_i \frac{Y_{\cdots\cdot}^2}{N_{i\cdot\cdot}}$	$B. - a$	$\frac{SS_{\text{B in }A}}{B. - a}$
Between C in B and A	$SS_{\text{C in }B\text{ and }A} = \sum_{i,j,k} \frac{Y_{ijk.}^2}{n_{ijk}} - \sum_{i,j} \frac{Y_{\text{ij}\cdot\cdot}^2}{N_{ij\cdot}}$	$C.. - B.$	$\frac{SS_{\text{C in }B\text{ and }A}}{C.. - B.}$
Residual	$SS_{\text{res}} = \sum_{i,j,k,l} Y_{ijkl}^2 - \sum_{i,j,k} \frac{Y_{\text{ijk.}}^2}{n_{ijk}}$	$N - C..$	$\frac{SS_{\text{res}}}{N - C..}$

Table 7.11 Expected mean squares for the balanced case of model III.

Source of variation	Model III $E(MS)$
Between A levels	$\sigma^2 + bcn\sigma_a^2$
Between B levels within A levels	$\sigma^2 + \frac{cn}{a(b-1)} \cdot \sum_{ij} b_{j(i)}^2$
Between C levels within B levels	$\sigma^2 + \frac{n}{ab(c-1)} \cdot \sum_{ijk} c_{k(i,j)}^2$
Residual	σ^2

The expected mean squares gives Table 7.11.

From this table, we can derive the test statistics by using steps 1–7 from Section 7.3.1. We find for testing

$$H_{B0} : b_{j(i)} = 0 \forall j, i; H_{BA} \text{ at least one } b_{j(i)} \neq 0$$

$$F_B = \frac{MS_{B \text{ in } A}}{MS_{\text{res}}} \tag{7.23}$$

and this is under H_{B0} $bj(i) = 0 \, \forall j, i$ $F[a(b-1); abc(n-1)]$-distributed with a $(b-1)$ and $abc(n-1)$ degrees of freedom.

To test $H_{C0}: c_{k(i,j)} = 0, \forall i, j, k$ against $H_{C0}: c_{k(i,j)} \neq 0$ for at least one combination of i,j,k we find, using steps 1–7 from Section 7.3.1, that

$$F_C = \frac{MS_{C \text{ in } B \text{ in } A}}{MS_{\text{res}}} \tag{7.24}$$

can be used, which, under $H_{C0}: c_{k(i,j)} = 0, \forall i, j, k$, is $F[a(b-1); abc(n-1)]$-distributed with $ab[c-1]$ and $abc(n-1)$ degrees of freedom.

Problem 7.23 Calculate the minimin and the maximin sub-class number n to test H_{B0}: $b_{j(i)} = 0 \, \forall j, i$; H_{BA} at least one $b_{j(i)} \neq 0$ for model III of the three-way analysis of variance – nested classification.

Solution

Use the R OPDOE-command

```
> size_n.three_way_nested.model_3_b(α,β,δ/σ,a,b,c, "cases")
```

By systematic search using different values of a we look for the best solution minimizing an.

Example

We use $\alpha = 0.05$, $\beta = 0.1$, $\delta/\sigma = 0.5$, $b = 5$, $c = 4$ and start with $a = 2$.

```
> size_n.three_way_nested.model_3_b(0.05,0.1,0.5,2,5,4, "minimin")
[1] 8
> size_n.three_way_nested.model_3_b(0.05,0.1,0.5,2,5,4, "maximin")
[1] 39
```

and the size of the experiment is for $a = 2$ between 320 and 1560.

With $a = 3$ we get

```
> size_n.three_way_nested.model_3_b(0.05,0.1,0.5,3,5,4, "minimin")
[1] 7
> size_n.three_way_nested.model_3_b(0.05,0.1,0.5,3,5,4, "maximin")
[1] 44
```

and the size of the experiment is for $a = 3$ between 420 and 2640.

Larger values of a lead to higher size of the experiment. A small number of levels of the factor A lead to smaller sizes of the experiment for testing hypotheses about the fixed effects, but for the variance component estimation $\boldsymbol{a}$ must be chosen larger.

Problem 7.24 Estimate the variance components for model III of the three-way analysis of variance – nested classification.

Solution

From Table 7.11 we obtain the following equations:

$$MS_A = \sigma^2 + bcn\sigma_a^2,$$

$$MS_{\text{res}} = \sigma^2,$$

and from this the estimators

$$s_a^2 = \frac{1}{bcn}(MS_A - MS_{\text{res}}),$$

$$s^2 = MS_{\text{res}}.$$

7.3.2.2 Three-Way Analysis of Variance – Nested Classification – Model IV – Balanced Case

The model equation is:

$$\boldsymbol{y}_{ijkl} = \mu + a_i + \boldsymbol{b}_{j(i)} + c_{k(ij)} + \boldsymbol{e}_{ijkl} \quad i = 1, \dots, a, j = 1, \dots b, k = 1, \dots, c, l = 1, \dots, n.$$

The model becomes complete under the conditions that the random variables on the right-hand side of the equation are uncorrelated and have with the same suffixes equal variances, and all fixed effects besides μ over j and k sum up to zero.

The expected mean squares are given in Table 7.12.

From this table, we can derive the test statistic for testing

$H_{0A}: a_i = 0\,\forall\, i$ against H_{AA}: at least one $a_i \neq 0$.

by using steps 1–7 from Section 7.3.1. From this we find

$\boldsymbol{F}_A = \frac{\boldsymbol{MS}_A}{\boldsymbol{MS}_{BinA}}$ is, under H_{0A}, $F((a-1); ab(c-1))$-distributed with $a-1$ and $ab(c-1)$ degrees of freedom.

Problem 7.25 Calculate the minimin and the maximin sub-class number n to test $H_{A0}: a_i = 0\,\forall\, i$ for model IV of the three-way analysis of variance – nested classification.

Table 7.12 Expected mean squares for balanced case of model IV.

Source of variation	Model IV $E(MS)$
Between A levels	$\sigma^2 + cn\sigma^2_{b(a)} + \frac{bcn}{a-1}\sum_i a_i^2$
Between B levels within A levels	$\sigma^2 + cn\sigma^2_{b(a)}$
Between C levels within B levels	$\sigma^2 + \frac{n}{ab(c-1)} \cdot \sum_{ijk} c^2_{k(i,j)}$
Residual	σ^2

Solution
Use the R OPDOE-command

```
> size_n.three_way_nested.model_4_a(α,β,δ,a,b,c, "minimin")
```

this gives:

```
> size_n.three_way_nested.model_4_a(0.05,0.1,0.4,6,5,4, "minimin")
[1] 6
```

and

```
> size_n.three_way_nested.model_4_a(0.05,0.1,0.4,6,5,4, "maximin")
[1] 13
```

Example
Determine the minimin and the maximin sub-class number n to test $H_{0A}: a_i = 0\,\forall\, i$ with the values $\alpha = 0.05$, $\beta = 0.1$, $\delta/\sigma = 0.4$, $a = 6$, $b = 5$, $c = 6$. We obtain

```
> size_n.three_way_nested.model_4_a(0.05,0.1,0.4,6,5,4, "minimin")
[1] 5
```

and

```
> size_n.three_way_nested.model_4_a(0.05,0.1,0.4,6,5,4, "maximin")
[1] 13
```

We can also derive the test statistic for testing $H_{0C}: c_k = 0 \forall k, j, i;$ against H_{AC} : at least one $c_k \neq 0$ using $\boldsymbol{F}_{C0} = \frac{\boldsymbol{MS}_{C \text{ in } AB}}{\boldsymbol{MS}_{\text{res}}}$, which, under H_{0C} $c = 0\,\forall\, k, j, i;$, is $F[ab(c-1); abc(n-1)]$-distributed with $ab(c-1)$ and $abc(n-1)$ degrees of freedom.

Problem 7.26 Calculate the minimin and the maximin sub-class number n to test $H_0: c_{k(ij)} = 0\,\forall\, k, j, i;$ H_A at least one $c_{k(ij)} \neq 0$ for model IV of the three-way analysis of variance – nested classification.

Solution
Use the R OPDOE-command

```
> size_n.three_way_nested.model_4_c(α,β,δ/σ,a,b,c, "cases").
```

By systematic search using different values of a we look for the best solution minimizing an.

Example

We use $\alpha = 0.05$, $\beta = 0.1$, $\delta/\sigma = 0.5$, $b = 5$, $c = 4$ and start with $a = 2$.

```
> size_n.three_way_nested.model_4_a(0.05,0.1,0.5,2,5,4, "minimin")
[1] 6
> size_n.three_way_nested.model_4_a(0.05,0.1,0.5,2,5,4, "maximin")
[1] 6
```

The size of the experiment for $a = 2$ is 240.

With $a = 3$ we get

```
> size_n.three_way_nested.model_4_a(0.05,0.1,0.5,3,5,4, "minimin")
[1] 5
> size_n.three_way_nested.model_4_a(0.05,0.1,0.5,3,5,4, "maximin")
[1] 7
```

and the size of the experiment is for $a = 3$ is between 300 and 420, hence larger than for $a = 2$.

Problem 7.27 Estimate the variance components for model IV of the three-way analysis of variance – nested classification.

Solution

From Table 7.12 we derive the estimators

$$s_b^2 = (1/(cn))(\boldsymbol{MS}_{B \text{ in } A} - \boldsymbol{MS}_{\text{res}}),$$

$$s^2 = \boldsymbol{MS}_{\text{res}}.$$

7.3.2.3 Three-Way Analysis of Variance – Nested Classification – Model V – Balanced Case

The model equation is:

$$\boldsymbol{y}_{ijkl} = \mu + a_i + \boldsymbol{b}_{j(i)} + \boldsymbol{c}_{k(ij)} + \boldsymbol{e}_{ijkl} \quad i = 1, \dots, a, j = 1, \dots b, k = 1, \dots, c, l = 1, \dots, n.$$

The model becomes complete under the conditions that the random variables on the right-hand side of the equation are uncorrelated and have with the same suffixes equal variances, and all fixed effects besides μ over j and k sum up to zero.

The expected mean squares are given in Table 7.13

From this table, we can derive the test statistic for testing $H_0 : a_i = 0 \,\forall\, i$; against H_A: at least one $a_i \neq 0$ in the usual way.

From this we find

$\boldsymbol{F}_A = \frac{\boldsymbol{MS}_A}{\boldsymbol{MS}_{C \text{ in } AB}}$ is, under H_0, $F((a-1); ab(c-1))$-distributed with $a-1$ and $ab(c-1)$ degrees of freedom.

Problem 7.28 Calculate the minimin and maximin number c of C – levels for testing the null hypotheses $H_0 : a_i = 0 \,\forall\, i$; against H_A: at least one $a_i \neq 0$ for model V of the three-way analysis of variance – nested classification.

Table 7.13 Expected mean squares for model V.

Source of variation	Model V $E(MS)$
Between A levels	$\sigma^2 + n\sigma^2_{c(ab)} + \frac{bcn}{a-1}\sum_i a_i^2$
Between B levels within A levels	$\sigma^2 + n\sigma^2_{c(ab)} + \frac{cn}{a(b-1)}\sum_{ij} b_{j(i)}^2$
Between C levels within B levels	$\sigma^2 + n\sigma^2_{c(ab)}$
Residual	σ^2

Solution

Use `>size_c.three_way_nested.model_5_a(`α`,`β`,`δ/σ`,a,b,n, "cases")`. We choose n as small as possible because the test statistic does not depend on n.

Example

We use $\alpha = 0.05$, $\beta = 0.1$, $\delta/\sigma = 0.75$, $a = 6$, $b = 5$, and $n = 2$.

```
> size_c.three_way_nested.model_5_a(0.05,0.1,0.75,6,5,2, "minimin")
[1] 3
> size_c.three_way_nested.model_5_a(0.05,0.1,0.75,6,5,2, "maximin")
[1] 7
```

From Table 7.13, we can also derive the test statistic for testing

$$H_0 : b_{j(i)} = 0 \forall j, i; H_A \text{ at least one } b_{j(i)} \neq 0.$$

We find that $\boldsymbol{F} = \frac{MS_{B \text{ in } A}}{MS_{C \text{ in } AB}}$ is under H_0 $F(a(b-1); ab\,(c-1))$-distributed with $a(b-1)$ and $abc - 1)$ degrees of freedom.

Problem 7.29 Calculate the minimin and maximin number c of C levels for testing the null hypotheses $H_0 : b_{j(i)} = 0 \,\forall\, j, i$; against H_A: at least one $b_{j(i)} \neq 0$ for model V of the three-way analysis of variance – nested classification.

Solution

Use the R `OPDDOE`-command

`> size_c.three_way_nested.model_5_b(`α`,`β`,`δ/σ`,a,b,n,"cases")`.

We choose n as small as possible because the test statistic does not depend on n.

Example

We use $\alpha = 0.05$, $\beta = 0.1$, $\delta/\sigma = 1$, $a = 6$, $b = 5$, and $n = 2$.

```
> size_c.three_way_nested.model_5_b(0.05,0.1,1,6,5,2, "minimin")
[1] 3

> size_c.three_way_nested.model_5_b(0.05,0.1,1,6,5,2, "maximin")
[1] 29
```

Problem 7.30 Estimate the variance components for model V of the three-way analysis of variance – nested classification.

Solution

$$s^2_{c(ab)} = \frac{1}{n}(MS_{C\text{ in }AB} - MS_{\text{res}}),$$

$$s^2 = MS_{\text{res}}.$$

7.3.2.4 Three-Way Analysis of Variance – Nested Classification – Model VI – Balanced Case

The model equation is given by:

$$y_{ijkl} = \mu + a_i + b_{j(i)} + c_{k(ij)} + e_{ijkl} \quad i = 1, \dots, a, j = 1, \dots b, k = 1, \dots, c, l = 1, \dots, n.$$

The model becomes complete under the conditions that the random variables on the right-hand side of the equation are uncorrelated and have with the same suffixes equal variances, and all fixed effects besides μ over j and k sum up to zero.

The expected mean squares are give in Table 7.14.

From this table, we can derive the test statistic for testing

$$H_0 : a_i = 0\forall i; H_A \text{ at least one } a_i \neq 0$$

We find

$F_A = \frac{MS_A}{MS_{B\text{ in }A}}$ is, under H_0, $F((a-1); ab(c-1))$-distributed with $a-1$ and $ab(c-1)$ degrees of freedom.

Problem 7.31 Calculate the minimin and maximin number b of B levels for testing the null hypotheses $H_0 : a_i = 0\,\forall\, i$; against H_A: at least one $a_i \neq 0$ for model VI of the three-way analysis of variance – nested classification.

Solution

Use the R OPDOE-command

```
> size_b.three_way_nested.model_6_a(α,β,δ/σ,a,c,n,"cases").
```

We choose n as small as possible because the test statistic does not depend on n.

Table 7.14 Expected mean squares for model VI.

Source of variation	Model VI $E(MS)$
Between A levels	$\sigma^2 + n\sigma^2_{c(ab)} + cn\sigma^2_{b(a)} + \frac{bcn}{a-1}\sum_i a_i^2$
Between B levels within A levels	$\sigma^2 + n\sigma^2_{c(ab)} + cn\sigma^2_{b(a)}$
Between C levels within B levels	$\sigma^2 + n\sigma^2_{c(ab)}$
Residual	σ^2

Example

We use $\alpha = 0.05$, $\beta = 0.1$, $\delta/\sigma = 1$, $a = 5$, $c = 4$, and $n = 2$.

```
> size_b.three_way_nested.model_6_a(0.05,0.1,1,5,4,2, "minimin")
[1] 3
> size_b.three_way_nested.model_6_a(0.05,0.1,1,5,4,2, "maximin")
[1] 5
```

Problem 7.32 Estimate the variance components for model VI of the three-way analysis of variance – nested classification.

Solution

The estimators can be derived from Table 7.14 and this gives the estimators

$$s^2_{b(a)} = \frac{1}{cn}(\boldsymbol{MS}_{B \text{ in } A} - \boldsymbol{MS}_{C \text{ in } AB}),$$

$$s^2_{c(ab)} = \frac{1}{n}(\boldsymbol{MS}_{C \text{ in } AB} - \boldsymbol{MS}_{\text{res}}),$$

$$s^2 = \boldsymbol{MS}_{\text{res}}.$$

7.3.2.5 Three-Way Analysis of Variance – Nested Classification – Model VII – Balanced Case

$$\boldsymbol{y}_{ijkl} = \mu + \boldsymbol{a}_i + b_{j(i)} + \boldsymbol{c}_{k(ij)} + \boldsymbol{e}_{ijkl}.$$

The model becomes complete under the conditions that the random variables on the right-hand side of the equation are uncorrelated and have with the same suffixes equal variances, and all fixed effects besides μ over j and k sum up to zero.

The expected mean squares are given in Table 7.15.

From this table we can derive the test statistic for testing

$$H_0 : b_{j(i)} = 0 \forall j, i; \text{ against } H_A : \text{ at least one } b_{j(i)} \neq 0$$

We find $\boldsymbol{F} = \frac{\boldsymbol{MS}_{B \text{ in } A}}{\boldsymbol{MS}_{C \text{ in } AB}}$ is, under H_0, $F(a(b-1); ab\ (c-1))$-distributed with $a(b-1)$ and $abc - 1)$ degrees of freedom.

Table 7.15 Expected mean squares for model VII.

Source of variation	Model VII E(MS)
Between A levels	$\sigma^2 + n\sigma^2_{c(ab)} + bcn\sigma^2_a$
Between B levels within A levels	$\sigma^2 + n\sigma^2_{c(ab)} + \frac{cn}{a(b-1)} \sum_{i,j} b^2_{j(i)}$
Between C levels within B levels	$\sigma^2 + n\sigma^2_{c(ab)}$
Residual	σ^2

Problem 7.33 Calculate the minimin and maximin number b of B levels for testing the null hypotheses $H_0: b_{j(i)} = 0\,\forall\, j,\, i$; against H_A: at least one $b_{j(i)} \neq 0$ for model VII of the three-way analysis of variance – nested classification.

Solution

Use `>size_c.three_way_nested.model_7_b(α,β,δ/σ,a,b,n, "cases")`.
We choose n as small as possible because the test statistic does not depend on n.

Example

We use $a = 6$, $b = 4$, $\alpha = 0.05$; $\beta = 0.1$ and $\delta/\sigma = 1$, and $n = 2$.

```
> size_c.three_way_nested.model_7_b(0.05,0.1,1,6,4,2, "minimin")
[1] 3
> size_c.three_way_nested.model_7_b(0.05,0.1,1,6,4,2, "maximin")
[1] 26
```

Problem 7.34 Estimate the variance components for model VII of the three-way analysis of variance – nested classification.

Solution

$$s_a^2 = \frac{1}{bcn}(\boldsymbol{MS}_A - \boldsymbol{MS}_{C \text{ in } AB}),$$

$$s_{c(ab)}^2 = \frac{1}{n}(\boldsymbol{MS}_{C \text{ in } AB} - \boldsymbol{MS}_{\text{res}}),$$

$$s^2 = \boldsymbol{MS}_{\text{res}}.$$

7.3.2.6 Three-Way Analysis of Variance – Nested Classification – Model VIII – Balanced Case

The model equation is:

$$\boldsymbol{y}_{ijkl} = \mu + \boldsymbol{a}_i + \boldsymbol{b}_{j(i)} + c_{k(ij)} + \boldsymbol{e}_{ijkl} \quad i = 1, \dots, a, j = 1, \dots b, k = 1, \dots, c, l = 1, \dots, n.$$

The model becomes complete under the conditions that the random variables on the right-hand side of the equation are uncorrelated and have with the same suffixes equal variances, and all fixed effects besides μ over j and k sum up to zero.

The expected mean squares are given in Table 7.16.

From this table, we can derive the test statistic for testing

$$H_0 : c_{k(ij)} = 0 \forall k, j, i; H_A \text{ at least one } c_{k(ij)} \neq 0$$

We find $\boldsymbol{F} = \frac{\boldsymbol{MS}_{C \text{ in } AB}}{\boldsymbol{MS}_R}$ is, under H_0, $F(ab(c-1); abc(n-1))$-distributed with $ab(c-1)$ and $abc(n-1)$ degrees of freedom.

Problem 7.35 Calculate the minimin and maximin sub-class numbers for testing the null hypotheses $H_0: c_{k(ij)} = 0\,\forall\, k,\, j,\, i$; against H_A: at least one $c_{k(ij)} \neq 0$ for model VIII of the three-way analysis of variance – nested classification.

Table 7.16 Expected mean squares for model VIII.

Source of variation	Model VIII $E(MS)$
Between A levels	$\sigma^2 + cn\sigma^2_{b(a)} + bcn\sigma^2_a$
Between B levels within A levels	$\sigma^2 + cn\sigma^2_{b(a)}$
Between C levels within B levels	$\sigma^2 + \frac{n}{ab(c-1)} \cdot \sum_{ijk} c^2_{k(i,j)}$
Residual	σ^2

Solution
Use `>size_n.three_way_nested.model_8_c(α,β,δ/σ,a,b,c, "cases")`.
We choose n as small as possible because the test statistic does not depend on n.

Example
We use $a = 6$, $b = 5$, $c = 4$, $\alpha = 0.05$; $\beta = 0.1$, and $\delta/\sigma = 0.5$, and $n = 2$.

```
> size_n.three_way_nested.model_8_c(0.05,0.1,0.5,6,5,4, "minimin")
[1] 7
> size_n.three_way_nested.model_8_c(0.05,0.1,0.5,6,5,4, "maximin")
[1] 378
```

Problem 7.36 Estimate the variance components for model VIII of the three-way analysis of variance – nested classification.

Solution

$$s^2_a = \frac{1}{bcn}(MS_A - MS_{B \text{ in } A}),$$

$$s^2_{b(a)} = \frac{1}{cn}(MS_{B \text{ in } A} - MS_{\text{res}}),$$

$$s^2 = MS_{\text{res}}.$$

7.3.3 Three-Way Analysis of Variance – Mixed Classification – $(A \times B) \succ C$

We have four mixed models in this classification. The ANOVA table is independent of the models and given in Table 7.17 for the balanced case.

We now consider the four models of this classification.

7.3.3.1 Three-Way Analysis of Variance – Mixed Classification – $(A \times B) \succ C$ Model III

The model equation is:

$$\boldsymbol{y}_{ijkl} = \mu + a_i + \boldsymbol{b}_j + (\boldsymbol{a}, \boldsymbol{b})_{ij} + c_{k(ij)} + \boldsymbol{e}_{ijkl}$$
$$i = 1, \dots, a, j = 1, \dots b, k = 1, \dots, c, l = 1, \dots, n.$$

Table 7.17 ANOVA table for the balanced three-way analysis of variance – mixed classification – $(A \times B) \succ C$.

Source of variation	SS	df	MS
Between A levels	$SS_A = \frac{1}{bcn}\sum_i Y^2_{\underline{i}\cdots} - \frac{1}{N}Y^2_{\underline{\cdots\cdot}}$	$a-1$	$\frac{SS_A}{a-1}$
Between B levels	$SS_B = \frac{1}{acn}\sum_j Y^2_{\underline{\cdot j}\cdot\cdot} - \frac{1}{N}Y^2_{\underline{\cdots\cdot}}$	$b-1$	$\frac{SS_B}{b-1}$
Between C levels within $A \times B$ levels	$SS_{C \text{ in } AB} = \frac{1}{n}\sum_{i,j,k} Y^2_{\underline{ijk}\cdot} - \frac{1}{cn}\sum_{i,j} Y^2_{\underline{ij}\cdot\cdot}$	$ab(c-1)$	$\frac{SS_{C \text{ in } AB}}{ab(c-1)}$
Interaction $A \times B$	$SS_{AB} = \frac{1}{cn}\sum_{i,j} Y^2_{\underline{ij}\cdot\cdot} - \frac{1}{bcn}\sum_i Y^2_{\underline{i}\cdots} - \frac{1}{acn}\sum_j Y^2_{\underline{\cdot j}\cdot\cdot} + \frac{1}{N}Y^2_{\underline{\cdots\cdot}}$	$(a-1)\,(b-1)$	$\frac{SS_{AB}}{(a-1)(b-1)}$
Residual	$SS_{\text{res}} = \sum_{i,j,k,r} y^2_{\underline{ijkr}} - \frac{1}{n}\sum_{i,j,k} Y^2_{\underline{ijk}\cdot}$	$N-abc$	$\frac{SS_{\text{res}}}{N-abc}$

Table 7.18 Expected mean squares for model III.

Source of variation	Model III $E(MS)$
Between A levels	$\sigma^2 + \frac{bcn}{a-1}\sum_i a_i^2 + cn\sigma_{ab}^2$
Between B levels	$\sigma^2 + cn\sigma_{ab}^2 + acn\sigma_b^2$
Between C levels within $A \times B$	$\sigma^2 + \frac{n}{ab(c-1)} \cdot \sum_{i,j,k} c_{k(ij)}^2$
Interaction $A \times B$	$\sigma^2 + cn\sigma_{ab}^2$
Residual	σ^2

The model becomes complete under the conditions that the random variables on the right-hand side of the equation are uncorrelated and have with the same suffixes equal variances, and all fixed effects besides μ over j and k sum up to zero.

The expected mean squares are given in Table 7.18.

From this table, we can derive the test statistic for testing

$$H_0 : a_i = 0 \forall i; H_A : \text{ at least one } a_i \neq 0.$$

We find

$\boldsymbol{F}_A = \frac{MS_A}{MS_{AB}}$ is, under H_0, $F(a-1;\ (a-1)(b-1))$-distributed with $a-1$ and $(a-1)(b-1)$ degrees of freedom.

Problem 7.37 Calculate the minimin and maximin sub-class numbers for testing the null hypotheses $H_0 : a_i = 0 \forall i$; against H_A: at least one $a_i \neq 0$ for model III of the three-way analysis of variance – mixed classification $(A \times B) \succ C$.

Solution

Use `>size_b.three_way_mixed_ab_in_c.model_3_a(α,β,δ/σ,a,c,n, "cases")`. We choose n as small as possible because the test statistic does not depend on n.

Example

We use $a = 6$, $c = 4$, $n = 2$, $\alpha = 0.05$; $\beta = 0.1$, and $\delta/\sigma = 0.5$, and $n = 2$.

```
> size_b.three_way_mixed_ab_in_c.model_3_a(0.05,0.1,0.5,6,4,2, "minimin")
[1] 7
> size_b.three_way_mixed_ab_in_c.model_3_a(0.05,0.1,0.5,6,4,2, "maximin")
[1] 18
```

From Table 7.18 we can also derive the test statistic for testing

$$H_0 : c_{k(ij)} = 0 \forall k, j, i; H_A : \text{ at least one } c_{k(ij)} \neq 0$$

$\boldsymbol{F} = \frac{MS_{C \text{ in } AB}}{MS_R}$ is, under H_0 $F[ab(c-1);\ abc(n-1)]$-distributed with $ab(c-1)$ and $abc(n-1)$ degrees of freedom.

Problem 7.38 Calculate the minimin and maximin sub-class numbers for testing the null hypotheses $H_0: c_{k(ij)} = 0 \forall k, j, i$; H_A: at least one $c_{k(ij)} \neq 0$ for model III of the three-way analysis of variance – mixed classification $(A \times B) \succ C$.

Solution

Use `>size_n.three_way_mixed_ab_in_c.model_3_c(α,β,δ/σ,a,b,c, "cases")`.

Example

We use $a = 6$, $b = 4$, $c = 2$, $\alpha = 0.05$; $\beta = 0.1$ and $\delta/\sigma = 0.5$.

```
> size_n.three_way_mixed_ab_in_c.model_3_c(0.05,0.1,0.5,6,4,2, "minimin")
[1] 10
> size_n.three_way_mixed_ab_in_c.model_3_c(0.05,0.1,0.5,6,4,2, "maximin")
[1] 224
```

Problem 7.39 Estimate the variance components for model III of the three-way analysis of variance – mixed classification $(\boldsymbol{A} \times \boldsymbol{B}) \succ \boldsymbol{C}$.

Solution

$$s_b^2 = \frac{1}{acn}(MS_B - MS_{AB}),$$

$$s_{ab}^2 = \frac{1}{cn}(MS_{AB} - MS_{\text{res}}),$$

$$s^2 = MS_{\text{res}}.$$

7.3.3.2 Three-Way Analysis of Variance – Mixed Classification – $(A \times B) \succ C$ Model IV

The model equation is:

$$\boldsymbol{y}_{ijkl} = \mu + \boldsymbol{a}_i + \boldsymbol{b}_j + (\boldsymbol{a}, \boldsymbol{b})_{ij} + c_{k(ij)} + \boldsymbol{e}_{ijkl}$$
$$i = 1, \dots, a, j = 1, \dots b, k = 1, \dots, c, l = 1, \dots, n.$$

The model becomes complete under the conditions that the random variables on the right-hand side of the equation are uncorrelated and have, with the same suffixes, equal variances, and all fixed effects besides μ over j and k sum up to zero.

The expected mean squares for the balanced case are given in Table 7.19.

From this table, we can derive the test statistic for testing

$$H_0 : c_{k(ij)} = 0 \forall k, j, i; H_A : \text{ at least one } c_{k(ij)} \neq 0.$$

We find

$\boldsymbol{F} = \frac{MS_{C \text{ in } AB}}{MS_R}$ is, under H_0, $F[ab(c-1);\ abc(n-1)]$-distributed with $ab(c-1)$ and $abc(n-1)$ degrees of freedom.

Problem 7.40 Calculate the minimin and maximin sub-class numbers for testing the null hypotheses $H_0: c_{k(ij)} = 0 \forall k, j, i$; H_A: at least one $c_{k(ij)} \neq 0$ for model IV of the three-way analysis of variance – mixed classification $(A \times B) \succ C$.

Table 7.19 Expected mean squares for model IV.

Source of variation	Model IV *E*(*MS*)
Between *A* levels	$\sigma^2 + bcn\sigma_a^2 + cn\sigma_{ab}^2$
Between *B* levels	$\sigma^2 + acn\sigma_b^2 + cn\sigma_{ab}^2$
Between *C* levels within $A \times B$	$\sigma^2 + \frac{n}{ab(c-1)} \cdot \sum_{ijk} c_{k(ij)}^2$
Interaction $A \times B$	$\sigma^2 + cn\sigma_{ab}^2$
Residual	σ^2

Solution

Use `>size_n.three_way_mixed_ab_in_c.model_4_c(α,β,δ/σ,a,b,c,"cases")`.

Example

We use $a = 6$, $b = 5$, $c = 4$, $\alpha = 0.05$; $\beta = 0.1$, and $\delta/\sigma = 0.5$.

```
> size_n.three_way_mixed_ab_in_c.model_4_c(0.05,0.1,0.5,6,5,4, "minimin")
[1] 7
> size_n.three_way_mixed_ab_in_c.model_4_c(0.05,0.1,0.5,6,5,4, "maximin")
[1] 378
```

Problem 7.41 Estimate the variance components for model IV of the three-way analysis of variance – mixed classification $(\boldsymbol{A} \times \boldsymbol{B}) \succ \boldsymbol{C}$.

Solution

$$s_a^2 = \frac{1}{bcn}(MS_A - MS_{AB}),$$

$$s_b^2 = \frac{1}{acn}(MS_B - MS_{AB}),$$

$$s_{ab}^2 = \frac{1}{cn}(MS_{AB} - MS_{\text{res}}),$$

$$s_b^2 = MS_{\text{res}}.$$

7.3.3.3 Three-Way Analysis of Variance – Mixed Classification – $(A \times B) \succ C$ Model V

The model equation is:

$$\boldsymbol{y}_{ijkl} = \mu + \boldsymbol{a}_i + b_j + (\boldsymbol{a}, b)_{ij} + \boldsymbol{c}_{k(ij)} + \boldsymbol{e}_{ijkl}$$
$$i = 1, \ldots, a, j = 1, \ldots b, k = 1, \ldots, c, l = 1, \ldots, n.$$

The model becomes complete under the conditions that the random variables on the right-hand side of the equation are uncorrelated and have, with the same suffixes, equal variances, and all fixed effects besides μ over j and k sum up to zero.

The expected mean squares for the balanced case are given in Table 7.20.

Table 7.20 Expected mean squares for the balanced model V.

Source of variation	Model V *E*(*MS*)
Between *A* levels	$\sigma^2 + \frac{bcn}{a-1}\sum_i a_i^2 + n\sigma_{c(ab)}^2$
Between *B* levels	$\sigma^2 + \frac{acn}{b-1}\sum_j b_j^2 + n\sigma_{c(ab)}^2$
Between *C* levels within $A \times B$	$\sigma^2 + n\sigma_{c(ab)}^2$
Interaction $A \times B$	$\sigma^2 + n\sigma_{c(ab)}^2 + \frac{cn}{(a-1)(b-1)}\sum_{i,j}(ab)_{ij}^2$
Residual	σ^2

From this table, we can derive the test statistic for testing

$$H_0 : a_i = 0 \forall i; H_A : \text{ at least one } a_i \neq 0$$

We find
$\boldsymbol{F}_A = \frac{MS_A}{MS_{C\,\text{in}\,AB}}$ is, under H_0, $F(a-1; ab\ (c-1))$-distributed with $a-1$ and $ab(c-1)$ degrees of freedom.

Problem 7.42 Calculate the minimin and maximin sub-class numbers for testing the null hypotheses $H_0 : a_i = 0 \,\forall\, i$; H_A: at least one $a_i \neq 0$ for model V of the three-way analysis of variance – mixed classification $(\boldsymbol{A} \succ \boldsymbol{B}) \times \boldsymbol{C}$.

Solution
Use `>size_c.three_way_mixed_ab_in_c.model_5_a(α,β,δ/σ,a,b,n, "cases")`.

Example
We use $a = 6$, $b = 5$, $n = 2$, $\alpha = 0.05$; $\beta = 0.1$ and $\delta/\sigma = 0.5$.

```
> size_c.three_way_mixed_ab_in_c.model_5_a(0.05,0.1,0.5,6,5,2, "minimin")
[1] 5
> size_c.three_way_mixed_ab_in_c.model_5_a(0.05,0.1,0.5,6,5,2, "maximin")
[1] 14
```

From Table 7.20 we can derive the test statistic for testing
$H_{B0} : b_j = 0 \,\forall\, j$ against H_{BA}: at least one $b_j \neq 0$.
We find $\boldsymbol{F}_B = \frac{MS_B}{MS_{C\,\text{in}\,AB}}$ is, under H_0, $F(b-1; ab\ (c-1))$-distributed with $b-1$ and $ab(c-1)$ degrees of freedom.

Problem 7.43 Calculate the minimin and maximin sub-class numbers for testing the null hypotheses $H_{B0} : b_j = 0 \,\forall\, j$ against H_{BA}: at least one $b_j \neq 0$ for model V of the three-way analysis of variance – mixed classification $(\boldsymbol{A} \succ \boldsymbol{B}) \times \boldsymbol{C}$.

Solution
Use `>size_c.three_way_mixed_ab_in_c.model_5_b(α,β,δ/σ,a,b,n, "cases")`.

Example
We use $a = 6$, $b = 5$, $n = 2$, $\alpha = 0.05$; $\beta = 0.1$ and $\delta/\sigma = 0.5$.

```
> size_c.three_way_mixed_ab_in_c.model_5_b(0.05,0.1,0.5,6,5,2, "minimin")
[1] 5
> size_c.three_way_mixed_ab_in_c.model_5_b(0.05,0.1,0.5,6,5,2, "maximin")
[1] 11
```

From Table 7.20 we can derive the test statistic for testing $H_{AB0}: (a,b)_{ij} = 0\,\forall j$ against H_{ABA}: at least one $(a,b)_{ij} \neq 0$.
We find
$\boldsymbol{F}_B = \frac{\boldsymbol{MS}_{AB}}{\boldsymbol{MS}_{C \text{ in } AB}}$ is, under H_0, $F((a-1)(b-1);\ ab\ (c-1))$-distributed with $(a-1)(b-1)$ and $ab(c-1)$ degrees of freedom.

Problem 7.44 Calculate the minimin and maximin number of levels of factor C for testing the null hypotheses $H_{AB0}: (a,b)_{ij} = 0\,\forall j$ against H_{ABA}: at least one $(a,b)_{ij} \neq 0$ for model V of the three-way analysis of variance – mixed classification $(\boldsymbol{A} \times \boldsymbol{B}) \succ \boldsymbol{C}$.

Solution
Use `>size_c.three_way_mixed_ab_in_c.model_5_axb(α,β,δ/σ,a,b, n, "cases")`.

Example
We use $a = 6$, $b = 5$, $n = 2$, $\alpha = 0.05$; $\beta = 0.1$ and $\delta/\sigma = 0.5$.

```
> size_c.three_way_mixed_ab_in_c.model_5_axb(0.05,0.1,0.5,6,5,2, "minimin")
[1] 8
> size_c.three_way_mixed_ab_in_c.model_5_axb(0.05,0.1,0.5,6,5,2, "maximin")
[1] 106
```

Problem 7.45 Estimate the variance components for model V of the three-way analysis of variance – mixed classification $(\boldsymbol{A} \times \boldsymbol{B}) \succ \boldsymbol{C}$.

Solution

$$s^2_{c(ab)} = \frac{1}{n}(\boldsymbol{MS}_{C \text{ in } AB} - \boldsymbol{MS}_{\text{res}}),$$

$$s^2 = \boldsymbol{MS}_{\text{res}}.$$

7.3.3.4 Three-Way Analysis of Variance – Mixed Classification – $(A \times B) \succ C$ Model VI

The model equation is:

$$\boldsymbol{y}_{ijkl} = \mu + \boldsymbol{a}_i + b_j + (\boldsymbol{a}, \boldsymbol{b})_{ij} + \boldsymbol{c}_{k(ij)} + \boldsymbol{e}_{ijkl}$$
$$i = 1, \dots, a, j = 1, \dots b, k = 1, \dots, c, l = 1, \dots, n.$$

Table 7.21 Expected mean squares for model VI.

Source of variation	Model VI $E(MS)$
Between A levels	$\sigma^2 + bcn\sigma_a^2 + n\sigma_{c(ab)}^2 + cn\sigma_{ab}^2$
Between B levels	$\sigma^2 + \frac{acn}{b-1}\sum_j b_j^2 + n\sigma_{c(ab)}^2 + cn\sigma_{ab}^2$
Between C levels within $A \times B$	$\sigma^2 + n\sigma_{c(ab)}^2$
Interaction $A \times B$	$\sigma^2 + n\sigma_{c(ab)}^2 + cn\sigma_{ab}^2$
Residual	σ^2

The model becomes complete under the conditions that the random variables on the right-hand side of the equation are uncorrelated and have, with the same suffixes, equal variances, and all fixed effects besides μ over j and k sum up to zero.

The expected mean squares are given Table 7.21.

From this table, we can derive the test statistic for testing

$$H_0 : b_j = 0\,\forall j; H_A : \text{ at least one } b_j \neq 0.$$

We find

$F_A = \frac{MS_B}{MS_{AB}}$ is, under H_0, $F(b-1;\ (a-1)\ (b-1))$-distributed with $a-1$ and $(a-1)$ $(b-1)$ degrees of freedom.

Problem 7.46 Calculate the minimin and maximin sub-class numbers for testing the null hypotheses $H_0 : b_j = 0\,\forall j$; against H_A: at least one $b_j \neq 0$ for model VI of the three-way analysis of variance – mixed classification$(A \times B) \succ C$.

Solution

Use `>size_c.three_way_mixed_ab_in_c.model_6_b(α,β,δ/σ,a,b,n,"cases")`.

Example

We use $a = 6$, $b = 5$, $n = 2$, $\alpha = 0.05$; $\beta = 0.1$ and $\delta/\sigma = 0.5$.

```
> size_c.three_way_mixed_ab_in_c.model_6_b (0.05,0.1,0.5,6,5,2, "minimin")
[1] 6

> size_c.three_way_mixed_ab_in_c.model_6_b (0.05,0.1,0.5,6,5,2, "maximin")
[1] 14
```

Problem 7.47 Estimate the variance components for model VI of the three-way analysis of variance – mixed classification $(\boldsymbol{A} \times \boldsymbol{B}) \succ \boldsymbol{C}$.

Solution

$$s_a^2 = \frac{1}{bcn}(MS_A - MS_{AB}),$$

$$s_{ab}^2 = \frac{1}{cn}(MS_{AB} - MS_{C \text{ in } AB}),$$

$$s_{c(ab)}^2 = \frac{1}{n}(MS_{C \text{ in } AB} - MS_{\text{res}}),$$

$$s^2 = MS_{\text{res}}.$$

7.3.4 Three-Way Analysis of Variance – Mixed Classification – $(A \succ B) \times C$

We have six mixed models in this classification. The ANOVA table is independent of the models and given in Table 7.22 for the balanced case.

7.3.4.1 Three-Way Analysis of Variance – Mixed Classification – $(A \succ B) \times C$ Model III

The model equation for the balanced case is:

$$y_{ijkl} = \mu + \boldsymbol{a}_i + b_{j(i)} + c_k + (\boldsymbol{a}, c)_{ik} + (\boldsymbol{a}, \boldsymbol{b}, \boldsymbol{c})_{ijk} + \boldsymbol{e}_{ijkl};$$
$$i = 1, \dots, a, j = 1, \dots, b, k = 1, \dots, c, l = 1, \dots, n.$$

The model becomes complete under the conditions that the random variables on the right-hand side of the equation are uncorrelated and have, with the same suffixes, equal variances, and all fixed effects besides μ over j and k sum up to zero.

The expected mean squares shows Table 7.23.

From Table 7.23, we can derive the test statistic for testing: $\boldsymbol{F} = \frac{MS_{B \text{ in } A}}{MS_R}$ is, under H_0, $F[a(b-1); abc(n-1)]$-distributed with $a(b-1)$ and $abc(n-1)$ degrees of freedom.

Problem 7.48 Calculate the minimin and maximin sub-class numbers for testing the null hypotheses $H_0: b_{j(i)} = 0 \, \forall j, i$; H_A: at least one $b_{j(i)} \neq 0$ for model III of the three-way analysis of variance – mixed classification $(\boldsymbol{A} \succ \boldsymbol{B}) \times \boldsymbol{C}$.

Solution

Use `>size_n.three_way_mixed_cxbina.model_3_b(α,β,δ/σ,a,b,c, "cases").`

Example

We use $a = 6$, $b = 5$, $c = 4$, $\alpha = 0.05$; $\beta = 0.1$ and $\delta/\sigma = 0.5$.

```
> size_n.three_way_mixed_cxbina.model_3_b (0.05,0.1,0.5,6,5,4, "minimin")
[1] 4

> size_n.three_way_mixed_cxbina.model_3_b (0.05,0.1,0.5,6,5,4, "maximin")
[1] 57
```

From Table 7.23 we can also derive the test statistic for testing $H_0: c_k = 0 \, \forall k$ against $H_A: c_k \neq 0$ for at least one k.

$\boldsymbol{F} = \frac{MS_C}{MS_{AC}}$ is, under H_0, $F(c-1; (a-1)(c-1))$-distributed with $c-1$ and $(a-1)(c-1)$ degrees of freedom.

Problem 7.49 Calculate the minimin and maximin number sub-class numbers for testing the null hypotheses $H_0: c_{k(ij)} = 0 \, \forall k, j, i$; H_A: at least one $c_{k(ij)} \neq 0$ for model III of the three-way analysis of variance – mixed classification $(\boldsymbol{A} \succ \boldsymbol{B}) \times \boldsymbol{C}$.

Table 7.22 ANOVA table for the three-way balanced analysis of variance – mixed classification $(A \succ B) \times C$.

Source of variation	SS	df	*MS*
Between A levels	$SS_A = \frac{1}{bcn}\sum_i Y^2_{i\cdots} - \frac{1}{N}Y^2_{\cdots\cdot}$	$a-1$	$\frac{\mathbf{SS}_A}{a-1}$
Between B levels within A levels	$SS_B = \frac{1}{cn}\sum_{i,j} Y^2_{ij\cdot\cdot} - \frac{1}{bcn}\sum Y^2_{i\cdots}$	$a(b-1)$	$\frac{\mathbf{SS}_{B\text{ in }A}}{a(b-1)}$
Between C levels	$SS_C = \frac{1}{abn}\sum_k Y^2_{\cdot\cdot k\cdot} - \frac{1}{N}Y^2_{\cdots\cdot}$	$c-1$	$\frac{\mathbf{SS}_C}{c-1}$
Interaction $A \times C$	$SS_{AC} = \frac{1}{bn}\sum_{i,k} Y^2_{i\cdot k\cdot} - \frac{1}{bcn}\sum_i Y^2_{i\cdots} - \frac{1}{abn}\sum_k Y^2_{\cdot\cdot k\cdot} + \frac{1}{N}Y^2_{i\cdots}$	$(a-1)(c-1)$	$\frac{SS_{AC}}{(a-1)(c-1)}$
Interaction $B \times C$ within A	$SS_{BCinA} = \frac{1}{n}\sum_{i,j,k} Y^2_{ijk\cdot} - \frac{1}{cn}\sum_{i,j} Y^2_{ij\cdot\cdot} - \frac{1}{bn}\sum_{i,k} Y^2_{i\cdot k\cdot} + \frac{1}{bcn}\sum_i Y^2_{i\cdots}$	$a(b-1)(c-1)$	$\frac{\mathbf{SS}_{BC\text{ in }A}}{a(b-1)(b-1)}$
Residual	$SS_R = \sum_{i,j,k,r} y^2_{ijkr} - \frac{1}{n}\sum_{i,j,k} Y^2_{ijk\cdot}$	$N-abc$	$\frac{SS_{\text{res}}}{N-abc}$

Table 7.23 Expected mean squares for balanced model III.

Source of variation	*E*(*MS*)
Between *A* levels	$bcn\sigma_a^2 + bn\sigma_{ac}^2 + \sigma^2$
Between *B* levels within *A* levels	$\frac{cn}{a(b-1)}\sum_{i,j} b_{j(i)}^2 + \sigma^2$
Between *C* levels	$\frac{abn}{c-1}\sum_{k} c_k^2 + bn\sigma_{ac}^2 + \sigma^2$
Interaction $A \times C$	$bn\sigma_{ac}^2 + \sigma^2$
Interaction $B \times C$ within *A*	$\frac{n}{a(b-1)(c-1)} \cdot \sum_{i,j,k} (bc)_{jk(i)}^2 + \sigma^2$
Residual	σ^2

Solution

Use `>size_n.three_way_mixed_cxbina.model_3_c(α,β,δ/σ,a,b,c, "cases").`

Example

We use $b = 5$, $c = 4$, $n = 2$, $\alpha = 0.05$; $\beta = 0.1$ and $\delta/\sigma = 0.5$.

```
> size_n.three_way_mixed_cxbina.model_3_bxc (0.05,0.1,0.5,5,4,2, "minimin")
[1] 10

> size_n.three_way_mixed_cxbina.model_3_bxc (0.05,0.1,0.5,5,4,2, "maximin")
[1] 189
```

Problem 7.50 Estimate the variance components for model III of the three-way analysis of variance – mixed classification $(\boldsymbol{A} \succ \boldsymbol{B}) \times \boldsymbol{C}$.

Solution

$$s_a^2 = \frac{1}{bcn}(MS_A - MS_{AC}),$$

$$s_{ac}^2 = \frac{1}{bn}(MS_{AC} - MS_{\text{res}}),$$

$$s^2 = MS_{\text{res}}.$$

7.3.4.2 Three-Way Analysis of Variance – Mixed Classification – $(A \succ B) \times C$ Model IV

The model equation for the balanced case is:

$$\boldsymbol{y}_{ijkl} = \mu + \boldsymbol{a}_i + \boldsymbol{b}_{j(i)} + c_k + (\boldsymbol{a}, c)_{ik} + (\boldsymbol{b}, \boldsymbol{c})_{jk(i)} + \boldsymbol{e}_{ijkl};$$
$$i = 1, \ldots, a, j = 1, \ldots, b, k = 1, \ldots, c, l = 1, \ldots, n.$$

The model becomes complete under the conditions that the random variables on the right-hand side of the equation are uncorrelated and have, with the same suffixes, equal variances, and all fixed effects besides μ over *j* and *k* sum up to zero.

Table 7.24 Expected mean squares for balanced model IV.

Source of variation	*E*(*MS*) for model IV
Between *A* levels	$\frac{bcn}{a-1}\sum_i a_i^2 + cn\sigma_{b(a)}^2 + n\sigma_{bc(a)}^2 + \sigma^2$
Between *B* levels within *A* levels	$cn\sigma_{b(a)}^2 + n\sigma_{bc(a)}^2 + \sigma^2$
Between *C* levels	$\frac{abn}{c-1}\sum_k c_k^2 + n\sigma_{bc(a)}^2 + \sigma^2$
Interaction $A \times C$	$\frac{bn}{(a-1)(c-1)}\sum_{i,k}(ac)_{ik}^2 + n\sigma_{bc(a)}^2 + \sigma^2$
Interaction $B \times C$ within *A*	$n\sigma_{bc(a)}^2 + \sigma^2$
Residual	σ^2

The expected mean squares are given in Table 7.24.

Problem 7.51 Calculate the minimin and maximin number of levels of the factor *B* for testing the null hypotheses $H_0: a_i = 0\,\forall\, i$ against $H_A: a_i \neq 0$ for at least one *i* calculated for model IV of the three-way analysis of variance – mixed classification $(\boldsymbol{A} \succ \boldsymbol{B}) \times \boldsymbol{C}$.

Solution

Use `>size_b.three_way_mixed_cxbina.model_4_a(α,β,δ/σ,a,c,n, "cases")`.

Example

We use $a = 6$, $c = 4$, $\alpha = 0.05$; $\beta = 0.1$ and $\delta/\sigma = 0.5$ and $n = 2$.

```
> size_b.three_way_mixed_cxbina.model_4_a(0.05,0.1,0.5,6,4,2, "minimin")
[1] 7
```

```
> size_b.three_way_mixed_cxbina.model_4_a(0.05,0.1,0.5,6,4,2, "maximin")
[1] 18
```

From Table 7.24, we also can derive the test statistic for testing $H_0: c_k = 0\,\forall\, k$ against $H_A: c_k \neq 0$ for at least one *k*.

$\boldsymbol{F} = \frac{MS_C}{MS_{BC\text{ in }A}}$ is, under H_0, $F(c-1; a(b-1)(c-1))$ – distributed with $c-1$ and $a(b-1)(c-1)$ degrees of freedom.

Problem 7.52 Calculate the minimin and maximin number of levels of the factor *B* for testing the null hypotheses $H_0: c_{k(ij)} = 0\,\forall\, k, j, i$; H_A: at least one $c_{k(ij)} \neq 0$ for model IV of the three-way analysis of variance – mixed classification $(\boldsymbol{A} \succ \boldsymbol{B}) \times \boldsymbol{C}$.

Solution

Use `>size_b.three_way_mixed_cxbina.model_4_c(α,β,δ/σ,a,c,n, "cases")`.

Example

We use $a = 6$, $c = 4$, $\alpha = 0.05$; $\beta = 0.1$ and $\delta/\sigma = 0.5$ and $n = 2$.

```
> size_b.three_way_mixed_cxbina.model_4_c(0.05,0.1,0.5,6,4,2, "minimin")
[1] 5
> size_b.three_way_mixed_cxbina.model_4_c(0.05,0.1,0.5,6,4,2, "maximin")
[1] 10
```

From Table 7.24, we also can derive the test statistic for testing $H_{AC0}:(a,c)_{jk} = 0\,\forall\, j,k$ against H_{ACA}: at least one $(ac)_{jk} \neq 0$.
$F_{AC} = \frac{MS_{AC}}{MS_{BC \text{ in } A}}$ is, under $H_{AC0}:(ac)_{jk} = 0\,\forall\, j,\ k$, $F((a-1)(c-1),\ a(b-1)(c-1))$ – distributed with $(a-1)(c-1)$, and $a(b-1)(c-1)$ degrees of freedom.

Problem 7.53 Calculate the minimin and maximin number of levels of the factor B for testing the null hypotheses $H_{AC0}:(a,c)_{jk} = 0\,\forall\, j,k$ against H_{ACA}: at least one $(a,c)_{jk} \neq 0$ for model IV of the three-way analysis of variance – mixed classification $(\boldsymbol{A} \succ \boldsymbol{B}) \times \boldsymbol{C}$.

Solution

Use `>size_b.three_way_mixed_cxbina.model_4_axc(α,β,δ/σ,a,c,n, "cases")`.

Example

We use $a = 6$, $c = 4$, $\alpha = 0.05$; $\beta = 0.1$ and $\delta/\sigma = 0.5$ and $n = 2$.

```
> size_b.three_way_mixed_cxbina.model_4_axc(0.05,0.1,0.5,6,4,2, "minimin")
[1] 9
> size_b.three_way_mixed_cxbina.model_4_axc(0.05,0.1,0.5,6,4,2, "maximin")
[1] 96
```

Problem 7.54 Estimate the variance components for model IV of the three-way analysis of variance – mixed classification $(\boldsymbol{A} \succ \boldsymbol{B}) \times \boldsymbol{C}$.

Solution

$$s^2_{b(a)} = \frac{1}{cn}(MS_{B \text{ in } A} - MS_{BC \text{ in } A}),$$

$$s^2_{bc(a)} = \frac{1}{n}(MS_{BC \text{ in } A} - MS_{\text{res}}),$$

$$s^2 = MS_{\text{res}}.$$

7.3.4.3 Three-Way Analysis of Variance – Mixed Classification – $(A \succ B) \times C$ Model V

The model equation for the balanced case is:

$$\boldsymbol{y}_{ijkl} = \mu + a_i + \boldsymbol{b}_{j(i)} + c_k + (a, c)_{ik} + (\boldsymbol{b}, \boldsymbol{c})_{jk(i)} + \boldsymbol{e}_{ijkl};$$
$$i = 1, \ldots, a, j = 1, \ldots, b, k = 1, \ldots, c, l = 1, \ldots, n.$$

The model becomes complete under the conditions that the random variables on the right-hand side of the equation are uncorrelated and have, with the same suffixes, equal variances, and all fixed effects besides μ over j and k sum up to zero.

The expected mean squares are given in Table 7.25.

Table 7.25 Expected mean squares for the balanced model V.

Source of variation	**E(*MS*) for model V**
Between *A* levels	$\frac{bcn}{a-1}\sum_i a_i^2 + bn\sigma_{ac}^2 + n\sigma_{bc(a)}^2 + \sigma^2$
Between *B* levels within *A* levels	$\frac{cn}{a-1}\sum_{ij} b_{j(i)}^2 + n\sigma_{bc(a)}^2 + \sigma^2$
Between *C* levels	$abn\sigma_c^2 + (bn\sigma_{ac}^2 + n\sigma_{bc(a)}^2) + \sigma^2$
Interaction $A \times C$	$bn\sigma_{ac}^2 + n\sigma_{bc(a)}^2 + \sigma^2$
Interaction $B \times C$ within *A*	$n\sigma_{bc(a)}^2 + \sigma^2$
Residual	σ^2

From Table 7.25, we can derive the test statistic for testing

$$H_0 : a_i = 0 \forall i; H_A : \text{ at least one } a_i \neq 0$$

$F_A = \frac{MS_A}{MS_{AC}}$ is, under H_0, $F((a-1); (a-1)(c-1))$-distributed with $a-1$ and $(a-1)(c-1)$ degrees of freedom.

Problem 7.55 Calculate the minimin and maximin number of levels of the factor *C* for testing the null hypotheses $H_0 : a_i \ \forall\, i$ against H_A: at least one a_i is unequal 0 for model V of the three-way analysis of variance – mixed classification $(\boldsymbol{A} \succ \boldsymbol{B}) \times \boldsymbol{C}$.

Solution

Use `>size_c.three_way_mixed_cxbina.model_5_a(α,β,δ/σ,a,b,n, "cases")`.

Example

We use $a = 6$, $b = 5$, $n = 2$, $\alpha = 0.05$; $\beta = 0.1$ and $\delta/\sigma = 0.5$.

```
> size_c.three_way_mixed_cxbina.model_5_a (0.05,0.1,0.5,6,5,2, "minimin")
[1] 6
```

```
> size_c.three_way_mixed_cxbina.model_5_a (0.05,0.1,0.5,6,5,2, "maximin")
[1] 15
```

From Table 7.25, we also can derive the test statistic for testing $H_0 : b_{j(i)} = 0 \,\forall\, j, i$ against H_A: at least one $b_{j(i)} \neq 0$.

$F = \frac{MS_{B \text{ in } A}}{MS_{B\times C \text{ in } A}}$ is, under H_0, $F[a(b-1); a(b-1)(c-1)]$-distributed with $a(b-1)$ and $a(b-1)(c-1)$ degrees of freedom.

Problem 7.56 Calculate the minimin and maximin number of levels of the factor *C* for testing the null hypotheses $H_0 : b_{j(i)} = 0 \,\forall\, j, i$ against H_A: at least one $b_{j(i)} \neq 0$ for model V of the three-way analysis of variance – mixed classification $(\boldsymbol{A} \succ \boldsymbol{B}) \times \boldsymbol{C}$.

Solution

Use `> size_c.three_way_mixed_cxbina.model_5_b(α,β,δ/σ,a,b,n, "cases")`.

Example

We use $a = 6$, $b = 5$, $n = 2$, $\alpha = 0.05$; $\beta = 0.1$ and $\delta/\sigma = 0.5$.

```
> size_c.three_way_mixed_cxbina.model_5_b (0.05,0.1,0.5,6,5,2, "minimin")
[1] 9
> size_c.three_way_mixed_cxbina.model_5_b (0.05,0.1,0.5,6,5,2, "maximin")
[1] 113
```

Problem 7.57 Estimate the variance components for model V of the three-way analysis of variance – mixed classification $(\boldsymbol{A} \succ \boldsymbol{B}) \times \boldsymbol{C}$.

Solution

$$s_c^2 = \frac{1}{abn}(MS_C - MS_{AC}),$$

$$s_{ac}^2 = \frac{1}{bn}(MS_{AC} - MS_{BC \text{ in } A}),$$

$$s_{bc(a)}^2 = \frac{1}{n}(MS_{BC \text{ in } A} - MS_{\text{res}}),$$

$$s^2 = MS_{\text{res}}.$$

7.3.4.4 Three-Way Analysis of Variance – Mixed Classification – $(A \succ B) \times C$ Model VI

The model equation for the balanced case is given by:

$$y_{ijkl} = \mu + a_i + \boldsymbol{b}_{j(i)} + c_k + (\boldsymbol{a}, \boldsymbol{c})_{ik} + (\boldsymbol{b}, \boldsymbol{c})_{jk(i)} + \boldsymbol{e}_{ijkl};$$
$$i = 1, \dots, a, j = 1, \dots, b, k = 1, \dots, c, l = 1, \dots, n.$$

The model becomes complete under the conditions that the random variables on the right-hand side of the equation are uncorrelated and have, with the same suffixes, equal variances, and all fixed effects besides μ over j and k sum up to zero.

The expected mean squares are given in Table 7.26.

From Table 7.26 we cannot derive an exact test statistic for testing $H_0: a_i = 0 \; \forall \, i$ against H_A: at least one $a_i \neq 0$, and this means that for this hypothesis no exact F-test exists.

We can only, analogously to Section 7.3.1, derive a test statistic for an approximate F-test and obtain

$$F = \frac{MS_A}{MS_{\text{B in A}} + MS_{A\times C} - MS_{B\times C \text{ in } A}}$$

and have, under $H_0: a_i = 0$ for all i, approximately $a - 1$ and

$$f = \frac{(MS_{B \text{ in } A} + MS_{A\times C} - MS_{B\times C \text{ in } A})^2}{\frac{MS_{B \text{ in } A}{}^2}{a(b-1)} + \frac{MS_{A\times C}{}^2}{(a-1)(b-1)} + \frac{MS_{B\times C \text{ in } A}{}^2}{a(b-1)(c-1)}}$$

degrees of freedom for this approximate F-statistics.

In addition, variance component estimation is not easy – at least with the ANOVA method and we drop this topic here. How to obtain minimal sample sizes for this model is shown in Spangl et al. (2020).

Table 7.26 Expected mean squares for model VI.

Source of variation	E(MS) for model VI
Between A levels	$\frac{bcn}{a-1}\sum_i a_i^2 + cn\sigma^2_{b(a)} + bn\sigma^2_{ac} + n\sigma^2_{bc(a)} + \sigma^2$
Between B levels within A levels	$cn\sigma^2_{b(a)} + n\sigma^2_{bc(a)} + \sigma^2$
Between C levels	$abn\sigma^2_c + n\sigma^2_{bc(a)} + bn\sigma^2_{ac} + \sigma^2$
Interaction $A \times C$	$bn\sigma^2_{ac} + n\sigma^2_{bc(a)} + \sigma^2$
Interaction $B \times C$ within A	$n\sigma^2_{bc(a)} + \sigma^2$
Residual	σ^2

7.3.4.5 Three-Way Analysis of Variance – Mixed Classification – ($A \succ B$) × C model VII

The model equation for the balanced case is:

$$\boldsymbol{y}_{ijkl} = \mu + \boldsymbol{a}_i + b_{j(i)} + \boldsymbol{c}_k + (\boldsymbol{a}, \boldsymbol{c})_{ik} + (\boldsymbol{b}, \boldsymbol{c})_{jk(i)} + \boldsymbol{e}_{ijkl}$$
$$i = 1, \dots, a, j = 1, \dots b, k = 1, \dots, c, l = 1, \dots, n.$$

The model becomes complete under the conditions that the random variables on the right-hand side of the equation are uncorrelated and have, with the same suffixes, equal variances, and all fixed effects besides μ over j and k sum up to zero.

The expected mean squares are given in Table 7.27.

From Table 7.27, we can derive the test statistic for testing $H_0: b_{j(i)} = 0\,\forall j, i$ against H_A: at least one $b_{j(i)} \neq 0$

$\boldsymbol{F} = \frac{\boldsymbol{MS}_{B\,in\,A}}{\boldsymbol{MS}_{B\times C\,in\,A}}$ is, under H_0, $F[a(b-1); a(b-1)(c-1)]$-distributed with $a(b-1)$ and $a(b-1)(c-1)$ degrees of freedom.

Table 7.27 Expected mean squares for the balanced model VII.

Source of variation	E(MS) for model VII
Between A levels	$bcn\sigma^2_a + bn\sigma^2_{ac} + n\sigma^2_{bc(a)} + \sigma^2$
Between B levels within A levels	$\frac{cn}{a(b-1)}\sum_{ij} b^2_{j(i)} + n\sigma^2_{bc(a)} + \sigma^2$
Between C levels	$abn\sigma^2_c + bn\sigma^2_{ac} + n\sigma^2_{bc(a)} + \sigma^2$
Interaction $A \times C$	$bn\sigma^2_{ac} + n\sigma^2_{bc(a)} + \sigma^2$
Interaction $B \times C$ within A	$n\sigma^2_{bc(a)} + \sigma^2$
Residual	σ^2

Problem 7.58 Calculate the minimin and maximin number of levels of the factor C for testing the null hypotheses $H_0: b_{j(i)} = 0\ \forall j, i$ against H_A: at least one $b_{j(i)} \neq 0$ for model VII of the three-way analysis of variance – mixed classification $(\boldsymbol{A} \succ \boldsymbol{B}) \times \boldsymbol{C}$.

Solution

Use `>size_c.three_way_mixed_cxbina.model_7_b(α,β,δ/σ,a,b,n, "cases")`.

Example

We use $a = 6$, $b = 5$, $n = 2$, $\alpha = 0.05$; $\beta = 0.1$ and $\delta/\sigma = 0.5$.

```
> size_c.three_way_mixed_cxbina.model_7_b (0.05,0.1,0.5,6,5,2, "minimin")
[1] 9
> size_c.three_way_mixed_cxbina.model_7_b (0.05,0.1,0.5,6,5,2, "maximin")
[1] 113
```

Problem 7.59 Estimate the variance components for model VII of the three-way analysis of variance – mixed classification $(\boldsymbol{A} \succ \boldsymbol{B}) \times \boldsymbol{C}$.

Solution

$$s_a^2 = \frac{1}{bcn}(MS_A - MS_{AC}),$$

$$s_c^2 = \frac{1}{abn}(MS_C - MS_{AC}),$$

$$s_{ac}^2 = \frac{1}{bn}(MS_{AC} - MS_{BC \text{ in } A}),$$

$$s_{bc(a)}^2 = \frac{1}{n}(MS_{BC \text{ in } A} - MS_{\text{res}}),$$

$$s^2 = MS_{\text{res}}.$$

7.3.4.6 Three-Way Analysis of Variance – Mixed Classification – $(A \succ B) \times C$ Model VIII

The model equation is:

$$\boldsymbol{y}_{ijkl} = \mu + \boldsymbol{a}_i + \boldsymbol{b}_{j(i)} + c_k + (\boldsymbol{a}, c)_{ik} + (\boldsymbol{b}, c)_{jk(i)} + \boldsymbol{e}_{ijkl},$$
$$i = 1, \dots, a, j = 1, \dots b, k = 1, \dots, c, l = 1, \dots, n.$$

The model becomes complete under the conditions that the random variables on the right-hand side of the equation are uncorrelated and have, with the same suffixes, equal variances, and all fixed effects besides μ over j and k sum up to zero.

The expected mean squares are given in Table 7.28.

From Table 7.28, we can derive the test statistic for testing $H_0: c_k = 0\ \forall k$ against H_A: at least one $c_k \neq 0$.

$\boldsymbol{F} = \frac{MS_C}{MS_{AC}}$ is, under H_0, $F(c-1; (a-1)(c-1))$-distributed with $c-1$ and $(a-1)(c-1)$ degrees of freedom.

Problem 7.60 Estimate the variance components for model VIII of the three-way analysis of variance – mixed classification $(\boldsymbol{A} \succ \boldsymbol{B}) \times \boldsymbol{C}$.

Table 7.28 Expected mean squares for model VIII.

Source of variation	*E*(*MS*) for model VIII
Between *A* levels	$bcn\sigma_a^2 + cn\sigma_{b(a)}^2 + \sigma^2$
Between *B* levels within *A* levels	$cn\sigma_{b(a)}^2 + \sigma^2$
Between *C* levels	$\frac{abn}{c-1}\sum_k c_k^2 + bn\sigma_{ac}^2 + n\sigma_{bc(a)}^2 + \sigma^2$
Interaction $A \times C$	$bn\sigma_{ac}^2 + n\sigma_{bc(a)}^2 + \sigma^2$
Interaction $B \times C$ within *A*	$n\sigma_{bc(a)}^2 + \sigma^2$
Residual	σ^2

Solution

$$s_a^2 = \frac{1}{bcn}(MS_A - MS_{B\text{ in }A}),$$

$$s_{b(a)}^2 = \frac{1}{cn}(MS_{B\text{ in }A} - MS_{\text{res}}),$$

$$s_{ac}^2 = \frac{1}{bn}(MS_{AC} - MS_{BC\text{ in }A}),$$

$$s_{bc(a)}^2 = \frac{1}{n}(MS_{BC\text{ in }A} - MS_{\text{res}}),$$

$$s^2 = MS_{\text{res}}.$$

References

Hartley, H.O. and Rao, J.N.K. (1967). Maximum likelihood estimation for the mixed analysis of variance model. *Biometrika* 54: 92–108.

Ott, R.L. and Longnecker, M. (2001). *Statistical Methods and Data Analysis*, 5e. Pacific Grove, CA USA: Duxbury.

Rasch, D., Spangl, B., and Wang, M. (2012). Minimal Experimental Size in the Three Way ANOVA Cross Classification Model with Approximate *F*-Tests. *Commun. Stat.– Simul. Comput.* 41: 1120–1130.

Rasch, D. and Schott, D. (2018). *Mathematical Statistics*. Oxford: Wiley.

Spangl, B., Kaiblinger, N., Ruckdeschel, P., and Rasch, D. (2019). Minimum experimental size in three-way ANOVA models with one fixed and two random factors exemplified by the mixed classification $(A \succ B) \times C$, working paper.

Searle, S.R. (1971, 2012). *Linear Models*. New York: Wiley.

Searle, S.R., Casella, G., and McCulloch, C.R. (1992). *Variance Components*. New York, Chichester, Brisbane, Toronto, Singapore: Wiley.

8

Regression Analysis

8.1 Introduction

The term regression stems from Galton (1885) who described biological phenomena. Later Yule (1897) generalised this term. It describes the relationship between two (or more) characters, which at this early stage were considered as realisations of random variables. Nowadays regression analysis is a theory within mathematical statistics with broad applications in empirical research.

Partially, we take notations from those of functions in mathematics. A mathematical function describes a deterministic relation between variables. So is the circumference c of a circle dependent on the radius r of this circle, $c = 2r\pi$.

Written in this way the circumference is called the dependent variable and the radius is called the independent variable or more generally in a function

$$y = f(x)$$

where x is called the independent variable and y the dependent variable. The role of these variables can be interchanged under some mathematical assumptions by using the (existence assumed) inverse function

$$x = f^{-1}(y).$$

In contrast to mathematics, in empirical sciences such functional relationships seldom exist. For instance, let us consider height at withers and age, or height at withers and chest girth of cattle. Although there is obviously no formula by which one can calculate the chest girth or the age of cattle from the height at withers, nevertheless there is obviously a connection between both. When both measurements are present and a point represents the value pair of each animal in a coordinate system. All these points are not, as in the case of a functional dependency, on a curve; it is rather a point cloud or as we say a scatter plot. In such a cloud, a clear trend is frequently recognisable, which suggests the existence of a relationship. We demonstrate this by examples.

Example 8.1 In Table 8.1, we find data from Barath et al. (1996).

The scatter-plot for these data is shown in Figure 8.1. It seems that there exists a relatively strong but non-linear relationship.

Example 8.2 In Table 8.2 data collected from 25 students on a statistics course in 1996 at Wageningen Agricultural University (The Netherlands) are shown.

Applied Statistics: Theory and Problem Solutions with R, First Edition.
Dieter Rasch, Rob Verdooren, and Jürgen Pilz.

Table 8.1 The height of hemp plants (y in centimetres) during growth (x age in weeks).

x_i	y_i	x_i	y_i
1	8.30	8	84.40
2	15.20	9	98.10
3	24.70	10	107.70
4	32.00	11	112.00
5	39.30	12	116.90
6	55.40	13	119.90
7	69.00	14	121.10

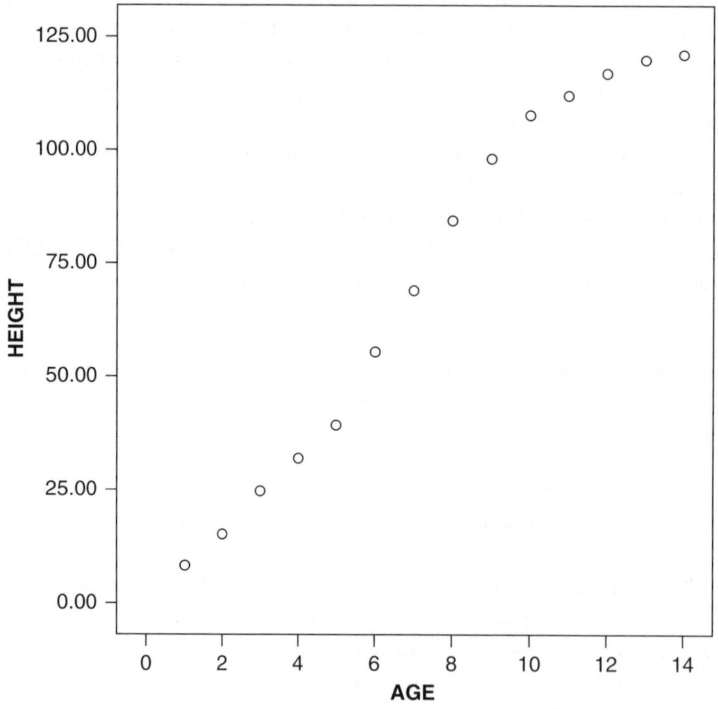

Figure 8.1 Scatter-plot for the association between age and height of hemp plants.

The scatter-plot for this group of students is shown in Figure 8.2. Here there is no strong relationship, but a linear relationship possibly exists.

Example 8.3 We consider data from Rasch (1968). Reported are the average withers heights (in centimetres) of 112 female cattle (Deutsches Frankenrind [German Frankonian Cattle]) measured at six-monthly intervals between birth (age = 0) and 60 months. These are shown in Table 8.3.

Table 8.2 Shoe sizes (x in centimetres) and body heights (y in centimetres) from 25 students.

Shoe size	Height	Shoe size	Height
42	165	40	162
43	185	41	163
43	178	40	160
41	170	38	151
39	157	42	170
44	170	41	170
36	159	43	178
40	180	42	163
40	168	38	160
39	165	39	166
42	172	40	170
40	158	42	178
43	180		

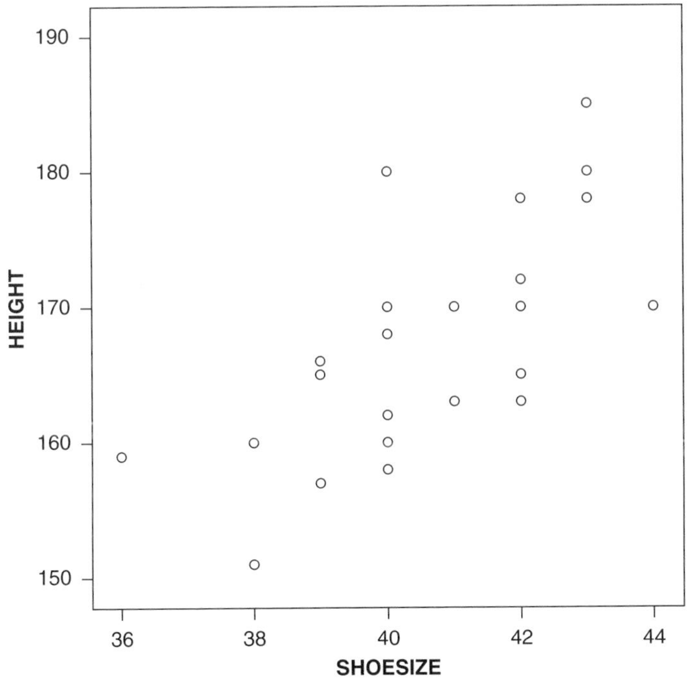

Figure 8.2 Scatter-plot for the observations of Example 8.2.

Table 8.3 Average withers heights of 112 cows in the first 60 months of life.

i	Age (months)	Height (cm)
1	0	77.20
2	6.00	94.50
3	12.00	107.20
4	18.00	116.00
5	24.00	122.40
6	30.00	126.70
7	36.00	129.20
8	42.00	129.90
9	48.00	130.40
10	54.00	130.80
11	60.00	131.20

Problem 8.1 Draw the scatter plot of Example 8.3.

Solution

```
> age <- c( 0,6,12,18,24,30,36,42,48,54,60)
> height <- c(77.2,94.5,107.2,116.0,122.4,126.7,129.2,129.9,130.4,130.8,131.2)
> plot(age,height,main = "Example 8.3")
```

Let us discuss the figures and examples. The scatter-plot in Figure 8.3 shows us that the relation between height and age is not linear. We shall therefore fit quasilinear and intrinsically non-linear functions to these data.

In Example 8.1 the age when hemp plants were measured was chosen by the experimenter – a situation discussed in more detail in Section 8.2 and of course the age is plotted on the x-axis. The corresponding scatter plot in Figure 8.1 shows a strong relation between both variables and it seems that they lie on an S-shaped (sigmoid) curve.

In Example 8.2 both variables are collected at the same time – a situation discussed in more detail in Section 8.3 and here any of the two variables could be plotted on the x-axis. The corresponding scatter plot in Figure 8.2 shows no clear relation between both variables and it is uncertain which curve could be drawn through the scatter plot.

Relationships that are not strictly functional are stochastic, and their investigation is the main object of a regression analysis. We call the variable(s) used to explain another variable the regressor(s) and the dependent variable the regressand. In Example 8.1, clearly, the age is the regressor, but in Example 8.2 the shoe size as well as the body height can be used as regressor.

In the style of mathematics, the terms dependent and independent variables are also used for stochastic relations in regression analysis – even if this sometimes makes no sense. If we consider the body height and the shoe size of students in Example 8.2 neither is independent of the other. In contrast, in the pair (height of hemp plants, age) in Example 8.1 the first depends on the second and not the other way around. We have here two very different situations. In the first case, we could model each of the two

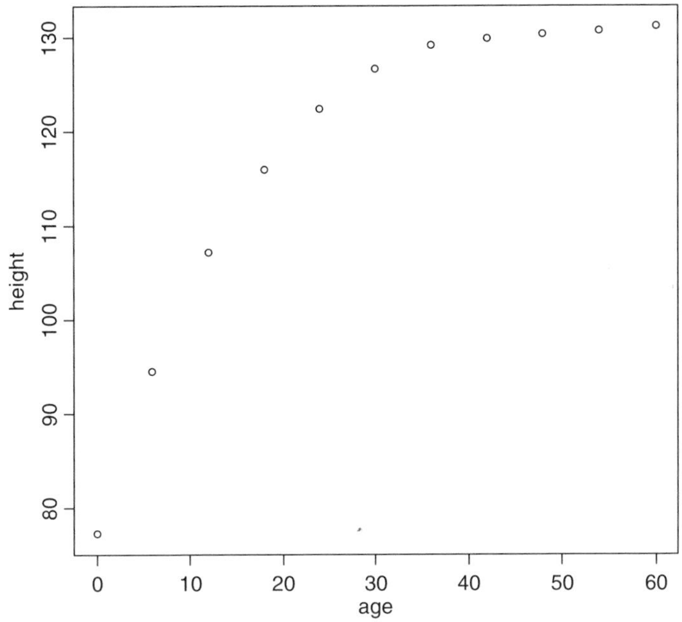

Figure 8.3 Scatter-plot of the data in Example 8.3.

characteristics by a random variable; both measured on the animal at the same time. In the second case, the experimenter determines at which ages the height of plants should be measured. As we see later the choice of age leads to a part of optimal experimental designs.

We use two different models for such situations. In regression model I the regressor is not a stochastic variable; its values are fixed before observation (given by the experimenter). In regression model II both variables are observed together and only in this case may we interchange the regressor – regressand – role of both variables.

In all cases, a function of the regressor variables is considered the regression function, and its parameters are estimated. In a narrower sense, regression may refer specifically to the estimation of continuous response variables, as opposed to the discrete response variables used in Chapter 11. The sample is representative of the population for inference prediction. The basic of regression analysis is the regression function; this is a mathematical function within a regression model. Let the regression function depend on a fixed or random vector of regressors x or $\boldsymbol{x}$ and a parameter vector β. In addition, the regression model contains a random error term $\boldsymbol{e}$. We write either the model for the regressand $\boldsymbol{y}$ as

$$\boldsymbol{y} = f(x, \beta) + \boldsymbol{e} \text{ and for each } n \text{ observation } \boldsymbol{y}_i = f(x_i, \beta) + \boldsymbol{e}_i, i = 1, \dots, n$$

for model I or as

$$\boldsymbol{y} = f(\boldsymbol{x}, \beta) + \boldsymbol{e} \text{ and for each } n \text{ observation } \boldsymbol{y}_i = f(\boldsymbol{x}_i, \beta) + \boldsymbol{e}_i, i = 1, \dots, n$$

for model II respectively.

For both models, the following assumptions must be fulfilled.

The error term is a random variable with expectation zero.

1. The regressor variables are measured without error. (If this is not so, modelling may be done instead using error-in-variables models not discussed in this book).
2. The regressor variables are linearly independent.
3. The error terms are uncorrelated for several observations $\boldsymbol{y}_i$.
4. The variance of the error is equal for all observations.

Regressor and regressand variables often refer to values measured at point locations. There may be spatial trends and spatial autocorrelation in the variables that violate statistical assumptions of regression. We discuss such problems in Chapter 12.

We start with linear and non-linear regression models with non-random regressors in Section 8.2 and continue with regression models with random regressors in Section 8.3.

8.2 Regression with Non-Random Regressors – Model I of Regression

In this section, we use the regression model

$$\boldsymbol{y}_i = f(x_i, \beta) + \boldsymbol{e}_i, i = 1, \dots, n \tag{8.1}$$

with n larger than the number of unknown components of β. The x_i may be vectors $x_i^T = (x_{1i}, x_{2i}, \dots, x_{ki})$, if $k = 1$ we speak about simple regression, and if $k > 1$ about multiple regression. For (8.1) the assumptions 1–4 are given above.

8.2.1 Linear and Quasilinear Regression

Definition 8.1 Let X be a $[n \times (k+1)]$ matrix of rank $k + 1 < n$ and $\Omega = R[X]$ the rank space of X. Further, let $\beta \in \Omega$ be a vector of parameters β_j $(j = 0, \dots, k)$ and $\boldsymbol{Y} = \boldsymbol{Y}_n$ an n-dimensional random variable. If the relations $E(\boldsymbol{e}) = 0_n$ and var$(\boldsymbol{e}) = \sigma^2 I_n$ for the error term e are valid, and the $\boldsymbol{e}_i$'s are uncorrelated, then

$$\mathrm{Y} = X\beta + \boldsymbol{e} \quad (Y \in R^n, \;\; \beta \in \Omega = R[X]) \tag{8.2}$$

using $X = (e_n, X^*)$ is called model I of the linear regression with k regressors in standard form; X^* is the matrix of the k column vectors of the regressors.

If some of the x_{ji}, $j = 1, \dots, k$ are nonlinear functions of some of the remaining x_{ji} we speak about quasiliner regression if these x_{ji} do not depend on β or on any other unknown parameters. We have var$(\boldsymbol{Y}) = $ var$(\boldsymbol{e}) = \sigma^2 I_n$.

We write the regression equation in matrix form as (8.2) with

$$\boldsymbol{Y}^T = (\boldsymbol{y}_1, \dots, \boldsymbol{y}_n); n > k + 1, \boldsymbol{e}^T = (\boldsymbol{e}_1, \dots, \boldsymbol{e}_n), X = \begin{pmatrix} 1 & x_{11} & \dots & x_{k1} \\ \dots & \dots & \dots & \dots \\ 1 & x_{1n} & \dots & x_{kn} \end{pmatrix}$$

and

$$\beta^T = (\beta_0, \beta_1 \dots, \beta_k).$$

Problem 8.2 Write down the general model of quasilinear regression.

Solution
We write $x_i^T = [g_1(x_{1i}, x_{2i}, \ldots, x_{ki}), \ldots, g_k(x_{1i}, x_{2i}, \ldots, x_{ki})\ ]$.

Example
We use $k = 2$ and

1. $x_i^T = (x_{1i}, x_{2i})$.

We use $k = 3$ and

2. $x_i^T = [x_{1i}, {x_{1i}}^2, {x_{1i}}^3]$
3. $x_i^T = [x_{1i}, \ln(x_{1i}), \sin(x_{1i})\ ]$
4. $x_i^T = (x_{1i}, x_{1i} \cdot x_{2i}, x_{1i} \cdot x_{3i})$.

Model (8.1) in case 2 is written with $\beta^T = (\beta_0, \beta_1, \beta_2, \beta_3\ )$ and

$$X = \begin{pmatrix} 1 & x_{11} & {x_{11}}^2 & {x_{11}}^3 \\ 1 & x_{12} & {x_{12}}^2 & {x_{12}}^3 \\ \cdots & \cdots & \cdots & \cdots \\ 1 & x_{1n} & {x_{1n}}^2 & {x_{1n}}^3 \end{pmatrix}, \qquad n > 4.$$

Problem 8.3 What are the conditions in the cases 1, …, 4 in the example of Problem 8.2 to make sure that $\text{rank}(X) = k + 1 \leq n$?

Solution
We have to assume that at least some values of the regressor variable are different from each other

1 + the number k of regressor variables.

Example
We must have at least

1. 3
2. 4
3. 4
4. 4

different values of the regressor variables.

Example 8.4 We use data from Steger and Püschel (1960) to demonstrate some calculations (Table 8.4).

8.2.1.1 Parameter Estimation

When we know nothing about the distribution of the error terms, we use the least squares method, which gives the same results as the maximum likelihood method for normally distributed $\boldsymbol{y}$. An estimator $\hat{\boldsymbol{\beta}}$ of the regression coefficient β using the least squares method is an estimator where its realisations $\hat{\beta}$ fulfil

$$\|\hat{e}\|^2 = \|Y - X\hat{\beta}\|^2 = \inf_{\theta \in \Omega} \|Y - X\beta\|^2.$$

Table 8.4 Carotene content (in mg/100 g dry matter) y of grass in dependency of the time of storage x (in days) for two kinds of storage from Steger and Püschel (1960).

Time	Sack	Glass
1	31.2500	31.2500
60	28.7100	30.4700
124	23.6700	20.3400
223	18.1300	11.8400
303	15.5300	9.4500

This leads to the estimators ($(X^TX)^{-1}$ existing because rank $(X) = k+1 \leq n$)

$$\boldsymbol{b} = \widehat{\boldsymbol{\beta}} = (X^TX)^{-1}X^T\boldsymbol{Y}. \tag{8.3}$$

The variance var($\boldsymbol{b}$) of the vector $\boldsymbol{b}$ is

$$\text{var}(\boldsymbol{b}) = \sigma^2(X^TX)^{-1}X^TX(X^TX)^{-1} = \sigma^2(X^TX)^{-1} = \sigma^2(c_{ij}); i, j = 0, 1, \dots, k. \tag{8.4}$$

From the Gauss–Markow theorem (Rasch and Schott (2018)), Theorem 4.3) we know that the least squares method gives in linear models linearly unbiased estimators with minimal variance – so-called best linear unbiased estimators (BLUE).

We write $(X^TX)^{-1} = (c_{ij})$, $i, j = 0, 1, \dots, k$ and use this later in the testing part.

For those not familiar with matrix notation we can use the following way.

Determine the minimum of

$$\sum_{i=1}^{n} [y_i - f(x_i, \beta)]^2. \tag{8.5}$$

We obtain the minimum by deriving to β and zeroising the derivatives, and checking that the solution gives a minimum. The solution we call $\widehat{\beta} = b$, the least squares estimate and switching to random variables $\boldsymbol{y}_i$ gives the least squares estimator $\boldsymbol{b} = \widehat{\boldsymbol{\beta}}$.

$$\boldsymbol{b} = \widehat{\boldsymbol{\beta}} = (X^TX)^{-1}X^T\boldsymbol{Y}.$$

The components of β in (8.5) are called regression coefficients, the components of (8.3) are the estimated regression coefficients.

We consider the special case of simple linear regression

$$\boldsymbol{y}_i = \beta_0 + \beta_1 x_i + \boldsymbol{e}_i, i = 1, \dots, n, \text{var}(\boldsymbol{y}_i) = \text{var}(\boldsymbol{e}_i) = \sigma^2 \tag{8.6}$$

Zeroising these derivatives gives the so-called normal equations.

The values of β_0 and β_1, minimising $S = \sum_{i=1}^{n} [y_i - \beta_0 - \beta_1 x_i]^2$, are denoted by b_0 and b_1. We obtain the following equations by putting the partial derivatives with respect to β_0 and β_1 above equal to zero and replacing all y_i by the random variable $\boldsymbol{y}_i$. (We check the fact that we really obtain a minimum by showing that the matrix of the second partial derivatives of S is positive definite). When S is a convex function, the solution of the first partial derivatives of S equal to zero gives a minimum.

We write with SS the sum of squares and SP the sum of products:

$$SS_x = \sum_{i=1}^{n} (x_i - \bar{x})^2 = \sum_{i=1}^{n} x_i^2 - \frac{\left(\sum_{i=1}^{n} x_i\right)^2}{n} \tag{8.7}$$

and

$$\boldsymbol{SP}_{xy} = \sum_{i=1}^{n} (x_i - \bar{x}) \cdot (y_i - \bar{y}) = \sum_{i=1}^{n} x_i \cdot y_i - \frac{\left(\sum_{i=1}^{n} x_i y_i\right)^2}{n} \tag{8.8}$$

and obtain

$$\boldsymbol{b}_1 = \frac{\boldsymbol{SP}_{xy}}{SS_x} \tag{8.9}$$

and

$$\boldsymbol{b}_0 = \bar{\boldsymbol{y}} - \boldsymbol{b}_1 \bar{x}. \tag{8.10}$$

Problem 8.4 Determine the least squares estimator of the simple linear regression (we assume that at least two of the x_i are different).

We call β_0 the intercept and β_1 the slope of the regression line.

Solution

Let $S = \sum_{i=1}^{n} [y_i - \beta_0 - \beta_1 x_i]^2$. The first partial derivatives of S are:

$$\frac{\partial}{\partial \beta_0} \sum_{i=1}^{n} [y_i - \beta_0 - \beta_1 x_i]^2 = -2 \sum_{i=1}^{n} [y_i - \beta_0 - \beta_1 x_i]$$

$$\frac{\partial}{\partial \beta_!} \sum_{i=1}^{n} [y_i - \beta_0 - \beta_1 x_i]^2 = -2 \sum_{i=1}^{n} x_i [y_i - \beta_0 - \beta_1 x_i].$$

Putting the first partial derivatives equal to zero, we find the solutions (8.9) and (8.10). Because S is a convex function, we really get a minimum.

Example

We use the sack-glass data from Example 8.4. The estimated regression coefficients in the linear model for the sack are:

$$b_1 = -0.0546 \text{ and } b_0 = 31.216,$$

the estimated regression function is with y = carotene content in sack and x = time:.

$$\hat{y}_{\text{sack}} = 31.216 - 0.0546x$$

```
> time <- c(1, 60, 124, 223, 303)
> sack <- c(31.25, 28.71, 23.67, 18.13, 15.53)
> lm( sack ~ time)
Call:
lm(formula = sack ~ time)
Coefficients:
(Intercept)          time
   31.21565      -0.05455
```

Figure 8.4 shows the estimated regression line and the scatter plot of the carotene example (sack) in Table 8.5.

When we use the carotene data for glass in Table 8.4 we receive analogously

```
> glass <- c(31.25 ,30.47,20.34, 11.84, 9.45)
> lm( glass ~time)
Call:
lm(formula = glass ~ time)
Coefficients:
(Intercept)          time
   32.18536      -0.08098
```

and the estimated regression line $\hat{y}_{\text{glass}} = 32.185 - 0.081x$

Both regression lines are shown in Figure 8.5.

With $k = 2$ and $f(x) = \beta_0 + \beta_1 x + \beta_2 x^2$ we obtain the following model equation for y_j .

$$y_j = \beta_0 + \beta_1 x_j + \beta_2 x_j^2 + e_j \quad (j = 1, \dots, n). \tag{8.11}$$

We assume that the e_j are mutually independent and distributed as $N(0; \sigma^2)$.

This is the model for a quadratic regression. Formally, we can handle this case like a multiple linear regression with two regressors. Let

$$x_1 = x.$$

$$x_2 = x^2.$$

Equation (8.11) gives us the following matrix expressions:

$$\beta^T = (\beta_0, \beta_1, \beta_2),$$

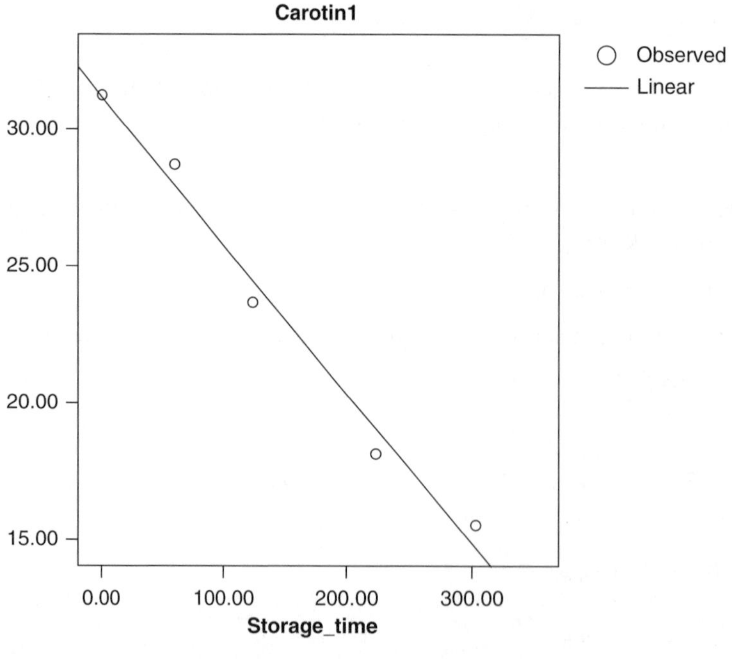

Figure 8.4 Scatter-plot and estimated regression line of the carotene example (sack) in Table 8.4.

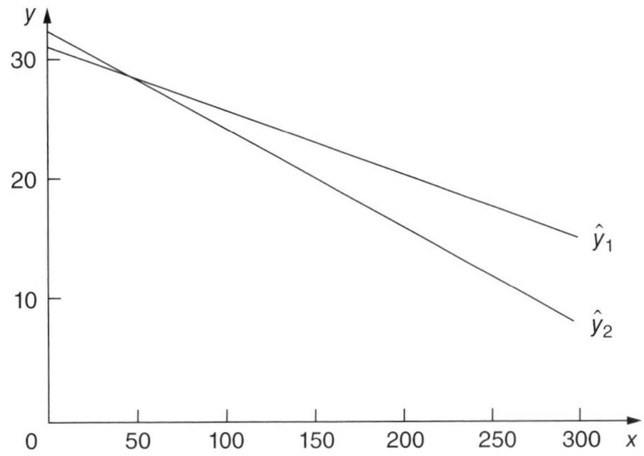

Figure 8.5 Estimated regression lines of the carotene example (sack $\hat{y}_1$, glass $\hat{y}_2$) of Table 8.4.

$$X = \begin{pmatrix} 1 & x_1 & x_1^2 \\ 1 & x_2 & x_2^2 \\ \vdots & \vdots & \vdots \\ 1 & x_n & x_n^2 \end{pmatrix}.$$

We obtain:

$$X^T X = \begin{pmatrix} n & \sum x_j & \sum x_j^2 \\ \sum x_j & \sum x_j^2 & \sum x_j^3 \\ \sum x_j^2 & \sum x_j^3 & \sum x_j^4 \end{pmatrix}$$

and

$$X^T y = \begin{pmatrix} \sum y_j \\ \sum x_j y_j \\ \sum x_j^2 y_j \end{pmatrix}.$$

In general, we write $\beta^T = (\beta_0, \beta_1, \dots, \beta_k), X^T X = \begin{pmatrix} n & \sum_{j=1}^{n} x_j & \dots & \sum_{j=1}^{n} x_j^k \\ \sum_{j=1}^{n} x_j & \sum_{j=1}^{n} x_j^2 & \dots & \sum_{j=1}^{n} x_j^{k+1} \\ \dots & \dots & \dots & \dots \\ \sum_{j=1}^{n} x_j^k & \sum_{j=1}^{n} x_j^{k+1} & \dots & \sum_{j=1}^{n} x_j^{2k} \end{pmatrix}$

and

$$X^T \boldsymbol{y} = \begin{pmatrix} \sum_{j=1}^{n} \boldsymbol{y}_j \\ \sum_{j=1}^{n} x_j \boldsymbol{y} \\ \dots \\ \sum_{j=1}^{n} x_j^k \boldsymbol{y} \end{pmatrix}.$$

Problem 8.5 Calculate the estimates of a linear, quadratic and cubic regression.

Solution
In R read y and x data and calculate then

```
> x2 <- x*x
> lm( y ~ x + x2)
> x3 <- x2*x
> lm( y ~ x + x2 + x3)
```

Example
We use the data of Example 8.4 y = sack and x = time.

For the linear regression we get:

```
> summary(lm( sack ~time))
Call:
lm(formula = sack ~ time)
Residuals:
      1       2       3       4       5
 0.0889  0.7676 -0.7809 -0.9200  0.8444
Coefficients:
            Estimate Std. Error t value Pr(>|t|)
(Intercept) 31.21565    0.70588   44.22 2.55e-05 ***
time        -0.05455    0.00394  -13.85 0.000815 ***
---
Signif. codes:  0 '***' 0.001 '**' 0.01 '*' 0.05 '.' 0.1 ' ' 1
Residual standard error: 0.9603 on 3 degrees of freedom
Multiple R-squared:  0.9846,    Adjusted R-squared:  0.9795
F-statistic: 191.8 on 1 and 3 DF,  p-value: 0.0008152
```

The estimate of the linear regression is: $y = 31.21565 - 0.05455x$

For the quadratic regression function, we get:

```
> time2 <- time*time
> summary(lm( sack ~time + time2))
Call:
lm(formula = sack ~ time + time2)
Residuals:
      1       2       3       4       5
-0.5104  0.9455 -0.1824 -0.5371  0.2844

Coefficients:
              Estimate Std. Error t value Pr(>|t|)
(Intercept)  3.183e+01  8.154e-01  39.039 0.000655 ***
time        -7.100e-02  1.369e-02  -5.188 0.035203 *
time2        5.367e-05  4.307e-05   1.246 0.338882
---
Signif. codes:  0 '***' 0.001 '**' 0.01 '*' 0.05 '.' 0.1 ' ' 1
Residual standard error: 0.8824 on 2 degrees of freedom
Multiple R-squared:  0.9913,    Adjusted R-squared:  0.9827
F-statistic: 114.3 on 2 and 2 DF,  p-value: 0.008671
```

The estimate for the quadratic regression is: $y = 31.83 - 0.071x + 0.00005367x^2$.

For the cubic regression we get:

```
> time3 <- time2*time
> summary(lm( sack ~time + time2 + time3))
Call:
lm(formula = sack ~ time + time2 + time3)
Residuals:
       1        2        3        4        5
-0.11843  0.38525 -0.41305  0.20105 -0.05483

Coefficients:
             Estimate Std. Error t value Pr(>|t|)
(Intercept)  3.141e+01  6.153e-01  51.044   0.0125 *
time        -3.934e-02  2.026e-02  -1.942   0.3027
time2       -2.400e-04  1.685e-04  -1.424   0.3897
time3        6.516e-07  3.680e-07   1.771   0.3272
---
Signif. codes:  0 '***' 0.001 '**' 0.01 '*' 0.05 '.' 0.1 ' ' 1
Residual standard error: 0.6136 on 1 degrees of freedom
Multiple R-squared:  0.9979,    Adjusted R-squared:  0.9916
F-statistic: 158.7 on 3 and 1 DF,  p-value: 0.05828
```

The estimate of the cubic regression function is:

$$\hat{y} = 31.41 - 0.03934x - 0.00024x^2 + 0.0000006517x^3.$$

Conclusion

The cubic regression gives no significant regression coefficients for x^2 and x^3. The quadratic regression has no significant regression coefficient for x^2.

The linear regression has a very significant regression coefficient for x.

Geometrically minimising $\sum_{i=1}^{n} [y_i - f(x_i, \beta)]^2$ in formula (8.5) means that the squares of the differences between the observed values y_i and the regression function $f(x_i, \hat{\beta})$ to be estimated are summarised and the sum must become a minimum. Be careful not to use the term distances because those are the orthogonal differences between the points in the scatter plot and the regression curve. However, the differences $y_i - f(x_i, \hat{\beta})$ are lines parallel to the x-axis in the scatter plot. This is the reason that in model II of regression analysis we have two (or more) estimated regression functions depending on which variable we choose as regressor.

Problem 8.6 Write down the normal equations of case 2 in the example of Problem 8.2.

Solution

In our case (8.5) has the form

$$S = \sum_{i=1}^{n} [y_i - \beta_0 - \beta_1 x_{1i} - \beta_2 x_{1i}{}^2 - \beta_3 x_{1i}{}^3]^2.$$

The first partial derivatives of S with respect to $\beta_0, \beta_1, \beta_2, \beta_3$ zeroing, leads to the normal equations

$$\sum_{i=1}^{n}[y_i - \beta_0 - \beta_1 x_{1i} - \beta_2 x_{1i}{}^2 - \beta_3 x_{1i}{}^3] = 0,$$

$$\sum_{i=1}^{n} x_{1i}[y_i - \beta_0 - \beta_1 x_{1i} - \beta_2 x_{1i}{}^2 - \beta_3 x_{1i}{}^3] = 0,$$

$$\sum_{i=1}^{n} x_{1i}^2[y_i - \beta_0 - \beta_1 x_{1i} - \beta_2 x_{1i}{}^2 - \beta_3 x_{1i}{}^3] = 0,$$

$$\sum_{i=1}^{n} x_{1i}^3[y_i - \beta_0 - \beta_1 x_{1i} - \beta_2 x_{1i}{}^2 - \beta_3 x_{1i}{}^3] = 0.$$

Example

We use the data from Table 8.3 and fit a third degree polynomial

```
> age <- c( 0,6,12,18,24,30,36,42,48,54,60)
> height <- c(77.2,94.5,107.2,116.0,122.4,126.7,129.2,129.9,130.4,130.8,131.2)
> age2 <- age*age
> age3 <- age2*age
> summary(lm( height ~age + age2 + age3))
Call:
lm(formula = height ~ age + age2 + age3)
Residuals:
      Min         1Q     Median         3Q        Max
-0.248485 -0.173485 -0.007459  0.194172  0.308392

Coefficients:
              Estimate Std. Error t value Pr(>|t|)
(Intercept)  7.743e+01  2.224e-01  348.17 4.26e-16 ***
age          3.159e+00  3.375e-02   93.61 4.18e-12 ***
age2        -6.342e-02  1.347e-03  -47.07 5.11e-10 ***
age3         4.289e-04  1.474e-05   29.11 1.45e-08 ***
---
Signif. codes:  0 '***' 0.001 '**' 0.01 '*' 0.05 '.' 0.1 ' ' 1
Residual standard error: 0.2502 on 7 degrees of freedom
Multiple R-squared:  0.9999,    Adjusted R-squared:  0.9998
F-statistic: 1.699e+04 on 3 and 7 DF,  p-value: 7.059e-14
```

The estimate of the third degree polynomial is:

$$\hat{y} = 77.43 + 3.159x - 0.06342x^2 + 0.0004289x^3.$$

We can derive the variances and the covariance of the estimators of the simple linear regression and obtain them using

$$SS_x = \sum_{i=1}^{n}(x_i - \bar{x})^2$$

$$\begin{cases} \sigma_0^2 = \mathrm{var}(\boldsymbol{b}_0) = \dfrac{\sigma^2 \sum_{i=1}^{n} x_i^2}{nSS_x} \\ \sigma_1^2 = \mathrm{var}(\boldsymbol{b}_1) = \dfrac{\sigma^2}{SS_x} \end{cases} \tag{8.12}$$

and

$$\sigma_{12} = \text{cov}(\boldsymbol{b}_0, \boldsymbol{b}_1) = -\frac{\overline{x}\sigma^2}{SS_x} \tag{8.13}$$

Thus the covariance matrix of $\begin{pmatrix} \boldsymbol{b}_0 \\ \boldsymbol{b}_1 \end{pmatrix}$ is given by

$$\text{var}\begin{pmatrix} \boldsymbol{b}_0 \\ \boldsymbol{b}_1 \end{pmatrix} = \begin{pmatrix} \sigma_0^2 & \sigma_{12} \\ \sigma_{12} & \sigma_1^2 \end{pmatrix} = \begin{pmatrix} \frac{\sigma^2 \sum_{i=1}^n x_i^2}{nSS_x} & -\frac{\overline{x}\sigma^2}{SS_x} \\ -\frac{\overline{x}\sigma^2}{SS_x} & \frac{\sigma^2}{SS_x} \end{pmatrix}. \tag{8.14}$$

From the estimators and the estimates of the regression coefficients, we can estimate the regression function (8.6). For this, we replace the parameters by their estimators and estimates respectively. The result is either the estimator of the regression function

$$\hat{\boldsymbol{y}}_i = \hat{\boldsymbol{\beta}}_0 + \hat{\boldsymbol{\beta}}_1 x_i + \hat{\boldsymbol{e}}_i, i = 1, \dots, n \tag{8.15}$$

or its estimate

$$\hat{y}_i = \hat{\beta}_0 + \hat{\beta}_1 x_i + \hat{e}_i, i = 1, \dots, n. \tag{8.16}$$

The variance $\text{var}(\boldsymbol{e}_i) = \sigma^2$ in (8.6) is called the residual variance and is estimated by the estimator

$$\boldsymbol{s}^2 = \frac{(\hat{\boldsymbol{y}}_i - \hat{\boldsymbol{\beta}}_0 - \hat{\boldsymbol{\beta}}_1 x_i)^2}{n-2} = \frac{\boldsymbol{SS}_{\text{res}}}{n-2}. \tag{8.17}$$

The estimate of the residual variance is

$$s^2 = \frac{(\hat{y}_i - \hat{\beta}_0 - \hat{\beta}_1 x_i)^2}{n-2} = \frac{SS_{\text{res}}}{n-2} \tag{8.18}$$

Problem 8.7 Calculate the determinant D of the covariance matrix of $\begin{pmatrix} \boldsymbol{b}_0 \\ \boldsymbol{b}_1 \end{pmatrix}$.

Solution

A determinant D of a quadratic matrix $\begin{pmatrix} a & b \\ c & d \end{pmatrix}$ is given by $D = ad - bc$.

Example

The determinant of $\begin{pmatrix} \frac{\sigma^2 \sum_{i=1}^n x_i^2}{nSS_x} & -\frac{\overline{x}\sigma^2}{SS_x} \\ -\frac{\overline{x}\sigma^2}{SS_x} & \frac{\sigma^2}{SS_x} \end{pmatrix}$ is therefore

$$D = \begin{vmatrix} \frac{\sigma^2 \sum_{i=1}^n x_i^2}{nSS_x} & -\frac{\overline{x}\sigma^2}{SS_x} \\ -\frac{\overline{x}\sigma^2}{SS_x} & \frac{\sigma^2}{SS_x} \end{vmatrix} = \frac{\sigma^4}{nSS_x} \tag{8.19}$$

Problem 8.8 Show that the matrix of second derivatives with respect to β_0 and β_1 in Problem 8.4 is positive definite.

Solution
A matrix is positive definite if all its eigenvalues are positive. Now we first determine the second derivatives in Problem 8.4.

$$\frac{\partial^2}{\partial {\beta_0}^2}\sum_{i=1}^{n}[y_i-\beta_0-\beta_1x_i]^2=-2\frac{\partial}{\partial\beta_0}\sum_{i=1}^{n}[y_i-\beta_0-\beta_1x_i]=2$$

$$\frac{\partial^2}{\partial {\beta_1}^2}\sum_{i=1}^{n}[y_i-\beta_0-\beta_1x_i]^2=-2\frac{\partial}{\partial\beta_1}\sum_{i=1}^{n}x_i[y_i-\beta_0-\beta_1x_i]=2\sum_{i=1}^{n}x_i^2$$

and

$$\frac{\partial^2}{\partial\beta_0\partial\beta_1}\sum_{i=1}^{n}[y_i-\beta_0-\beta_1x_i]^2=2\sum_{i=1}^{n}x_i.$$

The matrix of second derivatives the so-called Hesse matrix is therefore

$$A=\begin{pmatrix}2 & 2\sum_{i=1}^{n}x_i\\ 2\sum_{i=1}^{n}x_i & 2\sum_{i=1}^{n}x_i^2\end{pmatrix}.$$

The two eigenvalues are solutions of a quadratic function of λ:

$$\begin{vmatrix}2-\lambda & 2\sum_{i=1}^{n}x_i\\ 2\sum_{i=1}^{n}x_i & 2\sum_{i=1}^{n}x_i^2-\lambda\end{vmatrix}=0.$$

The corresponding quadratic equation has positive roots and therefore is A positive definite and we really obtained a minimum by zeroising the first derivatives.

Problem 8.9 Estimate the elements of the covariance matrix of the vector of estimators of all regression coefficients from data.

Solution
The covariance matrix of the vector of estimators of the regression coefficients depends on the values of the regressor variables and σ^2. We obtain for models with k regressors the estimators and estimates of the residual variance respectively the formulae

$$\boldsymbol{s}^2=\frac{\sum_{i=1}^{n}[\boldsymbol{y}_i-f(x_i,\widehat{\boldsymbol{\beta}})]^2}{n-k-1}=\frac{\boldsymbol{SS}_{\text{res}}}{n-k-1} \tag{8.20}$$

and

$$s^2=\frac{\sum_{i=1}^{n}[y_i-f(x_i,\widehat{\beta})]^2}{n-k-1}=\frac{SS_{\text{res}}}{n-k-1}$$

When we replace in the formula of the covariance matrix of the vector of estimators of all regression coefficients the residual variance σ^2 by its estimate s^2 from (8.18) we obtain the estimated covariance matrix

$$\widehat{\text{var}}\begin{pmatrix} \boldsymbol{b}_0 \\ \boldsymbol{b}_1 \end{pmatrix} = \begin{pmatrix} s_0^2 & s_{12} \\ s_{12} & s_1^2 \end{pmatrix} = \begin{pmatrix} \frac{s^2 \sum_{i=1}^{n} x_i^2}{nSS_x} & -\frac{\bar{x}s^2}{SS_x} \\ -\frac{\bar{x}s^2}{SS_x} & \frac{s^2}{SS_x} \end{pmatrix}. \tag{8.21}$$

Example
We use the data from the Example 8.4 (sack) and obtain

$$s^2 = \frac{2.7663}{3} = 0.9221.$$

Because $SS_x = 59410.8$, $\sum_{i=1}^{n} x_i^2 = 160515$ and $\bar{x} = 142.2$ we obtain

$$\widehat{\text{var}}\begin{pmatrix} \boldsymbol{b}_0 \\ \boldsymbol{b}_1 \end{pmatrix} = 0.9221 \cdot \begin{pmatrix} \frac{160515}{5 \cdot 59410.8} & -\frac{142.2}{59410.8} \\ -\frac{142.2}{59410.8} & \frac{1}{59410.8} \end{pmatrix} = \begin{pmatrix} 0.498 & -0.0022 \\ -0.0022 & 0.0000155 \end{pmatrix}.$$

Problem 8.10 Show how parameters of a general linear regression function $y = f(x_i, \beta) = \beta_0 + \beta_1 x_1 + \cdots + \beta_k x_k$ can be estimated.

Solution

```
> lm( y ~ x1 + .... + xk )
```

Example
As an example, we consider the quadratic (quasi-linear) regression function and the data of Example 8.3 with y = height and x = age

$$f(x, \beta) = \beta_0 + \beta_1 x + \beta_2 x^2.$$

```
> age <- c( 0,6,12,18,24,30,36,42,48,54,60)
> height <- c(77.2,94.5,107.2,116.0,122.4,126.7,129.2,129.9,130.4,130.8,131.2)
> age2 <- age*age
> summary(lm( height ~age + age2))
Call:
lm(formula = height ~ age + age2)
Residuals:
    Min      1Q  Median      3Q     Max
-3.5615 -1.8837 -0.0075  1.8090  3.2203
Coefficients:
             Estimate Std. Error t value Pr(>|t|)
(Intercept) 80.761538   1.969299   41.01 1.38e-10 ***
age          2.276092   0.152706   14.90 4.05e-07 ***
age2        -0.024819   0.002451  -10.12 7.74e-06 ***
---
Signif. codes:  0 '***' 0.001 '**' 0.01 '*' 0.05 '.' 0.1 ' ' 1
Residual standard error: 2.585 on 8 degrees of freedom
Multiple R-squared:  0.9832,    Adjusted R-squared:  0.9791
F-statistic: 234.8 on 2 and 8 DF,  p-value: 7.875e-08
```

The estimate of the quadratic regression function is:

$$\hat{y} = 80.761538 + 3.376092x - 0.024819x^2.$$

8.2.1.2 Confidence Intervals and Hypotheses Testing

For confidence estimation and hypotheses testing we must assume that the error terms $\boldsymbol{e}$ of the corresponding regression models are independent and normally distributed ($N(0, \sigma^2)$).We construct confidence intervals for the regression coefficients and confidence, and prediction intervals for the regression line for the simple linear regression only.

If $t(f;P)$ is the P-quantile of the t-distribution with f degrees of freedom and $CS(f;P)$ the P-quantile of the chi-squared distribution with f degrees of freedom we obtain the following $(1-\alpha)100\%$ confidence intervals for β_j, $j = 0, 1$ based on $n > 2$ data pairs

$$\left[\boldsymbol{b}_j - t\left(n-2, 1-\frac{\alpha}{2}\right)\boldsymbol{s};\quad \boldsymbol{b}_j + t\left(n-2, 1-\frac{\alpha}{2}\right)\boldsymbol{s}\right] \tag{8.22}$$

and σ^2

$$\left[\frac{\boldsymbol{s}^2(n-2)}{CS\left(n-2, 1-\frac{\alpha}{2}\right)};\quad \frac{\boldsymbol{s}^2(n-2)}{CS\left(n-2, \frac{\alpha}{2}\right)}\right]. \tag{8.23}$$

Equation (8.23) is not the best solution, there is a shorter interval obtained by $\alpha = \alpha_1 + \alpha_2$ as a sum of positive components α_1 and α_2 different from $\alpha_1 = \alpha_2 = \alpha/2$ and then replacing $\alpha/2$ by α_1 and $1 - \alpha/2$ by $1 - \alpha_2$ in (8.23). The reason for this is the asymmetry of the chi-squared distribution.

For $E(\boldsymbol{y}) = \beta_0 + \beta_1 x$ for any x in the experimental region we need the value of the following quantity K_x:

$$K_x = \sqrt{\frac{1}{n} + \frac{(x-\overline{x})^2}{SS_x}}$$

and then the $(1 - \alpha)$ confidence interval for $E(\boldsymbol{y}) = \beta_0 + \beta_1 x$ is:

$$\left[\boldsymbol{b}_0 + \boldsymbol{b}_1 x - t\left(n-2, 1-\frac{\alpha}{2}\right)\boldsymbol{s}\boldsymbol{K}_x;\quad \boldsymbol{b}_0 + \boldsymbol{b}_1 x + t\left(n-2, 1-\frac{\alpha}{2}\right)\boldsymbol{s}\boldsymbol{K}_x\right]. \tag{8.24}$$

If we calculate (8.24) for each x-value in the experimental region, and draw a graph of the realised upper and lower limits, the included region is called a confidence belt.

Problem 8.11 Draw a scatter plot with the regression line and a 95% confidence belt.

Solution

```
> summary(lm( y ~ x))
> plot(x, y)
> abline(lm(y ~ x ))
> pred.y <- predict( y ~x)
> pc <- predict(lm( y ~x ), interval = "confidence")
> matlines(x, pc, lty = c(1,2,2), col = "black")
```

Example

We consider in Example 8.4 storage in a sack where x = time and y = sack.

We choose $\alpha = 0.05$ and calculate using R

```
> qt(0.975,3)
```

```
[1] 3.182446
```

hence $t(3; 0.975) = 3.182$.

The lower and upper bounds of the realised 95% confidence belt for $\beta_0 + \beta_1 x$ can be found in Table 8.5.

Table 8.5 Lower and upper bounds of the realised 95%-confidence band for $\beta_0 + \beta_1 x$.

x	$\hat{y}$	K_x	Lower bound	Upper bound
1	31.16	0.73184	28.92	33.40
60	27.94	0.56012	26.23	29.65
124	24.45	0.45340	23.06	25.84
223	19.05	0.55668	17.35	20.75
303	14.69	0.79701	12.25	17.13

Using R:

```
> time <- c(1, 60, 124, 223, 303)
> sack <- c(31.25, 28.71, 23.67, 18.13, 15.53)
> plot( time , sack)
> abline(lm(sack ~time))
> pred.sack <- predict(lm(sack ~time))
> pred.sack
        1         2         3         4         5
31.16110 27.94238 24.45089 19.04999 14.68563
> pc <- predict(lm(sack ~time), interval = "confidence")
> pc
       fit      lwr      upr
1 31.16110 28.92463 33.39757
2 27.94238 26.23068 29.65408
3 24.45089 23.06530 25.83648
4 19.04999 17.34881 20.75118
5 14.68563 12.25001 17.12125
>  matlines(time, pc, lty=c(1,2,2), color= "black")
```

The scatter-plot and the estimated regression line with 95%-confidence bands are shown in Figure 8.6.

We can test the hypothesis that any regression coefficient in a simple or multiple or in a quasilinear regression is zero by a t-test.

To test the hypothesis that the element β_j of the vector β in (8.2) is a given value $\beta_j^{(0)}$ against a one- or two-sided alternative we write:

$$H_{0,j} : \beta_j = \beta_j^{(0)}, \quad j = 0, \ldots, k$$

against one of the alternatives

(a) $H_{0,j} : \beta_j < \beta_j^{(0)}, \; j = 0, \ldots, k$
(b) $H_{0,j} : \beta_j > \beta_j^{(0)}, \; j = 0, \ldots, k$
(c) $H_{0,j} : \beta_j \neq \beta_j^{(0)}, \; j = 0, \ldots, k.$

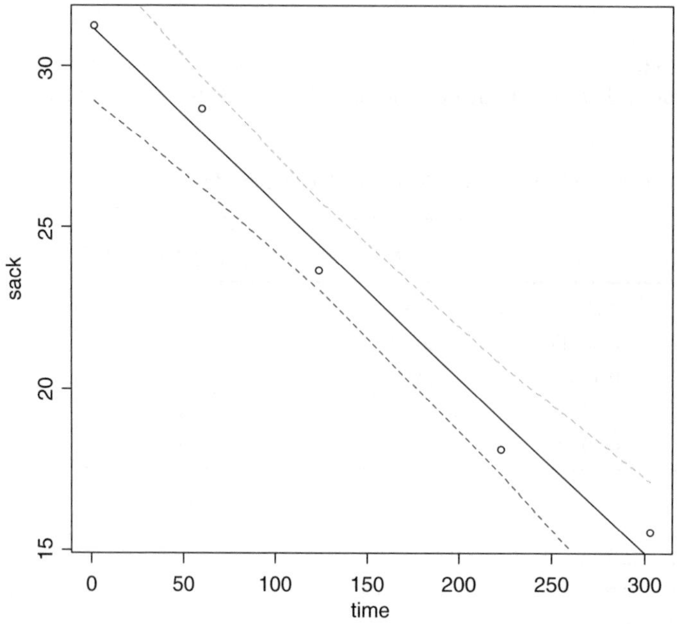

Figure 8.6 Scatter-plot and estimated regression line of the carotene example with 95%- confidence bands.

We use the test statistic

$$\boldsymbol{t}_j = \frac{\boldsymbol{b}_j - \beta_j^{(0)}}{\boldsymbol{s}_j}, \quad j = 0, \ldots, k \tag{8.25}$$

and reject H_0 with a type I error rate α if the realisation t_j of (8.25) is for

(a) $t_j < t(n-2, 1-\alpha)$
(b) $t_j > t(n-2, 1-\alpha)$
(c) $|t_j| > t\left(n-2, 1-\frac{\alpha}{2}\right)$.

We take $\boldsymbol{s}_j$ in analogy to (8.4).

Problem 8.12 Test for a simple linear regression function H_0: $\beta_1 = -0.1$ against H_A: $\beta_1 \neq -0.1$ with a significance level $\alpha = 0.05$.

Solution
From the linear regression with R command `lm()` we can find the estimate of the standard error of b_1. Insert the estimate b_1 and the standard error in (8.25) and obtain the observed t-value.

Example
We use the data of Example 8.4 with y = sack and x = time and test H_0: $\beta_1 = -0.1$.

For the linear regression we get:

```
> time <- c(1, 60, 124, 223, 303)
```

```
> sack <- c(31.25, 28.71, 23.67, 18.13, 15.53)
> summary(lm( sack ~time))
Call:
lm(formula = sack ~ time)
Residuals:
      1       2       3       4       5
 0.0889  0.7676 -0.7809 -0.9200  0.8444
Coefficients:
            Estimate Std. Error t value Pr(>|t|)
(Intercept) 31.21565    0.70588   44.22 2.55e-05 ***
time        -0.05455    0.00394  -13.85 0.000815 ***
---
Signif. codes:  0 '***' 0.001 '**' 0.01 '*' 0.05 '.' 0.1 ' ' 1
Residual standard error: 0.9603 on 3 degrees of freedom
Multiple R-squared:  0.9846,    Adjusted R-squared:  0.9795
F-statistic: 191.8 on 1 and 3 DF,  p-value: 0.0008152

> b1 <- -0.05455
> se1 <- 0.00394
> df <- 3
>  numt1 <- b1-(-0.1)
>  t1 <- numt1/se1
>  t1
[1] 11.53553
>  df <- 3
>  p_value <- 2*pt(abs(t1) , df, lower.tail = FALSE)
> p_value
[1] 0.001398722
```

Because the p-value $0.0014 < 0.05$ the H_0: $\beta_1 = -0.1$ is rejected.

Problem 8.13 To test the null hypothesis $H_{0,1}: \beta_1 = \beta_1^{(0)}$, the sample size n is to be determined so that for a given risk of the first kind α, the risk of the second kind has at least a given value β so long as, according to the alternative hypothesis, one of the following holds:

(a) $\beta_1 - \beta_1^{(0)} \leq \delta$,
(b) $\beta_1 - \beta_1^{(0)} \leq -\delta$, or
(c) $|\beta_1 - \beta_1^{(0)}| \leq \delta$.

Solution

The estimator $\boldsymbol{b}_1$ is under the normality assumption for $\boldsymbol{y}$ normally distributed with expectation β_1 and variance $\text{var}(\boldsymbol{b}_1) = \frac{\sigma^2}{SS_x}$. SS_x is the sum of squared deviations of the x – values. Therefore

$$\frac{\boldsymbol{b}_1}{\sigma}\sqrt{SS_x}$$

is $N\left(\frac{\beta_1}{\sigma}\sqrt{SS_x},\ 1\right)$ distributed.

If we replace σ by its estimator $\boldsymbol{s}^2 = \frac{(\hat{y}_i - \hat{\beta}_0 - \hat{\beta}_1 x_i)^2}{n-2} = \frac{\boldsymbol{SS}_{\text{res}}}{n-2}$ we obtain

$$\boldsymbol{t} = \frac{\boldsymbol{b}_1 - \beta_1}{\sqrt{\boldsymbol{SS}_{\text{res}}}} \sqrt{n-2}\sqrt{SS_x}$$

and this is centrally t-distributed with $n-2$ degrees of freedom. Then

$$\frac{\boldsymbol{b}_1}{\sqrt{\boldsymbol{SS}_{\text{res}}}} \sqrt{n-2}\sqrt{SS_x}$$

is non-centrally t-distributed with $n-2$ degrees of freedom and non-centrality parameter

$$\lambda = \frac{\beta_1}{\sigma}\sqrt{SS_x}.$$

We get the minimum sample size when we maximise λ and thus SS_x, and this is the case for the D-optimal design (see Rasch and Schott (2018) and Section 8.3.2).

Let us assume that the minimal sample size n is even, $n = 2t$, and that the experimental region is given by $[x_l, x_u]$. This means the D-optimal design is $\begin{pmatrix} x_l & x_u \\ t & t \end{pmatrix}$ and $SS_x = \frac{n}{4}(x_u - x_l)^2$. If n is odd, the D-optimal design is $\begin{pmatrix} x_l & x_u \\ \frac{n+1}{2} & \frac{n-1}{2} \end{pmatrix}$ and $SS_x = \left(n - \frac{1}{n}\right)\frac{1}{4}(x_u - x_l)^2$.

Now we solve the equation

$$t(n-2, P) = t(n-2, \lambda, 1-\beta) \tag{8.26}$$

for n with $P = 1 - \alpha$ for the cases (a) and (b) and $P = 1 - \frac{\alpha}{2}$ in case (c) using the relative effect size $\delta = \frac{\beta_1}{\sigma}$.

Example

We use for the two-sided case $\alpha = 0.05$, $\beta = 0.1$ and $\delta = \frac{\beta_1}{\sigma} = 0.5$ and obtain:

$t(n-2, 0.975) = t(n-2, \delta\sqrt{SS_x}, 0.9)$.

Problem 8.14 Test the two regression coefficients in a simple linear regression against zero with one- and two-sided alternatives.

Solution

The two pairs of hypotheses are

$$H_{0,0} : \beta_0 = \beta_0^{(0)}$$

against one of the alternatives

(a) $H_{a,0}: \beta_0 < \beta_0^{(0)}$
(b) $H_{a,0}: \beta_0 > \beta_0^{(0)}$
(c) $H_{a,0}: \beta_0 \neq \beta_0^{(0)}$

and

$$H_{0,1} : \beta_1 = \beta_1^{(0)}$$

against one of the alternatives

(a) $H_{a,1}: \beta_1 < \beta_1^{(0)}$
(b) $H_{a,1}: \beta_1 > \beta_1^{(0)}$
(c) $H_{a,1}: \beta_1 \neq \beta_1^{(0)}$.

The test statistics are:

$$\boldsymbol{t}_0 = \frac{\boldsymbol{b}_0 - \beta_0^{(0)}}{\boldsymbol{s}_0}$$

and

$$\boldsymbol{t}_1 = \frac{\boldsymbol{b}_1 - \beta_1^{(0)}}{\boldsymbol{s}_1}$$

respectively.

For $\boldsymbol{s}_0^2$ and $\boldsymbol{s}_1^2$ see (8.21) and the Example below it.

Example

We test the hypothesis that the slope of the simple linear regression in Example 8.4 for 'sack' is zero. Because the carotene content cannot increase during storage the pair of hypotheses is $H_{0,1}: \beta_1 = \beta_1^{(0)}$, $H_{0,1}: \beta_1 < \beta_1^{(0)}$. We choose $\alpha = 0.05$ and obtain using R:

```
> time <- c(1, 60, 124, 223, 303)
> sack <- c(31.25, 28.71, 23.67, 18.13, 15.53)
> summary(lm( sack ~time))
Call:
lm(formula = sack ~ time)
Residuals:
      1        2        3        4       5
 0.0889  0.7676 -0.7809 -0.9200  0.8444
Coefficients:
            Estimate Std. Error t value Pr(>|t|)
(Intercept) 31.21565    0.70588   44.22 2.55e-05 ***
time        -0.05455    0.00394  -13.85 0.000815 ***
---
Signif. codes:  0 '***' 0.001 '**' 0.01 '*' 0.05 '.' 0.1 ' ' 1

Residual standard error: 0.9603 on 3 degrees of freedom
Multiple R-squared:  0.9846,    Adjusted R-squared:  0.9795
F-statistic: 191.8 on 1 and 3 DF,  p-value: 0.0008152
```

In the output of R is the outcome of the two-sided test of $H_{0,1}$: $\beta_1 = 0$ against $H_{a,1}$: $\beta_1 \neq 0$ given. However, our alternative is $H_{a,1}$: $\beta_1 < 0$, hence we must use a left-sided critical region.

Because the estimate of β_1 is −0.05455 (negative), Pr(t(3) < −13.85) is 0.000815/2 < 0.05, hence $H_{0,1}$: $\beta_1 = 0$ is rejected. We also can check the p-value using R:

```
>  p_value <- pt(- 13.85 , 3)
> p_value
[1] 0.0004073806
```

Finally, we test the null hypothesis that two regression functions have equal parameters. We restrict ourselves on comparing two regression lines and use as an example the two kinds of storage (sack and glass) of Example 8.4

We consider two independent sets of measurements (y_{11}, ..., y_{1n1}) and (y_{21},..., y_{2n2}) of sizes n_1 and n_2 from two underlying populations, in order to test the null hypothesis

H_0: $\beta_1 = \beta_2$

against one of the following one- or two-sided alternatives

(a) H_A: $\beta_1 > \beta_2$
(b) H_A: $\beta_1 < \beta_2$
(c) H_A: $\beta_1 \neq \beta_2$

with the two regression models

$$y_{ji} = \beta^j{}_0 + \beta^j{}_1 x_{ji} + e_{ji}, j = 1, 2; i = 1, \dots, n_j > 2.$$

If we are sure that var(e_{1i}) = var(e_{1i}) = σ^2 we estimate σ^2 using

$$s^2 = \frac{1}{n_1 + n_2 - 4}[(n_1 - 2)s_1^2 + (n_2 - 2)s_2^2] \text{ and } \mathrm{var}(b_1 - b_2) = s_d^2.$$

The variance of the difference var($b_1 - b_2$) is $s_d^2 = s^2\left[\frac{1}{SS_{x_1}} + \frac{1}{SS_{x_2}}\right]$ and SS_{xj} (j = 1,2) analogous to (8.7) as $SS_{xj} = \sum_{i=1}^{n} (x_{ji} - \overline{x}_j)^2$.

The test statistic is

$$t = \frac{b_1 - b_2}{s_d} \tag{8.27}$$

with $n_1 + n_2 - 4$ degrees of freedom.

H_0 is rejected if:

(a) $t > t(n_1 + n_2 - 4; 1-\alpha)$,
(b) $t < -t(n_1 + n_2 - 4; 1-\alpha)$ and
(c) $|t| > t(n_1 + n_2 - 4; 1-\frac{\alpha}{2})$

and otherwise accepted.

If we are not sure that the two variances are equal, or if we know that they are unequal, we use the Welch test shown in Section 3.3.1.2.

We replace the test statistics (8.24) by

$$t^* = \frac{b_1 - b_2}{\sqrt{\frac{s_1^2}{SS_{x1}} + \frac{s_2^2}{SS_{x2}}}} \tag{8.28}$$

and H_0 is rejected if $|t^*|$ is larger than the corresponding quantile of the central t-distribution with approximate degrees of freedom f, the so-called Satterthwaite f,

$$f = \frac{\left(\frac{s_1^2}{n_1} + \frac{s_2^2}{n_2}\right)^2}{\frac{s_1^4}{n_1^2(n_1-1)} + \frac{s_2^4}{n_2^2(n_2-1)}}$$

Problem 8.15 Test the hypothesis that the slopes in the two types of storage in Example 8.4 are equal with significance level $\alpha = 0.05$ using the Welch test.

Solution and Example

```
> time <- c(1, 60, 124, 223, 303)
> sack <- c(31.25, 28.71, 23.67, 18.13, 15.53)
> glass <- c(31.25 ,30.47,20.34, 11.84, 9.45)
> ms <- lm( sack ~time)
> summary(ms)
Call:
lm(formula = sack ~ time)
Residuals:
      1       2       3       4       5
 0.0889  0.7676 -0.7809 -0.9200  0.8444
Coefficients:
            Estimate Std. Error t value Pr(>|t|)
(Intercept) 31.21565    0.70588   44.22 2.55e-05 ***
time        -0.05455    0.00394  -13.85 0.000815 ***
---
Signif. codes:  0 '***' 0.001 '**' 0.01 '*' 0.05 '.' 0.1 ' ' 1
Residual standard error: 0.9603 on 3 degrees of freedom
Multiple R-squared:  0.9846,    Adjusted R-squared:  0.9795
F-statistic: 191.8 on 1 and 3 DF,  p-value: 0.0008152

> mg <- lm ( glass ~ time)
> summary(mg)
Call:
lm(formula = glass ~ time)
Residuals:
      1       2       3       4       5
-0.8544  3.abs(Welch1434 -1.8038 -2.2868  1.8016
Coefficients:
            Estimate Std. Error t value Pr(>|t|)
(Intercept) 32.18536    2.00597  16.045 0.000527 ***
time        -0.08098    0.01120  -7.233 0.005450 **
---
Signif. codes:  0 '***' 0.001 '**' 0.01 '*' 0.05 '.' 0.1 ' ' 1
Residual standard error: 2.729 on 3 degrees of freedom
Multiple R-squared:  0.9458,    Adjusted R-squared:  0.9277
F-statistic: 52.32 on 1 and 3 DF,  p-value: 0.00545

> bs <-0.05455
```

```
> SEs<-0.00394
> dfs <- 3
> bg<-0.08098
> SEg <-0.01120
> dfg<-3
> Welch_t<-(bs-bg)/sqrt(SEs^2+SEg^2)
> Welch_t
[1]2.226095
> f <- (SEs^2 + SEg^2)^2/((SEs^4/dfs) + (SEg^4/dfg))
> f
[1] 3.731319
> p <- pt(abs(Welch_t),f)
> p_value <- 2*(1-p)
> p_value
[1] 0.09485371
```

Because the p-value $= 0.09485 > 0.05$, we cannot reject the hypothesis that the slopes in the two types of storage are equal.

8.2.2 Intrinsically Non-Linear Regression

In this section, we give estimates for parameters in such regression functions, which are non-linear in x and in the parameter vector β.

Definition 8.2 The regression function $f(x, \vec{\beta})$ in a regressor $x \in R$ and the parameter vector (we from now on use $\vec{\beta}$ for the parameter vector and β_j for one of its components, only for examples we use)

$$\vec{\beta}^T = (\beta_1, \ldots, \beta_k), \beta \in \Omega$$

which is non-linear in x and in at least one of the β_j, $j = 0, \ldots, k$ and cannot be made linear or quasi-linear by any continuous transformation of the non-linearity parameter is called an intrinsically non-linear regression function. We speak about k-parametric intrinsically non-linear regression functions.

We consider the intrinsically non-linear regression model

$$\boldsymbol{y}_i = f(x_i, \vec{\beta}) + \boldsymbol{e}_i \cdot i = 1, \ldots, n > k \tag{8.29}$$

and assume that the $\boldsymbol{e}_i$ are independently $N(0, \sigma^2)$ distributed. Normal distribution we need for the result found by Jennrich (1969).

One important application of intrinsically non-linear regression is growth research. Many special functions were first used as growth functions and researchers used a special notation for the regression coefficients. In place of $\beta_0, \beta_1, \ldots$ they used $\alpha, \beta, \ldots$. This leads to a conflict with our notation for the risks of the first and the second kind in hypotheses testing. Because the notation of growth research is so strongly accepted in applications, we in this section use the common notations and write our risks as α^* and β^*.

For linear models a general theory of estimating and tests was possible and by the Gauß–Markoff theorem we could verify optimal properties of the least squares estimator $\hat{\boldsymbol{\theta}}$. A corresponding theory for the general (non-linear) case does not exist.

For intrinsically non-linear problems, the situation can be characterised as follows.

- The existing theoretical results are not useful for solutions of practical problems; many results about distributions of the estimators are asymptotic.
- The practical approach leads to numerical problems for iterative solutions. We know scarcely anything about properties of the distributions of the estimators. The application of the methods of the intrinsically non-linear regression is more a problem of numerical mathematics than of mathematical statistics.
- A compromise could be to go over to quasilinear approximations for the non-linear problem and conduct the parameter estimation for the approximative model. However, by this we lose the interpretability of the parameters, desired in growth research and other applications by practitioners.

We do not repeat the general theory of the intrinsically non-linear regression – this can be found in chapter 9 of Rasch and Schott (2018) – but restrict ourselves to some important functions used in applications. In the theoretical part of this section we use β; only in the examples do we use $\alpha, \beta, \dots$.

The least squares estimate $\widehat{\beta}_n$ of β based on n observations is the magnitude minimises $\sum_{i=1}^{n} [y_i - f(x_i, \beta)]^2$; $\widehat{\boldsymbol{\beta}}$ is the corresponding least squares estimator and its distribution is unknown. We estimate σ^2 in analogy to the linear case by

$$s_n^2 = \frac{1}{n-k}\sum_{i=1}^{n} [\boldsymbol{y}_i - f(x_i, \widehat{\boldsymbol{\beta}})]^2 \tag{8.30}$$

and write $\widehat{\boldsymbol{\beta}} = (\widehat{\boldsymbol{\beta}}_1, \dots, \widehat{\boldsymbol{\beta}}_k)$. This estimator is biased.

8.2.2.1 The Asymptotic Distribution of the Least Squares Estimators

For the asymptotic distribution of the least squares estimators, i.e. the distribution of the estimators if the number of observation n tends to infinity, we use results of an important theorem of Jennrich (1969). We need some notation – readers with a less mathematical background may skip the next part.

We assume that the function $f(x, \beta)$ in Definition 8.2 is twice continuously differentiable with respect to β. For the n regressor values we obtain $f(x_i, \beta)$, $i = 1, \dots, n$. We consider the abbreviations:

$$\left.\begin{array}{l} f_j(x, \beta) = \frac{\partial f(x,\beta)}{\partial \beta_j}, j = 1, \dots, k \\ F_i = [f_1(x_i, \beta), \dots, f_k(x_i, \beta)] \\ F^T F = \sum_{i=1}^{n} F_i^T F_i \end{array}\right\}. \tag{8.31}$$

We assume that $F^T F$ is positive definite and $I(\beta) = \lim_{n\to\infty} \sum_{i=1}^{n} F_i{}^T F_i$ is the asymptotic information matrix.

Jennrich (1969) showed that $\sqrt{n}(\widehat{\boldsymbol{\beta}} - \beta)$ is asymptotically

$$N[0_k, \sigma^2 I^{-1}(\beta)]$$

distributed.

We call

$$\text{var}_{\text{as}}(\widehat{\boldsymbol{\beta}}) = \sigma^2\left(\sum_{i=1}^{n} F_i{}^T F_i\right)^{-1} = \sigma^2(F^T F)^{-1} \tag{8.32}$$

for each n the asymptotic covariance matrix of $\hat{\boldsymbol{\beta}}$ (a bit strange but generally in use).

We estimate $\text{var}_{\text{as}}(\hat{\boldsymbol{\beta}})$ by

$$\widehat{\text{var}_{\text{as}}(\hat{\boldsymbol{\beta}})} = \boldsymbol{s}_n^2 \sum_{i=1}^{n} (\widehat{\boldsymbol{F}_i^{*T} \boldsymbol{F}_i^{*}})^{-1} = \boldsymbol{s}_n^2 (\boldsymbol{v}_{jl}); j, l = 1, \ldots, k, n > k. \tag{8.33}$$

In $\boldsymbol{F}_i^*$ we replace the parameters by their estimators and $\boldsymbol{s}_n^2$ is given by (8.30).

Based on the estimated asymptotic covariance matrix we can derive test statistics to test the hypothesis that a component β_j of β equals β_{j0} as

$$\boldsymbol{t}_j = \frac{\hat{\boldsymbol{\beta}}_j - \beta_{j0}}{\boldsymbol{s}_n \sqrt{\boldsymbol{v}_{jj}}} \tag{8.34}$$

and this statistic is asymptotically t-distributed with $n - k$ degrees of freedom.

Analogous to (8.22) we obtain asymptotic confidence intervals for a component β_j as

$$\left[\boldsymbol{b}_j - t\left(n - 2, 1 - \frac{\alpha^*}{2}\right) \boldsymbol{s}_n \sqrt{\boldsymbol{v}_{jj}}; \boldsymbol{b}_j + t\left(n - 2, 1 - \frac{\alpha^*}{2}\right) \boldsymbol{s}_n \sqrt{\boldsymbol{v}_{jj}}\right]. \tag{8.35}$$

From now on we write in the special functions below $\hat{\boldsymbol{\theta}} = (\hat{\boldsymbol{\alpha}}, \hat{\boldsymbol{\beta}})$ or $\hat{\boldsymbol{\theta}} = (\hat{\boldsymbol{\alpha}}, \hat{\boldsymbol{\beta}}, \hat{\boldsymbol{\gamma}})$.

How good the corresponding tests (or confidence intervals) for small n are was investigated by simulation studies, which followed the scheme below:

For the role simulations play in statistics, see Pilz et al. (2018).

For simulations reported for special functions resulting from a research project of Rasch at the University of Rostock during the years 1980 to 1990, see Rasch and Schimke (1983) and Rasch (1990).

The hypothesis $H_{j0} : \beta_j = \beta_{j0}$ is tested against $H_{j0} : \beta_j \neq \beta_{j0}$ with the test statistic (8.32). For each of 10 000 runs we added pseudorandom numbers e_i from a normal distribution with expectation 0 and variance σ^2 to the values of the function $f(x_i, \theta)$ at n fixed support points inx_i $[x_l, x_u]$. Then for each i

$$y_i = f(x_i, \beta) + e_i \; (i = 1, \ldots, n; x_i \in [x_l, x_o])$$

is a simulated observation. We calculate then from the n simulated observation the least squares estimates $\hat{\beta}$, the estimate s_n^2 of σ^2 and the realisation $t_j = \frac{\hat{\beta}_j - \beta_{j0}}{s_n v_{jl}}$ of the test statistic (8.34). We then calculated 10 000 test statistic t_j and counted how often t_j fulfilled

$$t_j < -t\left(n - 1 \mid 1 - \frac{\alpha_{\text{nom}}}{2}\right); -t\left(n - 1 \mid 1 - \frac{\alpha_{\text{nom}}}{2}\right) \le t_j \le t\left(n - 1 \mid 1 - \frac{\alpha_{\text{nom}}}{2}\right)$$

where α_{nom} is the nominal confidence coefficient (risk of the first kind) and α_{act} is the actual confidence coefficient (risk of the first kind) reached by the simulation

$$t_j > t\left(n - 1 \mid 1 - \frac{\alpha_{\text{nom}}}{2}\right) \quad (j = 1, \ldots, 10, 000)$$

respectively (the null hypothesis in the simulation was always correct), divided by 10 000 giving an estimate of α_{act}. Further 10 000 runs to test $H_0 : \beta_j = \beta_j^* + \Delta_j$ with three Δ_j values have been performed to get information about the power.

We now consider a two-parametric and four three-parametric intrinsically non-linear regression functions important in applications and report the corresponding simulation results, changing the parameter symbols as mentioned above.

8.2.2.2 The Michaelis–Menten Regression

Michaelis–Menten kinetics is one of the best-known models of enzyme kinetics. It is named after German biochemist Michaelis and Canadian physician Menten. The model takes the form of an equation relating reaction rate y

$$\frac{d\sigma_m}{dh} = g\rho_b - \sigma_m \lambda_\sigma \frac{L_p}{A_b} \tan(\phi_{wi})$$

with the concentration x of a substrate. Michaelis and Menten (1913) published the data of Example 8.5 and an equation of the form

$$f_{\mathrm{M}}(x) = \frac{\alpha x}{1 + \beta x}. \tag{8.36}$$

Example 8.5 In Michaelis and Menten (1913) we find x (time in minutes) and y (rotation in degrees) to determine the effect of invertion on the inversion of sugar. Table 8.6 shows the values of experiment 5 in Michaelis and Menten (1913).

The model is

$$y_i = f_{\mathrm{M}}(x_i) = \frac{\alpha x_i}{1 + \beta x_i} + \boldsymbol{e}_i \cdot i = 1, \ldots, n > 2, \alpha \cdot \beta \neq 0. \tag{8.37}$$

To obtain estimates of the two parameters we minimise

$$\sum_{i=1}^{n} \left[y_i - \frac{\alpha x_i}{1 + \beta x_i} \right]^2$$

either directly or by zeroising the partial derivatives with respect to α and β, and check that the solution is a minimum.

Problem 8.16 Determine the estimates of the Michaelis–Menten regression and draw a plot of the estimated regression line.

Solution

The partial derivatives with respect to α and β of

$$S = \sum_{i=1}^{n} \left[y_i - \frac{\alpha x_i}{1 + \beta x_i} \right]^2$$

after zeroising putting $z_i = \frac{x_i}{1+\beta x_i}$ are after some manipulations:

$$\sum_{i=1}^{n} (y_i - a z_i^*) z_i^* = 0$$

Table 8.6 Time in minutes (x) and change of rotation in degrees (y) from experiment 5 of Michaelis and Menten (1913).

y	0	0.025	0.117	0.394	0.537	0.727	0.877	1.023	1.136	1.178
x	0	1	6	17	27	38	62	95	1372	1440

and

$$\sum_{i=1}^{n}(y_i - az_i^*)z_i^{*2} = 0$$

respectively. Here we put $z_i^* = \frac{x_i}{1+bx_i}$. Because S is a convex function, we really get a minimum.

The random residual variance is

$$\boldsymbol{s}_n^2 = \frac{1}{n-2}\sum_{i=1}^{n}\left[\boldsymbol{y}_i - \frac{\boldsymbol{a}x_i}{1+\boldsymbol{b}x_i}\right]^2 \tag{8.38}$$

with approximately df $= n - 2$.

Example

We use the data of Table 8.6 and calculate the estimates a and b, and the estimate

$$s_n^2 = \frac{1}{n-2}\sum_{i=1}^{n}\left[y_i - \frac{ax_i}{1+bx_i}\right]^2$$

of $\boldsymbol{s}_n^2$ in (8.38).

With R we must insert start values for the function $y = \frac{ax}{1+bx} = \frac{a}{\left(\frac{1}{x}+b\right)}$. For $x \to \infty$ we get $y = \frac{a}{b}$. For $x = 1$ we get $\hat{y} = \frac{a}{1+b}$. From the data we guess $1.2 = \frac{a}{b}$ and $0.025 = \frac{a}{1+b}$, hence $1.2b/(1+b) = 0.025$, $b = 0.0213$ and $a = 0.026$. The R-command `nls( )` gives a direct solution for the parameter estimates a and b.

```
> x <- c(0,1,6,17,27,38,62,95,1372,1440)
> y <- c(0,0.025,0.117,0.394,0.537,0.727,0.877,1.023,1.136,1.178)
> model <- nls(y~ a*x/(1 + b*x),start=list(a = 0.026, b = 0.0213))
> model
Nonlinear regression model
  model: y ~ a * x/(1 + b * x)
   data: parent.frame()
       a       b
0.04135 0.03392
 residual sum-of-squares: 0.02933
Number of iterations to convergence: 6
Achieved convergence tolerance: 5.338e-06
```

Hence the fitted function is $\hat{y} = \frac{0.04135x}{(1+0.03392x)}$ and $s_n^2 = 0.02923$.

The values of Table 8.3 and the corresponding regression curve gives Figure 8.7.

We now derive the asymptotic covariance matrix of the estimators. We first calculate

$$\left.\begin{aligned}\frac{\partial f_M(x_i)}{\partial \alpha} &= z_i\\ \frac{\partial f_M(x_i)}{\partial \beta} &= -\alpha z_i^2\end{aligned}\right\}$$

and

$$F_i = \begin{pmatrix} z_i & -\alpha z_i^2 \end{pmatrix}$$

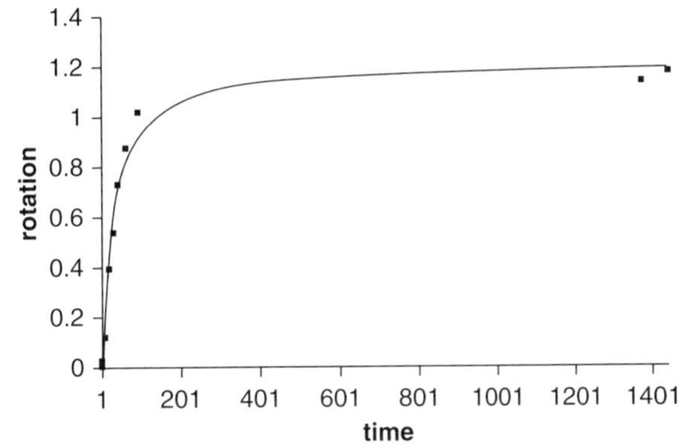

Figure 8.7 Scatter-plot of the observations from Table 8.3 and the fitted function.

and then obtain

$$F^T F = \sum_{i=1}^{n} F_i^T F_i = \begin{pmatrix} \sum_{i=1}^{n} z_i^2 & -\alpha \sum_{i=1}^{n} z_i^3 \\ -\alpha \sum_{i=1}^{n} z_i^3 & +\alpha^2 \sum_{i=1}^{n} z_i^4 \end{pmatrix} = \begin{pmatrix} c_{11} & c_{12} \\ c_{12} & c_{22} \end{pmatrix}.$$

The determinant is $\begin{vmatrix} c_{11} & c_{12} \\ c_{12} & c_{22} \end{vmatrix} = \alpha^2 \left(\sum_{i=1}^{n} z_i^2 \sum_{i=1}^{n} z_i^4 - \left[\sum_{i=1}^{n} z_i^3 \right]^2 \right) = \alpha^2 \Delta.$

The inverse of $F^T F$ is

$$(F^T F)^{-1} = \frac{1}{\Delta} \begin{pmatrix} \sum_{i=1}^{n} z_i^4 & \frac{1}{\alpha} \sum_{i=1}^{n} z_i^3 \\ \frac{1}{\alpha} \sum_{i=1}^{n} z_i^3 & \frac{1}{\alpha^2} \sum_{i=1}^{n} z_i^2 \end{pmatrix}. \tag{8.39}$$

We obtained asymptotic confidence estimations and tests as described in Section 8.2.2.1

We test

$$H_{0a} : a = a_0 \quad \text{against} \quad H_{Aa} : a \neq a_0$$

with the test statistic

$$\boldsymbol{t}_\alpha = \frac{(\boldsymbol{a} - \alpha_0)\sqrt{\hat{\boldsymbol{\Delta}}}}{\boldsymbol{s}_n \sqrt{\sum_{i=1}^{n} \boldsymbol{z}_i^{*4}}} \tag{8.40}$$

and

$$H_{0\beta} : \beta = \beta_0 \quad \text{against} \quad H_{A\beta} : \beta \neq \beta_0$$

is tested with

$$\boldsymbol{t}_\beta = \frac{(\boldsymbol{b} - \beta_0)\sqrt{\hat{\boldsymbol{\Delta}}}}{\boldsymbol{s}_n \sqrt{\frac{1}{\alpha^2} \sum_{i=1}^{n} \boldsymbol{z}_i^{*2}}}. \tag{8.41}$$

Problem 8.17 Test the hypotheses for the parameters of $f_M(x)$.

Solution

In R we continue the calculation for Problem 8.16 and calculate first $z_i^* = \frac{x_i}{1+\beta x_i}$ for $i = 1, \ldots, n$. Calculate then $Z_2 = \Sigma z_i^{*2}$, $Z_3 = \Sigma z_i^{*3}$ and $Z_4 = \Sigma z_i^{*4}$, then the estimate of Δ is $(Z_2 \cdot Z_4 - (Z_3)^2)$.

Further we insert these and the estimate $a = 0.04135$ and $s_n = \sqrt{0.02933/8} = 0.06055$ in t_α for (8.40) and the estimate $b = 0.03392$ and $s_n = \sqrt{0.02933/8} = 0.06055$ in t_β for (8.41).

Example

We test with the data of Example 8.5 with significance level $\alpha = 0.05$.

$$H_{0a} : a = 0.05 \quad \text{against} \quad H_{Aa} : a \neq 0.05,$$

$$H_{0\beta} : \beta = 0.05 \quad \text{against} \quad H_{A\beta} : \beta \neq 0.05,$$

Continuation of the R program of Problem 8.16:

```
> z <-c(x/(1 + 0.03392*x))
> z2 <- c(z*z)
> Z2 <- sum(z2)
> z3 <-c(z*z2)
> Z3 <- sum(z3)
> z4 <-c(z2*z2)
> Z4 <- sum(z4)
> Delta <- Z2*Z4 - (Z3)^2
> SSRes <- 0.02933
> n <- length(y)
> df <- n-2
> df
[1] 8
> sn <- sqrt(SSRes/df)
> sn
[1] 0.06054957
> numeratorta <- (0.04135 - 0.05)*sqrt(Delta)
> denominatorta <- sn*sqrt(Z4)
> ta <- numeratorta/denominatorta
> ta
[1] -1.937028
> p_valueta <- 2*(1-pt(abs(ta), df))
> p_valueta
[1] 0.08875258
> numeratortb <- (0.0392 - 0.05)*sqrt(Delta)
> denominatortb <- sn*sqrt(Z2)/(0.04135^2)
> tb <- numeratortb/denominatortb
> tb
[1] -0.1018714
```

```
> p_valuetb <- 2*(1-pt(abs(tb), df))
> p_valuetb
[1] 0.9213658
```

Because the p-value for the test statistic t_a of (8.40) is $0.08875 > 0.05$ H_{0a} is not rejected.

Because the p-value for the test statistic t_b of (8.41) is $0.92137 > 0.05$ H_{0b} is not rejected.

Problem 8.18 Construct $(1-\alpha^*)$ confidence intervals for the parameters of $f_M(x)$.

Solution

We obtain from (8.35)

$$\left[\boldsymbol{a} - t\left(n-2, 1-\frac{\alpha^*}{2}\right)\frac{s_n}{\sqrt{\hat{\Delta}}}\sqrt{\sum_{i=1}^{n} z_i^{*4}}; \boldsymbol{a} + \left(n-2, 1-\frac{\alpha^*}{2}\right)\frac{s_n}{\sqrt{\hat{\Delta}}}\sqrt{\sum_{i=1}^{n} z_i^{*4}}\right]$$

and

$$\left[\boldsymbol{b} - t\left(n-2, 1-\frac{\alpha^*}{2}\right)\frac{s_n}{\boldsymbol{a}\sqrt{\hat{\Delta}}}\sqrt{\sum_{i=1}^{n} z_i^{*2}}; \boldsymbol{b} + t\left(n-2, 1-\frac{\alpha^*}{2}\right)\frac{s_n}{\boldsymbol{a}\sqrt{\hat{\Delta}}}\sqrt{\sum_{i=1}^{n} z_i^{*2}}\right].$$

Continuing the R results of Problem 8.17 we calculate for a the Q_a with alpha = 0.05:

```
> Qa <- qt(0.025, df, lower.tail = FALSE)*(sn*sqrt(Z4))/sqrt(Delta)
```

The 0.95-confidence limits for a are then $a - Q_a$ and $a + Q_a$.

```
> Qb <- qt(0.025, df, lower.tail = FALSE)*sn*sqrt(Z4))/(a*sqrt(Delta))
```

The 0.95-confidence limits for b are then $b - Q_b$ and $b + Q_b$.

Example

We construct 0.95-confidence intervals with the data of Example 8.5 for α and β.

Continuing the R results of Problem 8.17.

```
> a<- 0.04135
> Qa <- qt(0.025, df, lower.tail = FALSE)* (sn*sqrt(Z4))/sqrt(Delta)
> Qa
[1] 0.0102977
>  LowerCLa <-  a-Qa
> LowerCLa
[1] 0.0310523
> UpperCLa <- a + Qa
> UpperCLa
[1] 0.0516477
> b <- 0.03392
> Qb <- qt(0.025, df, lower.tail = FALSE)* (sn*sqrt(Z2))/(a*sqrt(Delta))
> Qb
[1] 0.01010897
> LowerCLb <- b-Qb
> LowerCLb
[1] 0.02381103
> UpperCLb <- b + Qb
```

```
> UpperCLb
[1] 0.04402897
```

The 0.95-confidence interval for a is: [0.0310523; 0.0516477].
The 0.95-confidence interval for b is: [0.02381103; 0.04402897]

8.2.2.3 Exponential Regression

The model

$$y_i = f_E(x) = \alpha + \beta e^{\gamma x_i} + \boldsymbol{e}_i \cdot i = 1, \dots, n > k, \alpha \cdot \beta \cdot \gamma \neq 0 \tag{8.42}$$

is an exponential regression model where γ is the non-linearity parameter. $\alpha\beta\gamma \neq 0$ implies that no parameter is zero. Brody (1945) used it as a growth function – therefore model (8.42) is sometimes called Brody's model.

The least squares estimates a, b, c of α, β, γ are found by solving the first partial derivatives of y equal to zero and solve the simultaneous equations

$$\left.\begin{aligned} \frac{\partial SS}{\partial \alpha} &= -\sum_{i=1}^{n}(y_i - a - be^{cx_i}) = 0 \\ \frac{\partial SS}{\partial \beta} &= -\sum_{i=1}^{n} e^{cx_i}(y_i - a - be^{cx_i}) = 0 \\ \frac{\partial SS}{\partial \gamma} &= -\sum_{i=1}^{n} bx_i e^{cx_i}(y_i - a - be^{cx_i}) = 0 \end{aligned}\right\}. \tag{8.43}$$

For the solution of these simultaneous equations we check whether there is a minimum. This means that the matrix of the second partial derivatives must have positive eigenvalues.

From Rasch and Schott (2018, Chapter 9) we know that

$$F^T F : \begin{vmatrix} \sum_{i=1}^{n} 1 & \sum_{i=1}^{n} e^{\gamma x_i} & \sum_{i=1}^{n} \beta x_i e^{\gamma x_i} \\ \sum_{i=1}^{n} e^{\gamma x_i} & \sum_{i=1}^{n} e^{2\gamma x_i} & \sum_{i=1}^{n} \beta x_i e^{2\gamma x_i} \\ \sum_{i=1}^{n} \beta x_i e^{\gamma x_i} & \sum_{i=1}^{n} \beta x_i e^{2\gamma x_i} & \sum_{i=1}^{n} \beta^2 x_i^2 e^{2\gamma x_i} \end{vmatrix}. \tag{8.44}$$

We will not use this way in this section because R finds the minimum directly, as shown in Problem 8.19

Minimising SS $= \sum_{i=1}^{n} (y_i - \alpha - \beta e^{\gamma x_i})^2$ gives us the estimates a, b, c of α, β, γ.

Problem 8.19 Show how the parameters of $f_E(x)$ can be estimated.

Solution

To use the R-command `nls()` we need initial values. Because for $x = 0$ $\alpha + \beta e^{\gamma x_i} = \alpha + \beta$ we use $y(\min) = \alpha + \beta$, For $\gamma < 0$ is $\lim_{n\to\infty}[\alpha + \beta e^{\gamma x_i}] = \alpha$ we use it for $y(\max)$; γ we must find by systematic search. The R-command `nls( )` gives a direct solution. In R we use A, B, C for, respectively, a, b, c.

Example

Let us consider a numerical example with the growth of leaf surfaces of oil palms y in square metres and age x in years observed in Indonesia by Breure.

The data in Table 8.7 are available.

First we input the data and make a plot.

```
> age <- c(1,2,3,4,5,6,7,8,9,10,11,12)
> surface <-
c(2.02,3.62,5.71,7.13,8.33,8.29,9.81,11.3,12.18,12.67,12.62,13.01)
> growth <- cbind(age,surface)
```

The plot is given below and from the data we find for $x = 1$ $y = 2.02 \approx A + B\exp(C)$. From the plot we see that $C < 0$, hence we choose for C as start value a negative value such as -0.1.

Now $2.02 \approx A + B^{*}\ 0.905$. Further, $y(\max) = 13.01 \approx A$. We find $B \approx (2.02-13.01)/0.905 = -12.144$. We use as starting values $A = 13$, $B = -12$ and $C = -0.1$.

```
>  model <- nls(surface ~ A + B*exp(C*age), start=list(A = 13, B =-12,
          C = - 0.1))
> model
Nonlinear regression model
  model: surface ~ A + B * exp(C * age)
   data: parent.frame()
      A       B       C
 16.479 -16.650  -0.138
 residual sum-of-squares: 1.801
Number of iterations to convergence: 5
```

Table 8.7 Leaf surfaces of oil palms y in square metres and age x in years

Age	Surface
1	2.02
2	3.62
3	5.71
4	7.13
5	8.33
6	8.29
7	9.81
8	11.3
9	12.18
10	12.67
11	12.62
12	13.01

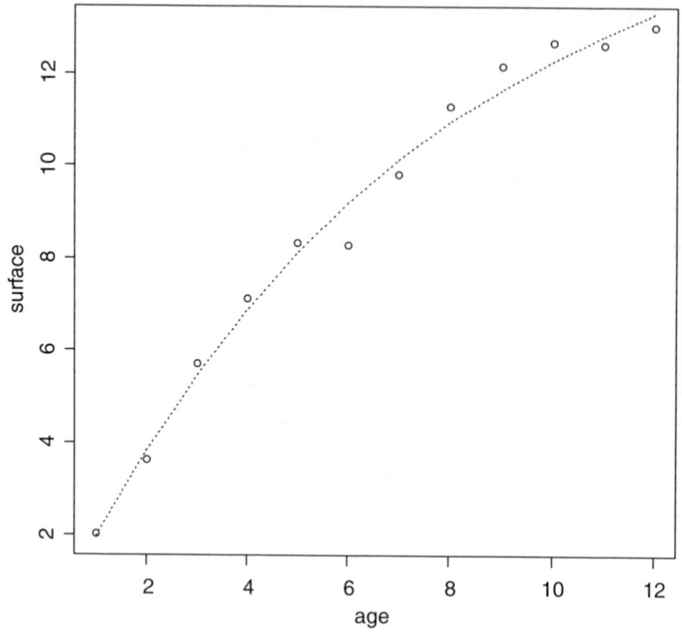

Figure 8.8 Scatter-plot of the data and the fitted exponential function in the example of Problem 8.19.

```
Achieved convergence tolerance: 3.579e-07
```

The fitted exponential function is $\widehat{\text{surface}}$ = 16.479–16.650*exp(–0.138*age) Figure 8.8.

```
> plot(age,surface)
> fittedsurf  <-  c(16.479-16.650*exp(-0.138*age))
```

Problem 8.20 Determine the determinant (8.44) of the matrix F^TF.

Solution

Note that the determinant $\left|\begin{pmatrix} c_{11} & c_{12} & c_{13} \\ c_{12} & c_{22} & c_{23} \\ c_{13} & c_{23} & c_{33} \end{pmatrix}\right|$ of a symmetric matrix C of order three $\begin{pmatrix} c_{11} & c_{12} & c_{13} \\ c_{12} & c_{22} & c_{23} \\ c_{13} & c_{23} & c_{33} \end{pmatrix}$ can be calculated as

$$c_{11}\begin{vmatrix} c_{22} & c_{23} \\ c_{23} & c_{33} \end{vmatrix} - c_{12}\begin{vmatrix} c_{12} & c_{13} \\ c_{23} & c_{33} \end{vmatrix} + c_{13}\begin{vmatrix} c_{12} & c_{13} \\ c_{22} & c_{23} \end{vmatrix}.$$

We obtain for the determinant

$$\begin{vmatrix} \sum_{i=1}^{n} 1 & \sum_{i=1}^{n} \mathrm{e}^{\gamma x_i} & \sum_{i=1}^{n} \beta x_i \mathrm{e}^{\gamma x_i} \\ \sum_{i=1}^{n} \mathrm{e}^{\gamma x_i} & \sum_{i=1}^{n} \mathrm{e}^{2\gamma x_i} & \sum_{i=1}^{n} \beta x_i \mathrm{e}^{2\gamma x_i} \\ \sum_{i=1}^{n} \beta x_i \mathrm{e}^{\gamma x_i} & \sum_{i=1}^{n} \beta x_i \mathrm{e}^{2\gamma x_i} & \sum_{i=1}^{n} \beta^2 x_i^2 \mathrm{e}^{2\gamma x_i} \end{vmatrix} = \Delta$$

$$= \beta^2 \{ n(c_{22}c_{33} - {c_{23}}^2) - c_{12}(c_{12}c_{33} - c_{23}c_{13}) + c_{13}(c_{12}c_{23} - c_{22}c_{13}) \}$$

$$= n\beta^2 \left[\sum_{i=1}^{n} \mathrm{e}^{2\gamma x_i} \sum_{i=1}^{n} x_i^2 \mathrm{e}^{2\gamma x_i} - \left(\sum_{i=1}^{n} x_i \mathrm{e}^{2\gamma x_i} \right)^2 \right]$$

$$- \beta^2 \sum_{i=1}^{n} \mathrm{e}^{\gamma x_i} \left[\sum_{i=1}^{n} \mathrm{e}^{\gamma x_i} \sum_{i=1}^{n} x_i^2 \mathrm{e}^{2\gamma x_i} - \sum_{i=1}^{n} x_i \mathrm{e}^{2\gamma x_i} \sum_{i=1}^{n} x_i \mathrm{e}^{\gamma x_i} \right]$$

$$+ \beta^2 \sum_{i=1}^{n} x_i \mathrm{e}^{\gamma x_i} \left[\sum_{i=1}^{n} \mathrm{e}^{\gamma x_i} \sum_{i=1}^{n} x_i \mathrm{e}^{2\gamma x_i} - \sum_{i=1}^{n} \mathrm{e}^{2\gamma x_i} \sum_{i=1}^{n} x_i \mathrm{e}^{\gamma x_i} \right]$$

$$= \beta^2.$$

The inverse of $F^T F$ has the form

$$(v_{jl}) = \frac{1}{\Delta} \begin{pmatrix} \sum_{i=1}^{n} \mathrm{e}^{2\gamma x_i} \sum_{i=1}^{n} x_i^2 \mathrm{e}^{2\gamma x_i} - \left(\sum_{i=1}^{n} x_i \mathrm{e}^{2\gamma x_i} \right)^2 & \sum_{i=1}^{n} x_i \mathrm{e}^{\gamma x_i} \sum_{i=1}^{n} x_i \mathrm{e}^{2\gamma x_i} - \sum_{i=1}^{n} \mathrm{e}^{\gamma x_i} \sum_{i=1}^{n} x_i^2 \mathrm{e}^{2\gamma x_i} & \frac{1}{\beta} \begin{pmatrix} \sum_{i=1}^{n} \mathrm{e}^{\gamma x_i} \sum_{i=1}^{n} x_i \mathrm{e}^{2\gamma x_i} - \\ \sum_{i=1}^{n} x_i \mathrm{e}^{\gamma x_i}, \sum_{i=1}^{n} \mathrm{e}^{2\gamma x_i} \end{pmatrix} \\ \sum_{i=1}^{n} x_i \mathrm{e}^{\gamma x_i} \sum_{i=1}^{n} x_i \mathrm{e}^{2\gamma x_i} - \sum_{i=1}^{n} \mathrm{e}^{\gamma x_i} \sum_{i=1}^{n} x_i^2 \mathrm{e}^{2\gamma x_i} & n \sum_{i=1}^{n} x_i^2 \mathrm{e}^{2\gamma x_i} - \left(\sum_{i=1}^{n} x_i \mathrm{e}^{\gamma x_i} \right)^2 & \frac{1}{\beta} \left(\sum_{i=1}^{n} \mathrm{e}^{\gamma x_i} \sum_{i=1}^{n} x_i \mathrm{e}^{\gamma x_i} - n \sum_{i=1}^{n} x_i \mathrm{e}^{2\gamma x_i} \right) \\ \frac{1}{\beta} \left(\sum_{i=1}^{n} \mathrm{e}^{\gamma x_i} \sum_{i=1}^{n} x_i \mathrm{e}^{2\gamma x_i} - \sum_{i=1}^{n} x_i \mathrm{e}^{\gamma x_i}, \sum_{i=1}^{n} \mathrm{e}^{2\gamma x_i} \right) & \frac{1}{\beta} \left(\sum_{i=1}^{n} \mathrm{e}^{\gamma x_i} \sum_{i=1}^{n} x_i \mathrm{e}^{\gamma x_i} - n \sum_{i=1}^{n} x_i \mathrm{e}^{2\gamma x_i} \right) & \frac{1}{\beta^2} \left(n \sum_{i=1}^{n} \mathrm{e}^{2\gamma x_i} - \left(\sum_{i=1}^{n} \mathrm{e}^{\gamma x_i} \right)^2 \right) \end{pmatrix} \tag{8.45}$$

Confidence estimations and tests can be obtained as described in Section 8.2.2.1 assuming that the errors $\boldsymbol{e}_{\mathrm{i}}$ are independent and normally distributed as $N(0, \sigma^2)$.

We test

$$H_{0a} : a = a_0 \quad \text{against} \quad H_{\mathrm{A}a} : a \neq a_0$$

with the test statistic

$$\boldsymbol{t}_\alpha = \frac{(\boldsymbol{a} - \alpha_0)\sqrt{\hat{\boldsymbol{\Delta}}}}{\boldsymbol{s}_n \sqrt{\sum_{i=1}^{n} \mathrm{e}^{2cx_i} \sum_{i=1}^{n} x_i^2 \mathrm{e}^{2cx_i} - \left(\sum_{i=1}^{n} x_i \mathrm{e}^{2cx_i} \right)^2}}. \tag{8.46}$$

Further

$$H_{0\beta} : \beta = \beta_0 \quad \text{against} \quad H_{\mathrm{A}\beta} : \beta \neq \beta_0$$

is tested with

$$t_\beta = \frac{(b - \beta_0)\sqrt{\hat{\Delta}}}{s_n\sqrt{n\sum_{i=1}^{n} x_i^2 e^{2cx_i} - \left(\sum_{i=1}^{n} x_i e^{cx_i}\right)^2}} \tag{8.47}$$

and

$$H_{0\gamma} : \gamma = \gamma_0 \quad \text{against} \quad H_{A\gamma} : \gamma \neq \gamma_0$$

with

$$t_\gamma = \frac{(c - c_0) \mid b \mid \sqrt{\hat{\Delta}}}{s_n\sqrt{\left(n\sum_{i=1}^{n} e^{2cx_i} - \left(\sum_{i=1}^{n} e^{cx_i}\right)^2\right)}} \tag{8.48}$$

and

$$s_n^2 = \frac{1}{n-3}\sum_{i=1}^{n}[y_i - a - be^{cx_i}]^2$$

approximately with $df = n - 3$.

Problem 8.21 Test the hypotheses for all three parameters of $f_E(x)$ with significance level $\alpha = 0.05$.

Solution
In R we first estimate the parameters of the exponential function. Then we construct the test statistic t_α for (8.46), t_β for (8.47) and t_γ for (8.48).

Example
We test in Example 8.3

$$H_{0a} : a = 130 \quad \text{against} \quad H_{Aa} : a \neq 130,$$

$$H_{0\beta} : \beta = -65 \quad \text{against} \quad H_{A\beta} : \beta \neq -65,$$

$$H_{0\gamma} : \gamma = -0.05 \quad \text{against} \quad H_{A\gamma} : \gamma \neq -0.05.$$

In Example 8.3 let we choose age to be x and height to be y. Because for $x = 0$ $\alpha + \beta e^{\gamma x_i} = \alpha + \beta$ we use $y(\min) = \alpha + \beta$, For $\gamma < 0$ is $\lim_{n\to\infty}[\alpha + \beta e^{\gamma x_i}] = \alpha$ we use it for $y(\max)$, γ we must find by systematic search. From the plot in Figure 8.3 we start with $y(\min) = a + b = 77$ and $y(\max) = a = 132$. Hence $b = -55$. The R-command `nls( )` gives a direct solution for the parameter estimates a and b. We start with a guess for c a negative value such as −0.1.

```
> x <- c( 0,6,12,18,24,30,36,42,48,54,60)
>  y <-c(77.2,94.5,107.2,116.0,122.4,126.7,129.2,129.9,130.4,130.8,131.2)
>  n <- length(y)
>  n
[1] 11
```

```
> model <- nls(y ~ a + b*exp(c*x), start=list(a = 132, b =-55, c = - 0.1))
> model
Nonlinear regression model
  model: y ~ a + b * exp(c * x)
   data: parent.frame()
       a        b        c
132.9622 -56.4210  -0.0677
 residual sum-of-squares: 6.084
Number of iterations to convergence: 6
Achieved convergence tolerance: 1.58e-06
```

The estimated exponential function is $y = 132.9622 - 56.4210e^{-0.677}$.
The estimate $s_n^2 = 6.084$ with approximately df $= n - 3 = 11-3 = 8$.

```
> SSRes <- 6.084
> df  <- n-3
> sn <- sqrt(SSRes/df)
> sn
[1] 0.8720665
> a <- 132.9622
> b <- -56.4210
> c <- -0.0677
> cc <- c(exp(2*c*x))
> C <- sum(cc)
> ee <- c(x^2*exp(2*c*x))
> E <- sum(ee)
> dd <- c(x*exp(2*c*x))
> D <- sum(dd)
> aa <- c(exp(c*x))
> A <- sum(aa)
> bb <- c(x*exp(c*x))
> B <- sum(bb)
>  Delta <- n*(C*E- D^2)-A*(A*E-D*B)+B*(A*D-C*B)
> Delta
[1] 358.7337
> num_ta <- (a-130)*sqrt(Delta)
> Sa <- C*E-D^2
> den_ta <- sn*sqrt(Sa)
> SEa <- den_ta/sqrt(Delta)
> SEa
[1] 0.5914927
> ta <- num_ta/den_ta
> ta
[1] 5.008008
> p_valuea <- 2*(1-pt(abs(ta),df))
> p_valuea
[1] 0.001042351
> num_tb <- (b-(-65))*sqrt(Delta)
> Sb <- n*E-B^2
> den_tb <- sn*sqrt(Sb)
> SEb <- den_tb/sqrt(Delta)
> SEb
[1] 0.8734449
> tb <- num_tb/den_tb
> tb
[1] 9.822027
```

```
> p_valueb <- 2*(1-pt(abs(tb),df))
> p_valueb
[1] 9.704446e-06
> num_tc <- (c-(-0.055))*abs(b)*sqrt(Delta)
> Sc <- n*C-A^2
> den_tc <- sn*sqrt(Sc)
> SEc <- den_tc/(sqrt(b^2)*sqrt(Delta))
> SEc
[1] 0.002707207
> tc <- num_tc/den_tc
> tc
[1] -4.691181
> p_valuec <- 2*(1-pt(abs(tc),df))
> p_valuec
[1] 0.001559142
```

Because the p-value for t_α is 0.001042351 < 0.05, H_{0a} is rejected.
Because the p-value for t_β is 9.704446e-06 < 0.05, H_{0b} is rejected.
Because the p-value for t_γ is 0.001559142 < 0.05, H_{0c} is rejected.

Confidence intervals with a nominal confidence coefficient $1\text{-}\alpha^*_{\text{nom}}$ are defined as follows:
Parameter α:

$$\left[\boldsymbol{a} - s_n \frac{\sqrt{\sum_{i=1}^{n} e^{2cx_i} \sum_{i=1}^{n} x_i^2 e^{2cx_i} - \left(\sum_{i=1}^{n} x_i e^{2cx_i} \right)^2}}{\sqrt{\widehat{\boldsymbol{\Delta}}}} t\left(n-3 \middle| 1 - \frac{\alpha^*_{\text{nom}}}{2} \right), \right.$$
$$\left. \boldsymbol{a} + s_n \frac{\sqrt{\sum_{i=1}^{n} e^{2cx_i} \sum_{i=1}^{n} x_i^2 e^{2cx_i} - \left(\sum_{i=1}^{n} x_i e^{2cx_i} \right)^2}}{\sqrt{\widehat{\boldsymbol{\Delta}}}} t\left(n-3 \middle| 1 - \frac{\alpha^*_{\text{nom}}}{2} \right) \right] \tag{8.49}$$

Parameter β:

$$\left[\boldsymbol{b} - s_n \frac{\sqrt{n \sum_{i=1}^{n} x_i^2 e^{2cx_i} - \left(\sum_{i=1}^{n} x_i e^{cx_i} \right)^2}}{\sqrt{\widehat{\boldsymbol{\Delta}}}} t\left(n-3 \middle| 1 - \frac{\alpha^*_{\text{nom}}}{2} \right), \right.$$
$$\left. \boldsymbol{b} + s_n \frac{\sqrt{n \sum_{i=1}^{n} x_i^2 e^{2cx_i} - \left(\sum_{i=1}^{n} x_i e^{cx_i} \right)^2}}{\sqrt{\widehat{\boldsymbol{\Delta}}}} t\left(n-3 \middle| 1 - \frac{\alpha^*_{\text{nom}}}{2} \right) \right] \tag{8.50}$$

Parameter γ:

$$\left[c - s_n \frac{\sqrt{\left(n \sum_{i=1}^{n} e^{2cx_i} - \left(\sum_{i=1}^{n} e^{cx_i} \right)^2 \right)}}{|\,b\,| \sqrt{\hat{\Delta}}} t\left(n-3 \middle| 1 - \frac{\alpha^*_{\text{nom}}}{2} \right), \right.$$

$$\left. c + s_n \frac{\sqrt{\left(n \sum_{i=1}^{n} e^{2cx_i} - \left(\sum_{i=1}^{n} e^{cx_i} \right)^2 \right)}}{|\,b\,| \sqrt{\hat{\Delta}}} t\left(n-3 \middle| 1 - \frac{\alpha^*_{\text{nom}}}{2} \right) \right] \tag{8.51}$$

Problem 8.22 Construct 0.95-confidence intervals for all three parameters of $f_E(x)$.

Solution
Continuation of the R program for Problem 8.21 we calculate in R (8.49)–(8.51).

Example
We use the data of Example 8.3 and continue the R program of Problem 8.21.

```
> t_value <- qt(0.025, df, lower.tail = FALSE)
> t_value
[1] 2.306004
> Qa <- ((sn*sqrt(C*E-D^2))/sqrt(Delta))*t_value
> Qa
[1] 1.363985
> lowerCLa <- a-Qa
> lowerCLa
[1] 131.5982
> upperCLa <- a + Qa
> upperCLa
[1] 134.3262
> Qb <- ((sn*sqrt(n*E-B^2))/sqrt(Delta))*t_value
> Qb
[1] 2.014168
> lowerCLb <- b-Qb
> lowerCLb
[1] -58.43517
> upperCLb <- b+Qb
> upperCLb
[1] -54.40683
> Qc <- (sn*sqrt(((n*C-A^2)/b^2))/sqrt(Delta))*t_value
> Qc
[1] 0.006242831
```

```
> lowerCLc <- c-Qc
> lowerCLc
[1] -0.07394283
> upperCLc <- c+Qc
> upperCLc
[1] -0.06145717
```

The 0.95-confidence interval for α is: [131.5982; 134.3262].
The 0.95-confidence interval for β is: [−58.43517; −54.40683].
The 0.95-confidence interval for γ is: [−0.07394283; −0.06145717].

Using the simulation experiment described in Section 8.2.2.1, we found that the empirical variances do not differ strongly from the main diagonal elements of the asymptotic covariance matrix for $n = 4$.

The choice of the denominator $n - 3$ in estimate s^2 of σ^2 is analogous to the linear case. There, $n - 3$ (or in general $n - p$) is the number degrees of freedom of the χ^2-distribution of the numerator of s^2. If we compare expectation, variance, skewness and kurtosis of a χ^2-distribution with $n - 3$ degrees of freedom with the corresponding empirical values from the simulation experiment, we see that even for the smallest possible $n = 4$ a good accordance is found. This means that $n - 3$ is a good choice for the denominator in the estimator of σ^2.

Table 8.8 shows the relative frequencies of confidence estimations and tests respectively (with $\alpha_{\text{nom}} = 0.05$ and $\alpha_{\text{nom}} = 0.1$ for a special parameter configuration of the exponential regression from 10 000 runs. As we can see already with $n = 4$ a sufficient alignment is found between α_{nom} and α_{act}. Therefore, the tests above can be used as approximative α_{nom} tests and the corresponding confidence intervals as approximative $(1-\alpha_{\text{nom}})$-confidence intervals.

Table 8.8 Relative frequencies of 10 000 simulated samples for the correct acception $1 - \alpha_{\text{act}}$ of $H_0: \alpha = 0.\ \beta = -50,\ \gamma = -0.05,\ n = 10(-1)4$, and $\alpha_{\text{nom}} = 0.05$ and $\alpha_{\text{nom}} = 0.1$ for the exponential regression.

	$H_{0\alpha}: \alpha = 0$		$H_{0\beta}: \beta = -50$		$H_{0\gamma}: \gamma = -0.05$	
n	$\alpha_{\text{nom}} = 0.05$	$\alpha_{\text{nom}} = 0.1$	$\alpha_{\text{nom}} = 0.05$	$\alpha_{\text{nom}} = 0.1$	$\alpha_{\text{nom}} = 0.05$	$\alpha_{\text{nom}} = 0.1$
10	95.26	90.63	95.25	90.55	95.24	90.53
9	94.76	88.33	95.13	90.04	94.72	88.38
8	95.46	90.66	95.33	90.61	95.07	90.21
7	95.38	90.22	95.25	90.28	95.39	90.41
6	94.98	90.22	94.89	88.76	94.93	90.24
5	95.01	90.21	95.14	90.10	94.94	90.23
4	94.93	88.92	95.14	88.86	95.09	88.86

8.2.2.4 The Logistic Regression

The model

$$y_i = f_L(x) = \frac{\alpha}{1 + \beta e^{\gamma x_i}} + e_i \cdot i = 1, \dots, n > k, \vec{\beta}^T = (\alpha\ \beta\ \gamma), \alpha \cdot \beta \cdot \gamma \neq 0 \tag{8.52}$$

is the model of a logistic regression where γ is the non-linearity parameter. $\alpha\beta\gamma \neq 0$ implies that no parameter is zero. The term logistic regression is not uniquely used – see Chapter 11 for another use of this term.

The first and second derivative to x are

$$\frac{\partial y}{\partial x} = -\frac{\alpha\beta\gamma e^{\gamma x}}{(1+\beta e^{\gamma x})^2}$$

and

$$\frac{\partial^2 y}{\partial x^2} = -\alpha\beta\gamma\frac{(1+\beta e^{\gamma x})^2\gamma e^{\gamma x} - 2e^{\gamma x}(1+\beta e^{\gamma x})\beta\gamma e^{\gamma x}}{(1+\beta e^{\gamma x})^4} = -\alpha\beta\gamma^2 e^{\gamma x}\frac{1-\beta e^{\gamma x}}{(1+\beta e^{\gamma x})^3}.$$

At the inflection point (x_ω, η_ω), the numerator of the second derivative has to be zero so that

$$1-\beta e^{\gamma x_\omega} = 0; \quad \beta = e^{-\gamma x_\omega} \text{ and } x_\omega = -\frac{1}{\gamma}\ln\beta,\ f_\omega = \frac{\alpha}{2} = \eta_\omega.$$

The second derivative $\frac{d^2 f(x,\ \theta)}{dx^2}$ changes at x_ω its sign and we really obtain an inflection point.

Now we show that the same curve is described by more than one function. The transition from one function to another is called reparametrisation.

Because $\beta = e^{-\gamma x_\omega}$ we have

$$\frac{\partial f_L(x)}{\partial x} = -\frac{\alpha\beta\gamma e^{\gamma x}}{(1+\beta e^{\gamma x})^2} = \frac{\alpha}{2}\left\{1+\tanh\left[-\frac{\gamma}{2}(x-x_\omega)\right]\right\}. \tag{8.53}$$

The logistic regression function is written as the three-parametric hyperbolic tangent-regression function.

Example 8.6 We consider an example. Table 8.9 shows the growth of leaf surfaces of oil palms y in square metres and age x in years observed in Indonesia by Breure.

Problem 8.23 Show how the parameters of $f_L(x)$ can be estimated.

Solution

We use the R-command `nls()` to minimise

$$SS = \sum_{i=1}^{n}\left(y_i - \frac{\alpha}{1+\beta e^{\gamma x_i}}\right)^2 \text{ directly.}$$

Example

We use the data of Example 8.1 and fit a logistic regression function to the hemp data.

To obtain initial valued for searching the minimum we see that for $x = 0$: $\frac{\alpha}{1+\beta e^{\gamma 0}} = \frac{\alpha}{1+\beta}$ and use $\frac{a}{1+b}$ for the value at the start of the growth. For $\gamma < 0$ we get $\lim_{x\to\infty}\frac{\alpha}{\beta e^{\gamma x}} = \alpha$ and we use the largest value of y as the initial value for a. For c we start as a guess after several

Table 8.9 Leaf surface y_i in m^2 of oil palms on a trial area in dependency of age x_i in years.

x_i	1	2	3	4	5	6	7	8	9	10	11	12
y_i	2.02	3.62	5.71	7.13	8.33	8.29	9.81	11.3	12.18	12.67	12.62	13.01

trials with the initial value −1. From the data we take as the largest value 121, hence the start value is $a = 121$. We take after several trials the start value $c = -1$; for $x = 1$ we obtain $8.3 = \frac{a}{1+be^{c\cdot 1}}$ hence $8.3 = \frac{121}{1+b\cdot 0.368}$, hence for b we find 36.9 and we take as the start value $b = 37$.

```
> x<-  c(1,2,3,4,5,6,7,8,9,10,11,12,13,14)
> y <- c(8.3,15.2,24.7,32,39.3,55.4,69,84.4,98.1,107.7,112,116.9,119.9,121.1)
> model <- nls(y ~ a/(1+b*exp(c*x)),start=list(a=121,b =37,c = -1))
> model
Nonlinear regression model
  model: y ~ a/(1 + b * exp(c * x))
   data: parent.frame()
       a         b         c
126.1910   19.7348   -0.4607
 residual sum-of-squares: 40.75
Number of iterations to convergence: 8
Achieved convergence tolerance: 7.927e-06
```

The fitted logistic function is $y = \frac{126.191}{1+19.7348\cdot e^{-0.4607x}}$.

Problem 8.24 Derive the asymptotic covariance matrix of the logistic regression function.

Solution

Because

$$\frac{\partial f_L(x)}{\partial \alpha} = \frac{1}{1+\beta e^{\gamma x}}, \qquad \frac{\partial f_L(x)}{\partial \beta} = \frac{-\alpha e^{\gamma x}}{(1+\beta e^{\gamma x})^2}, \qquad \frac{\partial f_L(x)}{\partial \gamma} = \frac{-\alpha\beta x e^{\gamma x}}{(1+\beta e^{\gamma x})^2}$$

we get

$$F(x_i) = \begin{pmatrix} \frac{1}{1+\beta e^{\gamma x_i}} & \frac{-\alpha e^{\gamma x_i}}{(1+\beta e^{\gamma x_i})^2} & \frac{-\alpha\beta x e^{\gamma x_i}}{(1+\beta e^{\gamma x_i})^2} \end{pmatrix}.$$

From this we obtain

$$F^T F = \begin{pmatrix} \sum_{i=1}^n \frac{1}{(1+\beta e^{\gamma x_i})^2} & \sum_{i=1}^n \frac{-\alpha e^{\gamma x_i}}{(1+\beta e^{\gamma x_i})^3} & \sum_{i=1}^n \frac{-\alpha\beta x_i e^{\gamma x_i}}{(1+\beta e^{\gamma x_i})^3} \\ \sum_{i=1}^n \frac{-\alpha e^{\gamma x_i}}{(1+\beta e^{\gamma x_i})^3} & \sum_{i=1}^n \frac{\alpha^2 e^{2\gamma x_i}}{(1+\beta e^{\gamma x_i})^4} & \sum_{i=1}^n \frac{\alpha^2\beta x_i e^{2\gamma x_i}}{(1+\beta e^{\gamma x_i})^4} \\ \sum_{i=1}^n \frac{-\alpha\beta x_i e^{\gamma x_i}}{(1+\beta e^{\gamma x_i})^3} & \sum_{i=1}^n \frac{\alpha^2\beta x_i e^{2\gamma x_i}}{(1+\beta e^{\gamma x_i})^4} & \sum_{i=1}^n \frac{\alpha^2\beta^2 x_i^2 e^{2\gamma x_i}}{(1+\beta e^{\gamma x_i})^4} \end{pmatrix} = \begin{pmatrix} c_{11} & c_{12} & c_{13} \\ c_{12} & c_{22} & c_{23} \\ c_{13} & c_{23} & c_{33} \end{pmatrix}. \tag{8.54}$$

The determinant of $F^T F$ is

$$\begin{aligned} |F^T F| = {} & \sum_{i=1}^n \frac{1}{(1+\beta e^{\gamma x_i})^2} \sum_{i=1}^n \frac{\alpha^2 e^{2\gamma x_i}}{(1+\beta e^{\gamma x_i})^4} \sum_{i=1}^n \frac{\alpha^2\beta^2 x_i^2 e^{2\gamma x_i}}{(1+\beta e^{\gamma x_i})^4} - 2\sum_{i=1}^n \frac{-\alpha e^{\gamma x_i}}{(1+\beta e^{\gamma x_i})^3} \\ & \times \sum_{i=1}^n \frac{-\alpha\beta x_i e^{\gamma x_i}}{(1+\beta e^{\gamma x_i})^3} \sum_{i=1}^n \frac{\alpha^2\beta x_i e^{2\gamma x_i}}{(1+\beta e^{\gamma x_i})^4} - \left(\sum_{i=1}^n \frac{-\alpha\beta x_i e^{\gamma x_i}}{(1+\beta e^{\gamma x_i})^3}\right)^2 \sum_{i=1}^n \frac{\alpha^2 e^{2\gamma x_i}}{(1+\beta e^{\gamma x_i})^4} \\ & - \sum_{i=1}^n \frac{1}{(1+\beta e^{\gamma x_i})^2} \left(\sum_{i=1}^n \frac{\alpha^2\beta x_i e^{2\gamma x_i}}{(1+\beta e^{\gamma x_i})^4}\right)^2 - \left(\sum_{i=1}^n \frac{-\alpha e^{\gamma x_i}}{(1+\beta e^{\gamma x_i})^3}\right)^2 \\ & \times \sum_{i=1}^n \frac{\alpha^2\beta^2 x_i^2 e^{2\gamma x_i}}{(1+\beta e^{\gamma x_i})^4} = \alpha^4\beta^2\Delta \end{aligned} \tag{8.55}$$

the inverse of F^TF has the form

$$(F^TF)^{-1} = \frac{1}{\alpha^4\beta^2\Delta}\begin{pmatrix} \alpha^2\beta^2(c_{22}c_{33} - c_{23}^2) & \alpha\beta^2(c_{13}c_{23} - c_{12}c_{33}) & \alpha\beta(c_{12}c_{23} - c_{13}c_{22}) \\ -\alpha\beta^2(c_{13}c_{23} - c_{12}c_{33}) & \beta^2(c_{11}c_{33} - c_{13}^2) & \beta(c_{12}c_{13} - c_{11}c_{23}) \\ \alpha\beta(c_{12}c_{23} - c_{13}c_{22}) & \beta(c_{12}c_{13} - c_{11}c_{23}) & c_{11}c_{22} - c_{12}^2 \end{pmatrix} \tag{8.56}$$

and the asymptotic covariance matrix is

$$\text{var}_{\text{as}}(\hat{\boldsymbol{\beta}}) = \sigma^2\left(\sum_{i=1}^{n} F_i^T F_i\right)^{-1}$$

$$= \frac{\sigma^2}{\alpha^2\beta^2\Delta}\begin{pmatrix} \alpha^2\beta^2(c_{22}c_{33} - c_{23}^2) & \alpha\beta^2(c_{13}c_{23} - c_{12}c_{33}) & \alpha\beta(c_{12}c_{23} - c_{13}c_{22}) \\ -\alpha\beta^2(c_{13}c_{23} - c_{12}c_{33}) & \beta^2(c_{11}c_{33} - c_{13}^2) & \beta(c_{12}c_{13} - c_{11}c_{23}) \\ \alpha\beta(c_{12}c_{23} - c_{13}c_{22}) & \beta(c_{12}c_{13} - c_{11}c_{23}) & c_{11}c_{22} - c_{12}^2 \end{pmatrix}. \tag{8.57}$$

We estimate $\text{var}_{\text{as}}(\hat{\boldsymbol{\beta}})$ by

$$\widehat{\text{var}_{\text{as}}(\hat{\boldsymbol{\beta}})} = \boldsymbol{s}_n^2 \sum_{i=1}^{n} (\widehat{\boldsymbol{F}_i^{*T}\boldsymbol{F}_i^{*}})^{-1}.$$

In $\boldsymbol{F}_i^*$ the parameters have been replaced by their estimators and $\boldsymbol{s}_n^2$ is given by (8.30). We obtain confidence estimations and tests as described in Section 8.2.2.1.

We test

$$H_{0a} : a = a_0 \quad \text{against} \quad H_{Aa} : a \neq a_0$$

with the test statistic

$$\boldsymbol{t}_\alpha = \frac{(\boldsymbol{a} - \alpha_0)\sqrt{\hat{\boldsymbol{\Delta}}}}{\boldsymbol{s}_n\sqrt{(\hat{\boldsymbol{c}}_{22}\hat{\boldsymbol{c}}_{33} - \hat{\boldsymbol{c}}_{23}^2)}}. \tag{8.58}$$

Further

$$H_{0\beta} : \beta = \beta_0 \quad \text{against} \quad H_{A\beta} : \beta \neq \beta_0$$

is tested with

$$\boldsymbol{t}_\beta = \frac{(\boldsymbol{b} - \beta_0) \mid \boldsymbol{a} \mid \sqrt{\hat{\boldsymbol{\Delta}}}}{\boldsymbol{s}_n\sqrt{(\hat{\boldsymbol{c}}_{11}\hat{\boldsymbol{c}}_{33} - \hat{\boldsymbol{c}}_{13}^2)}} \tag{8.59}$$

and

$$H_{0\gamma} : \gamma = \gamma_0 \quad \text{against} \quad H_{A\gamma} : \gamma \neq \gamma_0$$

with

$$\boldsymbol{t}_\gamma = \frac{(\boldsymbol{c} - c_0) \mid \boldsymbol{ab} \mid \sqrt{\hat{\boldsymbol{\Delta}}}}{\boldsymbol{s}_n\sqrt{\hat{\boldsymbol{c}}_{11}\hat{\boldsymbol{c}}_{22} - \hat{\boldsymbol{c}}_{12}^2}}. \tag{8.60}$$

The $\hat{c}_{ij}$, $i, j = 1, ..3$ are obtained from c_{ij} by replacing all parameters by their estimates. Further

$$s_n^2 = \frac{1}{n-3} \sum_{i=1}^{n} \left[y_i - \frac{a}{1 + be^{cx_i}} \right]^2$$

with approximately df $= n - 3$.

Problem 8.25 Test the hypotheses with significance $\alpha^* = 0.05$ for each of the three parameters of $f_L(x)$ with R. The null-hypotheses values are respectively $\alpha_0 = 15$, $\beta_0 = 7$ and $\gamma_0 = -0.05$.

Solution
Using the estimates *a*, *b* and *c* we calculate the estimates for c_{ij} and insert the values in (8.58)–(8.60).

Example
We use the data from Example 8.6 and estimate the logistic function with R.

We use the data of Example 8.6 and fit a logistic regression function to the oil palm data.

To obtain initial valued for searching the minimum we see that for $x = 0$: $\frac{\alpha}{1+\beta e^{\gamma 0}} = \frac{\alpha}{1+\beta}$ and we use $\frac{a}{1+b}$ for the value at the start of the growth. For $\gamma < 0$ we get $\lim_{x\to\infty} \frac{\alpha}{\beta e^{\gamma x}} = \alpha$ and we use the largest value of *y* as initial value for *a*. For *c* we start after several trials as a guess, with the initial value −1. From the data we take as the largest value 13.01, hence the start value for $a = 13$. We would take after several trials as a start value for $c = -1$; for $x = 1$ we obtain $2.02 = \frac{a}{1+be^c} = \frac{13}{1+b\cdot 0.368}$ hence $b = 14.77$ and we take as a start value $b = 15$.

```
> x <- c(1,2,3,4,5,6,7,8,9,10,11,12)
> y <- c(2.02,3.62,5.71,7.13,8.33,8.29,9.81,11.3,12.18,12.67,12.62,13.01)
> model <- nls(y ~ a/(1+b*exp(c*x)),start=list(a=13,b=15,c= -1))
> model
Nonlinear regression model
  model: y ~ a/(1 + b * exp(c * x))
   data: parent.frame()
      a       b       c
13.4470  5.8508 -0.4239
 residual sum-of-squares: 2.905

Number of iterations to convergence: 7
Achieved convergence tolerance: 1.518e-06
```

The estimate of the logistic function is $\hat{y} = \frac{13.447}{1+5.8508e^{-0.4239x}}$.

Now we calculate the estimate for the asymptotic covariance matrix of the estimators.

```
> a <- 13.4470
> b <- 5.8508
> c <- -0.4239
> z <- c(1/(1+b*exp(c*x)))
> aa <- c(z^2)
```

```
> A <- sum(aa)
> bb <-c((exp(c*x)*z^3))
> B <- sum(bb)
> cc <- c((z^3*x*exp(c*x)))
> C <- sum(cc)
> dd <-  c((z^4*exp(2*c*x)))
> D <- sum(dd)
> ee <- c((z^4*x*exp(2*c*x)))
> E <- sum(ee)
> gg <- c((z^4*x^2*exp(2*c*x)))
> G <- sum(gg)
> P1 <- A*D*G
> P2 <- 2*B*C*E
> P3 <- (C^2)*D
> P4 <- A*(E^2)
> P5 <- (B^2)*G
> Delta <- P1 + P2 - P3 - P4 - P5
> Delta
[1] 0.000582551
> SSRes <- 2.905
> n <- length(y)
> df <- n-3
> sn <- sqrt(SSRes/df)
> sn
[1] 0.5681354
> SEa <- (sn*sqrt((D*G-E^2))/sqrt(Delta))
> SEa
[1] 0.05991013
> SEb <- (sn*sqrt((A*G-C^2)/(a^2))/sqrt(Delta))
> SEb
[1] 0.102548
> SEc <- sn*sqrt((A*D-B^2)/((a^2)*(b^2)))/sqrt(Delta)
> SEc
[1] 8.531211e-05
```

Problem 8.26 Estimate the asymptotic covariance matrix for this example.

Solution
Use the estimates of the parameters and insert them into (8.56) and multiply this by SS_{res}/df.

Example
We continue with the R program of Problem 8.25.

```
> cova_a <- SEa^2
> cova_a
[1] 0.003589223
```

```
> covb_b <- SEb^2
> covb_b
[1] 0.0105161
> covc_c <- SEc^2
> covc_c
[1] 7.278157e-09
> sn2 <-sn^2
> sn2
[1] 0.3227778
> cova_b <- -(sn2*(E*C - B*G))/(a*Delta)
> cova_b
[1] 0.004791069
> cova_c <- -(sn2*(B*E - C*D))/(a*b*Delta)
> cova_c
[1] -2.260269e-08
> covb_c <- (sn2*(B*C - A*E))/(a^2*b*Delta)
> covb_c
[1] 1.259186e-08
```

This means the asymptotic covariance matrix is

$$\begin{pmatrix} 0.003589223 & 0.004791069 & -2.260269\text{e}-08 \\ 0.004791069 & 0.0105161 & \\ -2.260269\text{e}-08 & 1.259186\text{e}-08 & 7.278157\text{e}-09 \end{pmatrix}$$

Problem 8.27 Construct $(1-\alpha^*)$ confidence intervals for all three parameters of $f_L(x)$.

Solution

Confidence intervals with a nominal confidence coefficient $1-\alpha^*_{\text{nom}}$ are defined as follows.

Parameter α:

$$\left[\boldsymbol{a} - s_n\sqrt{(\hat{\boldsymbol{c}}_{22}\hat{\boldsymbol{c}}_{33} - \hat{\boldsymbol{c}}_{23}^2)}\, t\left(n-3\left|1-\frac{\alpha^*_{\text{nom}}}{2}\right.\right),\right.$$
$$\left.\boldsymbol{a} + s_n\sqrt{(\hat{\boldsymbol{c}}_{22}\hat{\boldsymbol{c}}_{33} - \hat{\boldsymbol{c}}_{23}^2)}\, t\left(n-3\left|1-\frac{\alpha^*_{\text{nom}}}{2}\right.\right)\right]. \tag{8.61}$$

Parameter β:

$$\left[\boldsymbol{b} - s_n\sqrt{(\hat{\boldsymbol{c}}_{11}\hat{\boldsymbol{c}}_{33} - \hat{\boldsymbol{c}}_{13}^2)}\, t\left(n-3\left|1-\frac{\alpha^*_{\text{nom}}}{2}\right.\right),\right.$$
$$\left[\boldsymbol{b} + s_n\sqrt{(\hat{\boldsymbol{c}}_{11}\hat{\boldsymbol{c}}_{33} - \hat{\boldsymbol{c}}_{13}^2)}\, t\left(n-3\left|1-\frac{\alpha^*_{\text{nom}}}{2}\right.\right)\right]. \tag{8.62}$$

Parameter γ:

$$\left[c - s_n\sqrt{\hat{c}_{11}\hat{c}_{22} - \hat{c}_{12}^2}\, t\left(n-3\left|1-\frac{\alpha_{\text{nom}}^*}{2}\right.\right),\right.$$
$$\left. c + s_n\sqrt{\hat{c}_{11}\hat{c}_{22} - \hat{c}_{12}^2}\, t\left(n-3\left|1-\frac{\alpha_{\text{nom}}^*}{2}\right.\right)\right]. \tag{8.63}$$

Example

We use the data from Example 8.6 and construct 0.95-confidence intervals for all parameters *a, b* and *c*.

```
> n <- length(y)
> df <- n-3
>  t_value <- qt(0.025, df, lower.tail = FALSE)
> t_value
[1] 2.262157
>  Qa <- SEa* t_value
>  lowerCLa <- a-Qa
>  lowerCLa
[1] 12.19065
> upperCLa <- a+Qa
> upperCLa
[1] 14.70335
>  Qb <- SEb*t_value
>  lowerCLb <- b-Qb
>  lowerCLb
[1] 3.55785
> upperCLb<- b+Qb
> upperCLb
[1] 8.14375
>  Qc <- SEc*t_value
>  lowerCLc <- c-Qc
>  lowerCLc
[1] -0.5411389
> upperCLc<- c+Qc
> upperCLc
[1] -0.3066611
```

The 0.95-confidence interval for α is: [12.19065; 14.70335].

The 0.95-confidence interval for β is: [3.55785; 8.14375].

The 0.95-confidence interval for γ is: [−0.5411389; −0.3066611].

In simulation experiments as described in Section 8.2.2.1 for 15 (α, β, γ) combinations, x_i values in [0, 65], normally distributed $\boldsymbol{e}_i$ and $\alpha_{\text{nom}} = 0.05$ and 0.1 were performed. For all parameter combinations, the result was that tests and confidence estimations based on the asymptotic covariance matrix already is recommended for $n > 3$.

8.2.2.5 The Bertalanffy Function

The regression function $f_B(x)$ of the model

$$y_i = (\alpha + \beta e^{\gamma x_i})^3 + e_i = f_B(x_i) + e_i, \quad i = 1, \dots, n, \quad n > 3, \tag{8.64}$$

is called the Bertalanffy function and was used by Bertalanffy (1929) to describe the growth of the body weight of animals. This function has two inflection points if α and β have different signs and are located at

$$x_{I1} = \frac{1}{\gamma}\ln\left(-\frac{\alpha}{\beta}\right) \quad \text{and} \quad x_{I2} = \frac{1}{\gamma}\ln\left(-\frac{\alpha}{3\beta}\right)$$

respectively

$$\text{with} \quad f_B(x_{I1}) = 0 \quad \text{and} \quad f_B(x_{I2}) = \left(\frac{2}{3}\alpha\right)^3.$$

Problem 8.28 Show how the parameters of $f_B(x)$ can be estimated.

Solution

We use the R-command `nls()` to minimise

$$SS = \sum_{i=1}^{n} (y_i - [\alpha + \beta e^{\gamma x_i}]^3)^2 \text{ directly.}$$

Example

We use the oil palm data of Example 8.6 and fit a Bertalanffy regression function. To obtain initial value a,b,c for the parameters we consider $[\alpha + \beta e^0]^3 = [\alpha + \beta]^3$ and use for $[a + b]^3$ the smallest y value. Because $\lim_{x\to\infty}[\alpha + \beta e^{\gamma x}]^3 = \alpha^3$ we use $a^3 = y_{max}$ and as an initial value for a we take $(y_{max})^{1/3}$. Further we use after several trials $c = -0.2$.

From the data we take as the largest value 13.01, hence the start value for $a^3 = 13$, hence $a \approx 2.3513$ and we take $a = 2.4$. For $x = 1$ we obtain $2.02 = [a + be^c]^3 = (13 + b\cdot 0.819)^3$ hence $b \approx 0.903$ and we take as start value $b = 1$.

```
> x <- c(1,2,3,4,5,6,7,8,9,10,11,12)
> y <- c(2.02,3.62,5.71,7.13,8.33,8.29,9.81,11.3,12.18,12.67,12.62,13.01)
> model <- nls(y ~ (a+b*exp(c*x))^3,start=list(a= 2.4,b=1,c= -0.2))
> model
Nonlinear regression model
  model: y ~ (a + b * exp(c * x))^3
   data: parent.frame()
      a       b       c
 2.4435 -1.4307 -0.2373
 residual sum-of-squares: 1.963
Number of iterations to convergence: 16
Achieved convergence tolerance: 1.493e-06
```

The estimated regression function is $[2.4435 - 1.4307e^{-0.2373x}]^3$.

In Figure 8.9 we find the graph of the fitted regression function.

With $\theta = (\theta_1, \theta_2, \theta_3)^T = (\alpha, \beta, \gamma)^T$ and with the notation of Definition 8.1 we obtain

$$F(x_i) = (3(\alpha + \beta e^{\gamma x_i})^2 \quad 3(\alpha + \beta e^{\gamma x_i})^2 e^{\gamma x_i} \quad 3\beta x_i(\alpha + \beta e^{\gamma x_i})^2 e^{\gamma x_i}).$$

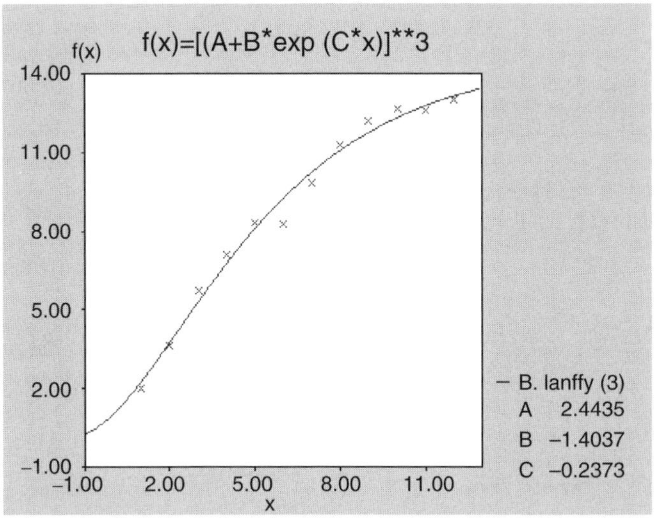

Figure 8.9 Fitted regression function of the example in Problem 8.28.

Again we use for writing

$$F^T(x_i)F(x_i) = 9\begin{pmatrix} c_{11} & c_{12} & c_{13} \\ c_{12} & c_{22} & c_{23} \\ c_{13} & c_{23} & c_{33} \end{pmatrix}$$

abbreviations like:

$$z_i = (\alpha + \beta e^{\gamma x_i})^4,$$

$$c_{11} = \sum_{i=1}^{n} z_i, c_{12} = \sum_{i=1}^{n} z_i e^{\gamma x_i}, c_{13} = \beta \sum_{i=1}^{n} x_i z_i e^{\gamma x_i}, c_{22} = \sum_{i=1}^{n} z_i e^{2\gamma x_i} \sum_{i=1}^{n} x_i z_i e^{2\gamma x_i}$$

$$c_{23} = \beta \sum_{i=1}^{n} x_i z_i e^{2\gamma x_i}, c_{33} = \beta^2 \sum_{i=1}^{n} x_i^2 z_i e^{2\gamma x_i}.$$

Now $|F^TF| = 9^3\{c_{11}(c_{22}c_{33} - c_{23}{}^2) - c_{12}(c_{12}c_{33} - c_{23}c_{13}) + c_{13}(c_{12}c_{23} - c_{22}c_{13})\} = 9^3\Delta$. The asymptotic covariance matrix is therefore

$$\text{var}_{\text{as}}(\hat{\beta}) = \sigma^2\left(\sum_{i=1}^{n} F_i{}^T F_i\right)^{-1}$$

$$= \frac{\sigma^2}{9\Delta}\begin{pmatrix} (c_{22}c_{33} - c_{23}^2) & (c_{13}c_{23} - c_{12}c_{33}) & (c_{12}c_{23} - c_{13}c_{22}) \\ (c_{13}c_{23} - c_{12}c_{33}) & (c_{11}c_{33} - c_{13}^2) & (c_{12}c_{13} - c_{11}c_{23}) \\ (c_{12}c_{23} - c_{13}c_{22}) & (c_{12}c_{13} - c_{11}c_{23}) & c_{11}c_{22} - c_{12}^2 \end{pmatrix}. \qquad (8.65)$$

We test

$$H_{0a} : a = a_0 \quad \text{against} \quad H_{Aa} : a \neq a_0$$

with the test statistic

$$t_\alpha = \frac{(a - \alpha_0)3\sqrt{\hat{\Delta}}}{s_n\sqrt{(\hat{c}_{22}\hat{c}_{33} - \hat{c}_{23}^2)}}. \tag{8.66}$$

Further

$$H_{0\beta} : \beta = \beta_0 \quad \text{against} \quad H_{A\beta} : \beta \neq \beta_0$$

is tested with

$$t_\beta = \frac{(b - \beta_0)3\sqrt{\hat{\Delta}}}{s_n\sqrt{(\hat{c}_{11}\hat{c}_{33} - \hat{c}_{13}^2)}} \tag{8.67}$$

and

$$H_{0\gamma} : \gamma = \gamma_0 \quad \text{against} \quad H_{A\gamma} : \gamma \neq \gamma_0$$

$$\text{with } t_\gamma = \frac{(c - c_0)3\sqrt{\hat{\Delta}}}{s_n\sqrt{\hat{c}_{11}\hat{c}_{22} - \hat{c}_{12}^2}}. \tag{8.68}$$

The $\hat{c}_{ij}, i, j = 1, ..3$ are obtained from the c_{ij} by replacing all parameters by their estimators. Further is

$$s_n^2 = \frac{1}{n-3}\sum_{i=1}^{n}[y_i - (a + be^{cx_i})^3]^2. \tag{8.69}$$

with approximately $df = n - 3$.

Problem 8.29 Test the hypotheses with significance level $\alpha^* = 0.05$ for each of the three parameters of $f_B(x)$ using R. The null-hypotheses values are respectively $\alpha_0 = 3$, $\beta_0 = -1$, and $\gamma_0 = -0.1$!

Solution

Using the estimates *a, b,* and c we calculate the estimates for c_{ij} and insert the values in (8.66)–(8.68).

Example

We use the oil palm data from Example 8.6 and we continue with the R commands of Problem 8.28 (Figure 8.10).

```
> a <- 2.4435
> b <-  -1.4307
> c <- -0.2373
> z <- c(( a + b*exp(c*x))^4)
> A <- sum(z)
> bb <- c(z*exp(c*x))
> B <- sum(bb)
> cc <- c(x*z*exp(c*x))
```

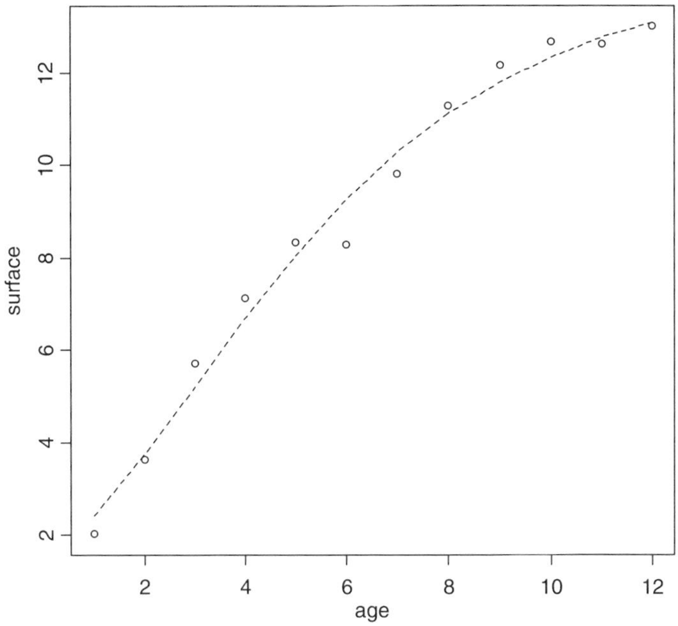

Figure 8.10 Fitted regression function of the example in Problem 8.32.

```
> C <- sum(cc)
> dd <- c(z*exp(2*c*x))
> D <- sum(dd)
> ee <- c(z*x*exp(2*c*x))
> E <- sum(ee)
> gg <- c(z*x^2*exp(2*c*x))
> G <- sum(gg)
> Delta <- A*D*G + 2*B*C*E -(C^2)*D - (E^2)*A -(B^2)*G
> Delta
[1] 13926.38
> n <- length(y)
> SSRes <- 1.963
> df <- n-3
> sn <- sqrt(SSRes/df)
> sn
[1] 0.4670237
> parta <- D*G-E^2
> SEa <- sn*sqrt((parta))/(sqrt(9)*sqrt(Delta))
> SEa
[1] 0.0434715
> partb <- A*G-C^2
> SEb <- sn*sqrt((partb))/(sqrt(9)*sqrt(Delta))
> SEb
[1] 0.08784467
```

```
> partc <- (A*D-B^2)/(b^2)
> SEc  <- sn*sqrt((partc))/(sqrt(9)*sqrt(Delta))
> SEc
[1] 0.03167163
> ta <- (a - 3)/SEa
> ta
> ta
[1] -12.80149
> p_valuea <- 2*(1-pt(abs(ta),df))
> p_valuea
[1] 4.431017e-07
> tb <- (b-(-1))/SEb
> tb
[1] -4.902973
> p_valueb <- 2*(1-pt(abs(tb),df))
> p_valueb
[1] 0.0008441792
> tc <- (c-(-0.1))/SEc
> tc
[1] -4.33511
> p_valuec <- 2*(1-pt(abs(tc),df))
> p_valuec
[1] 0.001890926
```

Because the *p*-value for t_a is 4.431017e-07 < 0.05 H_{0a} is rejected.
Because the *p*-value for t_b is 0.0008441792 < 0.05 H_{0b} is rejected.
Because the *p*-value for t_c is 0.001890926 < 0.05 H_{0c} is rejected.

Confidence intervals with a nominal confidence coefficient $1-\alpha^*_{nom}$ are defined as follows:

Parameter α:

$$\left[\boldsymbol{a} - \frac{\boldsymbol{s}_n}{3\sqrt{\Delta}}\sqrt{(\hat{\boldsymbol{c}}_{22}\hat{\boldsymbol{c}}_{33} - \hat{\boldsymbol{c}}_{23}^2)}\, t\left(n-3\left|1-\frac{\alpha^*_{\text{nom}}}{2}\right.\right),\right.$$
$$\left.\boldsymbol{a} + \frac{\boldsymbol{s}_n}{3\sqrt{\Delta}}\sqrt{(\hat{\boldsymbol{c}}_{22}\hat{\boldsymbol{c}}_{33} - \hat{\boldsymbol{c}}_{23}^2)}\, t\left(n-3\left|1-\frac{\alpha^*_{nom}}{2}\right.\right)\right] \tag{8.70}$$

Parameter β:

$$\left[\boldsymbol{b} - \frac{\boldsymbol{s}_n}{3\sqrt{\Delta}}\sqrt{(\hat{\boldsymbol{c}}_{11}\hat{\boldsymbol{c}}_{33} - \hat{\boldsymbol{c}}_{13}^2)}\, t\left(n-3\left|1-\frac{\alpha^*_{nom}}{2}\right.\right),\right.$$
$$\left.\boldsymbol{b} + \frac{\boldsymbol{s}_n}{3\sqrt{\Delta}}\sqrt{(\hat{\boldsymbol{c}}_{11}\hat{\boldsymbol{c}}_{33} - \hat{\boldsymbol{c}}_{13}^2)}\, t\left(n-3\left|1-\frac{\alpha^*_{\text{nom}}}{2}\right.\right)\right] \tag{8.71}$$

Parameter γ.

$$\left[c - \frac{s_n}{3\sqrt{\Delta}}\sqrt{\hat{c}_{11}\hat{c}_{22} - \hat{c}_{12}^2}\, t\left(n-3\left|1 - \frac{\alpha^*_{\text{nom}}}{2}\right.\right),\right.$$
$$\left. c + \frac{s_n}{3\sqrt{\Delta}}\sqrt{\hat{c}_{11}\hat{c}_{22} - \hat{c}_{12}^2}\, t\left(n-3\left|1 - \frac{\alpha^*_{\text{nom}}}{2}\right.\right)\right] \tag{8.72}$$

Problem 8.30 Construct $(1-\alpha^*)$-confidence intervals for all three parameters of $f_B(x)$.

Solution

Continuing with the R program of Problems 8.28 and 8.29 we insert the SEa into (8.70), the SEb into (8.71) and SEc into (8.72).

Example

We use the data from Example 8.6 and construct 0.95- confidence intervals for all parameters.

```
> t_value <- qt(0.025, df, lower.tail = FALSE)
> t_value
[1] 2.262157
>  Qa <- SEa*t_value
>  lowerCLa <- a-Qa
>  lowerCLa
[1] 2.345161
> upperCLa <- a+Qa
> upperCLa
[1] 2.541839
>  Qb <- SEb*t_value
>  lowerCLb <- b-Qb
>  lowerCLb
[1] -1.629418
> upperCLb<- b+Qb
> upperCLb
[1] -1.231982
>  Qc <- SEc*t_value
>  lowerCLc <- c-Qc
>  lowerCLc
[1] -0.3089462
> upperCLc<- c+Qc
> upperCLc
[1] -0.1656538
```

The 0.95-confidence interval for α is: [2.345161; 2.541839].
The 0.95-confidence interval for β is: [−1.629418; −1.231982].
The 0.95-confidence interval for $\gamma$is: [−0.3089462; −0.1656538].

Schlettwein (1987) did the simulation experiments described in Section 8.2.2.1 with normally distributed $\boldsymbol{e}_i$ and for several parameter combinations and n-values. They led to the conclusion that for normally distributed $\boldsymbol{e}_i$ in model (8.64) the asymptotic tests and confidence intervals for all $n \geq 4$ are appropriate.

8.2.2.6 The Gompertz Function

The regression function $f_G(x, \theta)$ of the model

$$\boldsymbol{y}_i = \alpha \cdot e^{\beta \cdot e^{\gamma x_i}} + \boldsymbol{e}_i = f_G(x, \theta) + \boldsymbol{e}_i, i = 1, \dots, n > 3, \alpha \neq 0, \gamma \neq 0, \beta < 0 \tag{8.73}$$

is the Gompertz-Function. In Gompertz (1825) it was used to describe the population growth. The function has an inflection point at

$$x_{\text{infl}} = -\frac{\ln(-\beta)}{\gamma} \quad \text{with} \quad f_G(x_{\text{infl}}, \theta) = \alpha \cdot e.$$

Problem 8.31 Show how the parameters of $f_G(x)$ can be estimated with the least squares method.

Solution

We use the R-command `nls()` to minimise

$$SS = \sum_{i=1}^{n} (y_i - \alpha e^{\beta e^{\gamma x_i}})^2 \text{ directly.}$$

Example

We use the oil palm data of Example 8.6 and fit a Gompertz regression function. Analogously to Section 8.2.2.5 we find for $x = 0$ that $ae^b = y_{\min}$ and for $\gamma < 0$ that $a = y_{\max} \approx 13$. We try several values for c and use $c = -0.1$. For $x = 1$ we have $y = 2.02 = 13 \cdot e^{b \cdot 0.905}$ hence $b \cdot 0.905 = \ln(2.02/13)$ and $b = -2.06$, let us use $b = -2$.

```
> x <- c(1,2,3,4,5,6,7,8,9,10,11,12)
>  y <- c(2.02,3.62,5.71,7.13,8.33,8.29,9.81,11.3,12.18,12.67,12.62,13.01)
> model <- nls(y ~ a*exp(b*exp(c*x)),start=list(a= 13,b=-2,c=-0.1))
> model
Nonlinear regression model
  model: y ~ a * exp(b * exp(c * x))
   data: parent.frame()
     a      b      c
14.144 -2.347 -0.285
 residual sum-of-squares: 2.16
Number of iterations to convergence: 8
Achieved convergence tolerance: 5.094e-06
```

The fitted Gompertz function is $\hat{y} = 14.144e^{-2.347e^{-0.235x}}$.

With the notation of Definition 8.1 we obtain and by this all components of θ are non-linearity parameters. Again we use for writing

$$F^T F = \begin{pmatrix} c_{11} & c_{12} & c_{13} \\ c_{12} & c_{22} & c_{23} \\ c_{13} & c_{23} & c_{33} \end{pmatrix}$$

$$c_{11} = \sum_{i=1}^{n} e^{2\beta\gamma x_i}, c_{12} = \alpha \sum_{i=1}^{n} e^{2\beta\gamma x_i} e^{\gamma x_i}, c_{13} = \alpha\beta \sum_{i=1}^{n} x_i e^{2\beta\gamma x_i} e^{\gamma x_i}, \tag{8.74}$$

$$c_{22} = \alpha^2 \sum_{i=1}^{n} e^{2\beta\gamma x_i} e^{2\gamma x_i} \sum_{i=1}^{n} x_i z_i e^{2\gamma x_i} c_{23} = \alpha^2 \beta \sum_{i=1}^{n} x_i e^{2\beta\gamma x_i} e^{2\gamma x_i}, c_{33} = \alpha^2 \beta^2 \sum_{i=1}^{n} x_i^2 e^{2\beta\gamma x_i} e^{2\gamma x_i}.$$

Now

$$|F^T F| = \{c_{11}(c_{22}c_{33} - c_{23}{}^2) - c_{12}(c_{12}c_{33} - c_{23}c_{13}) + c_{13}(c_{12}c_{23} - c_{22}c_{13})\} = \Delta.$$

The asymptotic covariance matrix is given by

$$\text{var}_{\text{as}}(\hat{\beta}) = \sigma^2 \left(\sum_{i=1}^{n} F_i^T F_i \right)^{-1}$$

$$= \frac{\sigma^2}{\Delta} \begin{pmatrix} (c_{22}c_{33} - c_{23}^2) & -(c_{13}c_{23} - c_{12}c_{33}) & (c_{12}c_{23} - c_{13}c_{22}) \\ -(c_{13}c_{23} - c_{12}c_{33}) & (c_{11}c_{33} - c_{13}^2) & (c_{12}c_{13} - c_{11}c_{23}) \\ (c_{12}c_{23} - c_{13}c_{22}) & (c_{12}c_{13} - c_{11}c_{23}) & c_{11}c_{22} - c_{12}^2 \end{pmatrix}.$$

We test the three hypotheses using (8.66)–(8.68).

The $\hat{c}_{ij}$, $i, j = 1, ..3$ are obtained from the c_{ij} by replacing all parameters by their estimators. Further

$$s_n^2 = \frac{1}{n-3} \sum_{i=1}^{n} [y_i - a e^{b e^{c x_i}}]^2. \tag{8.75}$$

With approximately $df = n - 3$.

Problem 8.32 Test the hypotheses with significance level $\alpha^* = 0.05$ for each of the three parameters of $f_G(x)$ with R. The null-hypothese values are respectively $\alpha_0 = 15$, $\beta_0 = -2$, and $\gamma_0 = -0.3$.

Solution

Using the estimates *a*, *b* and c we calculate the estimates for c_{ij} and insert the values in (8.66)–(8.68).

Example

We use the oil palm data from Example 8.3 and continue the R commands of Problem 8.31.

```
> a <-  14.144
> b <- -2.347
> c <- -0.285
> aa <- c(exp(2*b*exp(c*x)))
> A <- sum(aa)
> bb <- c(exp(c*x)*exp(2*b*exp(c*x)))
> B <- sum(bb)
> cc <- c(x*exp(c*x)*exp(2*b*exp(c*x)))
> C <- sum(cc)
> dd <- c(exp(2*c*x)*exp(2*b*exp(c*x)))
> D <- sum(dd)
> ee <- c(x*exp(2*c*x)*exp(2*b*exp(c*x)))
> E <- sum(ee)
> gg <- c(x^2*exp(2*c*x)*exp(2*b*exp(c*x)))
> G <- sum(gg)
> Delta <- A*D*G + 2*B*C*E -(C^2)*D - (E^2)*A -(B^2)*G
```

```
> Delta
[1] 0.06390853
>  n <- length(y)
> df <- n-3
>  SSRes <-  2.16
> sn <- sqrt(SSRes/df)
> sn
[1] 0.4898979
> SEa <- sn*sqrt(D*G - E^2)/sqrt(Delta)
> SEa
[1] 0.6790962
> SEb <- sn*sqrt((A*G-C^2)/a^2)/sqrt(Delta)
> SEb
[1] 0.2136496
> SEc <- sn*sqrt((A*D-B^2)/(a^2*b^2))/sqrt(Delta)
> SEc
[1] 0.03601059
> ta <- (a-15)/SEa
> ta
[1] -1.260499
>  p_valuea <- 2*(1-pt(abs(ta),df))
>  p_valuea
[1] 0.2391875
> tb <- (b-(-2))/SEb
> tb
[1] -1.624155
> p_valueb  <- 2*(1-pt(abs(tb),df))
> p_valueb
[1] 0.1387897
> tc <- (c-(-0.3))/SEc
> tc
[1] 0.4165442
> p_valuec <- 2*(1-pt(abs(tc),df))
> p_valuec
[1] 0.6867704
> age <- c(1,2,3,4,5,6,7,8,9,10,11,12)
> surface <-
c(2.02,3.62,5.71,7.13,8.33,8.29,9.81,11.3,12.18,12.67,12.62,13.01)
> plot(age,surface, main="Example 8.6 with fitted Gompertz function")
> lines(fittedsurf, lty = 2)
```

Problem 8.33 Construct $(1-\alpha^*)$-confidence intervals for all three parameters of $f_G(x)$.

Solution

Continuation with the R program of Problem 8.31, 8.32, and use the SEa, the SEb and SEc for the confidence intervals .

Example

We use the data from Example 8.3 and construct 0.95 – confidence intervals for all parameters

```
>   t_value <- qt(0.025, df, lower.tail = FALSE)
> t_value
[1] 2.262157
>  Qa <- SEa*t_value
>  lowerCLa <- a-Qa
```

```
>  lowerCLa
[1] 12.60778
>  upperCLa <- a+Qa
>  upperCLa
[1] 15.68022
> Qb <- SEb*t_value
> lowerCLb <- b-Qb
> lowerCLb
[1] -2.830309
> upperCLb<- b+Qb
> upperCLb
[1] -1.863691
>  Qc <- SEc*t_value
>  lowerCLc <- c-Qc
>  lowerCLc
[1] -0.3664616
> upperCLc<- c+Qc
> upperCLc
[1] -0.2035384
```

The 0.95-confidence interval for α is: [12.60778; 15.68022].
The 0.95-confidence interval for β is: [−2.830309; −1.863691].
The 0.95-confidence interval for γ is: [−0.3664616; −0.2035384].

In our simulation experiments described in Section 8.2.2.1 with normally distributed $\boldsymbol{e}_i$ and for several parameter combinations and n-values we concluded that for normally distributed $\boldsymbol{e}_i$ in model (8.73) the asymptotic tests and confidence intervals for all $n \geq 4$ are appropriate.

8.2.3 Optimal Experimental Designs

In a model I of regression, we determine the values of the predictor variable in advance. Clearly, we would wish to do this so that a given precision can be reached using a minimal number of observations. In this context, any selection of the x-values is called an experimental design. The value of the precision requirement is called the optimality criterion, and an experimental design that minimises the criterion for a given sample size is called an optimal experimental design.

Most optimal experimental designs are concerned with estimation criteria. Thus, the x_i values are often chosen in such a way that the variance of an estimator is minimised (for a given number of measurements) amongst all possible allocations of x_i values for linear or quasilinear regression functions. In the case of an intrinsically non-linear regression function, the asymptotic variance of an estimator is minimised.

First we have to define the so-called experimental region (x_l, x_u) in which measurements can or will be taken. Here x_l is the lower bound and x_u the upper bound of an interval on the x-axis.

A big disadvantage of optimal designs is their dependence on the accuracy of the chosen model, they are only certain to be optimal if the assumed model is correct.

8.2.3.1 Simple Linear and Quasilinear Regression

Definition 8.3 A matrix

$$V_{n,m} = \begin{pmatrix} x_1, \dots, x_m \\ n_1, \dots, n_m \end{pmatrix}, x_i \in [x_l, x_o], n_i \text{ integer}, \sum_{i=1}^{m} n_i = n$$

is called an exact m-point-design or a m-point-design for short with the support $S_m = (x_1, \dots, x_m)$ and the allocation $N_m = (n_1, n_2, \dots, n_m)$. The interval $[x_l, x_o]$ is the experimental region.

For the simple linear regression, the following criteria are used:

- $\text{var}(\boldsymbol{b_0})$ – (C_1)-optimality: minimise $\text{var}(\boldsymbol{b_0})$ in (8.11)
- $\text{var}(\boldsymbol{b_1})$ – (C_2)-optimality: minimise $\text{var}(\boldsymbol{b_1})$ in (8.12)
- D-optimality: minimise the determinant D in (8.19) of the covariance matrix (8.14)
- G-optimality: minimise the maximum [over (x_l, x_u)] of the expected width of the confidence interval (8.24).

To minimise $\text{var}(\boldsymbol{b_1})$ in (8.12) we have to maximise SS_x in its denominator and at the same time this minimises D in (8.19). The two criteria give the same optimal design solution and also minimise the expected width of the confidence interval for β_1 in (8.22).

In the case of normally distributed errors, we interpret the D-optimality criterion as follows. If we construct a confidence region for the vector of coefficients in the simple linear regression model, we get an ellipse. The D-optimal experimental design minimises the volume of this ellipse, among all such ellipses arising from any design having the same number of observations. We restrict ourselves in this section to the D-optimality criterion, and thus also to $\text{var}(\boldsymbol{b})$-optimality and on the G-optimality.

If the experimental region is the interval (x_l, x_u) and if n is even ($n = 2r$), then the D-optimal and the G-optimal experimental designs are identical: – we take r measurements at each of the two boundaries of the region.

Thus, these optimal (exact) designs take the form

$$\begin{pmatrix} x_l & x_u \\ r & r \end{pmatrix}.$$

If n is odd, with $n = 2r + 1$, then there are two D-optimal designs with r measurements at one boundary and $r + 1$ measurements at the other. Therefore, the D-optimal designs are given by

$$\begin{pmatrix} x_l & x_u \\ r & r+1 \end{pmatrix} \text{ or } \begin{pmatrix} x_l & x_u \\ r+1 & r \end{pmatrix}$$

respectively.

The (unique) G-optimal design for $n = 2r + 1$ requires r readings at both ends of the interval and one at the interval mid-point. Therefore, this design has the form

$$\begin{pmatrix} x_l & \frac{1}{2}(x_l + x_u) & x_u \\ r & 1 & r \end{pmatrix}.$$

We will now determine the D- and G-optimal designs for the carotene Example 8.4.

Problem 8.34 Determine the exact D- and G- optimal design in $[x_l = 1, x_u = 303]$ with $n = 5$.

Solution

We see that

$$\begin{pmatrix} 1 & 303 \\ 2 & 3 \end{pmatrix} \text{ and } \begin{pmatrix} 1 & 303 \\ 3 & 2 \end{pmatrix}$$

are *D*-optimal designs respectively. In both designs SS_x(optimal) = 109444.8 and the value of the *D*-criterion is $D = \frac{\sigma^4}{5 \cdot 1094448} = 1.8274 \cdot 10^{-7} \cdot \sigma^4$.

In general we would prefer the second design, in which we begin with three readings and take two readings at the end, because of the possibility those parts of the experimental material may become corrupted during the experiment.

The *G*-optimal design is

$$\begin{pmatrix} 1 & 152 & 303 \\ 2 & 1 & 2 \end{pmatrix}$$

This design is not *D*-optimal. After obtaining SS_x (*G*-optimal) = 91204 we find a larger *D*-value $D = 2.1929 \cdot 10^{-6} \cdot \sigma^4$. However, this design minimises the maximum of the expected width

$$E\left[2t\left(n-2\ ;\ 1-\frac{\alpha^*}{2}\right)sK_x\right] \approx 2\sigma t\left(n-2\ ;\ 1-\frac{\alpha}{2}\right)K_x$$

of (8.24). This means that it minimises the maximum of $K_x = \sqrt{\frac{1}{n} + \frac{(x-\bar{x})^2}{SS_x}}$. K_x takes its maximum at the border of the experimental region. In our case this is either at $x = 1$ or at $x = 303$.

If we insert $x = 1$ or $x = 303$ in the formula for K_x and because

$$\Sigma x_i^2 = 2 \cdot 1^2 + 152^2 + 2 \cdot 303^2 = 206724,$$

$$\Sigma x_i = 2 \cdot 1 + 152 + 2 \cdot 303 = 760 \text{ and}$$

SS_x (*G*-optimal) = 91204, we obtain $K_1 = \sqrt{\frac{1}{5} + \frac{(1-152)^2}{91204}} = 0.6708$ and $K_{303} = \sqrt{\frac{1}{5} + \frac{(303-152)^2}{91204}} = 0.6708$.

The maximum expected width of (8.24) of the design $\begin{pmatrix} 1 & 60 & 124 & 223 & 303 \\ 1 & 1 & 1 & 1 & 1 \end{pmatrix}$ in Example 8.4 is larger because its maximum is 0.797.

Its *D* value is $D = 3.3664 \cdot 10^{-6} \cdot \sigma^4$.

For polynomial quasilinear regression models, Pukelsheim (1993) gave *D*- and *G*-optimal designs, which we report in Table 8.10.

8.2.3.2 Intrinsically Non-linear Regression

For intrinsically non-linear regression models, the covariance matrix cannot be used for defining optimal design because it is unknown. We therefore use the asymptotic covariance matrix, but it unfortunately depends on at least one of the unknown parameters. Even if Rasch (1993) could show that the position of the support points often only slightly depends on θ we must check this for each special regression function separately. We therefore can find optimal designs only depending on at least one of the function parameters, we call such designs therefore locally optimal. We denote the part of θ which occurs in the formula of $\text{var}_{as}(\hat{\theta})$ in (8.33) by θ_0 and write θ_0 as an argument

Table 8.10 *D*- and *G*-optimal designs for polynomial regression for x ∈ [*a*, *b*] and *n* = *m*(*k* + 1).

k	**Optimal design**
1	$\begin{pmatrix} a & b \\ m & m \end{pmatrix}$
2	$\begin{pmatrix} a & \frac{a+b}{2} & b \\ m & m & m \end{pmatrix}$
3	$\begin{pmatrix} a & 0.7236a + 0.2764b & 0.2764a + 0.7236b & b \\ m & m & m & m \end{pmatrix}$
4	$\begin{pmatrix} a & 0.82735a + 0.17265b & \frac{a+b}{2} & 0.17265a + 0.82735b & b \\ m & m & m & m & m \end{pmatrix}$
5	$\begin{pmatrix} a & 0.88255a + 0.11745b & 0.6426a + 0.3574b & 0.3574a + 0.6426b & 0.11745a + 0.88255b & b \\ m & m & m & m & m & m \end{pmatrix}$

of the asymptotic covariance matrix as $\text{var}_{\text{as}}(\hat{\boldsymbol{\theta}}, \theta_0)$. At first, some general results are presented and later locally optimal designs for the functions in Section 8.2.3.1 are given.

The asymptotic covariance matrix $\text{var}_{\text{as}}(\hat{\boldsymbol{\theta}}, \theta_0)$ can now be written in dependency on θ_0 and $V_{n,m} \in V$ (V is the set of all admissible designs) as

$$\text{var}_{\text{as}}(\hat{\boldsymbol{\theta}}, \theta_0, V_{n,m})$$

and $V^*_{n,m}$ is called a locally *D*-, *G*- or *C*-optimal m-point design at $\theta = \theta_0$, if

$$H\{\text{var}_{\text{as}}(\hat{\boldsymbol{\theta}}, \theta_0, V^*_{n,m})\} = \min_{V_{n,m} \in V} H\{\text{var}_{\text{as}}(\hat{\boldsymbol{\theta}}, \theta_0, V_{n,m})\}. \tag{8.76}$$

Here H means any of the criteria *D*, *G* or *C*.

If V_m is the set of concrete m-point designs, then $V^*_{n,m}$ is called concrete locally H-optimal m-point design, if

$$H\{\text{var}_{\text{as}}(\hat{\boldsymbol{\theta}}, \theta_0, V^*_{n,m})\} = \min_{V_{n,m} \in V_m} H\{\text{var}_{\text{as}}(\hat{\boldsymbol{\theta}}, \theta_0, V_{n,m})\}. \tag{8.77}$$

For some regression functions and optimality criteria analytical solutions in closed form of the problems could be found. Otherwise, search methods are applied.

The first analytical solution we find in Box and Lucas 1959 in Theorem 8.1.

Theorem 8.1 (Box and Lucas (1959). For the regression function

$$f(x, \theta) = \alpha + \beta e^{\gamma x}$$

with n = 3, $\theta = (\alpha, \beta, \gamma)^T$ and $x \in [x_l, x_o]$ the locally *D*-optimal concrete design $V^*_{3,3}$ depends only on the component γ_0 of

$$\theta_0 = (\alpha_0, \beta_0, \gamma_0)^T$$

and has the form

$$V^*_{3,3} = \begin{pmatrix} x_l & x_2 & x_u \\ 1 & 1 & 1 \end{pmatrix}$$

with

$$x_2 = -\frac{1}{\gamma_0} + \frac{x_l e^{\gamma_0 x_l} - x_u e^{\gamma_0 x_u}}{e^{\gamma_0 x_l} - e^{\gamma_0 x_u}}. \tag{8.78}$$

Atkinson and Hunter (1968) gave sufficient and for $n = kp$ sufficient and necessary conditions for the function $f(x, \theta)$ that the support of a locally D-optimal design of size n is equal to the number of parameters of the regression function. These conditions are difficult to verify for $p > 2$.

Theorem 8.2 (Rasch 1990.) The support of a concrete locally D-optimal p-point design of size n is independent of n and the n_i of this design are as equal as possible (i.e. is $n = ap$, then $n_i = a$ otherwise the n_i differ maximal by 1).

In Rasch (1990) further theorems concerning the D-optimality can be found.

If $n > 2p$, the D-optimal concrete designs are approximately G-optimal in the sense that the value of the G-criterion for the concrete D-optimal p-point design even for $n \neq tp$ (t integer) does nearly not differ from that of the concrete. G-optimal design. For the functions in (8.6) we found optimal designs by search methods and for $n > p+2$ we often found p-point designs.

For all models and optimality criteria an equidistant design in the experimental region with one observation at each support point is far from being optimal.

We show this by examples for our special functions in Section 8.2.2.

8.2.3.3 The Michaelis-Menten Regression

The locally D-optimum design of the function $f_M(x) = \frac{\alpha x}{1+\beta x}$ in $[a,b]$ is for even $n = 2t$ given by (Ermakov and Zigljavski 1987):

$$\begin{pmatrix} a + \frac{b-a}{2+b\theta_2} & b \\ t & t \end{pmatrix}.$$

For odd $n = 2t+1$ we have two locally D-optimum designs namely $\begin{pmatrix} a + \frac{b-a}{2+b\theta_2} & b \\ t+1 & t \end{pmatrix}$ and $\begin{pmatrix} a + \frac{b-a}{2+b\theta_2} & b \\ t & t+1 \end{pmatrix}$ respectively.

In the interval [0, 1440], which was used in Table 8.6 we had $n = 10$ and by this $t = 5$. With the parameter estimated the locally D-optimum design is given as

$$\begin{pmatrix} 28.32 & 1440 \\ 5 & 5 \end{pmatrix}$$

with the criterium value 6.445. The original design of Michaelis and Menten in the form

$$\begin{pmatrix} 0 & 1 & 6 & 17 & 27 & 38 & 62 & 95 & 1372 & 1440 \\ 1 & 1 & 1 & 1 & 1 & 1 & 1 & 1 & 1 & 1 \end{pmatrix}$$

has the criterion value 16.48.

Rasch (2008) found by search procedures the locally C_1-optimal design $\begin{pmatrix} 21.55 & 1440 \\ 8 & 2 \end{pmatrix}$ and the locally C_2-optimal design $\begin{pmatrix} 20.29 & 1440 \\ 7 & 3 \end{pmatrix}$.

8.2.3.4 Exponential Regression

By constructing locally optimal designs in the class of m-point designs ($m \geq 3$) only such optimal designs have been found which are three-point designs as derived by Box and Lucas (1959) for $n = 3$.

Table 8.11 Locally optimal designs for the exponential regression.

Criterion	$(\alpha,\beta,\gamma) = (14,-18,-0.2)$	$(\alpha,\beta,\gamma) = (16.5,-16.7,-0.14)$	$(\alpha,\beta,\gamma) = (18,-17,-0.097)$
D	$\begin{pmatrix} 1 & 4.63 & 12 \\ 4 & 4 & 4 \end{pmatrix}$	$\begin{pmatrix} 1 & 5.14 & 12 \\ 4 & 4 & 4 \end{pmatrix}$	$\begin{pmatrix} 1 & 2.03 & 12 \\ 4 & 4 & 4 \end{pmatrix}$
C_1	$\begin{pmatrix} 1 & 4.46 & 12 \\ 2 & 4 & 6 \end{pmatrix}$	$\begin{pmatrix} 1 & 5.082 & 12 \\ 2 & 5 & 5 \end{pmatrix}$	$\begin{pmatrix} 1 & 1.22 & 12 \\ 1 & 1 & 10 \end{pmatrix}$
C_2	$\begin{pmatrix} 1.48 & 11.29 & 12 \\ 6 & 1 & 3 \end{pmatrix}$	$\begin{pmatrix} 3.87 & 11.59 & 12 \\ 6 & 2 & 4 \end{pmatrix}$	$\begin{pmatrix} 1 & 2.085 & 12 \\ 4 & 6 & 2 \end{pmatrix}$
C_3	$\begin{pmatrix} 1 & 4.48 & 12 \\ 3 & 6 & 3 \end{pmatrix}$	$\begin{pmatrix} 1 & 5.02 & 12 \\ 3 & 6 & 3 \end{pmatrix}$	$\begin{pmatrix} 1 & 2.05 & 12 \\ 2 & 6 & 4 \end{pmatrix}$

In Table 8.11 we list for rounded estimated parameter values found in the example of Problem 8.19 (oil palm data) and smaller and larger values of these parameters the locally D- and $C_i (i = 1, 2, 3)$-optimal designs.

We compare the criterion value $0.00015381\sigma^6$ of the D-optimal design $\begin{pmatrix} 1 & 5.14 & 12 \\ 4 & 4 & 4 \end{pmatrix}$ with the criterion value of the design $\begin{pmatrix} 1 & 2 & 3 & 4 & 5 & 6 & 7 & 8 & 9 & 10 & 11 & 12 \\ 1 & 1 & & 1 & 1 & 1 & 1 & 1 & 1 & 1 & 1 & 1 \end{pmatrix}$ used in the experiment, which is $0.0004782\sigma^6$.

8.2.3.5 The Logistic Regression

By constructing locally optimal designs in the class of m-point designs ($m \geq 3$) only three-point designs have been found.

In Table 8.12 we list for rounded estimated parameter values found in the example of Problem 8.23 (hemp data) and smaller and larger values of these parameters the locally D- and C_i ($i = 1, 2, 3$)-optimal designs.

Table 8.12 Locally optimal designs for the logistic regression.

Criterion	$(\alpha,\beta,\gamma) = (123,16,-0.5)$	$(\alpha,\beta,\gamma) = (126,20,-0.46)$	$((\alpha,\beta,\gamma) = (130,23,-0.42)$
D	$\begin{pmatrix} 3.30 & 7.39 & 14 \\ 5 & 5 & 4 \end{pmatrix}$	$\begin{pmatrix} 3.93 & 8.29 & 14 \\ 5 & 5 & 4 \end{pmatrix}$	$\begin{pmatrix} 4.42 & 8.00 & 14 \\ 5 & 5 & 4 \end{pmatrix}$
C_α	$\begin{pmatrix} 1.73 & 7.34 & 14 \\ 2 & 2 & 10 \end{pmatrix}$	$\begin{pmatrix} 1.15 & 8,07 & 14 \\ 2 & 3 & 9 \end{pmatrix}$	$\begin{pmatrix} 2.81 & 8.20 & 14 \\ 3 & 3 & 8 \end{pmatrix}$
C_β	$\begin{pmatrix} 2.16 & 2.37 & 14 \\ 11 & 1 & 2 \end{pmatrix}$	$\begin{pmatrix} 2.70 & 8.95 & 14 \\ 11 & 2 & 1 \end{pmatrix}$	$\begin{pmatrix} 3.11 & 3.46 & 8.75 \\ 11 & 1 & 2 \end{pmatrix}$
C_γ	$\begin{pmatrix} 1.81 & 7.37 & 14 \\ 8 & 4 & 2 \end{pmatrix}$	$\begin{pmatrix} 2.38 & 8.31 & 14 \\ 8 & 4 & 2 \end{pmatrix}$	$\begin{pmatrix} 2.80 & 8.12 & 14 \\ 8 & 4 & 2 \end{pmatrix}$

Table 8.13 Locally optimal designs for the Bertalanffy regression.

Criterion	$(\alpha,\beta,\gamma) = (2.35, -1.63, -0.31)$	$(\alpha,\beta,\gamma) = (2.44, -1.43, -0.24)$	$(\alpha,\beta,\gamma) = (2.54, -1.23, -0.17)$
D	$\begin{pmatrix} 1.19 & 5.05 & 12 \\ 4 & 4 & 4 \end{pmatrix}$	$\begin{pmatrix} 1 & 5.39 & 12 \\ 4 & 4 & 4 \end{pmatrix}$	$\begin{pmatrix} 1 & 5.88 & 12 \\ 4 & 4 & 4 \end{pmatrix}$
C_1	$\begin{pmatrix} 1 & 5.09 & 8.07 \\ 2 & 3 & 7 \end{pmatrix}$	$\begin{pmatrix} 1 & 5.55 & 12 \\ 2 & 4 & 6 \end{pmatrix}$	$\begin{pmatrix} 1 & 5.72 & 12 \\ 2 & 5 & 5 \end{pmatrix}$
C_2	$\begin{pmatrix} 1 & 6.97 & 8.07 \\ 10 & 1 & 1 \end{pmatrix}$	$\begin{pmatrix} 1 & 10.27 & 12 \\ 9 & 1 & 2 \end{pmatrix}$	$\begin{pmatrix} 1.94 & 2.69 & 12 \\ 4 & 4 & 4 \end{pmatrix}$
C_3	$\begin{pmatrix} 1 & 5.23 & 12 \\ 4 & 5 & 3 \end{pmatrix}$	$\begin{pmatrix} 1 & 5.54 & 12 \\ 4 & 5 & 3 \end{pmatrix}$	$\begin{pmatrix} 1 & 5.92 & 12 \\ 4 & 5 & 3 \end{pmatrix}$

Table 8.14 Locally optimal designs for the Gompertz regression.

Criterion	$(\beta,\gamma) = (-17,-0.14)$	$(\beta,\gamma) = (-19,-0.2)$	$(\beta,\gamma) = (-14,-0.08)$
D	$\begin{pmatrix} 1 & 5.14 & 12 \\ 4 & 4 & 4 \end{pmatrix}$	$\begin{pmatrix} 1 & 4.63 & 12 \\ 4 & 4 & 4 \end{pmatrix}$	$\begin{pmatrix} 1 & 5.7 & 12 \\ 4 & 4 & 4 \end{pmatrix}$
C_α	$\begin{pmatrix} 1 & 5.08 & 12 \\ 2 & 5 & 5 \end{pmatrix}$	$\begin{pmatrix} 1 & 4.61 & 12 \\ 2 & 4 & 6 \end{pmatrix}$	$\begin{pmatrix} 1 & 5.69 & 12 \\ 2 & 6 & 4 \end{pmatrix}$
C_β	$\begin{pmatrix} 3.87 & 11.59 & 12 \\ 6 & 2 & 4 \end{pmatrix}$	$\begin{pmatrix} 1.48 & 11.29 & 12 \\ 6 & 1 & 5 \end{pmatrix}$	$\begin{pmatrix} 1 & 5.44 & 12 \\ 2 & 6 & 4 \end{pmatrix}$
C_γ	$\begin{pmatrix} 1 & 5.02 & 12 \\ 3 & 6 & 3 \end{pmatrix}$	$\begin{pmatrix} 1 & 4.48 & 12 \\ 3 & 6 & 3 \end{pmatrix}$	$\begin{pmatrix} 1 & 5.63 & 12 \\ 3 & 6 & 3 \end{pmatrix}$

8.2.3.6 The Bertalanffy Function

By constructing locally optimal designs in the class of m-point designs ($m \geq 3$) only three-point designs have been found.

In Table 8.13 we list for rounded estimated parameter values found in the example of Problem 8.28 (oil palm data) and smaller and larger values of these parameters the locally D- and C_i ($i = 1, 2, 3$)-optimal designs.

8.2.3.7 The Gompertz Function

By constructing locally optimal designs in the class of m-point designs ($m \geq 3$) only three-point designs have been found.

In Table 8.14 we list for rounded estimated parameter values found in the example of Problem 8.31 and smaller and larger values of these parameters the locally D- and C_i ($i = 1, 2, 3$)-optimal designs.

8.3 Models with Random Regressors

We consider mainly the linear or quasilinear case for model II of the regression analysis where not only the regressand but also the regressor is a random variable. As mentioned at the top of this chapter we use the regression model

$$\boldsymbol{y}_i = f(\boldsymbol{x}_i, \beta) + \boldsymbol{e}_i, i = 1, \ldots, n$$

where the following assumptions must be fulfilled:

1. The error terms $\boldsymbol{e}_i$ are random variables with expectation zero.
2. The variance of the error terms $\boldsymbol{e}_i$ is equal for all observations.
3. For hypotheses testing and confidence estimation we assume that the $\boldsymbol{e}_i$ are $N(0, \sigma^2)$ distributed.
4. The error terms $\boldsymbol{e}_i$ are independent of the regressor variables $\boldsymbol{x}_i$.
5. $\text{cov}(\boldsymbol{x}_j, \boldsymbol{x}_k = 0, j \neq k), \text{var}(\boldsymbol{x}_i) = \sigma_x^2 \ \forall i$.

Intrinsically non-linear regression models with random regressors are more the exception than the rule and play a minor role. An exception is the relative growth of a part of a body to the whole body or another part, we speak about allometry and consider this in Section 8.3.2.

8.3.1 The Simple Linear Case

The model

$$\boldsymbol{y}_i = \beta_0 + \sum_{j=1}^{k} \beta_j \boldsymbol{x}_{ji} + \boldsymbol{e}_i, i = 1, \ldots n > k + 1 \tag{8.79}$$

with the additional assumptions (1) to (5) is called a model II of the (if $k > 1$, multiple) linear regression. Correlation coefficients are defined, as long as (8.79) holds and the distribution has finite second moments.

We first discuss the case with $k = 1$ and

$$\boldsymbol{y}_i = \beta_0 + \beta_1 \boldsymbol{x}_i + \boldsymbol{e}_i, i = 1, \ldots n > 2 \tag{8.80}$$

is the simple linear model II. Of course, we may interchange regressor and regressand and consider the model

$$\boldsymbol{x}_i = \beta_0^* + \beta_1^* \boldsymbol{y}_i + \boldsymbol{e}_i^*, i = 1, \ldots n > 2.$$

Instead of discussing both models we simply rename the two variables and come back to (8.80).

The two-dimensional density distribution of $\begin{pmatrix} \boldsymbol{x}_i \\ \boldsymbol{y}_i \end{pmatrix}$ may have existing second moments so that

$$E\begin{pmatrix} \boldsymbol{x}_i \\ \boldsymbol{y}_i \end{pmatrix} = \begin{pmatrix} \mu_x \\ \mu_y \end{pmatrix} \text{ and cov} \begin{pmatrix} \boldsymbol{x}_i \\ \boldsymbol{y}_i \end{pmatrix} = \begin{pmatrix} \sigma_x^2 & \text{cov}(\boldsymbol{x}, \boldsymbol{y}) \\ \text{cov}(\boldsymbol{x}, \boldsymbol{y}) & \sigma_y^2 \end{pmatrix}. \tag{8.81}$$

In model I we had $E(\boldsymbol{y}_i) = \beta_0 + \beta_1 x_i$ but now from (8.81) we see that $E(\boldsymbol{y}_i) = \beta_0 + \beta_1 \mu_x \, \forall \, i$ and does not depend on the regressor variable. The estimator of (8.81) is

$$\begin{pmatrix} \boldsymbol{s}_x^2 & \boldsymbol{s}_{xy} \\ \boldsymbol{s}_{xy} & \boldsymbol{s}_x^2 \end{pmatrix}.$$

We consider the conditional expectation (i.e. the expectation of the conditional distribution of $\boldsymbol{y}$ for given $\boldsymbol{x} = x$).

$$E(\boldsymbol{y}_i \mid \boldsymbol{x} = x_i) = \beta_0 + \beta_1 x_i, i = 1, \dots n > 2 \tag{8.82}$$

and the right-hand side of (8.82) equals (8.6) for model I.

Because the parameter estimation in model II is done for this conditional expectation (8.82) the formulae for the estimates are identical for both models. This is a reason why program packages like IBM SPSS-Statistics and SAS in regression analysis do not ask which model is assumed for the data as in the analysis of variance.

The estimators are

$$\boldsymbol{b}_1 = \frac{\boldsymbol{s}_{xy}}{\boldsymbol{s}_x^2} \tag{8.83}$$

and

$$\boldsymbol{b}_0 = \bar{\boldsymbol{y}} - \boldsymbol{b}_1 \bar{\boldsymbol{x}}. \tag{8.84}$$

Graybill (1961) showed with lemma 10.1 that $E\begin{pmatrix} \boldsymbol{b}_0 \\ \boldsymbol{b}_1 \end{pmatrix} = \begin{pmatrix} \beta_0 \\ \beta_1 \end{pmatrix}$. Further

$$\operatorname{var}(\boldsymbol{b}_1) = \frac{\sigma^2}{n\sigma_x^2}.$$

Even if the $\begin{pmatrix} \boldsymbol{x}_i \\ \boldsymbol{y}_i \end{pmatrix}$ are (independently) normally distributed, $\begin{pmatrix} \boldsymbol{b}_0 \\ \boldsymbol{b}_1 \end{pmatrix}$ is not normally distributed.

The ratio

$$\rho_{x,y} = \frac{\sigma_{xy}}{\sigma_x \sigma_y} \tag{8.85}$$

is called the correlation coefficient between $\boldsymbol{x}$ and $\boldsymbol{y}$.

By replacing $\sigma_{x_i y}, \sigma_{x_i}^2$ and σ_y^2 by their unbiased estimators $\mathrm{s}_{x_i y}, \mathrm{s}_{x_i}^2$ and s_y^2 we get the (biased) estimator

$$\mathrm{r}_{x_i y} = \frac{\mathrm{s}_{x_i y}}{\mathrm{s}_{x_i} \mathrm{s}_y} = \frac{\boldsymbol{SP}_{x_i y}}{\sqrt{\boldsymbol{SS}_{x_i} \boldsymbol{SS}_y}} \tag{8.86}$$

of the correlation coefficient.

To test the hypothesis H_0: $\rho \le \rho_0$ against the alternative hypothesis H_A: $\rho > \rho_0$ (respectively H_0: $\rho \ge \rho_0$ against H_A: $\rho < \rho_0$). we replace $\boldsymbol{r}$ by the modified Fisher transform

$$\boldsymbol{z} = \ln \frac{1 + \boldsymbol{r}}{1 - \boldsymbol{r}} \tag{8.87}$$

which is approximately normally distributed even if n is rather small. Cramér (1946) proved that for $n = 10$ the approximation is actually sufficient as long as $-0.8 \leq \rho \leq 0.8$. The expectation E of $\mathbf{z}$, being a function ζ of ρ, amounts to

$$E(\mathbf{z}) = \varsigma(\rho) = \ln\left(\frac{1+\rho}{1-\rho}\right) + \frac{\rho}{n-1}, \text{ the variance to } \operatorname{var}(\mathbf{z}) = \frac{4}{n-3}.$$

The statistic z in (8.87) can be used to test the hypothesis H_0: $\rho \leq \rho_0$ against the alternative hypothesis H_A: $\rho > \rho_0$ (respectively H_0: $\rho \geq \rho_0$ against H_A: $\rho < \rho_0$). H_0: $\rho \leq \rho_0$ is rejected with a first kind risk α, if $z \geq \zeta(\rho_0) + z_{1\text{-}\alpha} \cdot \frac{2}{\sqrt{n-3}}$ (respectively for H_0: $\rho \geq \rho_0$, if $z \geq \zeta(\rho_0) - z_{1\text{-}\alpha} \cdot \frac{2}{\sqrt{n-3}}$); $z_{1\text{-}\alpha}$ is the $(1 - \alpha)$-quantile of the standard normal distribution).

We use a sequential triangular test (Whitehead 1992; Schneider 1992). We split the sequence of data pairs into sub-samples of length, say $k > 3$ each. For each sub-sample j ($j = 1, 2, \ldots m$) we calculate a statistic of which distribution is only a function of ρ and k.

A triangular test must be based on a statistic with expectation 0, given the null-hypothesis, therefore we transform z into a realisation of a standardised variable

$$z^* = \left[z - \ln\left(\frac{1+\rho_0}{1-\rho_0}\right) - \frac{\rho_0}{k-1}\right]\frac{\sqrt{k-3}}{2}$$

which has the expectation 0 if the null-hypothesis is true.

The parameter

$$\theta == \left[\ln\frac{1+\rho}{1-\rho} - \ln\frac{1+\rho_0}{1-\rho_0} + \frac{\rho-\rho_0}{k-1}\right]\frac{\sqrt{k-3}}{2}$$

is used as a test parameter. For $\rho = \rho_0$ the parameter θ is zero (as demanded). For $\rho = \rho_1$ we obtain:

$$\theta_1 = \left[\ln\frac{1+\rho_1}{1-\rho_1} - \ln\frac{1+\rho_0}{1-\rho_0} + \frac{\rho_1-\rho_0}{k-1}\right]\frac{\sqrt{k-3}}{2}.$$

The difference $\delta = \rho_1 - \rho_0$ is the practical relevant difference which should be detected with power $1 - \beta$.

From each sub-sample j we now calculate the sample correlation coefficient r_j as well as its transformed values $z_j^* = \left[z_j - \ln\left(\frac{1+\rho_0}{1-\rho_0}\right) - \frac{\rho_0}{k-1}\right]\frac{\sqrt{k-3}}{2}$ ($j = 1, 2, \ldots, m$).

Now by $Z_{m\theta=0} = \sum_{j=1}^{m} z_j^*$ and $V_m = m$ the sequential path is defined by points (V_m, Z_m) for $m = 1, 2, \ldots$ up to the maximum of V below or exactly at the point where a decision can be made. The continuation region is a triangle whose three sides depend on α, β, and θ_1 via

$$a = \frac{\left(1 + \frac{z_{1-\beta}}{z_{1-\alpha}}\right)\ln\left(\frac{1}{2\alpha}\right)}{\theta_1}$$

and

$$c = \frac{\theta_1}{2\left(1 + \frac{z_{1-\beta}}{z_{1-\alpha}}\right)},$$

with the percentiles z_p of the standard normal distribution. That is, one side of the looked-for triangle lies between $-a$ and a on the ordinate of the (V, Z) plane $(V = 0)$. The two other borderlines are defined by the lines L_1: $Z = a + cV$ and L_2: $Z = -a + 3cV$, which intersect at

$$\left(V_{\max} = \frac{a}{c}, Z_{\max} = 2a\right).$$

The maximum sample size is of course $k \cdot V_{\max}$. The decision rule now is: Continue sampling as long as $-a + 3cV_m < Z_m < a + cV_m$ if $\theta_1 > 0$ or $-a + 3cV_m > Z_m > a + cV_m$ if $\theta_1 < 0$. Given $\theta_1 > 0$, accept H_A in case that Z_m reaches or exceeds L_1 and accept H_0 in case that Z_m reaches or underruns L_2, Given $\theta_1 < 0$, accept H_A in the case Z_m reaches or underruns L_1 and accept H_0 in the case Z_m reaches or exceeds L_2. If the point $\left(V_{\max} = \frac{a}{c}, Z_{\max} = 2a\right)$ is reached, H_A to be accepted.

What values must be chosen for k? The answer was found by a simulation experiment in Rasch et al. (2018) where

(a) the optimal size of subsamples (k_{opt}), where the actual type-I-risk (α_{act}) is below but as close as possible to the nominal type-I-risk (α_{nom}) was determined and
(b) the optimal nominal type-II-risk (β_{opt}), where the corresponding actual type-II-risk (β_{act}) is below but as close as possible to the nominal type-II-risk (β_{nom}) was determined.

Starting from $k = 4$, the size of the subsample was systematically increased with an increment of 1 for each parameter combination until the actual type-I-risk (α_{act}) fell below the nominal type-I-risk (α_{nom}). This optimal size of subsample (k_{opt}) was found in the next step to determine the optimal nominal type-II-risk (β_{opt}). That is, the nominal type-II-risk (β_{nom}) was systematically increased with an increment of 0.005 until the actual type-II-risk (β_{act}) fell below the nominal type-II-risk (α_{nom}).

Paths (Z, V) were generated by bivariate normally distributed random numbers x and y with means $\mu_x = \mu_y = 0$, variances $\sigma_x^2 = \sigma_y^2 = 1$, and a correlation coefficient $\sigma_{xy} = \rho$.

Using the `seqtest` package version 0.1-0 (Yanagida 2016) simulations can be performed for any α_{nom}, β_{nom}, and $\delta = \rho_1 - \rho_0$. We present here results for nominal risks $\alpha_{\text{nom}} = 0.05$ and 0.01, $\beta_{\text{nom}} = 0.1$ and 0.2, values of ρ_0 ranging 0.1–0.9 with an increment of 0.1, and $\delta = \rho_1 - \rho_0 = 0.05$, 0.10, 0.15, and 0.20.

For each parameter, combination 1 000 000 runs (paths) were generated. As criteria, we calculated:

(a) the relative frequency of wrongly accepting H_1, given $\rho = \rho_0$, which is an estimate of the actual type-I-risk (α_{act}),
(b) the relative frequency of keeping H_0, given $\rho = \rho_1$ which is an estimate of the actual type-II-risk (β_{act}),
(c) the average number of sample pairs (x, y), i.e. average sample number (ASN), is the mean (Table 8.15).

Table 8.15 Optimal size of sub-samples (k) and optimal nominal type-II-risk (β_{opt}) values for $\alpha_{nom} = 0.05$.

Given values of the test problem			Optimal values		
ρ_0	ρ_1	β	k_{opt}	β_{opt}	ASN | ρ_1
0.2	0.25	0.1	78	0.125	1909
		0.2	65	0.235	1532
	0.30	0.1	33	0.130	491
		0.2	27	0.245	396
	0.35	0.1	20	0.135	224
		0.2	16	0.250	183
	0.40	0.1	14	0.140	129
		0.2	12	0.260	106
0.3	0.35	0.1	76	0.125	1710
		0.2	79	0.240	1364
	0.40	0.1	38	0.135	428
		0.2	32	0.250	348
	0.45	0.1	23	0.140	193
		0.2	19	0.255	158
	0.50	0.1	16	0.150	109
		0.2	13	0.270	90
0.5	0.55	0.1	102	0.140	1122
		0.2	83	0.245	916
	0.60	0.1	41	0.145	278
		0.2	33	0.265	224
	0.65	0.1	23	0.155	120
		0.2	19	0.275	98
	0.70	0.1	16	0.165	66
		0.2	13	0.285	54
0.7	0.75	0.1	73	0.145	503
		0.2	61	0.265	403
	0.80	0.1	28	0.165	115
		0.2	23	0.285	94
	0.85	0.1	15	0.180	46

Example 8.7 We use results from Rasch et al. (2018).

The sequential triangular test for testing a correlation coefficient is implemented in the R package `seqtest` (Yanagida 2016), which is available on CRAN (The Comprehensive R Archive Network) and can be installed via command line `install.packages("seqtest")`. This package offers a simulation function to determine the optimal size of subsamples (k_{opt}) and the optimal nominal type-II-risk

(β_{opt}) for a user-specified parameter combination. In the following example we determine k_{opt} and β_{opt} for H_0: $\rho_0 \leq 0.3$ and H_1: $\rho_1 > 0.3$ with $\delta = 0.25$ and $\alpha_{nom} = 0.01$ and $\beta_{nom} = 0.05$.

After installing the package, the package is loaded using the function `library()`. We type

```
> library(seqtest)
```

In the first step, we determine the optimal size of subsamples (k_{opt}). We type

```
> sim.cor.seqtest(rho.sim = 0.3, rho.0 = 0.3, rho.1 = 0.55,
        k = seq(4, 16, by = 1), alpha = 0.05, beta = 0.05,
        runs = 10000)
```

i.e. we apply the function `sim.cor.seqtest()` using the first argument `rho.sim` to specify the simulated correlation coefficient ρ and the arguments `rho.0` and `rho.1` to specify ρ_0 and ρ_1. The argument `k` is used to specify a sequence for k, i.e. from 4 to 16 by increment of 1, for which the simulation is conducted, and the arguments `alpha` and `beta` are used to specify α_{nom} and β_{nom}. Last, we specify 10 000 runs using argument `runs` for each simulation condition.

As a results, we obtain:

```
Statistical Simulation for the Sequential Triangular Test

   H0: rho.0 <= 0.3  versus  H1: rho.1 > 0.3

   Nominal type-I-risk (alpha):        0.05
   Nominal type-II-risk (beta):        0.05
   Practical relevant effect (delta): 0.25

   Simulated data based on rho:        0.3
   Simulation runs:                    10000

  Estimated empirical type-I-risk (alpha):
   k = 4:   0.144
   k = 5:   0.107
   k = 6:   0.084
   k = 7:   0.068
   k = 8:   0.059
   k = 9:   0.063
   k = 10:  0.056
   k = 11:  0.054
   k = 12:  0.053
   k = 13:  0.051
   k = 14:  0.049
   k = 15:  0.047
   k = 16:  0.043
```

Simulation results indicate that $k = 14$ is the optimal value, where α_{act} is below but close to α_{nom}.

In the next step, we determine the optimal nominal type-II-risk (β_{opt}) based on the optimal size of subsamples $k_{opt} = 14$. We type

```
> sim.cor.seqtest(rho.sim = 0.55, rho.0 = 0.3, rho.1 = 0.55,
         k = 16, alpha = 0.05, beta = seq(0.05, 0.15, by = 0.01),
         runs = 10000)
```

i.e. again we apply the function `>sim.cor.seqtest()`. This time, we specify `rho.sim = 0.55` to simulate the H_1 condition and use the argument `beta` to specify a sequence for β_{nom}, i.e. from 0.05 to 0.15 by increment of 0.01, for which the simulation is conducted. As a result we obtain:

```
Statistical Simulation for the Sequential Triangular Test

   H0: rho.0 <= 0.3   versus   H1: rho.1 > 0.3

   Nominal type-I-risk (alpha):         0.05
   Practical relevant effect (delta): 0.25
   n in each sub-sample (k):            14

   Simulated data based on rho:         0.55
   Simulation runs:                     10000

  Estimated empirical type-II-risk (beta):
   Nominal beta = 0.05: 0.024
   Nominal beta = 0.06: 0.031
   Nominal beta = 0.07: 0.037
   Nominal beta = 0.08: 0.047
   Nominal beta = 0.09: 0.049
   Nominal beta = 0.10: 0.056
   Nominal beta = 0.11: 0.061
   Nominal beta = 0.12: 0.076
   Nominal beta = 0.13: 0.084
   Nominal beta = 0.14: 0.084
   Nominal beta = 0.15: 0.095
```

Simulation results indicate that $\beta_{nom} = 0.09$ is the optimal value, where β_{act} is below but close to β_{nom}.

The optimal values k_{opt} *and* β_{opt} determined by the simulation function are used for the sequential triangular test for testing a correlation coefficient. Let us assume that the first correlation coefficient calculated from a sample of 14 pairs is $r_1 = 0.46$. We type

```
> seq.obj <- cor.seqtest(x = 0.46, k = 14, rho.0 = 0.3, rho.1 = 0.55,
                 alpha = 0.05, beta = 0.09)
```

i.e. we apply the function `sim.cor.seqtest()`, using the first argument `x` to specify the sampled correlation coefficient 0.46. We specify $\rho_{0,}$ ρ_1 and α using arguments `rho.0`, `rho.1` and `alpha`, and specify k_{opt} and β_{opt} using functions `k` and `beta`. Results is assigned to the object `seq.obj`. As a result we obtain:

```
Sequential triangular test for Pearson's correlation coefficient

  H0: rho.0 <= 0.3   versus   H1: rho.1 > 0.3
  alpha: 0.05   beta: 0.09   delta: 0.25   k: 14
```

```
  Step 1
    V.m:       1.000          Z.m:        0.585
    Continuation range | V.m: [-3.084, 4.248]

  Test not finished, continue by adding data via update()
  Current sample size for 1 correlation coefficient: 1 x 14 = 14
```

Results show that the test statistic Z_m is within the continuation range conditioned on V_m. Hence, no final decision is achievable and for that reason we continue our study. Next, let us assume we sampled $r_2 = 0.57$, $r_3 = 0.49$, $r_4 = 0.69$, and $r_5 = 0.63$ from $k = 14$ pairs each. We type

```
> update(seq.obj, x = c(0.57, 0.49, 0.69, 0.63))
```

i.e. we apply the function `update()` to update results in the `seq.obj` object. As a result we obtain:

```
Sequential triangular test for Pearson's correlation coefficient

  H0: rho.0 <= 0.3  versus  H1: rho.1 > 0.3
  alpha: 0.05  beta: 0.09  delta: 0.25  k: 14

  Step 2
    V.m:       2.000          Z.m:        1.667
    Continuation range | V.m: [-2.211, 4.539]

  Step 3
    V.m:       3.000          Z.m:        2.380
    Continuation range | V.m: [-1.338, 4.830]

  Step 4
    V.m:       4.000          Z.m:        4.128
    Continuation range | V.m: [-0.465, 5.121]

  Step 5
    V.m:       5.000          Z.m:        5.522
    Continuation range | V.m: [0.408, 5.412]

  Test finished: Accept alternative hypothesis (H1)
  Final sample size for 5 correlation coefficients: 5 x 14 = 70
```

Results show that the cumulated test statistic Z_m is leaves the continuation range conditioned on V_m at step 5. Hence, the test is finished and the alternative hypothesis is accepted (see Figure 8.11).

In a simulation study, Rasch et al. (2018):

(a) Determined the optimal size of sub-samples (k_{opt}), where the actual type-I-risk (α_{act}) is below but as close as possible to the nominal type-I-risk (α_{nom}).
(b) Determined the optimal nominal type-II-risk (β_{opt}), where the corresponding actual type-II-risk (β_{act}) is below but as close as possible to the nominal type-II-risk (β_{nom}).

Starting from $k = 4$, the size of the sub-sample was systematically increased with an increment of 1 for each parameter combination until the actual type-I-risk (α_{act}) fell below the nominal type-I-risk (α_{nom}). This optimal size of subsample (k_{opt}) was found in

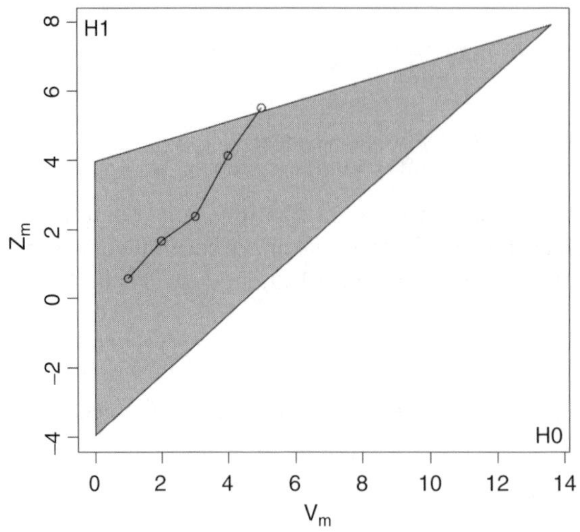

Figure 8.11 Graph of the triangle of the test of Example 8.7.

the next step to determine the optimal nominal type-II-risk (β_{opt}). That is, the nominal type-II-risk (β_{nom}) was systematically increased with an increment of 0.005 until the actual type-II-risk (β_{act}) fell below the nominal type-II-risk (α_{nom}).

Paths (Z, V) were generated by bivariate normally distributed random numbers x and y with means $\mu_x = \mu_y = 0$, variances $\sigma_x{}^2 = \sigma_y{}^2 = 1$, and a correlation coefficient $\sigma_{xy} = \rho$.

Using the `seqtest` package version 0.1–0 (Yanagida 2016) simulations can be performed for any α_{nom}, β_{nom}, and $\delta = \rho_1 - \rho_0$. We present here results for nominal risks $\alpha_{nom} = 0.05$ and 0.01, $\beta_{nom} = 0.1$ and 0.2, values of ρ_0 ranging 0.1–0.9 with an increment of 0.1, and $\delta = \rho_1 - \rho_0 =$ 0.05, 0.10, 0.15, and 0.20.

For each parameter, combination 1 000 000 runs (paths) were generated. As criteria, we calculated:

(a) the relative frequency of wrongly accepting H_1, given $\rho = \rho_0$, which is an estimate of the actual type-I-risk (α_{act}),
(b) the relative frequency of keeping H_0, given $\rho = \rho_1$, which is an estimate of the actual type-II-risk (β_{act}),
(c) the average number of sample pairs (x, y), i.e. *ASN*, is the mean number of sample pairs over all 1 000 000 paths runs of the simulation study (Table 8.16).

8.3.2 The Multiple Linear Case and the Quasilinear Case

In the case $k = 2$ random variable $(\boldsymbol{x}_1, \boldsymbol{x}_2, \boldsymbol{x}_3)$ is assumed to be three-dimensional normally distributed with existing second moments. Any of these three random variables may be use as regressand $\boldsymbol{y}$ in the regression model and the other renamed as $\boldsymbol{x}_1$, $\boldsymbol{x}_2$

$$\boldsymbol{y}_i = \beta_0 + \beta_1 \boldsymbol{x}_{1i} + \beta_2 \boldsymbol{x}_{2i} + \boldsymbol{e}_i, i = 1, \dots n > 3 \tag{8.88}$$

as special case of (8.79) [with assumptions (1) to (5)].

Table 8.16 Optimal size of subsamples (k) and optimal nominal type-II-risk (β_{opt}) values for $\alpha_{nom} = 0.05$.

Given values of the test problem			Optimal values		
ρ_0	ρ_1	β	k_{opt}	β_{opt}	ASN | ρ_1
0.1	0.15	0.1	53	0.120	2064
		0.2	43	0.225	1675
	0.20	0.1	22	0.125	543
		0.2	19	0.235	444
	0.25	0.1	13	0.130	257
		0.2	11	0.245	212
	0.30	0.1	10	0.135	152
		0.2	8	0.255	129
0.2	0.25	0.1	78	0.125	1909
		0.2	65	0.235	1532
	0.30	0.1	33	0.130	491
		0.2	27	0.245	396
	0.35	0.1	20	0.135	224
		0.2	16	0.250	183
	0.40	0.1	14	0.140	129
		0.2	12	0.260	106
0.3	0.35	0.1	76	0.125	1710
		0.2	79	0.240	1364
	0.40	0.1	38	0.135	428
		0.2	32	0.250	348
	0.45	0.1	23	0.140	193
		0.2	19	0.255	158
	0.50	0.1	16	0.150	109
		0.2	13	0.270	90
0.4	0.45	0.1	105	0.130	1436
		0.2	85	0.240	1162
	0.50	0.1	41	0.140	358
		0.2	34	0.255	290
	0.55	0.1	25	0.150	157
		0.2	20	0.265	129
	0.60	0.1	17	0.160	87
		0.2	14	0.275	73
0.5	0.55	0.1	102	0.140	1122
		0.2	83	0.245	916
	0.60	0.1	41	0.145	278
		0.2	33	0.265	224

(Continued)

Table 8.16 (Continued)

Given values of the test problem			Optimal values		
ρ_0	ρ_1	β	k_{opt}	β_{opt}	ASN | ρ_1
	0.65	0.1	23	0.155	120
		0.2	19	0.275	98
	0.70	0.1	16	0.165	66
		0.2	13	0.285	54
0.6	0.65	0.1	92	0.140	810
		0.2	76	0.255	652
	0.70	0.1	36	0.155	193
		0.2	29	0.270	158
	0.75	0.1	21	0.165	82
		0.2	17	0.285	68
	0.80	0.1	13	0.180	43
		0.2	11	0.305	36
0.7	0.75	0.1	73	0.145	503
		0.2	61	0.265	403
	0.80	0.1	28	0.165	115
		0.2	23	0.285	94
	0.85	0.1	15	0.180	46
		0.2	13	0.300	39
	0.90	0.1	10	0.200	23
		0.2	8	0.330	20
0.8	0.85	0.1	49	0.160	235
		0.2	40	0.275	192
	0.90	0.1	17	0.185	50
		0.2	15	0.305	42
	0.95	0.1	9	0.210	17
		0.2	7	0.345	15
0.9	0.95	0.1	19	0.190	53
		0.2	16	0.310	44

Unbiased estimators of β_0, β_1, β_2 can be received analogous to (8.83) and (8.84).

The three conditional two-dimensional distributions are two-dimensional distributions with correlation coefficients

$$\rho_{ij.k} = \frac{\rho_{ij} - \rho_{ik}\rho_{jk}}{\sqrt{(1-\rho_{ik}^2)(1-\rho_{jk}^2)}} \quad (i \neq j \neq k;\ \ i,j,k = 1,2,3). \tag{8.89}$$

Here ρ_{ij},ρ_{ik} and ρ_{jk} are the correlation coefficients of the three two-dimensional marginal distributions of $(\boldsymbol{x}_i, \boldsymbol{x}_j, \boldsymbol{x}_k)$. It can easily be shown that these marginal distributions are two-dimensional normal distributions if $(\boldsymbol{x}_1,\boldsymbol{x}_2,\boldsymbol{x}_3)$ is normally distributed.

The correlation coefficient (8.89) of the conditional two-dimensional normal distribution of (x_i,x_j) for given x_k, $i, j, k = 1, 2, 3$ but different is called the partial correlation coefficient between (x_i,x_j) after fixing $\boldsymbol{x}_k$ at x_k.

We obtain estimators $\boldsymbol{r}_{ij\,\cdot\,k}$ for partial correlation coefficients by replacing the simple correlation coefficients in (8.87) by their estimators and obtain the biased estimators

$$\boldsymbol{r}_{ij\,\cdot\,k} = \frac{\boldsymbol{r}_{ij} - \boldsymbol{r}_{ik}\boldsymbol{r}_{jk}}{\sqrt{(1-\boldsymbol{r}_{ik}^2)(1-\boldsymbol{r}_{jk}^2)}}, i, j, k = 1, 2, 3 \text{ but different} \tag{8.90}$$

For the general parameter estimation, we use the formulae in Section 8.2

8.3.2.1 Hypotheses Testing - General

Bartlett (1933) showed that all the tests in 8.2.1.2 could be applied also for model II, but the power of the tests differ between both models.

Problem 8.35 Estimate the parameters of a linear and a quadratic regression for the data of Example 8.2 and test the hypothesis that the factor of the quadratic term is zero.

Solution

We use the estimates from (8.9) and (8.10) and the formula for the factor of the quadratic term from the solution of Problem 8.5. We test the null hypothesis $H_0: \beta_2 = 0$ against $H_A: \beta_2 \neq 0$ with the test statistic (8.25) but now $\boldsymbol{x}$ is random.

Example

Further we can test $H_0: \rho_{x,y} = 0$ by (8.25) for $j = 1$ and $\beta_1 = 0$ because the hypotheses $\rho_{x,y} = \frac{\sigma_{xy}}{\sigma_x\sigma_y} = 0$ and $\beta_1 = \frac{\sigma_{xy}}{\sigma_x^2}$=0 both imply $\sigma_{xy} = 0$.

Analogously we could test $H_0: \varrho_{ij\,\cdot\,k} = 0$ by

$$t = \frac{\boldsymbol{r}_{ij\,\cdot\,k}\sqrt{n-3}}{\sqrt{1-\boldsymbol{r}_{ij\,\cdot\,k}^2}} \tag{8.91}$$

which is $t(n-3)$-distributed.

8.3.2.2 Confidence Estimation

An approximate $(1-\alpha)$ 100%-confidence interval for ϱ is

$$\left\langle \tanh\left(\text{z} - \frac{z_{1-\frac{\alpha}{2}}}{\sqrt{n-3}}\right), \tanh\left(\text{z} + \frac{z_{1-\frac{\alpha}{2}}}{\sqrt{n-3}}\right)\right\rangle, \tag{8.92}$$

with the $\left(1-\frac{\alpha}{2}\right)$-quantile $z_{1-\frac{\alpha}{2}}$ of the standard normal distribution $\left[P\left(\mathbf{z} > z_{1-\frac{\alpha}{2}}\right) = \frac{\alpha}{2}\right]$.

To interpret the value of ϱ (and also of r), we again consider the regression function $f(\text{x}) = E(\text{y}|\text{x}) = \beta_0 + \beta_1\text{x}$.

ϱ^2 can now be explained as a measure of the proportion of the variance of $\boldsymbol{y}$, explainable by the regression on $\boldsymbol{x}$. The conditional variance of $\boldsymbol{y}$ is

$$\text{var}(\text{y} \mid x) = \sigma_y^2(1 - \rho^2),$$

and

$$\frac{\text{var}(\text{y} \mid x)}{\sigma_y^2} = 1 - \rho^2$$

is the proportion of the variance of $\boldsymbol{y}$, not explainable by the regression on x, and by this the statement above follows. We call $\varrho^2 = B$ measure of determination.

To construct confidence intervals for β_0 and β_1 or to test hypotheses about these parameters seems to be difficult but the methods for model I can also be applied for model II. We demonstrate this for the example of the confidence interval for β_0. The argumentation for confidence intervals for other parameters and for the statistical tests is analogous.

The probability statement

$$P\left[\boldsymbol{b}_0 - t\left(n-2 \;\middle|\; 1-\frac{\alpha}{2}\right) \text{s}_0 \leqq \beta_0 \leqq \boldsymbol{b}_0 + t\left(n-2 \;\middle|\; 1-\frac{\alpha}{2}\right) \text{s}_0\right] = 1 - \alpha,$$

leading to the confidence interval (8.58) for $j = 0$ is true, if for fixed values $x_1, \ldots, x_n$ samples of y values are selected repeatedly. Using the frequency interpretation, β_0 is covered in about $(1 - \alpha)$ 100% of the cases by the interval (8.58). This statement is valid for each arbitrary n-tuple $x_{i1}, \ldots, x_{in}$, also for an n-tuple $x_{i1}, \ldots, x_{in}$, randomly selected from the distribution because (8.58) is independent of $x_1, \ldots, x_n$, if the conditional distribution of the $\boldsymbol{y}_j$ is normal. However, this is the case, because $(\boldsymbol{y}, \boldsymbol{x}_1, \ldots, \boldsymbol{x}_k)$ was assumed to be normally distributed. By this the construction of confidence intervals and testing of hypotheses can be done by the methods and formulae given above. However, the expected width of the confidence intervals and the power function of the tests differ for both models.

That $\left\langle \boldsymbol{b}_i - t\left(n-2 \middle| 1-\frac{\alpha}{2}\right) \text{s}_i, \boldsymbol{b}_i + t\left(n-2 \middle| 1-\frac{\alpha}{2}\right) \text{s}_i \right\rangle$ is really a confidence interval with a confidence coefficient $1 - \alpha$ also for model II, can of course be proven exactly, using a theorem of Bartlett (1933) by which

$$\boldsymbol{t}_i = \frac{\text{s}_x\sqrt{n-2}}{\sqrt{\text{s}_y^2 - \boldsymbol{b}_i^2\text{s}_x^2},}(\boldsymbol{b}_i - \beta_i)$$

is $t(n-2)$-distributed.

8.3.3 The Allometric Model

In model II we discuss only one intrinsically non-linear model, the allometric model

$$\boldsymbol{y}_i = \alpha\boldsymbol{x}_i^{\beta} + \boldsymbol{e}_i, i = 1, \ldots n > 2 \tag{8.93}$$

first used by Snell (1892) to describe the dependency between body mass and brain mass.

The allometric model is mainly used to describe relative growth of one part of a growing individual (plant or animal) to the total mass or the mass of another part see Huxley (1972). Nowadays this model is also used in technical applications.

8.3.4 Experimental Designs

The experimental design for model II of the regression analysis differs fundamentally from that of model I. Because $\boldsymbol{x}$ in model II is a random variable, the problem of the optimal choice of x does not occur. Experimental design in model II means only the optimal choice of n in dependency of given precision requirements. A systematic description about that is given in Rasch et al. (2008). We repeat this in the following.

At first we restrict in (8.59) on $k = 1$, and consider the more general model of the regression within of $a \geq 1$ groups with the same slope β_1

$$\boldsymbol{y}_{hj} = \beta_{h0} + \beta_1 x_{hj} + \boldsymbol{e}_{hj} \quad (h = 1, \dots, a; j = 1, \dots, n_h \geq 2). \tag{8.94}$$

We estimate β_1 for $a > 1$ not by (8.9), but by

$$\boldsymbol{b}_{I1} = \frac{\sum_{h=1}^{a} \boldsymbol{SP}_{x,y}^{(h)}}{\sum_{h=1}^{a} \boldsymbol{SS}^{(h)}} = \frac{\boldsymbol{SP}_{Ixy}}{\boldsymbol{SS}_{Ix}}, \tag{8.95}$$

with $\boldsymbol{SP}_{xy}^{(h)}$ and $\boldsymbol{SS}_x^{(h)}$ for each of the a groups as defined in Example 8.1.

If we look for a minimal $n = \sum_{h=1}^{a} n_h$ so that $V(\boldsymbol{b}_{11}) \leq C$ we find in Bock (1998)

$$n - a - 2 = \left\lceil \frac{\sigma^2}{C\sigma_k^2} \right\rceil.$$

If $k = 1$ for the expectation $E(\boldsymbol{y}|x) = \beta_0 + \beta_1 x$ $(1-\alpha)-$ confidence interval should be given so that the expectation of the square of the half width of the interval does not exceed δ, then

$$n - 3 = \left\lceil \frac{\sigma^2}{\delta^2} \left[1 - \frac{2}{n\sigma_x^2} \max[(x_0 - \mu_0)^2, (x_1 - \mu_1)^2]\, t^2 \left(n - 2 \middle| 1 - \frac{\alpha}{2} \right) \right] \right\rceil \tag{8.96}$$

must be chosen.

References

Atkinson, A.C. and Hunter, W.G. (1968). The design of experiments for parameter estimations. *Technometrics* 10: 271–289.

Barath, C.S., Rasch, D., and Szabo, T. (1996). Összefügges a kiserlet pontossaga es az ismetlesek szama kozött. *Allatenyesztes es takarmanyozas* 45: 359–371.

Bartlett, M.S. (1933). On the theory of statistical regression. *Proc. Royal Soc. Edinburgh* 53: 260–283.

Bertalanffy, L. von (1929). Vorschlag zweier sehr allgemeiner biologischer Gesetze. *Biol. Zbl.* 49: 83–111.

Bock, J. (1998). *Bestimmung des Stichprobenumfanges für biologische Experimente und kontrollierte klinische Studien.* München Wien: R. Oldenbourg Verlag.

Box, G.E.P. and Lucas, H.L. (1959). Design of experiments in nonlinear statistics. *Biometrics* 46: 77–96.

Brody, S. (1945). *Bioenergetics and Growth*. N.Y: Rheinhold Pub. Corp.
Cramér, H. (1946). *Mathematical Methods of Statistics*. Princeton: Princeton University Press.
Ermakov, S.M. and Zigljavski, A.A. (1987). Математическаь тероиь рптималньв экспеоиметрв *(Mathematical Theory of Optimal Experiments)*. Moskwa: Nauka.
Galton, F. (1885). Opening address as President of the Anthropology Section of the British Association for the Advancement of Science, September 10th, 1885, at Aberdeen. *Nature* 32: 507–510.
Gompertz, B. (1825). On the nature of the function expressive of the law of human mortality, and on a new method determining the value of life contingencies. *Phil. Trans. Roy. Soc. (B), London* 513–585.
Graybill, A.F. (1961). *An Introduction to Linear Statistical Models*. New York: Mc Graw Holl.
Huxley, J.S. (1972). *Problems of Relative Growth*, 2e. New York: Dover.
Jennrich, R.J. (1969). Asymptotic properties of nonlinear least squares estimation. *Ann. Math. Stat.* 40: 633–643.
Michaelis, L. and Menten, M.L. (1913). Die Kinetik der Invertinwirkung. *Biochem. Z.* 49: 333–369.
Pilz, J., Rasch, D., Melas, V.B., and Moder, K. (eds.) (2018). *Statistics and Simulation:* IWS 8, Vienna, Austria, September 2015, *Springer Proceedings in Mathematics & Statistics*, vol. 231. Heidelberg: Springer.
Pukelsheim, F. (1993). *Optimal Design of Experiments*. New York: Wiley.
Rasch, D. (1968). *Elementare Einführung in die Mathematische Statistik*. Berlin: VEB Deutscher Verlag der Wissenschaften.
Rasch, D. (1990). Optimum experimental design in nonlinear regression. *Commun. Stat. Theory and Methods* 19: 4789–4806.
Rasch, D. (1993). The robustness against parameter variation of exact locally optimum experimental designs in growth models - a case study. In: *Techn. Note 93-3*. Department of Mathematics, Wageningen Agricultural University.
Rasch, D. (2008) Versuchsplanung in der nichtlinearen Regression - Michaelis-Menten gestern und heute. Vortrag an der Boku Wien, Mai 2008.
Rasch, D. and Schimke, E. (1983). Distribution of estimators in exponential regression – a simulation study. *Skand. J. Stat.* 10: 293–300.
Rasch, D. and Schott, D. (2018). *Mathematical Statistics*. Oxford: Wiley.
Rasch, D., Herrendörfer, G., Bock, J. et al. (eds.) (2008). *Verfahrensbibliothek Versuchsplanung und - auswertung, 2. verbesserte Auflage in einem Band mit CD*. München Wien: R. Oldenbourg Verlag.
Rasch, D., Yanagida, T., Kubinger, K.D., and Schneider, B. (2018). Determination of the optimal size of subsamples for testing a correlation coefficient by a sequential triangular test. In: *Statistics and Simulation*: IWS 8, Vienna, Austria, September 2015, Springer Proceedings in Mathematics & Statistics, vol. 231 (ed. J. Pilz, D. Rasch, V.B. Melas and K. Moder).
Schlettwein, K. (1987). *Beiträge zur Analyse von vier speziellen Wachstumsfunktionen*. Rostock: Dipl.Arbeit, Sektion Mathematik, Univ.
Schneider, B. (1992). An interactive computer program for design and monitoring of sequential clinical trials. In *Proceedings of the XVIth international biometric conference* (pp. 237–250), Hamilton, New Zealand.

Snell, O. (1892). Die Abhängigkeit des Hirngewichts von dem Körpergewicht und den geistigen Fähigkeiten. *Arch. Psychiatr.* 23: 436–446.

Steger, H. and Püschel, F. (1960). Der Einfluß der Feuchtigkeit auf die Haltbarkeit des Carotins in künstlich getrocknetem Grünfutter. *Die Deutsche Landwirtschaft* 11: 301–303.

Whitehead, J. (1992). *The Design and Analysis of Sequential Clinical Trials*, 2e. Chichester: Ellis Horwood.

Yanagida, T. (2016). seqtest: Sequential triangular test, R package version 0.1-0. http://CRAN.R-project.org/package=seqtest.

Yule, G.U. (1897). On the theory of correlation. *J. R. Stat. Soc.* 60 (4): 812–854, Blackwell Publishing.

9

Analysis of Covariance (ANCOVA)

9.1 Introduction

Analysis of covariance combines elements of the analysis of variance and of regression analysis. Because the analysis of variance can be seen as a multiple linear regression analysis, the analysis of covariance (ANCOVA) can be defined as a multiple linear regression analysis in which there is at least one categorical explanatory variable and one quantitative variable. Usually the categorical variable is a treatment of primary interest measured at the experimental unit and is called response y. The quantitative variable x, which is also measured in experimental units in anticipation that it is associated linearly with the response of the treatment. This quantitative variable x is called a covariate or covariable or concomitant variable.

Before we explain this in general, we give some examples performed in a completely randomised design.

1. In a completely randomised design treatments were applied on tea bushes where the yields y_{ij} are the yields in kilograms of the tea bushes. An important source of error is that, by the luck of the draw by randomisation, some treatments will be allotted to a more productive set of bushes than others. Fisher described in 1925 in his first edition of "Statistical Methods for Research Workers" the application of the covariate x_{ij}, which was the yield in kilograms of the tea bushes in a period before treatments were applied. Since the relative yields of the tea bushes show a good deal of stability from year to year, x_{ij} serves as a linear predictor of the inherent yielding stabilities of the bushes. The regression lines of y_{ij} on x_{ij} are parallel regression lines for the treatments. By adjusting the treatment yields so as to remove these differences in yielding ability, we obtain a lower variance of the experimental error and more precise comparisons amongst the treatments. See Fisher (1935) section 49.1.
2. In variety trials with corn (*Zea mays*) or sugar beet (*Beta vulgaris*) usually plots with 30 plants are used. The weight per plot in kilograms is the response y_{ij}, but often some plants per plot are missing. As covariate, x_{ij} is taken as the number of plants per plot. The regression lines of y_{ij} on x_{ij} are parallel regression lines for the treatments.
   One could perhaps analyse the yield per plant (y/x) in kilograms as a means of removing differences in plant numbers. This is satisfactory if the relation between y and x is a straight line through the origin. However, the regression line of y on x is a straight line not through the origin and the estimated regression coefficient b is often substantially less than the mean yield per plant because when plant numbers are high,

Applied Statistics: Theory and Problem Solutions with R, First Edition.
Dieter Rasch, Rob Verdooren, and Jürgen Pilz.

competition between plants reduces the yield per plant. If this happens, the use of y/x overcorrects for the stand of plants. Of course, the yield per plant should be analysed if there is direct interest in this quantity.

3. A common clinical method to evaluate an individual's cardiovascular capacity is through treadmill exercise testing. One of the measures obtained during treadmill testing, maximal oxygen uptake, is the best index of work capacity and maximal cardiovascular function. As subjects, e.g., 12 healthy males who did not participate in a regular exercise program were chosen. Two treatments selected for the study were a 12-week step aerobics training program and a 12-week outdoor running regimen on a flat terrain. Six men were randomly assigned to each group in a completely randomised design. Various respiratory measurements were made on the subjects while on the treadmill before the 12-week period. There were no differences in the respiratory measurements of the two groups of subjects prior to treatment. The measurement of interest y_{ij} is the change in of maximal ventilation (litres/minute) of oxygen for the 12-week period. The relationship between maximal ventilation change and age (years) is linear and the regression lines for the two treatments are parallel regression lines. Hence as covariate x_{ij} is taken as the age of the subjects.
4. One wants to assess the strength of threads y_{ij} in pounds made by three different machines. Each thread is made from a batch of cotton, and some batches tend to form thicker thread than other batches. There is no way to know how thick it will be until you make it. Regardless of how the machines may affect thread strength, thicker threads are stronger. Thus, we record the diameter x_{ij} in 10^{-3} in. as a covariate. The regression lines of of y_{ij} on x_{ij} are parallel regression lines for the machines.

When one wants to use the ANCOVA it is a good idea to check whether the linear regression lines are parallel for the different treatments. If the regression lines are not parallel then we have an interaction between the treatments and the covariate. Hence it is a good idea to run first an ANCOVA model with interaction of treatments and covariate; if the slopes are not statistically different (no significant interaction), then we can use an ANCOVA model with parallel lines, which means that there is a separate intercepts regression model. The main use of ANCOVA is for testing a treatment effect while using a quantitative control variable as covariate to gain power.

Note that if needed the ANCOVA model can be extended with more covariates. The R program can handle this easily. If the regression of y is quadratic on the covariate x, we have then an ANCOVA model with two covariates, $x_1 = x$ and $x_2 = x^2$.

9.2 Completely Randomised Design with Covariate

We first discuss the balanced design.

9.2.1 Balanced Completely Randomised Design

We assume that we have a balanced completely randomised design for a treatment A with a classes. Assuming further that there is a linear relationship between the response y and the covariate x, we find that an appropriate statistical model is

$$\boldsymbol{y}_{ij} = \mu + a_i + \beta x_j + \boldsymbol{e}_{ij}, i = 1, \dots, a; j = 1, \dots, n, \text{var}(\boldsymbol{y}_{ij}) = \text{var}(\boldsymbol{e}_{ij}) = \sigma^2 \quad (9.1)$$

where μ is a constant, a_i is the treatment effect with the side condition $\sum_{i=1}^{a} a_i = 0$, β is the coefficient for the linear regression of $\boldsymbol{y}_{ij}$ on x_{ij}, and the $\boldsymbol{e}_{ij}$ are random independent normally distributed experimental errors with expectation 0 and variance σ^2.

Two additional key assumptions for this model are that the regression coefficient β is the same for all treatment groups (parallel regression lines) and the treatments do not influence the covariate x.

The first objective of the covariance analysis is to determine whether the addition of the covariate has reduced the estimate of experimental error variance. This means that the test of the null hypothesis $H_{\beta 0}$: $\beta = 0$ against the alternative hypothesis $H_{\beta A}$: $\beta \neq 0$ results in rejection of $H_{\beta 0}$. If the reduction of the estimate of the experimental error variance is significant then we obtain estimates of the treatment group means $\mu + a_i$ adjusted to the same value of the covariate x for each of the treatment groups and determine the significance of treatment differences on the basis of the adjusted treatment means. Usually the statistical packages estimate the adjusted treatment means for the overall mean $\bar{x}_{..}$ of the covariate x.

The least squares estimates of the parameters of model (9.1) are:

$$\hat{\mu} = \bar{y}_{..} + \hat{\beta}\bar{x}_{..}, \hat{a}_{\mathrm{i}} = \bar{y}_{\mathrm{i.}} - \bar{y}_{..} - \hat{\beta}(\bar{x}_{\mathrm{i.}} - \bar{x}_{..}),$$

$$\hat{\beta} = E_{xx}/E_{yy} \text{ with } E_{xx} = \sum_{i=1}^{a}\sum_{j=1}^{n}(x_{ij} - \bar{x}_{i.})^2 \text{ and}$$

$$E_{xy} = \sum_{i=1}^{a}\sum_{j=1}^{n}(x_{ij} - \bar{x}_{i.})(y_{ij} - \bar{y}_{i.})$$

see Montgomery (2013), section 15.3. For the derivation of the least squares estimates of the parameters see the general unbalanced completely randomised design in Section 9.2.2.

For the balanced case we have then that n_i for $i = 1, \ldots, a$ is equal to n.

The nested ANOVA table (note the sequence of the source of variation) for the test of the null hypothesis $H_{\beta 0}$: $\beta = 0$ is given in Table 9.1.

The test of the null hypothesis $H_{\beta 0}$: $\beta = 0$ against the alternative hypothesis $H_{\beta A}$: $\beta \neq 0$ is done with the test-statistic

$$F_b = (SS_b/1)/(SS_E/(a(n-1)-1)) \tag{9.2}$$

Table 9.1 Nested ANOVA table for the test of the null hypothesis $H_{\beta 0}$: $\beta = 0$.

Source of variation	*df*	*SS*
Treatments	$(a-1)$	$T_{yy} = n\sum_{i=1}^{a}(y_{i.} - \bar{y}_{...})^2$
Regression coefficient $\hat{\beta}$	1	$SS_b = (E_{xy})^2/E_{xx}$
Error	$a(n-1)-1$	$SS_E = SS_{yy} - T_{yy} - SS_b$
Corrected total	$an-1$	$SS_{yy} = \sum_{i=1}^{a}\sum_{j=1}^{n}(y_{ij} - \bar{y}_{i.})^2$

Table 9.2 Nested ANOVA table for the test HA0: 'all a_i are equal'.

Source of variation	*df*	*SS*
Regression coefficient $\hat{\beta}$	1	$S_b = \left[\sum_{i=1}^{a}\sum_{j=1}^{n}\left(x_{ij}-\bar{x}_{i.}\right)\left(y_{ij}-\bar{y}_{i.}\right)\right]^2 \Big/ \sum_{i=1}^{a}\sum_{j=1}^{n}\left(x_{ij}-\bar{x}_{i.}\right)^2$
Treatments	$a-1$	$S_T = SS_{yy} - S_b - SS_E$
Error	$a(n-1)-1$	SS_E
Corrected Total	$an-1$	$SS_{yy} = \sum_{i=1}^{a}\sum_{j=1}^{n}(y_{ij}-\bar{y}_{i.})^2$

which has under $H_{\beta 0}$: $\beta = 0$ the F-distribution with $df = 1$ for the numerator and $df = a(n\text{-}1)\text{-}1$ for the denominator.

For the test of the null hypothesis H_{A0}: 'all a_i are equal' against the alternative hypothesis H_{AA}: 'at least one a_i is different from the other' we use the nested ANOVA table given in Table 9.2.

The test of the null hypothesis H_{A0}: 'all a_i are equal' against the alternative hypothesis H_{AA}: 'at least one a_i is different from the other' is done with the test-statistic

$$F_A = (S_T/(a-1))/(SS_E/(a(n-1)-1) \tag{9.3}$$

which has under H_{0A}: 'all a_i are equal' the F-distribution with $df = a - 1$ for the numerator and $df = a(n-1) - 1$ for the denominator.

The estimate of the treatment mean adjusted for the covariate at $x =$ overall mean $\bar{x}_{..}$ is

$$\bar{y}_{i.} - \hat{\beta}(\bar{x}_{i.} - \bar{x}_{..}) \text{ for } i = 1, \dots, a. \tag{9.4}$$

The estimate of the standard error of this estimate (9.2) is

$$MS_E\left[\left(\frac{1}{n} + \frac{(\bar{x}_{i.} - \bar{x}_{..})^2}{E_{xx}}\right]\text{ with } MS_E = \frac{SS_E}{a(n-1)-1}. \tag{9.5}$$

The estimate for the difference between two adjusted treatment means at overall mean $\bar{x}_{..}$ is

$$\sqrt{MS_E\left[\left(\frac{2}{n} + \frac{(\bar{x}_{i.} - \bar{x}_{j.})^2}{E_{xx}}\right]}\text{ with } MS_E = \frac{SS_E}{a(n-1)-1}. \tag{9.6}$$

However, we first want to check if we can use an ANCOVA model with parallel lines, which means that there is a separate intercepts regression model. Therefore, we run first an ANCOVA model with interaction of treatments and covariate.

An appropriate statistical model is

$$\boldsymbol{y}_{ij} = \mu + a_i + \beta_i x_j + \boldsymbol{e}_{ij}, i = 1, \dots, a; j = 1, \dots, n, \text{var}(\boldsymbol{y}_{ij}) = \text{var}(\boldsymbol{e}_{ij}) = \sigma^2 \tag{9.7}$$

where μ is a constant, a_i is the treatment effect with the side condition $\sum_{i=1}^{a} a_i = 0$, β_i is the coefficient for the linear regression for treatment A with class a_i of $\boldsymbol{y}_{ij}$ on x_{ij}, and the

Table 9.3 Strength of a monofilament fibre produced by three different machines *M*.

M_1		M_2		M_3	
y	*x*	*y*	*x*	*y*	*x*
36	20	40	22	35	21
41	25	48	28	37	23
39	24	39	22	42	26
42	25	45	30	34	21
49	32	44	28	32	15

Note: This experiment is described in Section 9.1 as example 4.

$\boldsymbol{e}_{ij}$ are random independent normally distributed experimental errors with mean 0 and variance σ^2.

In R we would make an ANOVA table with this model and look for the test of the interaction effect of treatment and covariate whether we must reject the null hypothesis $H_{0\beta i}$: $\beta_1 = \ldots = \beta_a$ and accept the alternative hypothesis $H_{a\beta i}$: 'there is at least one β_i different from another β_j with $i \neq j$'. If this is the case we cannot use the ANCOVA model.

Example 9.1 From Montgomery (2013) we take his Example 15.5. Consider a study performed to determine if there is a difference in the strength of a monofilament fibre produced by three different machines *M*. The response is the breaking strength of the fibre *y* (in pounds) and the diameter of the fibre *x* (in 10^{-3} in.). The data from this experiment are shown in Table 9.3.

Problem 9.1 Test the null hypothesis $H_{\beta 0}$: $\beta_1 = \ldots = \beta_a$ against the alternative hypothesis H_{BA}: 'there is at least one β_i different from another β_j with $i \neq j$' with a significance level $\alpha = 0.05$ using model (9.7) in Example 9.1.

Solution

In R we use the command

```
>fit.seplines <- lm(strength ~ machine * diameter, data=example9_1)
```

Example

```
> strength <- c(36,41,39,42,49,40,48,39,45,44,35,37,42,34,32)
> diameter <- c(20,25,24,25,32,22,28,22,30,28,21,23,26,21,15)
> machine <- factor(rep(1:3, each = 5))
> plot(diameter, strength, pch = as.character(machine))
See Figure 9.1.
> example9_1 <- data.frame(machine, strength,diameter)
> fit.seplines <- lm(strength ~ machine * diameter , data = example9_1)
> anova(fit.seplines)
Analysis of Variance Table
Response: strength
                 Df  Sum Sq Mean Sq F value    Pr(>F)
```

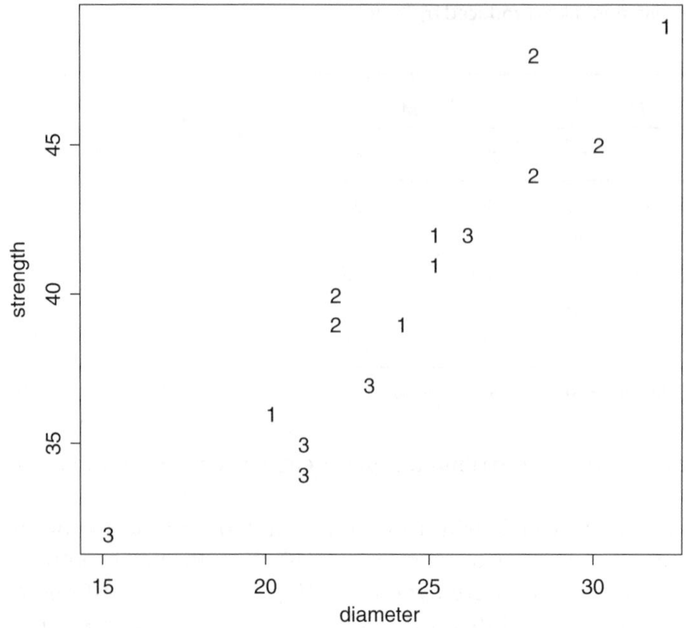

Figure 9.1 Scatter-plot of the example in Problem 9.1 with M1 as 1, M2 as 2, M3 as 3.

```
machine           2 140.400   70.200 25.0231 0.0002107 ***
diameter          1 178.014 178.014 63.4538 2.291e-05 ***
machine:diameter  2   2.737   1.369  0.4878 0.6292895
Residuals         9  25.249   2.805
---
Signif. codes:  0 '***' 0.001 '**' 0.01 '*' 0.05 '.' 0.1 ' ' 1
```

Because the interaction machine × diameter has the p-value $\Pr(>F) = 0.6292895 > 0.05$ we cannot reject the null hypothesis H_{B0}: $\beta_1 = \ldots = \beta_a$ and we can use the ANCOVA model (9.1) with parallel lines for the regression lines of strength on diameter for the machines.

Problem 9.2 Test in Example 9.1 with the ANCOVA model (9.1) the null hypothesis $H_{\beta 0}$: $\beta = 0$ against the alternative hypothesis H_{BA}: $\beta \neq 0$ with significance level $\alpha = 0.05$.

Solution

In R we use the command

```
> testbeta0 <- lm ( strength ~machine  + diameter , data = example9_1)
```

Example

We continue with the R program of Problem 9.1.

```
> testbeta0 <- lm ( strength ~machine  + diameter , data = example9_1)
> anova(testbeta0)
Analysis of Variance Table
Response: strength
          Df  Sum Sq Mean Sq F value    Pr(>F)
```

```
machine    2 140.400  70.200  27.593 5.170e-05 ***
diameter   1 178.014 178.014  69.969 4.264e-06 ***
Residuals 11  27.986   2.544
---
Signif. codes:  0 '***' 0.001 '**' 0.01 '*' 0.05 '.' 0.1 ' ' 1
> summary(testbeta0)
Call:
lm(formula = strength ~ machine + diameter, data = example9_1)
Residuals:
    Min      1Q  Median      3Q     Max
-2.0160 -0.9586 -0.3841  0.9518  2.8920
Coefficients:
            Estimate Std. Error t value Pr(>|t|)
(Intercept)   17.360      2.961   5.862 0.000109 ***
machine2       1.037      1.013   1.024 0.328012
machine3      -1.584      1.107  -1.431 0.180292
diameter       0.954      0.114   8.365 4.26e-06 ***
---
Signif. codes:  0 '***' 0.001 '**' 0.01 '*' 0.05 '.' 0.1 ' ' 1
Residual standard error: 1.595 on 11 degrees of freedom
Multiple R-squared:  0.9192,    Adjusted R-squared:  0.8972
F-statistic: 41.72 on 3 and 11 DF,  p-value: 2.665e-06
```

Because in the ANOVA table with diameter as the last model variable we find the p-value $\Pr(>F) = 4.264\text{e-}06 < 0.05$ we reject the null hypothesis H_{B0}: $\beta = 0$.

Of course this can also be concluded from the t-test for diameter where we find the same p-value $\Pr(>|t|) = 4.26\text{e-}06 < 0.05$.

Problem 9.3 Test in Example 9.1 with the ANCOVA model (9.1) the null hypothesis H_{A0}: 'all a_i are equal' against the alternative hypothesis H_{AA}: 'at least one a_i is different from the other' with significance level $\alpha = 0.05$.

Solution

In R we use the command

```
> machine0 <- lm ( strength ~ diameter+ machine , data = example9_1)
```

Example

We continue with the R program of Problem 9.1.

```
> machine0 <- lm ( strength ~ diameter+ machine , data = example9_1)
> anova(machine0)
Analysis of Variance Table
Response: strength
          Df  Sum Sq Mean Sq  F value    Pr(>F)
diameter   1 305.130 305.130 119.9330 2.96e-07 ***
machine    2  13.284   6.642   2.6106   0.1181
Residuals 11  27.986   2.544
---
Signif. codes:  0 '***' 0.001 '**' 0.01 '*' 0.05 '.' 0.1 ' ' 1
> summary(machine0)

Call:
lm(formula = strength ~ diameter + machine, data = example9_1)
Residuals:
```

```
    Min      1Q  Median      3Q     Max
-2.0160 -0.9586 -0.3841  0.9518  2.8920
Coefficients:
            Estimate Std. Error t value Pr(>|t|)
(Intercept)   17.360      2.961   5.862 0.000109 ***
diameter       0.954      0.114   8.365 4.26e-06 ***
machine2       1.037      1.013   1.024 0.328012
machine3      -1.584      1.107  -1.431 0.180292
---
Signif. codes:  0 '***' 0.001 '**' 0.01 '*' 0.05 '.' 0.1 ' ' 1
Residual standard error: 1.595 on 11 degrees of freedom
Multiple R-squared:  0.9192,    Adjusted R-squared:  0.8972
F-statistic: 41.72 on 3 and 11 DF,  p-value: 2.665e-06
```

From the ANOVA table with machine as the last model variable we find the p-value $\Pr(>F) = 0.1181 > 0.05$ we cannot reject the null hypothesis H_{A0}: 'all a_i are equal'.

Problem 9.4 Estimate the adjusted machine means at the overall mean of the diameter.

Solution

Look at the outcomes of `summary(beta0)` in Problem 9.2; from this we find the estimate of β as `diameter 0.954`. Apply then formula (9.4).

Example

We calculate first the means of strength and diameter per machine and the overall mean of strength and diameter.

```
> tapply(strength, machine, mean)
   1    2    3
41.4 43.2 36.0
> tapply(diameter, machine, mean)
   1    2    3
25.2 26.0 21.2
> mean.strength <- mean(strength)
> mean.strength
[1] 40.2
> mean.diameter <- mean(diameter)
> mean.diameter
[1] 24.13333
> mean.strength1 <- 41.4
> mean.strength2 <- 43.2
> mean.strength3 <- 36.0
> mean.diameter1 <- 25.2
> mean.diameter2 <- 26.0
> mean.diameter3 <- 21.2
> estimate.beta <- 0.954
> adj.meanM1 <- mean.strength1-estimate.beta*(mean.diameter1-mean.diameter)
> adj.meanM1
[1] 40.3824
> adj.meanM2 <- mean.strength2-estimate.beta*(mean.diameter2-mean.diameter)
> adj.meanM2
[1] 41.4192
```

```
> adj.meanM3 <- mean.strength3-estimate.beta*(mean.diameter3-mean.diameter)
> adj.meanM3
[1] 38.7984
```

Hence the adjusted mean of M_1 is 40.3824; the adjusted mean of M_2 is 41.4192, and the adjusted mean of M_3 is 38.7984.

Note that the difference in the adjusted means of $M_2 - M_1$ is 41.4192–4140.3824 = 1.0368 and this is given in Problem 9.3 in the `summary(machine0)` as the `Estimatemachine2 1.037` with the estimate of the standard error 1.013. The difference in the adjusted means of $M_3 - M_1$ is 38.7984–7940.3824 = −1.584 and this isgiven in Problem 9.3 in the `summary(machine0)` as the `Estimate machine3 −1.584` with the estimate of the standard error 1.107.

Problem 9.5 Draw the estimated regression lines for the machines M_1, M_2, and M_3.

Solution

The regression line for machine M_i goes through the point $(\overline{x}_{i..}, \overline{y}_{i.})$, the sample mean of the x-values and the y-values of M_i. The estimate of β is $b = 0.954$ as given in Problem 9.2 in the `summary(testbeta0)` as diameter 0.954 with the estimate of its standard error as 0.114.

The estimated regression line for M_i. is $y_{ij} = I_i + bx_{ij}$ hence the intercepts are $I_i = \overline{y}_{i.} - b\,\overline{x}_{i.}$.

Example

Continuation with the results of the R program of Problems 9.3 and 9.4 we find the intercepts I_i.

```
> intercept_I1 <- mean.strength1 - estimate.beta*mean.diameter1
> intercept_I1
[1] 17.3592
> intercept_I2 <- mean.strength2 - estimate.beta*mean.diameter2
> intercept_I2
[1] 18.396
> intercept_I3 <- mean.strength3 - estimate.beta*mean.diameter3
> intercept_I3
[1] 15.7752
```

Using y for strength and x for diameter we have the regression lines for the machines: the regression line for M_1 is $y = 17.3592 + 0.954x$; the regression line for M_2 is $y = 18.396 + 0.954x$; the regression line for M_3 is $y = 15.7752 + 0.954x$.

Note that in the output of Problem 9.3 the output of `summary(machine0)` we find the `estimate(intercept)` 17.360, which is the intercept I_1; further we find `machine2 1.037` and `machine3−1.584`.

The intercept I_2 is `(intercept) +machine2=17.360+1.037=18.397` and the intercept I_3 is `(intercept) +machine3=17.360+ (−1.584) =15.776.`

In the rationale in Problem 9.4 we have found that the difference in the adjusted means of $M_2 - M_1$ is 41.4192 − 40.3824 = 1.0368 and this is given in Problem 9.3 in the `summary(machine0 as the Estimate machine2 1.037)`.

But the difference of the adjusted means of $M_2 - M_1$ is according to (9.4) $[\bar{y}_{2.} - \hat{\beta}(\bar{x}_{2.} - \bar{x}_{..})] - [\bar{y}_{1.} - \hat{\beta}(\bar{x}_{1.} - \bar{x}_{..})] = [\bar{y}_{2.} - \hat{\beta}(\bar{x}_{2.})] - [\bar{y}_{1.} - \hat{\beta}(\bar{x}_{1.})] = I_2 - I_1$.

Analogous to the difference of the adjusted means of $M_3 - M_1$ is 38.7984 − 40.3824 = −1.584 and this is given in Problem 9.3 in the summary (`machine0`) as the `Estimate machine3 -1.584`. However, the difference in the adjusted means of $M_3 - M_1$ is, according to (9.4), $I_3 - I_1$.

Problem 9.6 Give the difference in the adjusted means $M_1 - M_2$, $M_1 - M_3$ and $M_2 - M_3$. Give for these expected differences the (1–0.05)-confidence limits.

Solution

Look at the outcomes of `summary(beta0)` in Problem 9.2; from this we find in the `summary(testbeta0)` the estimate of β as diameter 0.954 and the estimate of its standard error as standard error = 0.114.

Look further at the outcomes of `summary(machine0)` in Problem 9.3. From this we find `Estimatemachine2 1.037` with standard error 1.013; further, `Estimate machine3 -1.584` with standard error 1.107. In Problem 9.5 we have already seen that the difference om the adjusted means of $M_2 - M_1$ is 1.037, hence the difference in the adjusted means of $M_1 - M_2$ is −1.037 with the estimate of its standard error 1.013. Analogous to the difference in the adjusted means of $M_3 - M_1$ is −1.584, hence the difference of the adjusted means of $M_1 - M_3$ is 1.584 with the estimate of its standard error 1.107. For the difference in the adjusted means of $M_2 - M_3$ we must calculated it from the results of Problem 9.4 and the estimate of its standard error with (9.6).

The mean square error and its *df* is given in Problem 9.2 in `anova(testbeta0)` as residuals 11 27.986 2.544 hence MSE = 2.544 with *df* = 11. Of course this result is also given in Problem 9.3 in `anova(machine0)`. It is better to use more decimals with MS = 27.986/11.

Example

```
> df <- 11
> MSE <- 27.986/11
> MSE
[1] 2.544182
> tvalue <- qt(0.975, df)
> tvalue
[1] 2.200985
> adj.meanM1 <- 40.3824
> adj.meanM2 <- 41.4192
> adj.meanM3 <- 38.7984
> diffM1M2 <- adj.meanM1 - adj.meanM2
> diffM1M2
[1] -1.0368
> diffM1M3 <- adj.meanM1-adj.meanM3
> diffM1M3
[1] 1.584
> diffM2M3 <- adj.meanM2 - adj.meanM3
> diffM2M3
```

```
[1] 2.6208
> SEM1M2 <- 1.013
> SEM1M3 <- 1.107
>  diameter <- c(20,25,24,25,32,22,28,22,30,28,21,23,26,21,15)
> machine <- factor(rep(1:3, each = 5))
>  tapply(diameter, machine, mean)
   1    2    3
25.2 26.0 21.2
> mean.diameter1 <- 25.2
> mean.diameter2 <- 26.0
> mean.diameter3 <- 21.2
>  tapply(diameter, machine, sum)
  1   2   3
126 130 106
> sum.diameter1 <- 126
> sum.diameter2 <- 130
> sum.diameter3 <- 106
> A <- sum(diameter^2)
> A
[1] 8998
> B <- (sum.diameter1^2 + sum.diameter2^2 + sum.diameter3^2)/5
> B
[1] 8802.4
> Exx <- A-B
> Exx
[1] 195.6
> n <- 5
> SEM2M3 <- sqrt(MSE*(2/n + (mean.diameter2-mean.diameter3)^2/Exx))

> SEM2M3
 [1] 1.147761
> CLlowerM1M2 <- diffM1M2 - SEM1M2*tvalue
> CLlowerM1M2
[1] -3.266398
> CLupperM1M2 <- diffM1M2 + SEM1M2*tvalue
> CLupperM1M2
[1] 1.192798
> CLlowerM1M3 <- diffM1M3 - SEM1M3*tvalue
> CLlowerM1M3
[1] -0.8524906
> CLupperM1M3 <- diffM1M3 + SEM1M3*tvalue
> CLupperM1M3
[1] 4.020491
> CLlowerM2M3 <- diffM2M3 - SEM2M3*tvalue
> CLlowerM2M3
[1] 0.0945949
> CLupperM2M3 <- diffM2M3 + SEM2M3*tvalue
> CLupperM2M3
[1] 5.147005
```

The 0.95-confidence interval for $\alpha_1 - \alpha_2$ is: [−3.266398; 1.192798]

The 0.95-confidence interval for $\alpha_1 - \alpha_3$ is: [−0.8524906; 4.020491]
The 0.95-confidence interval for $\alpha_2 - \alpha_3$ is: [0.0945949; 5.147005]

9.2.2 Unbalanced Completely Randomised Design

Now we will give the ANCOVA for an unbalanced completely randomised design for a treatment A with a classes. Assuming further that there is a linear relationship between the response y and the covariate x, we find that an appropriate statistical model is

$$\boldsymbol{y}_{ij} = \mu + a_i + \beta x_{ij} + \boldsymbol{e}_{ij}, i = 1, \dots, a; j = 1, \dots, n_i, \text{var}(\boldsymbol{y}_{ij}) = \text{var}(\boldsymbol{e}_{ij}) = \sigma^2 \tag{9.8}$$

where μ is a constant, a_i are the treatment effects with the side condition $\sum_{i=1}^{a} a_i = 0$, β is the coefficient for the linear regression of $\boldsymbol{y}_{ij}$ on x_{ij}, and the $\boldsymbol{e}_{ij}$ are random independent normally distributed experimental errors with mean 0 and variance σ^2.

The least squares estimator of the parameters in (9.8) are as follows.

Deriving the first partial derivatives of

$$S = \sum_{i=1}^{a} \sum_{j=1}^{n_i} (y_{ij} - \mu - a_i - \beta x_j)^2$$

with respect to the parameters and zeroing the result at which the parameter values in the equations are replaced by their estimates leads to the equations ($b = \hat{\beta}$):

$$\begin{cases} N\hat{\mu} + \sum_{i=1}^{a} n_i \hat{a}_i + bX_{..} = Y_{..}, \\ n_i\hat{\mu} + n_i\hat{a}_i + bX_{..} = Y_{i.}, i = 1, \dots, a \\ X_{..}\hat{\mu} + \sum_{i=1}^{a} X_{i.}\hat{a}_i + b\sum_{j=1}^{n_i} x_{ij}^2 = \sum_{i=1}^{a}\sum_{j=1}^{n_i} x_{ij}y_{ij} \end{cases} . \tag{9.9}$$

From this, we obtain the estimates explicitly as:

$$\begin{cases} \hat{\mu} + \hat{a}_i = \frac{1}{n_i}(Y_{i.} - bX_{i.}), i = 1, \dots, a \\ b = \frac{\sum_{i=1}^{a}\sum_{j=1}^{n_i}(x_{ij} - \bar{x}_{i.})(y_{ij} - \bar{y}_{i.})}{\sum_{i=1}^{a}\sum_{j=1}^{n_i}(x_{ij} - \bar{x}_{i.})^2} = \frac{\sum_{i=1}^{a} SP_{xy}^{(i)}}{\sum_{i=1}^{a} SS_x^{(i)}} \end{cases} . \tag{9.10}$$

Because

$$S = \sum_{i=1}^{a}\sum_{j=1}^{n_i} \left(y_{ij} - \mu - a_i - \beta x_j\right)^2$$

is a convex function the solution of the equation to set the partial derivatives equal to zero gives a minimum of S.

Replacing the realisations of the random variables in (9.8) by the corresponding random variables results in the least squares estimators

$$\begin{cases} \hat{\mu} + \hat{a}_i = \frac{1}{n_i}(Y_{i.} - bX_{i.}), i = 1, \dots, a \\ b = \dfrac{\sum_{i=1}^{a}\sum_{j=1}^{n_i}(x_{ij} - \bar{x}_{i.})(y_{ij} - \bar{y}_{i.})}{\sum_{i=1}^{a}\sum_{j=1}^{n_i}(x_{ij} - \bar{x}_{i.})^2} = \dfrac{\sum_{i=1}^{a} SP_{xy}^{(i)}}{\sum_{i=1}^{a} SS_x^{(i)}} \end{cases}. \tag{9.11}$$

The estimators in (9.11) are best linear unbiased estimators (BLUE) and normally distributed.

The analysis for the unbalanced ANCOVA goes with R with the same commands, but the differences are in the estimator of β, the standard errors of the estimate of the adjusted treatment means and the standard errors of the differences between the estimates of the adjusted treatment means.

The estimate of β is $\hat{\beta} = E_{xx} / E_{yy}$ with

$$E_{xx} = \sum_{i=1}^{a}\sum_{j=1}^{n}(x_{ij} - \bar{x}_{i.})^2 \text{ and } E_{xy} = \sum_{i=1}^{a}\sum_{j=1}^{n}(x_{ij} - \bar{x}_{i.})(y_{ij} - \bar{y}_{i.})$$

which is also given in (9.10) with $\hat{\beta} = b$.

The estimates of the treatment means adjusted for the covariate at $x =$ overall mean $\bar{x}_{..}$ is

$$\bar{y}_{i.} - \hat{\beta}(\bar{x}_{i.} - \bar{x}_{..}) \text{ for } i = 1, \dots, a. \tag{9.12}$$

The estimate of the standard error of this estimate (9.2) is

$$\sqrt{MS_E\left[\frac{1}{n_i} + \frac{(\bar{x}_{i.} - \bar{x}_{..})^2}{E_{xx}}\right]} \text{ with } MS_E = \frac{SS_E}{a(n-1)-1}. \tag{9.13}$$

The estimate for the difference between two adjusted treatment means is

$$\sqrt{MS_E\left[\frac{1}{n_i} + \frac{1}{n_i} + \frac{(\bar{x}_{i.} - \bar{x}_{j.})^2}{E_{xx}}\right]} \text{ with } MS_E = \frac{SS_E}{a(n-1)-1}. \tag{9.14}$$

Example 9.2 From Walker (1997) we take his example 9.1 "Triglyceride changes adjusted for glycemic control". A new cholesterol-lowering supplement Fibralo, was studied in a randomised double-blind study against a marketed reference agent, Gemfibrozil, in 34 non-insulin dependent diabetic (NIDDM) patients. One of the study objectives was to compare the mean decrease in triglyceride levels (y in percent change) between groups. Degree of glycemic control, measured by haemoglobin A_{1c} levels (HbA_{1c})x (in ng/ml) was thought to be an important factor in the responsiveness to the treatment. This covariate x was measured at the start of the study, and is shown with the percent changes in triglycerides from pretreatment to the end of the 10 week trial for the patients in the data given in Table 9.4.

Problem 9.7 Test in Example 9.2 the null hypothesis H_{B0}: $\beta_1 = \dots = \beta_a$ against the alternative hypothesis H_{BA}: 'there is at least one β_i different from another β_j with $i \neq j$' with a significance level $\alpha = 0.05$ using the model of (9.8).

Table 9.4 Data of a randomised double-blind study.

Fibralo = group 1			Gemfibrozil = group 2		
Patient No	*x*	*y*	**Patient No**	*x*	*y*
7.0	5		1	5.1	10
4	6.0	10	3	6.0	15
7	7.1	−5	5	7.2	−15
8	8.6	−20	6	6.4	5
11	6.3	0	9	5.5	10
13	7.5	−15	10	6.0	−15
16	6.6	10	12	5.6	−5
17	7.4	−10	14	5.5	−10
19	5.3	20	15	6.7	−20
21	6.5	−15	18	8.6	−40
23	6.2	5	20	6.4	−5
24	7.8	0	23	6.0	−10
27	8.5	−40	25	9.3	−40
28	9.2	−25	26	8.5	−20
30	5.0	25	29	7.9	−35
33	7.0	−10	31	7.4	0
			32	5.0	0
			34	6.5	−10

Solution

In R we use the command

```
> fit.seplines <- lm(y ~ group + x + group * x , data = example9_2)
```

Example

```
> x1 <- c(7.0,6.0,7.1,8.6,6.3,7.5,6.6,7.4,5.3,6.5,6.2,7.8,8.5,9.2,5.0,7.0)
> y1 <- c(5,10,-5,-20,0,-15,10,-10,20,-15,5,0,-40,-25,25,-10)
> x2 <- c(5.1,6.0,7.2,6.4,5.5,6.0,5.6,5.5,6.7,8.6,6.4,6.0,9.3,8.5,
        7.9,7.4,5.0,6.5)
> y2 <- c(10,15,-15,5,10,-15,-5,-10,-20,-40,-5,-10,-40,-20,-35,0,0,-10)
> group <- rep(1:2, times= c(16,18))
> x <- c(x1,x2)
> y <- c(y1,y2)
> Group <- factor(group)
> example9_2 <- data.frame(Group,x,y)
> fit.seplines <- lm(y ~ Group + x + Group * x , data = example9_2)
> anova(fit.seplines)
Analysis of Variance Table
Response: y
          Df Sum Sq Mean Sq F value    Pr(>F)
Group      1  327.2   327.2  3.4535   0.07296 .
x          1 5980.9  5980.9 63.1225 7.227e-09 ***
```

```
Group:x     1   61.2    61.2  0.6458    0.42794
Residuals 30 2842.5    94.7
---
Signif. codes:  0 '***' 0.001 '**' 0.01 '*' 0.05 '.' 0.1 ' ' 1
```

Because the interaction `Group:x` has the p-value $Pr(>F) = 0.42794 > 0.05$ we cannot reject the null hypothesis H_{B0}: $\beta_1 = \ldots = \beta_a$ and we can use the ANCOVA model (9.8) with parallel lines for the regression lines of y on x diameter for the groups.

Problem 9.8 Test in Example 9.2 with the ANCOVA model (9.8) the null hypothesis H_{B0}: $\beta = 0$ against the alternative hypothesis H_{BA}: $\beta \neq 0$ with significance level $\alpha = 0.05$.

Solution

In R we use the command

```
> testbeta0 <- lm ( y ~ Group  + x , data = example9_2)
```

Example

We continue with the R program of Problem 9.7.

```
> testbeta0 <- lm ( y ~ Group  + x , data = example9_2)
> anova(testbeta0)
Analysis of Variance Table
Response: y
          Df Sum Sq Mean Sq F value    Pr(>F)
Group      1  327.2   327.2  3.4934  0.07109.
x          1 5980.9  5980.9 63.8521 5.063e-09 ***
Residuals 31 2903.7    93.7
---
Signif. codes:  0 '***' 0.001 '**' 0.01 '*' 0.05 '.' 0.1 ' ' 1
```

Because in the ANOVA table with x as the last model variable we find the p-value $Pr(>F) = 5.063e\text{-}09 < 0.05$ and we reject the null hypothesis H_{B0}: $\beta = 0$.
Of course this can also concluded from the t-test for x where we find the same p-value $Pr(>|t|) = 5.06e\text{-}09 < 0.05$.

```
> summary(testbeta0)
Call:
lm(formula = y ~ Group + x, data = example9_2)
Residuals:
     Min       1Q   Median       3Q      Max
-19.0353  -6.8607   0.1951   6.1214  18.7915
Coefficients:
            Estimate Std. Error t value Pr(>|t|)
(Intercept)   74.814     10.163   7.361 2.75e-08 ***
Group2       -10.222      3.363  -3.040  0.00478 **
x            -11.268      1.410  -7.991 5.06e-09 ***
---
Signif. codes:  0 '***' 0.001 '**' 0.01 '*' 0.05 '.' 0.1 ' ' 1
```

```
Residual standard error: 9.678 on 31 degrees of freedom
Multiple R-squared:  0.6848,    Adjusted R-squared:  0.6644
F-statistic: 33.67 on 2 and 31 DF,  p-value: 1.692e-08
```

Problem 9.9 Test in Example 9.2 with the ANCOVA model (9.8) the null hypothesis H_{A0}: 'all a_i are equal' against the alternative hypothesis H_{AA}: 'at least one a_i is different from the other' with significance level $\alpha = 0.05$.

Solution

In R we use the command

```
> Group0 <- lm ( y ~ x + Group , data = example9_2)
```

Example

We continue with the R program of Problem 9.7.

```
> Group0 <- lm ( y ~ x + Group , data = example9_2)
> anova(Group0)
Analysis of Variance Table
Response: y
          Df Sum Sq Mean Sq F value    Pr(>F)
x          1 5442.7  5442.7 58.1068 1.353e-08 ***
Group      1  865.4   865.4  9.2387  0.004783 **
Residuals 31 2903.7    93.7
---
Signif. codes:  0 '***' 0.001 '**' 0.01 '*' 0.05 '.' 0.1 ' ' 1
> summary(Group0)
Call:
lm(formula = y ~ x + Group, data = example9_2)
Residuals:
     Min       1Q   Median       3Q      Max
-19.0353  -6.8607   0.1951   6.1214  18.7915
Coefficients:
            Estimate Std. Error t value Pr(>|t|)
(Intercept)   74.814     10.163   7.361 2.75e-08 ***
x            -11.268      1.410  -7.991 5.06e-09 ***
Group2       -10.222      3.363  -3.040  0.00478 **
---
Signif. codes:  0 '***' 0.001 '**' 0.01 '*' 0.05 '.' 0.1 ' ' 1
Residual standard error: 9.678 on 31 degrees of freedom
Multiple R-squared:  0.6848,    Adjusted R-squared:  0.6644
F-statistic: 33.67 on 2 and 31 DF,  p-value: 1.692e-08
```

From the ANOVA table with the group as the last model variable we find the p-value $\mathrm{PR}(>F) = 0.004783 < 0.05$ and we reject the null hypothesis H_{A0}: 'all a_i are equal'.

Hence, the mean triglyceride response adjusted for glycemic control differs between the two treatment groups.

Problem 9.10 Estimate the adjusted group means at the overall mean of x.

Solution
Look at the outcomes of summary(beta0) in Problem 9.8; from this we find the estimate of β as x: − 11.268. Apply then formula (9.12).

Example
We continue with the R program of Problems 9.8 and 9.9.

We calculate first the means of y and x per group and the overall mean of y and x.

```
> tapply(y , Group, mean)
        1         2
 -4.06250 -10.27778
> tapply(x , Group, mean)
       1        2
7.000000 6.644444
> mean.y1 <- -4.06250
> mean.y2 <- -10.27778
> mean.x1 <- 7.000000
> mean.x2 <- 6.644444
> mean.y <- mean(y)
> mean.y
[1] -7.352941
> mean.x <- mean(x)
> mean.x
[1] 6.811765
> estimate.beta <- -11.268
> adj.Group1 <- mean.y1-estimate.beta*(mean.x1-mean.x)
> adj.Group1
[1] -1.941465
> adj.Group2 <- mean.y2-estimate.beta*(mean.x2-mean.x)
> adj.Group2
[1] -12.16315
```

The adjusted mean at the overall mean of x of group 1 is −1.941465 and the adjusted mean of group 2 at the overall mean of x is −12.16315.

Problem 9.11 Give the difference in the adjusted means at the overall mean of x of group 1 − group 2.

Give for this expected difference the (0.95)-confidence limits.

Solution
Look at the outcomes of summarybeta0) in Problem 9.8; from this we find in the summary(testbeta0) the estimate of β as x =−11.268 and the estimate of its standard error as Standard Error=1.410.

Look further at the outcomes of `summary(Group0)` in Problem 9.9. From this we find `(Intercept) 74.814` and `Estimate Group2−10.222` with `Standard Error 3.363.`

The value of intercept = 74.814 is the adjusted treatment mean of group 1 at the value $x = 0$, `mean.y1-estimate.beta*mean.x1 = −4.06250 − (−11.268)* 7.000000 = 74.8135.`

From Problem 9.10 we can calculate the difference in the adjusted means at the overall mean of x of group 2 – group 1 is −12.16315 (−1.941465), hence the difference of the adjusted means at the overall mean of x of group 2 – group 1 is −10.221685 with the estimate of its standard error 3.363. We see that this is already calculated in `summary(Group0)` in Problem 9.9. Hence, the difference in the adjusted means at the overall mean of x of group 1 – group 2 is 10.222 also with the estimate of its standard error 3.363.

We will control this estimate of its standard error using (9.14).

The nean square error and its df is given in Problem 9.8 in `anova(testbeta0)` as `Residuals 31 2903.7 93.7` hence MSE = 93.7 with $df = 31$. Of course this result is also given in Problem 9.9 in `anova(Group0)`. However, it is better to get more decimals to use MSE = 2903.7/31.

Example

```
> x1 <- c(7.0,6.0,7.1,8.6,6.3,7.5,6.6,7.4,5.3,6.5,6.2,7.8,8.5,9.2,5.0,7.0)
> x2 <- c(5.1,6.0,7.2,6.4,5.5,6.0,5.6,5.5,6.7,8.6,6.4,6.0,9.3,8.5,7.9,7.4,5.0,6.5)
> x <- c(x1,x2)
> n1 <- length(x1)
> n2 <- length(x2)
> n1
[1] 16
> n2
[1] 18
> mean.x1 <- mean(x1)
> mean.x2 <- mean(x2)
> sum.x1 <- sum(x1)
> sum.x2 <- sum(x2)
> MSE <- 2903.7/31
> MSE
[1] 93.66774
> df <- 31
> tvalue <- qt(0.975, df)
> tvalue
[1] 2.039513
> A <- sum(x^2)
> B <- (sum.x1^2)/n1 + (sum.x2^2)/n2
> Exx <- A-B
> Exx
[1] 47.10444
> SEG1G2 <- sqrt(MSE*(1/n1+1/n2+(mean.x1-mean.x2)^2/Exx))
> SEG1G2
[1] 3.362943
> diff.G1G2 <- 10.222
> CLlowerG1G2 <- diff.G1G2 - SEG1G2*tvalue
> CLlowerG1G2
[1] 3.363233
> CLupperG1G2 <- diff.G1G2 + SEG1G2*tvalue
> CLupperG1G2
[1] 17.08077
```

We see that the estimate of the standard error of the adjusted means at the overall mean of x of group 1 – group 2 is `SEG1G2 = 3.363`, which we have already found as 3.363 in `summary(Group0)` in Problem 9.9.

The 0.95-confidence interval for the expected means at the overall mean of x of group 1 – group 2 is: [3.363233; 17.08077].

Problem 9.12 Give the estimated regression lines for the groups Fibralo = group 1 and Gemfibrozil = group 2 and make a scatter plot of the data with the two regression lines.

Solution

The regression line for group i goes through the point $(\bar{x}_{i.,}\ \bar{y}_{i.})$, the sample mean of the x values and the y values of group i. The estimate of β is b as `x − 11.268`, given in Problem 9.8 in the `summary(testbeta0)` and with the estimate of its standard error as `1.410`.

The estimated regression line for group i is $y_{ij} = I_i + bx_{ij}$ hence the intercept is $I_i = \bar{y}_{i.} - b\,\bar{x}_{i..}$

Example

Continuing with the results of the R program of Problem 9.8 we find the intercepts a_i.

Note that in the output of Problem 9.8 the output of `summary(Group0)` we find the `Estimate (Intercept) 74.814` and `Group2 −10.222`.

The intercept a_1 is hence `(Intercept) = 74.814`

The intercept a_2 is hence `(Intercept) + Group2 = 74.814 + (−10.222) = 64.592`.

Using the results of Problem 9.10 we find:

```
> estimate.beta <- -11.268
> mean.y1 <- -4.06250
> mean.y2 <- -10.27778
> mean.x1 <- 7.000000
> mean.x2 <- 6.644444
> I1 <- mean.y1 - estimate.beta* mean.x1
> I1
[1] 74.8135
> I2 <- mean.y2 - estimate.beta* mean.x2
> I2
[1] 64.59181
```

Hence the regression line for Fibralo = group 1 is $y = 74.8135 - 11.268x$ and the regression line for Gemfibrozil = group 2 is $y = 64.59181 - 11.268x$.

The scatter plot with the regression lines is done as follows.

```
> x1 <- c(7.0,6.0,7.1,8.6,6.3,7.5,6.6,7.4,5.3,6.5,6.2,7.8,8.5,9.2,5.0,7.0)
> y1 <- c(5,10,-5,-20,0,-15,10,-10,20,-15,5,0,-40,-25,25,-10)
> x2 <- c(5.1,6.0,7.2,6.4,5.5,6.0,5.6,5.5,6.7,8.6,6.4,6.0,9.3,8.5,
        7.9,7.4,5.0,6.5)
> y2 <- c(10,15,-15,5,10,-15,-5,-10,-20,-40,-5,-10,-40,-20,-35,0,0,-10)
> Group <- rep(1:2, times= c(16,18))
```

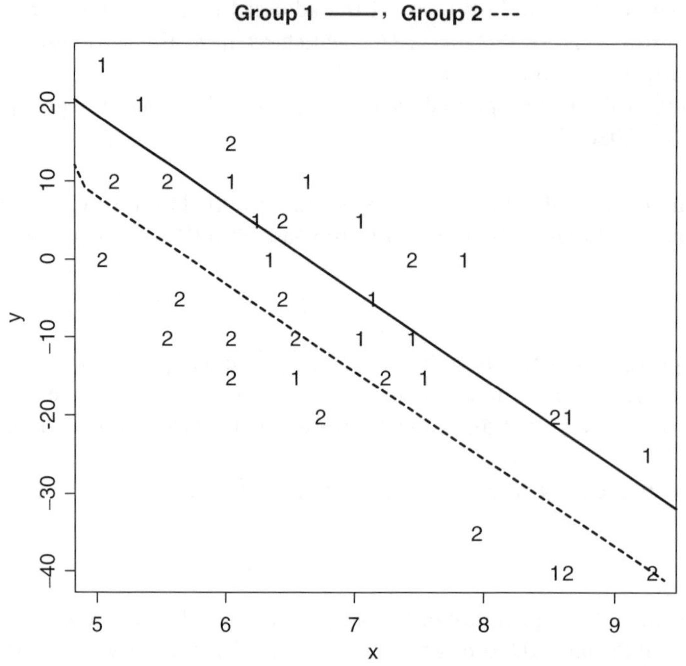

Figure 9.2 Scatter-plot with regression lines of the example in Problem 9.2.

```
> x <- c(x1,x2)
> y <- c(y1,y2)
> plot(x, y, main = "Group 1 ---, Group 2 ----", pch = as.character(Group))
> abline(I1,estimate.beta, lty=1, lwd=2)
> abline(I2,estimate.beta, lty=2, lwd=3)
```

See Figure 9.2.

9.3 Randomised Complete Block Design with Covariate

We assume that we have a balanced randomised complete block design for a treatment A with a classes and b blocks. Assuming further that there is a linear relationship between the response y and the covariate x, we find that an appropriate statistical model is

$$\boldsymbol{y}_{ijk} = \mu + a_i + c_j + \beta x_{ijk} + \boldsymbol{e}_{ijk}, i = 1, \dots, a; j = 1, \dots, b, k = 1, \dots, n \tag{9.15}$$

where μ is a constant, a_i is the treatment effect with the side condition $\sum_{i=1}^{a} a_i = 0$, c_j is the block effect with the side condition $\sum_{j=1}^{b} c_j = 0$, β is the coefficient for the linear regression of $\boldsymbol{y}_{ijk}$ on x_{ijk}, and the $\boldsymbol{e}_{ijk}$ are random independent normally distributed experimental errors with mean 0 and variance σ^2.

The analysis is analogous to Section 9.2.1 for the balanced completely randomised design; only the ANOVA table additionally includes the effect of blocks.

We want first to check if we can use an ANCOVA model with parallel lines, which means that there is a separate intercepts regression model. Therefore, we run first an ANCOVA model with interaction of treatments and covariate.

An appropriate statistical model is

$$y_{ijk} = \mu + a_i + c_j + \beta_i x_{ijk} + e_{ijk}, i = 1, \ldots, a; j = 1, \ldots, b, k = 1, \ldots, n \quad (9.16)$$

μ is a constant, a_i is the treatment effect with the side condition $\sum_{i=1}^{a} a_i = 0$, c_j is the block effect with the side condition $\sum_{j=1}^{b} c_j = 0$, β_i is the coefficient for the linear regression for treatment A with class A_i of y_{ijk} on x_{ijk}, and the e_{ijk} are random independent normally distributed experimental errors with mean 0 and variance σ^2.

In R make an ANOVA table for this model and look for the test of the interaction effect of treatment and covariate whether we must reject the null hypothesis H_{BO}: $\beta_1 = \ldots = \beta_a$ and accept the alternative hypothesis H_{BA}: 'there is at least one β_i different from another β_j with $i \neq j$'. If this is the case we cannot use the ANCOVA model.

Example 9.3 From Snedecor and Cochran (1989) we use example 18.4.2. In a randomised complete block design with four blocks (factor B), six varieties of corn (Zea mays) (factor V) are tested. The plots are planted with 30 plants of a variety, but during the growing season, several plants die. The yield of ear corn y (in pounds) per plot and the stand x (number of plants at harvesting per plot) are given in Table 9.5.

The experiment compared the yields y of six varieties of corn. There are some variation from plot to plot in number of plants (stand). If this variation is caused by differences in fertility in different plots and if higher plants result in higher yields per plot, increased precision will be obtained by adjusting for the covariance analysis of yield on plant number. The plant numbers in this event serve as an index of fertility levels of the plots. If some varieties characteristically have higher plant numbers than others through a greater ability to germinate or to survive when the plants are young, the adjustment for stand distorts the yields because it is trying to compare the varieties at some average plant number level that the varieties do not attain in practice.

With this in mind, look first at an ANOVA for varieties in x (stand). The ANOVA table give for varieties the MS(V) = 9.167 with df = 5 and for error the MS(E) = 7.589 with df = 15.

The F-ratio = 1.208 has a p-value = 0.352. The low value of F gives some assurance that the variations in stand are mostly random and that adjustment for stand will not introduce bias.

Note: this experiment is described in Section 9.1 as example 2.

Table 9.5 Data of a randomised complete block design with four blocks (factor B) and six varieties of corn (Zea mays) (factor V).

Block (B)	B_1		B_2		B_3		B_4	
Variety (V)	x	y	x	y	x	y	x	y
V_1	28	202	22	165	27	191	19	134
V_2	23	145	26	201	28	203	24	180
V_3	27	188	24	185	27	185	28	220
V_4	24	201	28	231	30	238	30	261
V_5	30	202	26	178	26	198	29	226
V_6	30	228	25	221	27	207	24	204

Problem 9.13 Test in Example 9.3 the null hypothesis H_{B0}: $\beta_1 = \ldots = \beta_a$ against the alternative hypothesis H_{BA}: 'there is at least one β_i different from another β_j with $i \neq j$' with a significance level $\alpha = 0.05$ using the model (9.16).

Solution
In R we use the command

```
> fit.seplines <- lm(y ~ B + V + x + V:x , data = example9_3)
```

Example

```
> x1 <- c(28,23,27,24,30,30)
> y1 <- c(202,145,188,201,202,228)
> x2 <- c(22,26,24,28,26,25)
> y2 <- c(165,201,185,231,178,221)
> x3 <- c(27,28,27,30,26,27)
> y3 <- c(191,203,185,238,198,207)
> x4 <- c(19,24,28,30,29,24)
> y4 <- c(134,180,220,261,226,204)
> x <- c(x1,x2,x3,x4)
> y <- c(y1,y2,y3,y4)
> b <- rep(1:4, each = 6)
> B <- factor(b)
> v1 <- c(1,2,3,4,5,6)
> v <- c(v1,v1,v1,v1)
> V <- factor(v)
> example9_3 <- data.frame(B, V, x, y)
> fit.seplines <- lm(y ~ B + V + x + V:x , data = example9_3)
> anova(fit.seplines)
Analysis of Variance Table
Response: y
          Df Sum Sq Mean Sq F value    Pr(>F)
B          3  436.2   145.4  1.3792 0.3106353
V          5 9490.0  1898.0 18.0049 0.0001896 ***
x          1 7391.0  7391.0 70.1131 1.535e-05 ***
V:x        5  412.5    82.5  0.7827 0.5867580
Residuals  9  948.7   105.4
---
Signif. codes:  0 '***' 0.001 '**' 0.01 '*' 0.05 '.' 0.1 ' ' 1
```

Because the interaction $V{:}x$ has the p-value $\Pr(>F) = 0.5867580 > 0.05$ we cannot reject the null hypothesis H_{B0}: $\beta_1 = \ldots = \beta_a$ and we can use the ANCOVA model (9.15) with parallel lines for the regression lines of y on x for V.

Problem 9.14 Test in Example 9.3 with the ANCOVA model (9.15) the null hypothesis H_{B0}: $\beta = 0$ against the alternative hypothesis H_{BA}: $\beta \neq 0$ with significance level $\alpha = 0.05$.

Solution
In R we use the command

```
> testbeta0 <- lm ( y ~ B + V  + x , data = example9_3)
```

Example

We continue with the R program of Problem 9.13.

```
> testbeta0 <- lm ( y ~ B + V + x , data = example9_3)
> anova(testbeta0)
Analysis of Variance Table
Response: y
          Df Sum Sq Mean Sq F value    Pr(>F)
B          3  436.2   145.4  1.4952     0.259
V          5 9490.0  1898.0 19.5198 7.313e-06 ***
x          1 7391.0  7391.0 76.0124 4.963e-07 ***
Residuals 14 1361.3    97.2
---
Signif. codes:  0 '***' 0.001 '**' 0.01 '*' 0.05 '.' 0.1 ' ' 1
> summary(testbeta0)
Call:
lm(formula = y ~ B + V + x, data = example9_3)
Residuals:
     Min       1Q   Median       3Q      Max
-15.3829  -6.3924  -0.1922   4.4051  16.5853
Coefficients:
            Estimate Std. Error t value Pr(>|t|)
(Intercept) -31.1765    23.5836  -1.322  0.20738
B2           17.2727     5.9399   2.908  0.01146 *
B3            5.3044     5.7118   0.929  0.36880
B4           20.5771     5.8250   3.533  0.00331 **
V2           -0.8223     7.0677  -0.116  0.90903
V3            1.3554     7.3455   0.185  0.85625
V4           27.5187     7.8920   3.487  0.00363 **
V5           -2.2169     7.7865  -0.285  0.78004
V6           21.8554     7.3455   2.975  0.01003 *
x             8.0578     0.9242   8.719 4.96e-07 ***
---
Signif. codes:  0 '***' 0.001 '**' 0.01 '*' 0.05 '.' 0.1 ' ' 1
Residual standard error: 9.861 on 14 degrees of freedom
Multiple R-squared:  0.9271,    Adjusted R-squared:  0.8803
F-statistic: 19.79 on 9 and 14 DF,  p-value: 1.761e-06
```

Because in the ANOVA table with x as the last model variable we find the p-value Pr(>F) = 4.963e-07 < 0.05, we reject the null hypothesis H_{B0}: $\beta = 0$.
Of course this can also concluded from the t-test for x where we find the same p-value Pr(>|t|) = 4.96e-07 < 0.05.

Problem 9.15 Test in Example 9.3 with the ANCOVA model (9.15) the null hypothesis H_{A0}: 'all a_i are equal' against the alternative hypothesis H_{AA}: 'at least one a_i is different from the other' with significance level $\alpha = 0.05$.

Solution

In R we use the command

```
> V0 <- lm ( y ~ x + B + V , data = example9_3)
```

Example
We continue with the R program of Problem 9.14.

```
> V0 <- lm ( y ~ x + B + V , data = example9_3)
> anova(V0)
Analysis of Variance Table
Response: y
          Df  Sum Sq Mean Sq  F value     Pr(>F)
x          1 12161.2 12161.2 125.0702 2.298e-08 ***
B          3  1928.8   642.9   6.6121  0.005207 **
V          5  3227.3   645.5   6.6381  0.002296 **
Residuals 14  1361.3    97.2
---
Signif. codes: 0 '***' 0.001 '**' 0.01 '*' 0.05 '.' 0.1 ' ' 1
> summary(V0)
Call:
lm(formula = y ~ x + B + V, data = example9_3)
Residuals:
     Min       1Q   Median       3Q      Max
-15.3829  -6.3924  -0.1922   4.4051  16.5853
Coefficients:
            Estimate Std. Error t value Pr(>|t|)
(Intercept) -31.1765    23.5836  -1.322  0.20738
x             8.0578     0.9242   8.719 4.96e-07 ***
B2           17.2727     5.9399   2.908  0.01146 *
B3            5.3044     5.7118   0.929  0.36880
B4           20.5771     5.8250   3.533  0.00331 **
V2           -0.8223     7.0677  -0.116  0.90903
V3            1.3554     7.3455   0.185  0.85625
V4           27.5187     7.8920   3.487  0.00363 **
V5           -2.2169     7.7865  -0.285  0.78004
V6           21.8554     7.3455   2.975  0.01003 *
---
Signif. codes: 0 '***' 0.001 '**' 0.01 '*' 0.05 '.' 0.1 ' ' 1
Residual standard error: 9.861 on 14 degrees of freedom
Multiple R-squared:  0.9271,    Adjusted R-squared:  0.8803
F-statistic: 19.79 on 9 and 14 DF,  p-value: 1.761
```

From the ANOVA table with V as the last model variable we find the p-value $\Pr(>F) = 0.002296 < 0.05$ and we reject the null hypothesis H_{A0}: 'all α_i are equal'.
Hence the mean yield y adjusted for stand x differs between the varieties.

Problem 9.16 Estimate the adjusted variety means at the overall mean of x.

Solution
Look at the outcomes of `summary(beta0)` in Problem 9.14; from this we find the estimate of β as x 8.0578. Apply then formula (9.12).

Example
We continue with the R program of Problems 9.14 and 9.15.

We calculate now first the means of y and x per variety and the overall mean of y and x.

```
>tapply(y, V, mean)
>tapply(y, V, mean)

      1      2      3      4      5      6
173.00 182.25 194.50 232.75 201.00 215.00
> mean.yV1 <- 173.00
> mean.yV2 <- 182.25
> mean.yV3 <- 194.50
> mean.yV4 <- 232.75
> mean.yV5 <- 201.00
> mean.yV6 <- 215.00
> tapply (x, V, mean)
    1     2     3     4     5     6
24.00 25.25 26.50 28.00 27.75 26.50
> mean.x1 <- 24.00
> mean.x2 <- 25.25
> mean.x3 <- 26.50
> mean.x4 <- 28.00
> mean.x5 <- 27.75
> mean.x6 <- 26.50
> mean.y <- mean(y)
> mean.y
[1] 199.75
> mean.x <- mean(x)
> mean.x
[1] 26.33333
> estimate.beta <- 8.0578
> adj.V1 <- mean.yV1-estimate.beta*(mean.x1-mean.x)
> adj.V1
[1] 191.8015
> adj.V2 <- mean.yV2-estimate.beta*(mean.x2-mean.x)
> adj.V2
[1] 190.9793
> adj.V3 <- mean.yV3-estimate.beta*(mean.x3-mean.x)
> adj.V3
[1] 193.157
> adj.V4 <- mean.yV4-estimate.beta*(mean.x4-mean.x)
> adj.V4
[1] 219.3203
> adj.V5 <- mean.yV5-estimate.beta*(mean.x5-mean.x)
> adj.V5
[1] 189.5848
> adj.V6 <- mean.yV6-estimate.beta*(mean.x6-mean.x)
> adj.V6
[1] 213.657
```

The adjusted means at the overall mean of x are for: V_1 191.8015; V_2 190.9793; V_3 193.157; V_4 219.3203; V_5 189.5848; V_6 213.657.

Problem 9.17 Give the difference of the adjusted means at the overall mean of x of $V_1 - V_2$.

Give for the expected difference of the adjusted means at the overall mean of x of $V_1 - V_2$ the (0.95)-confidence limits.

Solution

Look at the outcomes of `summary(beta0)` in Problem 9.14; from this we find in the `summary(testbeta0)` the estimate of β as x 8.0578 and the estimate of its standard error as standard error = 0.9242.

Look further at the outcomes of `summary(V0)` in Problem 9.15. From this we find estimate `V2−0.8223` with standard error 7.0677.

From Problem 9.16 we can calculate the difference of the adjusted means of $V_2 - V_1$ is 190.9793–191.8015 = –0.8222 which is given in `summary(V0)` with the estimate of its standard error 7.0677. Hence the difference of $V_1 - V_2$ is 0.8222 also with the estimate of its standard error 7.0677. We will control this using (9.14).

The mean square error and its df is given in Problem 9.14 in `anova(testbeta0)` as `Residuals 141361.3 97.2` hence MSE = 97.2 with df = 14. Of course this result is also given in Problem 9.15 in `anova(V0)`. But to have more decimals it is better to use MSE = 1361.3/14.

```
> x1 <- c(28,23,27,24,30,30)
> x2 <- c(22,26,24,28,26,25)
> x3 <- c(27,28,27,30,26,27)
> x4 <- c(19,24,28,30,29,24)
> x <- c(x1,x2,x3,x4)
> v1 <- c(1,2,3,4,5,6)
> v <- c(v1,v1,v1,v1)
> tableVX <- data.frame(v,x)
> vx1 <- tableVX[v==1,]
> vx2 <- tableVX[v==2,]
> vx3 <- tableVX[v==3,]
> vx4 <- tableVX[v==4,]
> vx5 <- tableVX[v==5,]
> vx6 <- tableVX[v==6,]
> n <- length(vx1$x)
> svx1 <- sum(vx1$x)
> svx2 <- sum(vx2$x)
> svx3 <- sum(vx3$x)
> svx4 <- sum(vx4$x)
> svx5 <- sum(vx5$x)
> svx5 <- sum(vx5$x)
> svx6 <- sum(vx6$x)
> df <- 14
> MSE <- 1361.3/df
```

```
> MSE
[1] 97.23571
> tvalue <- qt(0.975, df)
> tvalue
[1] 2.144787
> A <- sum(x^2)
> B <- (svx1^2+svx2^2+svx3^2+svx4^2+svx5^2+svx6^2)/n
> Exx <- A-B
> Exx
[1] 135.5
> mean.vx1 <- mean(vx1$x)
> mean.vx2 <- mean(vx2$x)
> SEV1V2 <- sqrt(MSE*(2/n +(mean.vx1-mean.vx2)^2/Exx))
> SEV1V2
[1] 7.052597
> diff.V1V2 <- 0.8222
> CLlowerV1V2 <- diff.V1V2 - SEV1V2*tvalue
> CLlowerV1V2
[1] -14.30412
> CLupperV1V2 <- diff.V1V2 + SEV1V2*tvalue
> CLupperV1V2
[1] 15.94852
```

The 0.95-confidence interval for the expected yield of the adjusted means at the overall mean of x of $V_1 - V_2$ is: [−14.30412; 15.94852].

The estimate of the standard error of the difference of the adjusted means at the overall mean of x of $V_1 - V_2$ is given by `anova(testb0)` as 7.0677 and the calculated standard error 7.052597 is due to the output of `anova(testb0)` not showing enough decimals for the `SS(Residuals)`. If we use the estimate of the standard error of the difference of the adjusted means at the overall mean of x of $V_1 - V_2$ as 7.0677 then the 0.95-confidence limits are:

```
> SEV1V2 <- 7.0677
> CLlowerV1V2 <- diff.V1V2 - SEV1V2*tvalue
> CLlowerV1V2
[1] -14.33651
> CLupperV1V2 <- diff.V1V2 + SEV1V2*tvalue
> CLupperV1V2
[1] 15.98091
```

Now the 0.95-confidence interval for the expected yield of $V_1 - V_2$ is: [−14.333651; 15.98091].

9.4 Concluding Remarks

If the covariate x as well as the primary response variable y is affected by the treatments the resultant response is multivariate and the covariance adjustments for treatments means is inappropriate. In these cases an analysis of the bivariate response $(\boldsymbol{x}, \boldsymbol{y})$ utilising multivariate methods is in order. Multivariate methods are not covered in this book.

Adjustment for the covariate is appropriate if it is measured prior to treatment administration since the treatments have not yet had the opportunity to affect its value. If the covariate is measured concurrently with the response variable, then it must be decided whether it could be affected by the treatments before the covariance adjustments are considered. See the rationale given in Example 9.3.

Practical application of the analysis of covariance has been demonstrated only with completely randomised designs and randomised complete block designs; however, the use of covariates can be extended to any treatment and experiment design as well as to comparative observational studies of complex structure and studies requiring the use of multiple covariates for adjustment. Using R, the analysis of multiple covariates is easy to do. Add the multiple covariates in the model equation.

For further information about topics in analysis of covariance, see Snedecor and Cochran (1989) sections 18.5–18.9.

Extensive discussions on the use and misuses of covariates in research studies were provided in two special issues of *Biometrics* (1957), Volume 13, No. 3; and *Biometrics* (1982), Volume 38, No. 3. Of particular interest are articles by Cochran (1957), Smith (1957), and Cox and McCullagh (1982). A number of issues arise relevant to the use of covariates. Amongst those concerns are the applicability in certain situations and the relationship between blocking and covariates.

Analysis of covariance for general random and mixed-effects models is considerably more difficult. Henderson and Henderson (1979) and Henderson (1982) discuss the problem and possible approaches.

References

Cochran, W.G. (1957). Analysis of covariance: its nature and uses. *Biometrics* 13: 261–281.

Cox, D.R. and McCullagh, P. (1982). Some aspects of analysis of covariance. *Biometrics* 38: 54–561.

Fisher, R.A. (1935). *Statistical Methods for Research Workers, 5th* e. Edinburgh: Oliver & Boyd.

Henderson, C.R. Jr., (1982). Analysis of covariance in the mixed model: higher-level, nonhomogeneous, and random regressions. *Biometrics* 38: 623–640.

Henderson, C.R. Jr., and Henderson, C.R. (1979). Analysis of covariance in mixed models with unequal subclass numbers. *Communications in Statistics A* 8: 751–788.

Montgomery, D.C. (2013). *Design and Analysis of Experiments, 8th* e. New York: John Wiley & Sons, Inc.

Smith, H.F. (1957). Interpretation of adjusted treatment means and regressions in analysis of covariance. *Biometrics* 13: 282–308.

Snedecor, G.W. and Cochran, W.G. (1989). *Statistical Methods, 8th* e. Ames: Iowa State University Press.

Walker, G.A. (1997). *Common Statistical Methods for Clinical Research, with SAS Examples.* Cary, NC: SAS Institute Inc.

10

Multiple Decision Problems

10.1 Introduction

We will make statements about $a \geq 2$ populations (distributions). We first discuss selection procedures and later multiple comparisons of means. Gupta and Huang (1981) give a good overview of multiple decision problems.

10.2 Selection Procedures

We start with a set G of a populations, i.e. $G = \{P_i, i = 1, \dots, a\}$. These populations correspond with random variables with parameter vectors for which at least one component is unknown. This general approach is described in Rasch and Schott (2018). In this book we restrict ourselves to a special case. In P_i we consider stochastically independent random variables with a $N(\mu_i^*, \sigma^2)$-distribution.

It is our aim to select the population with the largest expectation or the populations with the $t < a$ largest expectations. In Section 10.7 we discuss selection procedures for variances.

We first order the populations in G by magnitude of their expectations. For this, we need an order relation.

Definition 10.1 A population P_k is considered to be better than the population P_j $(j \neq k = 1, \dots, a)$, if $\mu_k^* > \mu_j^*$. P_k is not worse than P_j, if $\mu_k^* \geq \mu_j^*$.

The values $\mu_1^*, \dots, \mu_a^*$ and the a populations can be ordered; if $\mu_{(i)}^*$ is the ith ordered (by magnitude) expectation, then we have $\mu_{(1)}^* \leq \mu_{(2)}^* \leq \dots \leq \mu_{(a)}^*$.

Next we renumber the populations by permuting the indices 1,…, a. To avoid confusion between the original and the permuted indices we denote the populations, the random variables and the parameters afresh. The permutation $\begin{pmatrix} (1) & (2) & \dots & (a) \\ 1 & 2 & \dots & a \end{pmatrix}$ transforms the population P_j with its parameter μ_j^* into the population A_j, with the random variable $\boldsymbol{y}_j$ with expectation μ_j respectively, and A_i is not worse than A_{i^*} if $i \geq i^*$. Of course the permutation is unknown; we use it only to simplify writing.

Definition 10.2 If the set $G = \{A_1, \dots, A_a\} = \{P_1, \dots, P_a\}$ should be partitioned into at least two subsets, so that in one of the subsets the better elements of G following

Applied Statistics: Theory and Problem Solutions with R, First Edition.
Dieter Rasch, Rob Verdooren, and Jürgen Pilz.

Definition 10.1 are contained, we say, we have a selection problem. A decision function (−rule) performing such a partition is called a selection rule or a selection procedure.

The theory of selection procedures is only about 70 years old -see Miescke and Rasch (1996 a, b)- and is the youngest branch of statistical procedures.

We consider the case that G shall be partitioned exactly into two subsets G_1 and G_2 so that $G_1 = \{A_{t+1}, \dots A_a\}$ and $G_2 = \{A_1, \dots A_t\}$.

10.2.1 The Indifference Zone Formulation for Selecting Expectations

Selection Problem 10.1. (Bechhofer, 1954). For a given risk of wrong decision β with $\binom{a}{t}^{-1} < 1-\beta < 1$ and $\delta > 0$ a subset M_B of size t from G has to be selected. Selection is based on random samples $(y_{i1}, \dots y_{in_i})$ from A_i with normally distributed components y_i. Select M_B in such a way that the probability $P(CS)$ of a correct selection is

$$P(CS) = P_C = P(M_B = G_1 | d(G_1, G_2) \geqq \delta) \geqq 1 - \beta. \tag{10.1}$$

In (10.1) $d(G_1, G_2)$ is the distance between A_{a-t+1} and A_{a-t}. the value δ, given in advance and, besides β, is part of the precision requirement.

The condition $\binom{a}{t}^{-1} < 1-\beta < 1$ is reasonable, because for $1-\beta \leq \binom{a}{t}^{-1}$ no real statistical problem exists. Without experimenting, one can denote any of the $\binom{a}{t}$ subsets of size t by M_B and with $1-\beta \leq \binom{a}{t}^{-1}$ it fulfils (10.1).

The region $[\mu_{a-t+1}, \mu_{a-t+1} - \delta]$ is the indifference zone and Selection Problem 10.1 is often called the indifference zone formulation of selection problems.

From Guiard (1994) a modified problem formulation is:

Selection Problem 10.1A. Select a subset M_B of size t corresponding to Selection Problem 10.1 in such a way that in place of (10.1)

$$P_{C^*} = P(M_B \subset G_1^*) \geq 1 - \beta. \tag{10.2}$$

is used. Here G_1^* is the set in G, containing all A_i with $\mu_i \geq \mu_{a-t+1} - \delta$.

For practical purposes, this formulation is easier to understand. Later, we show how results for Selection Problem 10.1 apply to Selection Problem 10.1A.

For the selection procedure, take from each of a normal populations with the random variable y_i with $E(y_i) = \mu_i; \text{var}(y_i) = \sigma^2$ a random sample $(y_{i1}, \dots, y_{in_i})$. These a random samples are assumed to be stochastically independent and the components y_{ij} are assumed to be distributed like y_i. We base decisions on the estimators $\overline{y}_i$ or its realisations of μ_i.

10.2.1.1 Indifference Zone Selection, σ^2 Known

We first assume that the variances of the a populations are equal and known. In practice this is rarely the case but this approach gives a better understanding.

Selection Rule 10.1. From the a independent random samples the sample means $\overline{y}_1, \dots, \overline{y}_a$ are calculated and then we select the t populations with the t largest means into the set M_B, see Bechhofer (1954).

Selection Rule 10.1 can only be applied, if σ^2 is known. If σ^2 is unknown, we apply multi-stage selection procedures in Section 10.2.1.2.

Theorem 10.1 (Bechhofer, 1954). Under the assumptions of this section, we put $\delta^* = \frac{\delta\sqrt{n}}{\sigma}$ and if $n_1 = \cdots = n_a = n$ for the probability of a correct selection P_C we obtain:

$$P_C = P\{\max(\overline{y}_1, \ldots, \overline{y}_{a-t}) < \min(\overline{y}_{a-t+1}, \ldots, \overline{y}_a)\} \tag{10.3}$$

$$P_C \geqq t \int_{-\infty}^{\infty} [\Phi(z+\delta^*)]^{a-t}[1-\Phi(z)]^{t-1}\varphi(z)dz \tag{10.4}$$

as long as $\mu_{a-t+1} - \mu_{a-t} > \delta^*$ always holds where Φ and φ are the distribution function and the density function of the standardised normal distribution, respectively.

Bechhofer (1954) gave the proof – see also Rasch and Schott (2018, theorem 11.1).

For the often occurring special case $t = 1$ formula (10.4) becomes

$$P_C \geq \int_{-\infty}^{\infty} [\Phi(z+\delta^*)]^{a-1}\varphi(z)dz. \tag{10.5}$$

Problem 10.1 Determine the minimal sample size n so that for given a and δ the probability of a correct selection of the largest expectation P_C in (10.5) is at least $1 - \beta$.

Solution

Using the R-command in the package `OPDOE`:

```
> size.selection.bechhofer(a=,beta=,delta=,sigma=)
```

we can calculate n.

Example

For $a = 10$, $P_C = 0.95$, $\delta = 0.5$, $\sigma = 1$ we get

```
> library(OPDOE)
> size.selection.bechhofer(a=10,beta=0.05,delta=0.5,sigma=1)
[1] 47
```

and therefore need in each sample 47 observations.

For $a = 10$, $P_C = 0.95$, $\delta = 1$, and $\sigma = 1$ we get

```
> size.selection.bechhofer(a=10,beta=0.05,delta=1,sigma=1)
[1] 12
```

Problem 10.2 Determine the minimal sample size n so that for given a and $\delta = c\sigma$ the probability of a correct selection of the t largest expectation P_C in (10.5) is at least $1 - \beta$.

Solution

At first add to R the command to make a function

```
 > Bech.snr = function(a,t, c, beta) # Bechhofer's optimum
sample number
  {
   K=(1-beta)/t
 BechInt = function(n,a,t,c) # Bechhofer-Integral
```

```
{d=c *sqrt(n)
        integrand=function(z)
        {(pnorm(z+d) ^ (a-t))*((1-pnorm(z)) ^ (t-1))*dnorm(z)}
        intnew=integrate(integrand,-Inf,Inf)$value
        intnew
       }
f = function(n,a,t,c, beta)}BechInt(n,a,t,c)- K}
k = uniroot(f,c(1,1000),a = a,t = t,c = c,beta = beta)$root
  # Finds the unique minimum integer k0 for which f(k0)>K
  k0=ceiling(k)
  print(paste("optimum sample number: n =",k0), quote=F)
  }
```

Then use the function

```
> Bech.snr(a=,t=,c =,beta=)
```

Example
We use:

1. $a = 20$, c $= 0.5$, $t = 4$, $\beta = 0.05$.
2. $a = 10$, c $= \delta = 0.5$, $t = 1$, $\beta = 0.05$, to show that we can solve the example of Problem 10.1 without `OPDOE`.
3. $a = 10$, c $= \frac{22}{\sqrt{300}}$, $t = 1$, $\beta = 0.05$.

The results are

1. `> Bech.snr(a = 20,t = 4, c=0.5,beta = 0.05)`
   `[1] optimum sample number: n = 75`
2. `> Bech.snr(a = 10,t = 1, c=0.5,beta = 0.05)`
   `[1] optimum sample number: n = 47`
3. `> Bech.snr(a = 10,t = 4,c=1.27,beta = 0.05)`
   `[1] optimum sample number: n = 10`

Of course, in (2) we obtain the same sample size as in the example of Problem 10.1.

Example 10.1 To select the four populations with the largest expectations from $a = 10$ populations with equal variances we observed ten data per population and received the sample means in Table 10.1.

Table 10.1 Sample means of Example 10.1.

Popula tion	P_1	P_2	P_3	P_4	P_5	P_6	P_7	P_8	P_9	P_{10}
$\bar{y}_{i\cdot}$	138.6	132.2	138.4	122.7	130.6	131.0	139.2	131.7	128.0	122.5

Using Selection Rule 10.1, we have to select the populations P_7, P_1, P_3, P_2.

10.2.1.2 Indifference Zone Selection, σ^2 Unknown

Bechhofer showed for Selection Rule 10.1 that P_C has the maximal lower bound of (10.4) if we have normal distributions with known and equal variances and use $n_i = n$; $i = 1, \ldots, a$ for fixed a, t, δ.

If σ^2 is unknown a two-stage selection rule is proposed.

Selection Rule 10.2. Calculate from observations (realisations of $\boldsymbol{y}_{ij}$) $y_{ij}(i = 1, \ldots, a$; $j = 1, \ldots, n_0)$ from the a populations $A_1, \ldots, A_a$ with

$$10 \leq n_0 \leq 30$$

the estimate

$$s_0^2 = MS_{\text{res}} \tag{10.6}$$

with $df = a(n_0 - 1)$ degrees of freedom, as in Table 5.2 with $N = an_0$. For given a, c, t, and β we calculate using the R-command of Problem 10.2 the sample size n. If $n > n_0$ we take from each of the a populations $n - n_0$ additional observations, otherwise n_0 is the final sample size. With n or n_0 we continue as in Selection Rule 10.1.

In place of Selection Problem 10.l, Selection Problem 10.1A can always be used. There are advantages in applications. The researcher could ask what can be said about the probability that we really selected the t best populations, if $(\mu_{a-t+1}, \mu_{a-t}) < \delta$. An answer to such a problem is not possible, it is better to formulate Selection Problem 10.1A, which can better be interpreted, and where we now know at least with probability $1 - \beta$ that we selected exactly t populations not being more than δ worse than A_{a-t+1}.

Guiard (1994) showed that the least favourable cases concerning the values of P_C and P_{C^*} for Selection Problem 10.1 and 10.1A are identical. By this, the lower bounds $1 - \beta$ in (10.2) are identical for Selection Problems 10.1 and 10.1A. We call P_{C^*} the probability of a δ-precise selection.

10.3 The Subset Selection Procedure for Expectations

Now it is again our aim to select the population with the largest expectations. The size $\boldsymbol{r}$ of the selected subset $\boldsymbol{M}_G$ is random. Selection is based on a stochastically independent random samples $(\boldsymbol{y}_{i1}, \ldots \boldsymbol{y}_{in_i})$; its components $\boldsymbol{y}_{ij}$ are assumed to be distributed like $\boldsymbol{y}_i$, which are $N(\mu_i, \sigma^2)$ distributed. We base decisions on the estimators $\overline{\boldsymbol{y}}_i$ or its realisations of μ_i.

Selection Problem 10.2 (Gupta 1956).

For a given risk β of incorrect decision with

$$\binom{a}{t}^{-1} < 1 - \beta < 1 \tag{10.7}$$

select from G a subset $\boldsymbol{M}_G$ of random size $\boldsymbol{r}$ so that

$$P\{G_2 \subset \mathrm{M}_G\} \geqq 1 - \beta. \tag{10.8}$$

Selection Problem 10.2 is the subset formulation of the selection problem.

The following selection rule stems from Gupta and Panchapakesan (1970, 1979).

Selection Rule 10.3. We use the estimators $\overline{\boldsymbol{y}}_i$ based on a samples of equal size n, which are $N\left(\mu_i, \frac{\sigma^2}{n}\right)$-distributed.

All the A_i are put into M_G, for which

$$\bar{y}_{i.} \geq \bar{y}_{a.} - \delta. \tag{10.9}$$

We have to choose $\delta = \frac{D\sigma}{\sqrt{n}}$ with a pre-given D so that

$$1 - \beta = \int_{-\infty}^{\infty} [\Phi(u + D)]^{a-1} \varphi(u) \mathrm{d}u \tag{10.10}$$

where Φ and φ are the distributions function and the density function of the standardised normal distribution, respectively.

If σ^2 is unknown we write approximately $\delta \approx \frac{Ds}{\sqrt{n}}$, where s^2 is an estimate of σ^2, based on f degrees of freedom. Then (10.10) is replaced by

$$1 - \beta = \int_{0}^{\infty} \int_{-\infty}^{\infty} [\Phi(u + Dy)]^{a-1} \varphi(u) h_f(y) \mathrm{d}u\mathrm{d}y, \tag{10.11}$$

where $h_f(y)$ is the density function of $\sqrt{\chi_f^2/f}$ and χ_f^2 is $CS(f)$-distributed.

Theorem 10.2 (Gupta and Panchapakesan [1970]) For a normal random variable $\boldsymbol{y}$ for $t = 1$ Selection Problem 10.2 is solvable with Selection Rule 10.3 for all β with $\frac{1}{a} < \beta < 1$.

Gupta and Panchapakesan [1970] further proved Theorem 10.3.

Theorem 10.3 (Gupta and Panchapakesan (1970)). Using Selection Rule 10.3 and with the assumptions of Theorem 10.2 the supremum of the expectation $E(\boldsymbol{r})$ is taken for $\mu_1 = \mu_2 = \ldots = \mu_a$.

Therefore $\mu_1 = \mu_2 = \ldots = \mu_a$ is the least favourable parameter constellation for Selection Problem 10.2.

10.4 Optimal Combination of the Indifference Zone and the Subset Selection Procedure

Which of the two problem formulations is better suited for practical purposes? Often experiments with a technologies, a varieties, medical treatments and others, have the purpose of selecting the best of them (i.e. $t = 1$). If we then have a huge list of candidates at the beginning – say about 500, such as in drug screening, then it is reasonable at first to reduce the number of candidates by a subset procedure down to let us say $r = 20$. In a second step we then use an indifference zone procedure with $a = r$. We will present an optimal combination of the two approaches taken from simulation results in Rasch and Yanagida (2019).

A Simulation Experiment. (Rasch and Yanagida, 2019)

We choose $t = 1$ and as final probability of a correct selection $P_C = 0.95$. Further we assume that the μ_i, $i = 1, \ldots a > 1$ are expectations of normal distributions with equal variances.

We start with Gupta's approach with a probability of correct selection $PC_{Gu} \geq 0.95$ and a sample size n_{Gu} as small as possible, both found by systematic search. This results in a subset of size r for each run. If A_a is not in that subset the simulation run was finished with the result 'incorrect selection'. If A_a is in the subset, and $r = 1$, the simulation run was finished with the result 'correct selection'. If, however, A_a is in the subset and $r > 1$ we continue with Bechhofer's approach with a sample size n_B from the R-program of Selection Problem 10.1.

How can the free parameters of both problems be combined in an optimal way so that the overall probability of correct selection is $P_C = 0.95$ and the total experimental size

$$N = an_G + r(n_B - n_{Gu}) \tag{10.12}$$

is as small as possible?

Rasch and Yanagida (2019) showed that for unknown variances and $a \geq 30$ nearly no difference occurs between known and unknown variances due to the large degrees of freedom of the t-distribution. Therefore, for a not too small the results are valid for unknown variances too.

In the simulation experiment, we used Gupta's approach with sample sizes n_{Gu} shown in Table 10.2.

The simulated observations in each sample and each run are x_{ij}, $i = 1, \ldots, a$; $j = 1, \ldots, n_{Gu}$. Each x_{ij} is a realisation of a normally distributed random variable with expectation μ_i; $i = 1, \ldots, a$; $\mu_a = 1$; $\mu_j = 0, j = 1, \ldots, a{-}1$ and variance $\sigma^2 = 1$.

Using Gupta's approach for the a realised sample means $\bar{x}_{i.}, i = 1, \ldots, a$ results in a subset of size r.

If $\bar{x}_a$ is in this subset and $r > 1$ we simulate from the r populations in this subset samples of size obtained by R.

Using Bechhofer's approach results in an a best-selected population; if it is not A_a, the selection is incorrect. We used in the simulation experiment besides the probability of correct selection $P_C = 0.95$ $a = 30, 50, 100, 200$. All simulations had 100 000 runs, which means that the relative frequencies of correct selection are not far from 0.95. When in N_f of the runs the selection was incorrect (either already after Gupta's approach or at the end) then

$$1 - \frac{N_f}{100000}$$

is a good estimate of P_C.

Table 10.3 shows the average size of the selected subset, which must be used as the number of populations in the indifference zone selection.

Table 10.4 shows the optimal values of P_B. Table 10.5 shows the average total number of observations needed for both selection procedures and Table 10.6 the estimated overall probability of correct selection.

Table 10.2 Values of n_{Gu} used in the simulation experiment.

	a			
δ	30	50	100	200
$\delta = \frac{\sigma}{2}$	19	20	20	23
$\delta = \sigma$	5	5	5	6

We learn from the simulation experiment that a combination of Gupta's and Bechhofer's approach leads to a smaller total sample size than the use of Bechhofer's approach alone.

In the mean time further results are obtaind by simulation for $t > 1$ and for non-normal distributions. For this see the special issue of JSTP and a proceedings volume (Springer) of the 10th International Workshop on Simulation and Statistics, Salzburg 2019.

Table 10.3 Values of average $\bar{r}$ found in the subset selection.

	a			
δ	30	50	100	200
$\delta = \frac{\sigma}{2}$	12.25	18.81	32.08	59.81
$\delta = \sigma$	12.48	18.70	31.91	62.07

Table 10.4 Optimal values of P_B used in the simulation experiment.

	a			
δ	30	50	100	200
$\delta = \frac{\sigma}{2}$	0.981	0.983	0.991	0.988
$\delta = \sigma$	0.975	0.981	0.991	0.982

Table 10.5 Average total size of the simulation experiment (upper entry) and the size needed for Bechhofer's approach only (lower entry).

	a			
δ	30	50	100	200
$\delta = \sigma/2$	1151.7	2013.70	4253.95	8836.00
	1830	3350	7500	16 400
$\delta = \sigma$	288.38	500.25	1071.70	2215.90
	480	850	1900	4200

Table 10.6 Relative frequencies of correct selection calculated from 100 000 runs.

	a			
δ	30	50	100	200
$d = \frac{\sigma}{2}$	0.9519	0.9512	0.9510	0.9504
$d = \sigma$	0.9504	0.9516	0.9507	0.9502

10.5 Selection of the Normal Distribution with the Smallest Variance

Let the random variable $\boldsymbol{x}$ in P_i be $N(\mu_i, \sigma_i^2)$-distributed with known μ_i. From n observations from each population P_i; $i = 1, \dots, a$ we calculate

$$y_i = \frac{1}{n} \sum_{j=1}^{n} \left(x_{ij} - \mu_i\right)^2.$$

In the case of unknown μ_i we use estimates $\overline{x}_i$ and calculate

$$y_i = \frac{1}{n-1} \sum_{j=1}^{n} (x_{ij} - \overline{x}_i)^2, (i = 1, \dots, a).$$

The y_i will be used to select the population with the smallest variance; each y_i has the same number f of degrees of freedom (if μ_i is known we have $f = n$ and if μ_i is unknown then $f = n-1$).

Selection for the smallest variance follows from Selection Rule 10.4.

Selection Rule 10.4.

Put into $\boldsymbol{M}_G$ all P_i for which

$s_i^2 \leq \frac{s_{(1)}^2}{z^*}$, where $s_{(1)}^2$ is the smallest sample variance. $z^* = z(f, a, \beta) \leq 1$ depends on the degrees of freedom f, the number a of populations and on β.

For z^* we choose the largest number, so that the right-hand side of (10.11) equals $1 - \beta$. We have to calculate $P(CS)$ for the least favourable case given by $s_{(2)}^2 = \dots = s_{(a)}^2$ (monotonicity of the likelihood ratio is given). We denote the estimates of σ_i^2 as usual by s_i^2 and formulate (for the proof see Rasch and Schott (2018, theorem 11.5)

Theorem 10.4 Let the $\boldsymbol{y}_i$ in a populations be $N(\mu_i, \sigma_i^2)$-distributed, there may be independent estimators $\boldsymbol{s}_i^2$ of σ_i^2 with f degrees of freedom each. Select from the a populations a subset $\boldsymbol{N}_G$ so that it contains the smallest variance σ_1^2 at least with probability $1 - \beta$. Using Selection Rule 10.4 with an appropriately chosen $z^* = z(f, a, \beta)$ the probability of a correct selection $P(CS)$ then is

$$P(CS) = \int_0^\infty \left[1 - G_f(z^*, v)\right]^{a-1} g_f(v) \mathrm{d}v \tag{10.13}$$

In (10.13) G_f and g_f are the distribution function and the density function of the central χ^2-distribution with f degrees of freedom, respectively.

10.6 Multiple Comparisons

Real multiple decision problems (with more than two decisions) are present; if we consider results of some tests simultaneously, their risks must be mutually evaluated. Of course, we cannot give a full overview about methods available. For more details, see Miller (1981) and Hsu (1996).

We consider the situation that a populations P_i, $i = 1, \dots, a$ are given in which random variables $\boldsymbol{y}_i$, $i = 1, \dots, a$ are independent of each other normally distributed with expectations μ_i and common but unknown variance σ^2. When the variances are unequal, some special methods are applied and we discuss only some of them. The situation is similar to that of the one-way ANOVA in Chapter 5; the a populations are the levels of a fixed factor A.

We assume that from a populations independent random samples $\boldsymbol{Y}_i^T = (\boldsymbol{y}_{i1}, \dots, \boldsymbol{y}_{in_i})$ $(i = 1, \dots, a)$ of size n_i are drawn. Some of the methods assume the balanced case with equal sample sizes. We consider several multiple comparison (MC) problems.

MC Problem 10.1. Test the null hypothesis

$$H_0: \quad \mu_1 = \mu_2 = \cdots = \mu_a$$

against the alternative hypothesis

H_A: there exists at least one pair (i,j) with $i \neq j$ so that $\mu_i \neq \mu_j$.

The first kind risk is defined not for a single pair (i,j) but for all possible pairs of the experiment and this risk is therefore called an experiment-wise risk α_e.

MC Problem 10.2. Test each of the $\binom{a}{2}$ null hypotheses

$$H_{0,ij}; \mu_i = \mu_j; i \neq j; i,j = 1, \dots, a$$

against the corresponding alternative hypothesis

$$H_{A,ij}; \mu_i \neq \mu_j; i \neq j; i,j = 1, \dots, a.$$

Each pair $(H_{0,ij}; H_{A,ij})$ of these hypotheses is independently tested of the others. This corresponds with the situation of Chapter 3 for which the two-sample t-test was applied.

The first kind risk α_{ij} is defined for the pair $(H_{0,ij}; H_{A,ij})$ of hypotheses and may differ for each pair. Often we choose all $\alpha_{ij} = \alpha_c$ and call it a comparison-wise first kind risk. If we perform all $\binom{a}{2}$ t-tests using the two-sample t-test or the Welch test than we speak about the multiple t-procedure or W-procedure.

MC Problem 10.3. One of the populations (without loss of generality P_a) is prominent (a standard method, a control treatment, and so on). Test each of the $a - 1$ null hypotheses

$$H_{0,i}; \mu_i = \mu_a; i \neq a; i = 1, \dots, a-1 \tag{10.14}$$

against the alternative hypothesis

$$H_{A,i}; \mu_i \neq \mu_a; i \neq a; i = 1, \dots, a-1. \tag{10.15}$$

The first kind risk α_i for the pair $(H_{0,i}; H_{A,i})$ of hypotheses is independent of that one of other pairs and therefore also called comparison-wise. Often we choose $\alpha_i = \alpha_c$.

If we use the term experiment-wise, risk of the first kind α_e in MC Problem 10.3 means that it is the probability that in at least one of the $\binom{a}{2}$ pairs $(H_{0,ij}; H_{A,ij})$ of hypotheses in MC Problem 10.2 or that in at least one of the $a - 1$ pairs $(H_{0,i}; H_{A,i})$ of hypotheses in MC Problem 10.3 the null hypothesis is erroneously rejected. However, make sure that such a risk is not really a risk of a statistical test but a probability in a multiple decision problem because when we consider all possible pairs (null hypothesis–alternative hypothesis) of

MC Problem 10.2 or 10.3, we have a multiple decision problem with more than two possible decisions if $a > 2$.

In general we cannot convert α_e and α_c into each other. The asymptotic (for known σ^2) relations for k orthogonal contrasts

$$\alpha_e = 1 - (1 - \alpha_c)^k, \tag{10.16}$$

$$\alpha_c = 1 - (1 - \alpha_e)^{1/k}, \tag{10.17}$$

follow from elementary rules of probability theory, because we can assign to the independent contrasts independent F-tests (transformed z-tests) with $f_1 = 1, f_2 = \infty$ degrees of freedom.

To solve MC Problems 10.2 and 10.3 we first construct confidence intervals for differences of expectations as well as for linear contrasts in these expectations. With these confidence intervals, the problems easily are handled.

As already mentioned at the start of Section 10.2, we assume that from a populations independent random samples $\boldsymbol{Y}_i^T = (\boldsymbol{y}_{i1}, \ \ldots, \ \boldsymbol{y}_{in_i}) \quad (i = 1, \ \ldots, \ a)$ of size n_i are drawn. We consider the a populations P_i as the a levels of a fixed factor A in a one-way analysis of variance as discussed in Section 5.3, therefore we write

$$\boldsymbol{y}_{ij} = \mu + a_i + \boldsymbol{e}_{ij} \ \ (i = 1, \ldots, a; j = 1, \ldots, n_i) \tag{10.18}$$

and call μ the overall expectation and $\mu + a_i$ the expectation of the ith level of population P_i. The total size of the experiment is $N = \sum_{i=1}^{a} n_i$. In applied statistics, we use the notation 'multiple comparison of means' synonymously with 'multiple comparison of expectations'.

10.6.1 The Solution of MC Problem 10.1

We can solve MC Problem 10.1 by several methods. At first, we use the F-test from Chapter 5.

10.6.1.1 The F-test for MC Problem 10.1

If MC Problem 10.1 is handled by the F-test, we use the notations of Table 5.2. $H_0: \mu_1 = \mu_2 = \cdots = \mu_a$ is rejected, if

$$\frac{MS_A}{\text{res}} > F(a-1, \ N - a \mid 1 - \alpha_e). \tag{10.19}$$

Problem 10.3 Determine in a balanced design the subclass number n for MC Problem 10.1 for a precision determined by $\alpha_e = 0.05$, $\beta = 0.05$, and $\delta/\sigma = 1$.

Solution

Use the design function of the R-package OPDOE for the one-way analysis of variance:

```
>size.anova(model="a", a= ,alpha= ,beta= ,delta= ,case= ).
```

Example

Determine n_{min} and n_{max} for $a = 10$, $\alpha_e = 0.05$, $\beta = 0.05$, and $\delta/\sigma = 1$. Attention, the δ in the command corresponds with δ/σ in our precision requirement.

```
> library(OPDOE)
> size.anova(model="a", a=10, alpha=0.05, beta=0.05, delta=1,
      case="minimin")
 n
11
> size.anova(model="a", a=10, alpha=0.05, beta=0.05, delta=1,
      case="maximin")
 n
49
```

We have to draw between 11 and 49 observations from each population.

10.6.1.2 Scheffé's Method for MC Problem 10.1

The method proposed in Scheffé (1953) allows the calculation of simultaneous confidence intervals for all linear contrasts of the $\mu + a_i$ in (10.17).

We now reformulate MC Problem 10.1 as an MC problem to construct confidence intervals. If H_0 is correct then all linear contrasts in the $\mu_i = \mu + \alpha_i$ equal zero. The validity of H_0 conversely follows from the fact that all linear contrasts $L_r = \sum_{j=1}^{a} c_{rj}\mu_j$ vanish. Therefore, confidence intervals $\boldsymbol{K}_r$ for all linear contrasts L_r can be constructed in such a way that the probability that $L_r \in \boldsymbol{K}_r$ for all r is at least $1 - \alpha_e$. We then reject H_0 with a first kind risk α_e, if at least one of the $\boldsymbol{K}_r$ does not cover L_r.

Problem 10.4 Use Scheffé's method to test the null hypothesis of MC Problem 10.1. We consider all linear contrasts L_r in the μ_i.

Solution

Construct

$$\left\langle \hat{\boldsymbol{L}}_r - \boldsymbol{s}\sqrt{(a-1)F(a-1,\ N-a|1-\alpha_e)}\sqrt{\sum_{i=1}^{a}\frac{c_{ri}^2}{n_i}},\right.$$

$$\left.\hat{\boldsymbol{L}}_r + \boldsymbol{s}\sqrt{(a-1)F(a-1,\ N-a|1-\alpha_e)}\sqrt{\sum_{i=1}^{a}\frac{c_{ri}^2}{n_i}},\right\rangle. \tag{10.20}$$

This confidence interval contains all L_r with $\sum_{i=1}^{a} c_{ri} = 0$ with probability $1 - \alpha_e$. Here

$$s = \sqrt{\boldsymbol{MS}_{\text{res}}} \tag{10.21}$$

(from Table 5.2). Using (10.19) to construct confidence intervals for all $\binom{a}{2}$ differences of expectations only, then the confidence interval in (10.20) has a too large expected width, it contains the differences of expectations with a probability $\geq 1 - \alpha_e$. We say in such cases that the confidence intervals and the corresponding tests are conservative. Therefore, the minimal sample sizes are too large.

Example

We construct confidence intervals (10.19) for the $\binom{10}{2}$ differences of $a = 10$ expectations for $\alpha_e = 0.05$ and means calculated with 20 observations each. In this case we have $\sqrt{\sum_{i=1}^{a} \frac{c_{ri}^2}{n_i}} = \sqrt{\frac{1}{n_i} + \frac{1}{n_j}}, i \neq j; i, j = 1, \dots, a$ and $\hat{\boldsymbol{L}}_r = \overline{\boldsymbol{y}}_i - \overline{\boldsymbol{y}}_j, i \neq j; i, j = 1, \dots, a$. At first we calculate $F(9,\ 200 - 10|0.95)$ with R as

```
>    qf(0.95,9,190)
[1] 1.929425
```

and from this $\sqrt{9F(9,\ 200 - 10|0.95)} = 4.16741$ and $\sqrt{\frac{1}{20} + \frac{1}{20}} = 0.3162$. Then we obtain $[\overline{\boldsymbol{y}}_i - \overline{\boldsymbol{y}}_j - 1.318\boldsymbol{s}, \overline{\boldsymbol{y}}_i - \overline{\boldsymbol{y}}_j + 1.318\boldsymbol{s}]$, $i \neq j; i, j = 1, \dots, a$ as confidence intervals for the Scheffé method. Pairs of sample means differing by more than 1.318 standard deviations lead to significant differences in the corresponding expectations.

Example 10.2 We use data simulated by Rasch and Schott (2018, example 11.4). A (pseudo-)random number generator has generated 10 samples of size five each. The values of the samples 1 to 8 are realisations of an $N(50,\ 64)$-normally distributed random variable; the two other samples differ only in expectations, we have $\mu_9 = 52$ and $\mu_{10} = 56$, respectively. Table 10.7 gives the simulated samples.

Table 10.8 contains the differences between the means.

10.6.1.3 Bonferroni's Method for MC Problem 10.1

Confidence intervals by Scheffé's method are not optimal if confidence intervals for k special but not for all contrasts are wanted. Sometimes we obtain shorter intervals using the Bonferroni inequality (Bonferroni, 1936).

Table 10.7 Simulated observations of Example 10.2.

	Number of sample									
y_{ij}	1	2	3	4	5	6	7	8	9	10
y_{i1}	63.4	49.6	50.3	55.5	62.5	30.7	56.7	64.5	44.4	55.7
y_{i2}	46.7	48.4	52.8	36.1	45.8	48.6	46.2	42.2	38.2	64.7
y_{i3}	59.1	49.3	52.5	54.0	52.8	45.8	41.9	49.6	64.8	61.8
y_{i4}	60.7	48.3	58.6	55.9	44.9	44.9	55.8	48.9	43.7	38.9
y_{i5}	54.9	51.5	48.0	52.9	51.3	52.9	48.9	40.7	61.3	61.8
$\overline{y}_{i.}$	56.96	49.42	52.44	50.88	51.46	44.58	49.90	49.18	50.48	56.58

Table 10.8 Differences $\overline{y}_{i.} - \overline{y}_{j.}$ between means of Example 10.2.

j / i	2	3	4	5	6	7	8	9	10
1	7.54	4.52	6.08	5.50	12.38	7.06	7.78	6.48	0.38
2		−3.02	−1.46	−2.04	4.84	−0.48	0.24	−1.06	−7.16
3			1.56	0.98	7.86	2.54	3.26	1.96	−4.14
4				−0.58	6.30	0.98	1.70	0.40	−5.70
5					6.88	1.56	2.28	0.98	−5.12
6						−5.32	4.60	−5.90	−12.00
7							0.72	−0.58	−6.68
8								−1.30	−7.40
9									−6.10

Theorem 10.5 If the components $\boldsymbol{y}_i$ of an a-dimensional random vector $\boldsymbol{y}^T = (\boldsymbol{y}_1, \ldots, \boldsymbol{y}_a)$ with distribution function $F(y_1, \ldots, y_a)$ have the same marginal distribution function $F(y)$ then the Bonferroni inequality

$$1 - F(y_1, \ldots, y_a) \leq \sum_{i=1}^{a} [1 - F(y_i)] \tag{10.22}$$

is valid with the distribution function $F(y_i)$ of $\boldsymbol{y}_i$.

The proof is given in Rasch and Schott (2018, Theorem 11.7).

If k special linear contrasts $L_r = \sum_{j=1}^{a} c_{rj}\mu_j \quad (r = 1, \ldots, k)$ are given, then the estimator $\hat{\boldsymbol{L}}_r = \sum c_{rj}\overline{\boldsymbol{y}}_j$ is $N(L_r,\ k_r\sigma^2)$-distributed for each r with $k_r = \sum_{i=1}^{a} \frac{c_{ri}^2}{n_i}$. Then

$$\boldsymbol{t}_r = \frac{\hat{\boldsymbol{L}}_r - L_r}{\boldsymbol{s}\sqrt{k_r}} \quad (r = 1, \ldots, k) \tag{10.23}$$

with $\boldsymbol{s} = \sqrt{\boldsymbol{MS}_{\text{res}}}$ (from Table 5.2) are components of a k-dimensional random variable. The marginal distributions are central t-distributions with $\nu = \sum_{i=1}^{a}(n_i - 1)$ degrees of freedom and the density $f(t,\ \nu)$.

The Bonferroni inequality allows us to find a lower bound of the probability that all t_r values $(r = 1,\ldots,k)$ lie between $-w$ and $w (w > 0)$. Due to the symmetry of the t-distribution and Theorem 10.5 we get

$$P = P\{-w \leqq \boldsymbol{t}_r < w, r = 1, \ldots, k\} \geqq 1 - 2/k \int_{w}^{\infty} f(t,\ \nu)\mathrm{d}t. \tag{10.24}$$

We choose w so that the right-hand side of (10.23) equals $(1 - \alpha_e)$ and obtain simultaneous $(1 - \alpha_e)$-confidence intervals for the L_r as

$$\left\langle \hat{\boldsymbol{L}}_r - w\sqrt{k_r}\boldsymbol{s},\ \hat{\boldsymbol{L}}_r + w\sqrt{k_r}\boldsymbol{s} \right\rangle. \tag{10.25}$$

This means we determine w so that

$$\int_w^\infty f(t,\ v)\mathrm{d}t = \frac{\alpha_e}{2k} = \alpha_c \tag{10.26}$$

and the Bonferroni inequality has the form

$$P > 1 - \alpha_e \geq 1 - 2k\alpha_c. \tag{10.27}$$

Problem 10.5 Write the confidence interval down for the $\binom{a}{2}$ multiple comparisons of the expectations of a normal distributions with equal variances.

Solution

We have $k = \binom{a}{2}$ and $\hat{\boldsymbol{L}}_r = \overline{\boldsymbol{y}}_i - \overline{\boldsymbol{y}}_j, i \neq j; i, j = 1, \ldots, a$ with means based on n observations and $f(t,\ v)$ is the density of the central t-distribution with $v = N - a$ degrees of freedom. The intervals are

$$\left[\overline{\boldsymbol{y}}_i - \overline{\boldsymbol{y}}_j - \boldsymbol{s}wk_c, \overline{\boldsymbol{y}}_i - \overline{\boldsymbol{y}}_j + \boldsymbol{s}wk_c\right] \tag{10.28}$$

with $k_c = \sqrt{\frac{1}{n} + \frac{1}{n}}$.

Example

As in the example of Problem 10.4 we use $a = 10$ expectations, $\alpha_e = 0.05$ and means calculated with 20 observations each. Then the realisation of (10.28) is with $k_c = \sqrt{\frac{1}{20} + \frac{1}{20}} = 0.3162$

$$\left[\overline{y}_i - \overline{y}_j - 0.3162sw, \overline{y}_i - \overline{y}_j + 0.3162sw\right].$$

The $\left(1 - \frac{0.05}{90}\right) = 0.9994$-quantile of the central t-distribution with 190 degrees of freedom is found by R as

```
>w<- qt(0.9994,190)
>w
3.288529
```

and the interval is

$$\left[\overline{y}_i - \overline{y}_j - 1.04s, \overline{y}_i - \overline{y}_j + 1.04s\right].$$

10.6.1.4 Tukey's Method for MC Problem 10.1 for $n_i = n$

Tukey's method (1953) is applied for equal numbers of observations in all a normally distributed samples.

Definition 10.3 If $X = (\boldsymbol{x}_1, \dots, \boldsymbol{x}_a)^T$ is a random sample with independent $N(\mu, \sigma^2)$-distributed components and if $\nu \boldsymbol{s}^2/\sigma^2$ is independent of $\boldsymbol{Y}$ $CQ(\nu)$-distributed and $\boldsymbol{w}$ is the range of $(\boldsymbol{x}_1, \dots, \boldsymbol{x}_a)$ we call the random variable

$$\boldsymbol{q}_{a,\nu} = \frac{\boldsymbol{w}}{\boldsymbol{s}}$$

the studentised range of $\boldsymbol{X}$ and $\boldsymbol{s}^2$. The studentised augmented range is the random variable

$$\boldsymbol{q}*_{a,\nu} = \frac{1}{\boldsymbol{s}}\left[\max\left(\boldsymbol{w}, \max\left\{|\boldsymbol{x}_i - \mu|: i = 1, \dots, a\right\}\right)\right].$$

Note that $\sqrt{2}\ |q^*|$ equals the critical value traditionally denoted by q, the upper α-quantile of the studentised range distribution with a treatments and ν degrees of freedom.

In applications, the $\boldsymbol{x}_i$ are typically the means of random samples each of size n, $\boldsymbol{s}^2$ is the pooled variance, and the degrees of freedom are $\nu = a(n-1)$.

Tukey's method is based on the distribution of $\boldsymbol{q}_{a,\nu}$. The distribution function of the studentised range $\boldsymbol{q}_{a,\nu}$ is given by formula (11.38) in Rasch and Schott (2018).

We denote by $q(a, \nu|1-\alpha)$ the $(1-\alpha)$-quantile of the distribution function of $\boldsymbol{q}_{a,\nu}$ in dependency on the number a of components of $\boldsymbol{Y}$and the degrees of freedom of $\boldsymbol{s}^2$ in Definition 10.3.

Tukey's method to construct confidence intervals for the differences $\mu_i - \mu_{i'}$ of the expectations of a independent $N(\mu_i, \sigma^2)$-distributed random variables $\boldsymbol{y}_i (i = 1, \dots, a)$ is based on the equivalence of the probabilities

$$P\left\{\frac{1}{\boldsymbol{s}}[(\boldsymbol{y}_i - \boldsymbol{y}_k) - (\mu_i - \mu_k)] \leqq K \quad \text{for all } i \neq k;\ i,\ k = 1, \dots, a\right\}$$

and

$$P\left\{\frac{1}{\boldsymbol{s}} \operatorname*{Max}_{i,\ k}[\boldsymbol{y}_i - \mu_i - (\boldsymbol{y}_k - \mu_k)] \leqq K\ (i,\ k = 1, \dots, a)\right\}.$$

This equivalence is a consequence of the fact that the validity of the inequality in the second term is necessary and sufficient for the validity of the inequality in the first term. The maximum of a set of random variables is understood as its largest order statistic and

$$\max_{\substack{i,k \\ i,k = 1, \dots, a}} [\boldsymbol{y}_i - \mu_i - (\boldsymbol{y}_k - \mu_k)]$$

is the range $\boldsymbol{w}$ of $N(0, \sigma^2)$-distributed random variables, if the $\boldsymbol{y}_i$ independently from each other are $N(\mu_i, \sigma^2)$-distributed. We know from theorem 11.8 in Rasch and Schott (2018) that

$$P\{|(\boldsymbol{y}_i - \boldsymbol{y}_k) - (\mu_i - \mu_k)| \leq q(a,\ f|1-\alpha_e)\boldsymbol{s}\ (i,\ k = 1, \dots, a)\} = 1 - \alpha_e. \tag{10.29}$$

(10.28) gives a class of simultaneous confidence intervals with confidence coefficient $1-\alpha_e$.

Problem 10.6 Construct confidence intervals for differences $\mu_i - \mu_j$ of a expectations of random variables $\boldsymbol{y}_i$; $i = 1, \dots, a$, which are independent of each other normally distributed with expectations μ_i and common but unknown variance σ^2. From a populations independent random samples $\boldsymbol{Y}_i^T = (\boldsymbol{y}_{i1}, \dots, \boldsymbol{y}_{in})$ $(i = 1, \dots, a)$ of equal size n are drawn.

Solution
First we estimate σ^2 by $s^2 = \boldsymbol{MS}_{\mathbf{res}}$ as in the methods above. The intervals

$$\left\langle \bar{y}_{i\cdot} - \bar{y}_{j\cdot} - q[a,\ a(n-1) \mid 1-\alpha_e]\frac{s}{\sqrt{n}},\ \bar{y}_{i\cdot} - \bar{y}_{l\cdot} + q[a,\ a(n-1) \mid 1-\alpha_e]\frac{s}{\sqrt{n}} \right\rangle$$

are $(1-\alpha_e)$-confidence intervals, $q[a,\ a(n-1) \mid 1-\alpha_e]$ is calculated using R.

Example
As in the example of Problem 10.4 we use $a = 10$ expectations, $\alpha_e = 0.05$ and means calculated with 20 observations each. Then we need $q[10,\ 190 \mid 0.95]$.

```
> qtukey(0.95,10, df = 190)
1] 4.527912
```

and obtain

$$\left[\bar{y}_i - \bar{y}_j - 4.5279\frac{s}{\sqrt{20}}, \bar{y}_i - \bar{y}_j + 4.5279\frac{s}{\sqrt{20}}\right] = \left[\bar{y}_i - \bar{y}_j - 1.0125s, \bar{y}_i - \bar{y}_j + 1.0125s\right].$$

10.6.1.5 Generalised Tukey's Method for MC Problem 10.1 for $n_i \neq n$

Spjøtvoll and Stoline (1973) generalised Tukey's method without assuming $n_i = n$.

If all conditions of Section 10.8.1.2 are fulfilled, all linear contrasts $L = \sum_{i=1}^{a} c_i\mu_i$ are simultaneously covered with probability $1-\alpha_e$ by intervals

$$\left\langle \hat{\boldsymbol{L}} - \frac{1}{2}\sum |c_i| q^*(a,\ f \mid 1-\alpha_e)\boldsymbol{s},\ \ \hat{\boldsymbol{L}} + \frac{1}{2}\sum |c_i| q^*(a,\ f \mid 1-\alpha_e)\boldsymbol{s} \right\rangle. \tag{10.30}$$

In (10.29) $\hat{\boldsymbol{L}} = \sum_{i=1}^{a} c_i \boldsymbol{y}_{i\cdot}$, and $\boldsymbol{q}^*(a,\ f \mid 1-\alpha)$ is the $(1-\alpha)$-quantile of the distribution of the augmented studentised range $\boldsymbol{q}^*(a,f)$ (see Rasch and Schott (2018)). These intervals, contrary to the Tukey method, depend on the degree of unbalancedness.

A further generalisation of the Tukey method can be found in Hochberg (1974) and Hochberg and Tamhane (1987) – see also Hochberg and Tamhane (2008).

Problem 10.7 Construct confidence intervals for differences $\mu_i - \mu_j$ of a expectations of random variables $\boldsymbol{y}_i$; $i = 1, \dots, a$, which are independent of each other normally distributed with expectations μ_i and common but unknown variance σ^2. From a populations independent random samples $\boldsymbol{Y}_i^T = (\boldsymbol{y}_{i1}, \ \dots,\ \boldsymbol{y}_{in})$ $(i = 1,\ \dots,\ a)$ of size n are drawn.

Solution
From (10.30) we receive $[\hat{y}_i - \hat{y}_j - sq^*(a,f,1-\alpha_e), \hat{y}_i - \hat{y}_j + sq^*(a,f,1-\alpha_e)]$.

Example
As in the example of Problem 10.4 we use $a = 10$ expectations, $\alpha_e = 0.05$ and means calculated with 20 observations each.

This leads to

$$\left[\hat{y}_i - \hat{y}_j - sq^*(10{,}190{,}0.95), \hat{y}_i - \hat{y}_j + sq^*(10{,}190{,}0.95)\right].$$

Using R we get $q^*(10{,}190{,}0.95)$:

```
> qtukey(.95,10, df =  190)/sqrt(2)
[1] 3.201717
```

10.6.2 The Solution of MC Problem 10.2 – the Multiple t-Test

In the case of equal sample sizes a multiple Welch test analogous to the two-sample Welch test is possible.

Each of the $\binom{a}{2}$ null hypotheses

$$H_{0,ij}; \mu_i = \mu_j; i \neq j; i,j = 1, \ldots, a$$

has to be tested against the corresponding alternative hypothesis

$$H_{A,ij}; \mu_i \neq \mu_j; i \neq j; i,j = 1, \ldots, a.$$

Each pair $(H_{0,ij}; H_{A,ij})$ of these hypotheses is tested independently of the others. This corresponds with the situation of Chapter 3 for which the two-sample t-test was applied.

By contrast to the two-sample t-test the multiple t-test uses in place of the variances of the two samples under comparison the overall variance estimator, which is equal to the residual mean square of the one-way ANOVA in Chapter 5. Notice that it is assumed that all a sample sizes are equal to $n > 1$. The test statistic for testing $H_{0,ij}$ is

$$\boldsymbol{t} = \frac{\bar{\boldsymbol{y}}_i - \bar{\boldsymbol{y}}_j}{\sqrt{\boldsymbol{MS}_{\text{res}}}} \sqrt{\frac{2}{n}}\ (i \neq j; i,j = 1, \ldots, a) \tag{10.31}$$

with $\boldsymbol{MS}_{\text{res}}$ from (5.8).

$H_{0,ij}$ is rejected if the realisation t of $\boldsymbol{t}$ in (10.31) is larger than the $(1-\alpha)$-quantile of the central t-distribution with $N - a$ degrees of freedom.

Problem 10.8 Determine the minimal sample size for testing $H_{0,ij}$; $\mu_i = \mu_j$, $i \neq j$, $i, j = 1, \ldots, a$ for a first kind risk α, a second kind risk β, an effect size δ, and a standard deviation sd.

Solution
Use the R-command in `OPDOE: >size.multiple_t.test(a=,alpha=, beta = Delta=,sd=).`

Example

We calculate the minimal sample size for testing $H_{0,ij}$; $\mu_i = \mu_j$ for $a = 10$, a first kind risk $\alpha = 0.01$, a second kind risk $\beta = 0.1$, and an effect size $\delta = 0.5$ sd using

```
> library(OPDOE)
```

`> size.multiple_t.test(a = 10,alpha = 0.01,beta = 0.1, delta = 0.5,sd = 1)` which gives

```
$size
[1] 121
$total
[1] 1210
```

Next we use lower precision and an effect size $\delta =$ sd to obtain

```
> size.multiple_t.test(a=10,alpha=0.05,beta=0.1,delta=1,sd=1)
$size
[1] 23
$total
[1] 230
```

10.6.3 The Solution of MC Problem 10.3 – Pairwise and Simultaneous Comparisons with a Control

Sometimes $a - 1$ treatments have to be compared with a standard procedure called control. We assume first that we have equal sample sizes n for the treatments.

10.6.3.1 Pairwise Comparisons – The Multiple t-Test

If one of the a populations P, without loss of generality let it be P_a, is considered as a standard or control, we use in case of a pairwise comparison in place of (10.31)

$$\boldsymbol{t} = \frac{\bar{y}_i - \bar{y}_a}{\sqrt{\boldsymbol{MS}_{\text{res}}}}\sqrt{\frac{2}{n}}\ (i = 1, \dots, a-1) \tag{10.32}$$

to test each of the $a - 1$ null hypotheses

$$H_{0,i};\ \mu_i = \mu_a;\ i = 1, \dots, a-1$$

against the corresponding alternative hypothesis

$$H_{A,i};\ \mu_i \neq \mu_a;\ i = 1, \dots, a-1.$$

When we use for the tests of the $a - 1$ differences an overall significance level α^* and we use for each test of the $a - 1$ differences a significance level α, then the Bonferroni inequality gives $\alpha \leq \alpha^* \leq (a - 1)\ \alpha$. When we use for the test of each of the $a - 1$ differences a significance level $\alpha = \alpha^*/(a - 1)$ we have an approximate overall significance level of α^*. To test with an exact overall significance level α^* we must use the multivariate Student distribution $t(a-1, f)$ with $f = df(MS_{\text{res}}) = a(n - 1)$. Dunnett (1955) solved this problem. Let us now derive the optimum choice of sample sizes for the multiple t-test with a control.

Problem 10.9 Determine the minimal sample size for testing $H_{0,i}$; $\mu_i = \mu_a$, $i = 1, \dots, a-1$ with an approximate overall significance level α^* [hence we use for the test a first kind risk $\alpha = \alpha^*/(a-1)$], a second kind risk β, an effect size δ and a standard deviation sd.

Solution

Use the R-command in OPDOE

```
>size.t.test(power =,sig.level =, delta =, sd=, type
="two.sample").
```

Example

We calculate the minimal sample size for testing $H_{0,i}$; $\mu_i = \mu_a$, $i = 1, \dots, a-1$, for $a = 10$, a first kind risk $\alpha = 0.15/9 = 0.0167$, a second kind risk $\beta = 0.05$, and an effect size $\delta = \text{sd}$ using

```
> library(OPDOE)
> size.t.test(power=0.95,sig.level=0.0167,delta=1,sd=1,type="two.sample")
```

which gives

```
[1] 35
```

The total sample size is then 10*35 = 350.

Next we use lower precision with $\beta = 0.1$ and obtain

```
> size.t.test(power=0.8,sig.level=0.0167,delta=1,sd=1,type="two.sample")
[1] 23
```

The total sample size is then 10*23 = 230.

Often we have a fixed set of resources and a budget that allows for only N observations. To maximise the power of the tests we want to minimise the total variance. If we use n_a observations for the control population P_a and for the other treatments an equal sample size of n observations then $N = (a-1)n + n_a$. The variance of $\bar{y}_i - \bar{y}_a$ is $\sigma^2 (1/n + 1/n_a)$. The total variance (TSV) is the sum of the $(a-1)$ parts, hence $\text{TSV} = (a-1)[\sigma^2 (1/n + 1/n_a)]$ and we have to minimise it subject to the constraint $N = (a-1)n + n_a$. This is a Lagrange multiplier problem in calculus where we must minimise $M = \text{TSV} + \lambda[N - (a-1)n - n_a]$. Setting the partial derivatives w.r.t. n and n_a equal to zero yields the equations:

(1) $\partial M/\partial n = -(a-1)\,\sigma^2/n^2 - \lambda(a-1) = 0$

(2) $\partial M/\partial n_a = -(a-1)\,\sigma^2/n_a^2 - \lambda = 0.$

Hence $n_a = N - (a-1)n$ and inserting this in TSV we have a function of n as $f(n)$; the first derivative of $f(n)$ to n, $f'(n) = 0$ gives as solution the stationary point $n = N/[(a-1) + \sqrt{(a-1)}]$. To decide what type of extreme it is calculate $f''(n) = 2/n^3 + 2(a-1)^2/[N-(a-1)n]^2$ and this is positive for all $n > 0$, hence the stationary point belongs to a minimum of $f(n)$ due to the constraint. Of course we must take integer numbers for n and n_a.

In the R package `OPDOE` we cannot solve the minimal sample size when we take the integer numbers for $n_a = n\sqrt{(a-1)}$ and $n = N/[(a-1)+\sqrt{(a-1)}]$.

10.6.3.2 Simultaneous Comparisons – The Dunnett Method

Simultaneous $1-\alpha_e$-confidence intervals for the $a-1$ differences

$$\mu_i - \mu_a \quad (i = 1, \dots, a-1)$$

are constructed for the equal sample sizes n for the a samples. (After renumbering μ_a is always the expectation of the control.)

We consider a independent $N(\mu_i, \sigma^2)$-distributed random variables $\boldsymbol{y}_i$, independent of a CS(f)-distributed random variable $\frac{f\boldsymbol{s}^2}{\sigma^2}$.

Dunnett (1955) derived the distribution of

$$\left(\frac{\boldsymbol{y}_1 - \boldsymbol{y}_a}{\boldsymbol{s}}, \dots, \frac{\boldsymbol{y}_{a-1} - \boldsymbol{y}_a}{\boldsymbol{s}}\right)$$

which is the multivariate Student distribution $t(a-1, f)$ with $f = df(MS_{\text{res}}) = a(n-1)$, which he called the distribution $d(a-1, f)$ with $f = df(MS_{\text{res}}) = a(n-1)$.

Dunnett (1964) and Bechhofer and Dunnett (1988) present the quantiles $d(a-1, f \mid 1-\alpha_e)$ of the distribution of

$$\boldsymbol{d} = \frac{\max\limits_{1\le i\le a-1} \left[\left|\boldsymbol{y}_i - \boldsymbol{y}_a - (\mu_i - \mu_a)\right|\right]}{\boldsymbol{s}\sqrt{2}}.$$

We see that $\boldsymbol{d} \le d(a-1, f \mid 1-\alpha_e)$ is necessary and sufficient for

$$\frac{1}{\boldsymbol{s}\sqrt{2}} \left|\boldsymbol{y}_i - \boldsymbol{y}_a - (\mu_i - \mu_a)\right| \le d(a-1, f \mid 1-\alpha_e).$$

For all i, by

$$\left\langle \boldsymbol{y}_i - \boldsymbol{y}_a - d(a-1, f \mid 1-\alpha_e)\boldsymbol{s}\sqrt{2},\ \boldsymbol{y}_i - \boldsymbol{y}_a + d(a-1, f \mid 1-\alpha_e)\boldsymbol{s}\sqrt{2} \right\rangle \quad (10.33)$$

a class of confidence intervals is given, covering all differences $\mu_i - \mu_a$ with probability $1-\alpha_e$.

For the one-way classification with the notation of Example 10.2 we receive, for equal sub-class numbers n, the class of confidence intervals

$$\left\langle \bar{\boldsymbol{y}}_{i\cdot} - \bar{\boldsymbol{y}}_{a\cdot} - d(a-1, a(n-1) \mid 1-\alpha_e)\, \boldsymbol{s}\frac{\sqrt{2n}}{n}, \right.$$
$$\left. \bar{\boldsymbol{y}}_{i\cdot} - \bar{\boldsymbol{y}}_{a\cdot} + d(a-1, a(n-1) \mid 1-\alpha_e)\boldsymbol{s}\frac{\sqrt{2n}}{n} \right\rangle$$
$$(i = 1, \dots, a-1). \quad (10.34)$$

If $n_i = n$, for $i = 1, \dots, a$, we use the Dunnett procedure, based on confidence intervals of the Dunnett method. Then H_{0i} is rejected, if

$$\sqrt{n}\frac{|\bar{\boldsymbol{y}}_{i\cdot} - \bar{\boldsymbol{y}}_{a\cdot}|}{\sqrt{2}\boldsymbol{s}} > d(a-1, a(n-1) \mid 1-\alpha_e). \quad (10.35)$$

If the n_i are not equal, we use a method proposed by Dunnett (1964) with modified quantiles. For example, when the optimal sample size of the control $n_a = n\sqrt{a-1}$ is taken and $n_i = n$ for $i = 1, 2, \ldots, a - 1$, the table of critical values are given in Dunnett (1964) and Bechhofer and Dunnett (1988).

Dunnett and Tamhane (1992) proposed a further modification, called a step up procedure, starting testing the hypothesis with the least significant of the $a - 1$ test statistics at the left-hand side of (10.33) and then going further with the next larger one and so on. Analogously, a step down procedure is defined. Dunnett and Tamhane (1992) showed that the step up procedure is uniformly more powerful than a method proposed by Hochberg (1988) and preferable to the step down procedure. As a disadvantage they mentioned the greater difficulty of calculating critical values and minimum sample sizes in the standard and other groups. This problem is of no importance if we use the R-package `DunnettTests` together with the R-package `mvtnorm`.

Problem 10.10 Determine the minimal sample size for simultaneously testing $H_{0,i}$; $\mu_i = \mu_a; i = 1, \ldots, a-1$ for a first kind risk α, a second kind risk β, an effect size δ, and a standard deviation sd.

Solution

We use the command of the R package `DunnettTests`:

```
> nvDT(ratio,power,r,k,mu,mu0,contrast,sigma=NA,dist,
       alpha=0.05,mcs=1e+05,testcall)
```

Here k equals our $a - 1$. Further are

`ratio`	The pre-specified ratio of sample size in each of the treatment groups to the sample size in the control group
`power`	The power required to be achieved
`r`	The least number of null hypotheses to be rejected, $r \le 1 \le k$
`k`	Number of hypotheses to be tested, $k \ge 2$ and $k \le 16$
`mu`	Assumed population mean in each of the k treatment groups
`mu0`	Assumed population mean in the control group.
`contrast`	Because in this chapter `mu` and `mu0` are concerned with means of continuous outcomes, specify contrast = 'means'
`sigma`	An (rough) estimate of the error standard deviation
`dist`	Whether the sample size is calculated for t-distributed test statistics (dist = 'tdist') or standard normally distributed test statistics (dist = 'zdist')
`alpha`	The pre-specified overall significance level, default = 0.05 must not be input
`mcs`	The number of Monte Carlo sample points to numerically approximate the power for a given sample size, refer to equations (4.3) and (4.5) in Dunnett and Tamhane (1992)
`testcall`	The applied Dunnett test procedure: 'SD' = step-down Dunnett test; 'SU' = step-up Dunnett test.

In Problem 3.3 it is shown what to do when σ is unknown to obtain an estimate of the error standard deviation.

Example

We use as expectation of the standard $\mu_0 = 2$ and as expectation of the other populations $\mu_i = 3$. We receive in the step up case for testing all $r = k = 9$ hypotheses assuming `ratio` = 1 (This is the case of equal sample sizes n for all a treatments).

```
> library(DunnettTests)
> nvDT(1, 0.95, r=9, 9, mu=3, mu0=2,
      sigma=1.,contrast="means",dist="zdist", testcall="SU")
$'least sample size required in each treatment groups'
[1] 34
$'least sample size required in the control group'
[1] 34
```

with a total experimental size $N = 10 \cdot 34 = 340$. If `ratio` is increased the total sample size increases as we see from

```
> nvDT(2, 0.95, r=9, 9, mu=3, mu0=2,
      sigma=1.,contrast="means",dist="zdist", testcall="SU")
$'least sample size required in each treatment groups'
[1] 48
$'least sample size required in the control group'
[1] 24
```

with a total experimental size $N = 9 \cdot 48 + 24 = 456$.

It can be shown that the best choice for the control is $n_a = n\sqrt{k}$, hence `ratio` $= \frac{1}{\sqrt{k}}$, which gives

```
> nvDT(0.3333, 0.95, r=9, 9, mu=3, mu0=2,
      sigma=1.,contrast="means",dist="zdist", testcall="SU")
$'least sample size required in each treatment groups'
[1] 23
$'least sample size required in the control group'
[1] 70
```

with a total experimental size $N = 9 \cdot 23 + 70 = 277$. A smaller `ratio`, let us say `ratio` = 0.2, leads again to a larger N.

```
> nvDT(0.2, 0.95, r=9, 9, mu=3, mu0=2,
      sigma=1.,contrast="means",dist="zdist", testcall="SU")
$'least sample size required in each treatment groups'
[1] 21
$'least sample size required in the control group'
[1] 105
```

with $N = 9 \cdot 21 + 105 = 294$.

A rough graphical representation of the relation between $a - 1$ and N is shown in Figure 10.1.

Finally, we compare the minimal sample sizes of the methods used for Example 10.2 in Table 10.9.

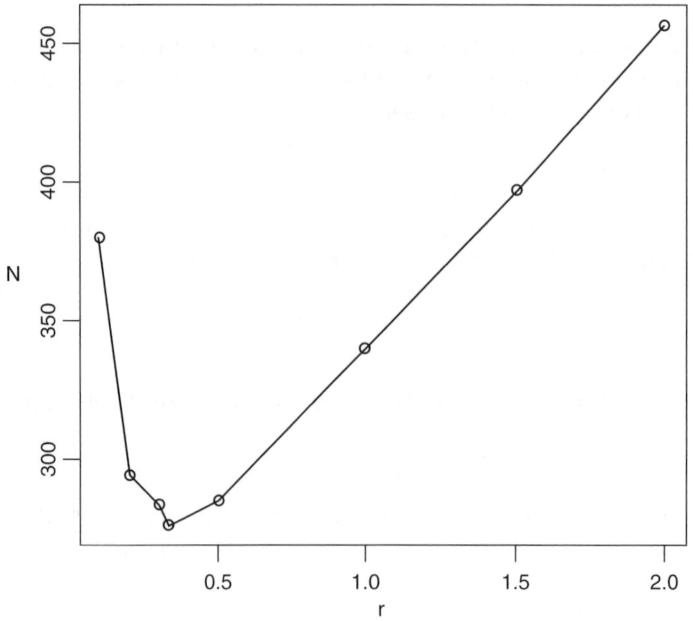

Figure 10.1 Relationship between the total experimental size N and $r = \frac{n}{n_0}$.

Table 10.9 Minimal sample sizes for several multiple decision problems.

Method	n **for** $a = 10, \alpha = 0.05,\ \alpha_e = 0.05$ $\beta = 0.05;\ d = \sigma$	**Remarks**
Selection Rule 10.1 ($t = 1$)	12	$t = 1$
Dunnett's stepwise	27.7 average	9
procedure	23	comparisons
Multiple t-procedure	$11 \le n \le 49$	$n_{10} = n\sqrt{9}$
F-test		45
		comparisons
		One test

Average means $\bar{n} = \frac{1}{10}(9n_1 + n_{10})$.

References

Bechhofer, R.E. (1954). A single sample multiple decision procedure for ranking means of Normal populations with known variances. *Ann. Math. Statist.* 25: 16–39.

Bechhofer, R.E. and Dunnett, C.W. (1988). *Percentage Points of Multivariate Student T Distributions, Selected Tables in Mathematical Statistics 11*. Providence: American Mathematical Society.

Bonferroni, C.E. (1936). Teoria statistica delle classi e calcolo delle probabilità. *Pubbl. d. R. Ist. Super. di Sci. Econom. e Commerciali di Firenze.*

Dunnett, C.W. (1955). A multiple comparison procedure for comparing several treatments with a control. *Jour. Amer. Stat. Ass.* 50: 1096–1121.

Dunnett, C.W. (1964). New tables for multiple comparisons with a control. *Biometrics* 20: 482–491.

Dunnett, C.W. and Tamhane, A.C. (1992). A step-up multiple test procedure. *Jour..Amer. Stat. Ass.* 87: 162–170.

Guiard, V. (1994). Different definitions of Δ-correct selection for the indifference zone formulation. *In Miescke and Rasch* 1994: 176–199.

Gupta, S.S. (1956). *On a Decision Rule for a Problem in Ranking Means*. Mim. Ser., no 150. *Univ. North Carolina.*

Gupta, S.S. and Huang, D.Y. (1981). *Multiple Statistical Decision Theory: Recent Develop-Ments*. New York: Springer.

Gupta, S.S. and Panchapakesan, S. (1970). *On a class of subset selection procedures*. Mim. Ser. 225. *Purdue Univ.* .

Gupta, S.S. and Panchapakesan, S. (1979). *Multiple Decision Procedures: Theory and Methodology of Selecting and Ranking Populations*. New York: Wiley.

Hochberg, Y. (1974). Some generalization of the *T*- method in simultaneous inference. *Jour. Multiv. Anal.* 4: 224–234.

Hochberg, Y. (1988). A sharper Bonferroni procedure for multiple tests of significance. *Biometrika* 75: 800–802.

Hochberg, Y. and Tamhane, A.C. (1987). *Multiple Comparison Procedures*. New York: Wiley.

Hochberg, Y. and Tamhane, A.C. (2008). *Some Theory of Multiple Comparison Procedures for Fixed-Effects Linear Models*. New York: Wiley.

Hsu, J.C. (1996). *Multiple Comparisons*. London: Chapman & Hall.

Miescke, K.J. and Rasch, D. (eds.) (1996a). Special issue on 40 years of statistical selection theory, part I. *J. Statist. Plann. Inference* 54: 2.

Miescke, K.J. and Rasch, D. (eds.) (1996b). Special issue on 40 years of statistical selection theory, part II. *J. Statist. Plann. Inference* 54: 3.

Miller, R.G. (1981). *Simultaneous Statistical Inference*. New York: Springer Verlag.

Rasch, D. and Schott, D. (2018). *Mathematical Statistics*. Oxford: Wiley.

Rasch, D. and Yanagida, T. (2019). An optimal two-stage procedure to select the best out of a Normal population. *Journal of Statistical Theory and Practice* 13: 1.

Scheffé, H. (1953). A method for judging all contrasts in the analysis of variance. *Biometrika* 40: 87–104.

Spjøtvoll, E. and Stoline, M.R. (1973). An extension of the *T*-method of multiple comparisons to include the cases with unequal sample size. *Jour. Amer. Stat. Ass.* 68: 975–978.

Tukey, J.W. (1953). Multiple Comparisons. *Jour. Amer. Stat. Ass.* 48: 624–625.

11

Generalised Linear Models

11.1 Introduction

Like in analysis of variance (ANOVA) or regression analysis a generalised linear model (GLM) describes the relation between a random regressand (response variable) $\boldsymbol{y}$ and a vector $x^T = (x_0, \dots, x_k)$ of regressor variables influencing it, and is a flexible generalisation of ordinary linear regression allowing for regressands that have error distribution models other than a normal one. The GLM generalises linear regression by writing the linear model to be related to the regressand via a link function of the corresponding exponential family and by allowing the magnitude of the variance of each measurement to be a function of its predicted value.

Possibly the first who introduced a GLM was Rasch (1960). The Rasch model is a psychometric model for analysing categorical data, such as answers to questions on a reading assessment or questionnaire responses, as a function of the trade-off between (i) the respondent's abilities, attitudes or personality traits and (ii) the item difficulty. For example, we may use it to estimate a student's reading ability, or the extremity of a person's attitude to capital punishment from responses on a questionnaire. In addition to psychometrics and educational research, the Rasch model and its extensions are used in other areas, including the health profession and market research because of their general applicability. The mathematical theory underlying Rasch models is a special case of a GLM. Specifically, in the original Rasch model, the probability of a correct response is modelled as a logistic function of the difference between the person and item parameter – see Section 11.5.

GLMs were formulated by Nelder and Wedderburn (1972) and later by McCullagh and Nelder (1989) as a way of unifying various other statistical models, including linear regression, logistic regression (Rasch model), and Poisson regression. They proposed an iteratively reweighted least squares method for maximum likelihood (ML) estimation of the model parameters. Maximum likelihood estimation remains popular and is the default method in many statistical computing packages. Other approaches, including Bayesian approaches and least squares fits to variance-stabilised responses, have also been developed. Possibly Nelder and Wedderburn did not know the Rasch model. Later Rasch (1980) extended his approach.

In GLMs, exponential families of distributions play an important role.

Applied Statistics: Theory and Problem Solutions with R, First Edition.
Dieter Rasch, Rob Verdooren, and Jürgen Pilz.

11.2 Exponential Families of Distributions

Definition 11.1 The distribution of a random variable $\boldsymbol{y}$ (may be a vector) with parameter vector $\theta = (\theta_1,\theta_2,\ldots,\theta_p)^T$ belongs to a k-parametric exponential family if its likelihood function can be written as

$$f(y,\theta) = h(y)\,\mathrm{e}^{\sum_{i=1}^{k} \phi_i(\theta)\cdot T_i(y) - B(\theta)} \tag{11.1}$$

where the following conditions hold:

- φ_i and B are real functions of θ and B does not depend on y,
- the function $h(y)$ is non-negative and does not depend on θ.

The exponential family with a random variable $\boldsymbol{y}$ (may be a vector) is in canonical form with the so-called natural parameters η_i, meaning that (11.1) can be reparameterised as

$$f(y,\eta) = h(y)\mathrm{e}^{\sum_{i=1}^{k} \eta_i(\theta)T_i(y) - A(\eta)} \tag{11.2}$$

with $\eta = (\eta_1, \ldots, \eta_k)^T$.

The exponential families include many of the most common distributions. The term $\eta(\theta)$ is called the canonical link function.

We present here some members of the family, which we use in this book.

Example 11.1 Let $(P^*(\theta),\ \theta \in \Omega)$ be the family of two-dimensional normal distributions with the random variable $\boldsymbol{y} = \begin{pmatrix} y_1 \\ y_2 \end{pmatrix}$, the expectation $\mu = \begin{pmatrix} \mu_1 \\ \mu_2 \end{pmatrix}$ and the covariance matrix $\Sigma = \begin{pmatrix} \sigma_1^2 & 0 \\ 0 & \sigma_2^2 \end{pmatrix}$. This is a four-parametric exponential family with the density (likelihood) function

$$f(y,\mu,\Sigma) = \frac{1}{2\pi\sigma_1\sigma_2}\mathrm{e}^{-\frac{1}{2}\left[\frac{(y_1-\mu_1)^2}{\sigma_1^2} + \frac{(y_2-\mu_2)^2}{\sigma_2^2}\right]}$$

and the natural parameters

$$\eta_1 = \frac{\mu_1}{\sigma_1^2}, \eta_2 = \frac{\mu_2}{\sigma_2^2}, \eta_3 = -\frac{1}{2\sigma_1^2}, \eta_4 = -\frac{1}{2\sigma_2^2}$$

and the factors

$$T_1(y) = y_1, T_2(y) = y_2,$$

$$T_3(y) = {y_1}^2, T_4(y) = {y_2}^2,$$

$$A(\eta) = \frac{1}{2}\left(\frac{\mu_1^2}{\sigma_1^2} + \frac{\mu_2^2}{\sigma_2^2}\right).$$

Example 11.2 Let P^* be the family of normal distributions $N(\mu, 1)$ with expectation $\mu = \theta$ and variance 1, i.e. it holds $\Omega = R^1$. P^* is a one-parametric exponential family with $\dim(\Omega) = 1$, the link function is $\eta = \mu$. P^* is complete using theorem 1.4 in Rasch and Schott (2018). If $\boldsymbol{Y} = (y_1, y_2, \ldots, y_n)^T$ is a random sample with components from P^*, then $T_1(Y) = \sum y_i/n$ is distributed as $N(\mu, \frac{1}{n})$. Consequently the family of distributions P^* is also complete. Because of theorem 1.3 in Rasch and Schott (2018) $\bar{\boldsymbol{y}}$ is minimal sufficient and therefore complete sufficient. The distribution family of CS($n-1$)-distributions (χ^2-distributions with $n-1$ degrees of freedom) induced by $(n-1)T_2(Y) = \sum \boldsymbol{y}_i^2 - n\bar{\boldsymbol{y}}^2$ is independent of μ. Hence $\boldsymbol{s}^2 = \frac{1}{n-1}\sum_{i=1}^{n}(\boldsymbol{y}_i - \bar{\boldsymbol{y}})^2$ is an ancillary statistic relative to $\mu = \theta$.

For instance, the following discrete distribution families can be written as exponential families:

Bernoulli distribution, $P(\boldsymbol{y} = y) = p^y(1-p)^{1-y}, 0 < p < 1, y = 0, 1$.

Binomial distribution with the probability (likelihood) function $P(\boldsymbol{y} = y) = \binom{n}{y} p^y(1-p)^{n-y}, 0 < p < 1, y = 0, 1, \ldots, n$.

Poisson distribution with the likelihood function $L_y(y, \lambda) = \frac{\lambda^y}{y!}\mathrm{e}^{-\lambda}, \lambda > 0, y = 0, 1, 2, \ldots$.

Multinomial distribution

$$P(\boldsymbol{y}_i = y_i, i = 1, \ldots, k) = \frac{n!}{y_1! \ldots y_k!} p_1^{y_1} \cdot \ldots \cdot p_k^{y_k}, 0 < p_i < 1, \sum_{i=1}^{k} p_i = 1, \sum_{i=1}^{k} y_i = n.$$

Problem 11.1 Show that the binomial distributions are an exponential family.

Solution

With the canonical notation of Definition 11.1 we get

$$\binom{n}{y} p^y(1-p)^{n-y} = \binom{n}{y} \mathrm{e}^{y \cdot \ln\left(\frac{p}{1-p}\right) + ln(1-p)^n}$$

and $\eta = \ln\left(\frac{p}{1-p}\right)$, $T(y) = y$, $h(y) = \binom{n}{y}$, $A(\eta) = -\ln\left(1 - \frac{\mathrm{e}^\eta}{1+\mathrm{e}^\eta}\right)$.

The function $B(\theta)$ in (11.1) has with $\theta = p$ the form $B(p) = -n\ln(1-p)$.

Problem 11.2 Show that the Poisson distributions are an exponential family.

Solution

With the canonical notation of Definition 11.1 we get

$$\frac{\lambda^y}{y!}\mathrm{e}^{-\lambda} = \frac{1}{y!}\mathrm{e}^{(y\ln\lambda - \lambda)}$$

and $\eta = \ln(\lambda)$, $T(y) = y$, $A(\eta) = \lambda = \mathrm{e}^\eta$, $h(y) = \frac{1}{y!}$.

The function $B(\theta)$ in (11.1) has with $\theta = \lambda$ the form $B(\lambda) = \lambda$.

Problem 11.3 Show that the gamma distributions are an exponential family.

Solution

A continuous random variable $\boldsymbol{y}$ is gamma-distributed with positive parameters $\lambda > 0$ und $\nu > 0$ if its density is given by

$$f(y) = \begin{cases} \dfrac{\lambda^\nu}{\Gamma(\nu)} y^{\nu-1}\mathrm{e}^{-\lambda y}, & \text{if } y \geq 0 \\ 0, & \text{if } y < 0 \end{cases}$$

and $\Gamma(x)$ is the gamma-function

$$\Gamma(x) = \int_0^\infty e^{-t} t^{x-1} dt.$$

The first two moments are $E(\boldsymbol{y}) = \frac{\nu}{\lambda}$ and $var(\boldsymbol{y}) = \frac{\nu}{\lambda^2}$.

For $y \geq 0$ we rewrite $f(y)$ as $f(y) = e^{\ln f(y)} = e^{-\lambda y + \nu \ln \lambda + (\nu - 1)\ln y - \ln \Gamma(\nu)}$.

With the components of the natural parameter $\eta_1 = -\lambda$ and $\eta_2 = \nu$ we see that the gamma distributions belong to the exponential family.

$$e^{\ln f(y)} = e^{-\lambda y + \nu \ln(\lambda) - \ln y + \nu \ln y - \ln \Gamma(\nu)}$$

with $\eta = \begin{pmatrix} \eta_1 \\ \eta_2 \end{pmatrix} = \begin{pmatrix} -\lambda \\ \nu \end{pmatrix}$, $T(y) = \begin{pmatrix} y \\ \ln y \end{pmatrix}$, $A(\eta) = [-\eta_2 \ln(-\eta_1) + \ln \Gamma(\eta_2)]$, $h(y) = \frac{1}{y}$.

As we see in Section 11.7 we can avoid the negative sign of η_1 by using a transformation of the natural parameter as the link function.

11.3 Generalised Linear Models – An Overview

The term GLM usually refers to conventional linear regression models for a continuous or discrete response variable $\boldsymbol{y}$ given continuous and/or categorical predictors. In this chapter we assume that the distribution of $\boldsymbol{y}$ belongs to the exponential families. A GLM does not assume a linear relationship between the regressand and the regressor, but the relation between the regressor variables $(x_0, x_1, \ldots, x_k)$ and the parameter(s) is described by a link function for which we usually use the natural parameters of the canonical form of the exponential family.

The GLMs are a broad class of models that include linear regression, ANOVA, ANCOVA, Poisson regression, log-linear models, etc. Table 11.1 provides a summary of GLMs following Agresti (2018, chapter 4) where Mixed means categorical, nominal or ordinal, or continuous.

In all GLM in this chapter we assume the following:

- All random components refer to the probability distribution of the regressand.
- We have systematic components specifying the regressors $(x_1, x_2, \ldots, x_k)$ in the model or their linear combination in creating the so-called linear predictor.
- We have a link function specifying the link between random and systematic components.
- $\boldsymbol{y}_1, \boldsymbol{y}_2, \ldots, \boldsymbol{y}_n$ are independently distributed.
- The regressand has a distribution from an exponential family.
- The relationship between regressor and regressand is linear via the link function.
- Homogeneity of variance is not assumed, overdispersion is possible as we can see later.
- Errors are independent but not necessarily normally distributed.

In GLM the deviance means sum of squares and the residual deviance means mean squares. It may happen that after fitting a GLM the residual deviance exceeds the value expected. In such cases we speak about overdispersion.

Table 11.1 Link function, random and systematic components of some GLMs.

Model	Random component	Link	Systematic component
Linear regression	Normal	Identity	Continuous
ANOVA	Normal	Identity	Categorical
ANCOVA	Normal	Identity	Mixed
Log-linear regression	Gamma	Log	Continuous
Logistic regression	Binomial	Logit	Mixed
Log-linear regression	Poisson	Log	Categorical
Multinomial response	Multinomial	Generalised logit	Mixed

Sources of this may be:

- The systematic component of the model is wrongly chosen (important regressors have been forgotten or were wrongly included, e.g. linear in place of quadratic).
- There exist outliers in the data.
- The data are not realisations of a random sample.

We demonstrate this for binary data.

Let $\boldsymbol{p}$ be a random variable with $E(\boldsymbol{p}) = \mu$ and $var(\boldsymbol{p}) = n\mu(1-\mu) = \sigma^2$. For a realisation p of $\boldsymbol{p}$ we assume that $\boldsymbol{k}$ is $B(n; p)$-distributed. Then by the laws of iterated expectation and variance

$$E(\boldsymbol{k}) = n\mu$$

and

$$\mathrm{var}(\boldsymbol{k}) = n\mu(1-\mu) + n(n-1)\sigma^2.$$

As we can see var($\boldsymbol{k}$) is larger than for a binomial distribution with parameter μ. For $n = 1$ overdispersion cannot be detected.

How to detect and handle overdispersion is demonstrated in Section 11.5.2.

Example 11.3 In the binary logistic regression the probability p of a success is transformed by the logit function and linearly related to the regressor vector $x_i^T = (x_{i0}, \dots, x_{ik})$ by

$$\mathrm{logit}(p_i) = \eta_i = \ln\left(\frac{p_i}{1-p_i}\right) = \sum_{j=0}^{k} \beta_j x_{ij}, i = 1, \dots, n$$

or especially for $k = 1$ by

$$\mathrm{logit}(p_i) = \eta_i = \ln\left(\frac{p_i}{1-p_i}\right) = \beta_0 + \beta_1 x_i, i = 1, \dots, n.$$

We use for calculations the R package `glm`.

11.4 Analysis – Fitting a GLM – The Linear Case

We demonstrate the analysis by fitting a GLM to the data of Example 5.9 but use now the R program glm2 – this means the linear case. The analysis of intrinsic GLM is shown in Sections 11.5, 11.6, and 11.7.

Problem 11.4 Fit a GLM with an identity link to the data of Example 5.9.

Solution and Example

We take the data of Table 5.8 which is now Table 11.2 and explain the procedure with these data.

For the analysis of this design we can use in Chapter 5 the R the commands >aov() or >glm().

While aov() can only be used for balanced two-way cross classifications glm() gives more detailed information and can also be used for unbalanced two-way cross classifications. With glm the ANOVA proceeds, in the command at family = "gaussian" with default the link identity, as follows (Table 11.3):

```
> loss <- c(8.39, 11.58, 5.42, 9.53, 9.44, 12.21, 5.56, 10.39)
> storage <- c(1,2,3,4,1,2,3,4)
> crop <- c(1,1,1,1,2,2,2,2)
> Table_11_2 <- data.frame(cbind(loss,storage,crop))
> STORAGE <- factor(storage)
> CROP <- factor(crop)
> library(glm2)
> glm.loss<- glm(loss ~STORAGE + CROP, family="gaussian", Table_11_2)
> summary(glm.loss)
Call:
glm(formula = loss ~ STORAGE + CROP, family = "gaussian", data =
Table_11_2)
Deviance Residuals:
1             2             3             4             5             6
    7          8
-0.190     0.020      0.265     -0.095      0.190     -0.020
 -0.265     0.095
Coefficients:
               Estimate      Std. Error        t value            Pr(>|t|)
(Intercept)    8.5800         0.2196          39.069          3.69e-05 ***
STORAGE2       2.9800         0.2778          10.728           0.00173 **
STORAGE3      -3.4250         0.2778         -12.330           0.00115 **
STORAGE4       1.0450         0.2778           3.762           0.03285 *
CROP2          0.6700        0.1964            3.411           0.04212 *
Signif. codes:  0 '***' 0.001 '**' 0.01 '*' 0.05 '.' 0.1 ' ' 1
(Dispersion parameter for gaussian family taken to be 0.07716667)
Null deviance:          44.3554  on 7  degrees of freedom
Residual deviance:  0.2315  on 3  degrees of freedom
AIC: 6.3621
Number of Fisher Scoring iterations: 2
> anova(glm.loss)
Analysis of Deviance Table
Model: gaussian, link: identity
Response: loss
Terms added sequentially (first to last)
```

Table 11.2 Observations (loss during storage in percent of dry mass during storage of 300 days) of the experiment of Example 5.9.

		Forage crop	
		Green rye	Lucerne
Kind of storage	Glass in refrigerator	8.39	9.44
	Glass in barn	11.58	12.21
	Sack in refrigerator	5.42	5.56
	Sack in barn	9.53	10.39

Table 11.3 Analysis of variance table of Problem 11.4.

Source of variation	SS	df	MS	F
Between the storages	43.2261	3	14.4087	186.7
Between the forage crops	0.8978	1	0.8978	11.63
Residual	0.2315	3	0.0772	
Total	44.3554	7		

```
           Df  Deviance   Resid. Df     Resid. Dev
NULL                          7         44.355
STORAGE     3    43.226       4          1.129
CROP        1     0.898       3          0.232
```

The dispersion parameter above equals the mean square of the residuals (= residual deviance): 0.0772 = 0.2315/3. In `glm` the deviance means sum of squares and the residual deviance means mean squares.

Also, the null deviance is equal to the ordinary total variance. A more detailed explanation of these facts and the necessary modifications needed for fitting GLMs are given below.

11.5 Binary Logistic Regression

Logistic regression is a special type of regression where the probability of 'success' is modelled through a set of predictors (regressands). The predictors may be categorical, nominal or ordinal, or continuous. Binary logistic regression is a special type of regression where a binary response (regressor) variable is related to explanatory variables, which can be discrete and/or continuous.

We consider for a fixed factor with $a \geq 1$ levels in each level a binary random variable $\boldsymbol{y}_i$ with possible outcomes 0 and 1. We assume $P(\boldsymbol{y}_i = 1) = p_i$; $0 < p_i < 1$, $i = 1, \dots a$. The statistics (absolute frequencies) $\boldsymbol{Y}_{i.} = \sum_{j=1}^{n_i} \boldsymbol{y}_{ij}$ of independent random samples $(\boldsymbol{y}_{i1}, \dots, \boldsymbol{y}_{in_i})^T, n_i > 1$ with independent components distributed as $\boldsymbol{y}_i$ are binomially distributed with parameters n_i (known) and p_i. The relation between the regressor variables $(x_{i0}, \dots, x_{ik})$ and the parameter is described by the link function $\eta_i = \sum_{j=0}^{k} \beta_j x_{ij}$.

11.5.1 Analysis

Analogous to Section 8.1.1 the vector $\boldsymbol{Y}_{i.}$ may depend on a vector of (fixed) regressor variables $x_i^T = (x_{i0}, \dots, x_{ik})$.

Problem 11.5 Show the stochastic and systematic component and the link function for the binary logistic regression.

Solution

The absolute frequencies $\boldsymbol{Y}_{i.} = \sum_{j=1}^{n_i} \boldsymbol{y}_{ij}$ are binomially distributed with parameters p_i.

The logarithm of the likelihood function is

$$\ln \prod_{i=1}^{a} L(Y_{i.}, p_i) = \sum_{i=1}^{a} \left[Y_{i.} \ln\left(\frac{p_i}{1-p_i}\right) + n_i \ln(1-p_i) + \text{const} \right] \tag{11.3}$$

with a negligible constant const $= \sum_{i=1}^{a} \ln \binom{n_i}{Y_{i.}}$.

Often we use the relative frequencies $\hat{\boldsymbol{p}}_i = \frac{Y_{i.}}{n_i}$ in place of the absolute frequencies $\boldsymbol{Y}_{i.}$. We know that

$E(\hat{\boldsymbol{p}}_i) = p_i$ and $var(\hat{\boldsymbol{p}}_i) = \frac{p_i(1-p_i)}{n_i}$ and the variance depends on the expectation p_i.

The logistic regression function is

$$\eta_i = \sum_{j=0}^{k} \beta_j x_{ij} \tag{11.4}$$

For the link function we use in this book mainly the canonical link function (which means the link defined by the canonical form of the exponential family given in the solution of Problem 11.1)

$$\eta_i = \eta(p_i) = \ln\left(\frac{p_i}{1-p_i}\right) \text{ with } p_i = \frac{1}{1+e^{-\eta_i}}. \tag{11.5}$$

Other possible link functions are the probit function

$$\eta(p) = \Phi^{-1}(p) = z_p$$

or the complementary log–log-function (cloglog)

$$\eta(p) = \ln[-\ln(1-p)].$$

In this section we only demonstrate the logistic function and the model (11.4).

After estimating the parameters of model (11.4) with less than k regressors (an unsaturated model) with the methods of Chapter 8 by

$$\hat{\boldsymbol{\eta}}_i = \sum_{j=0}^{g} \hat{\boldsymbol{\beta}}_j x_{ij}$$

and evaluating of this estimated regression model. This can be done by comparing it with the estimated saturated model, which includes all $k-1$ regressors, which provides the best fit to the data. The parameter p_i in (11.5) is estimated by the relative frequency $\hat{\boldsymbol{p}}_i = \frac{Y_{i.}}{n_i} = \bar{\boldsymbol{y}}_i$.

We can now formally test whether two models differ significantly. This can be done by either the Pearson's χ^2-statistic or by the deviance; both are explained below. We write now

$$\pi_i = \frac{e^{\eta_i}}{1 + e^{\eta_i}}.$$

Pearson's χ^2-statistic is the sum of the squared standardised residuals of a model with g components ($g = k$ for the saturated model).

$$\chi^2 = \sum_{i=1}^{g} \frac{n_i(\hat{p}_i - \hat{\pi}_i)^2}{\hat{p}_i(1 - \hat{p}_i)}. \tag{11.6}$$

The deviance is defined for a model with g components if l_i is the log-likelihood of component i by

$$D = 2\sum_{i=1}^{g}[l_i(\hat{p}_i) - l_i(\hat{\pi}_i)]. \tag{11.7}$$

χ^2 and D are asymptotically equivalent, which means that

$$\lim_{n_i \to \infty \text{ for all } i} D = \lim_{n_i \to \infty \text{ for all } i} \chi^2.$$

Problem 11.6 Analyse the proportion of damaged o-rings from the space shuttle challenger data.

The data can be found in Faraway (2016).

```
>library(faraway)
>data(orings)
```

Each shuttle had two boosters, each with three o-rings. For each mission, we know the number of o-rings out of six showing some damage and the launch temperature (in degrees Fahrenheit, °F). What is the probability of failure in a given o-ring related to the launch temperature? Compare the results for the three common choices of logit-link, probit-link and complementary log–log-link.

Predict the probability of damage at 31 °F (this was the actual temperature on January 28, 1986, the date of the Challenger catastrophe).

Solution

First of all we take a look at the raw data and try to fit it with a simple linear model, i.e. via the `lm-function`:

```
> library(faraway)
> data(orings)
> attach(orings)
> plot(damage/6~temp, ylim=c(0,1), xlim=c(25,85), xlab="Temperature in
  °F", ylab="Probability of damage", main="orings data - lm fit")
> lmtry<-lm(damage/6~temp)
> abline(lmtry, col="black")
```

When looking at the raw data, the damage variable is always divided by 6 to get the proportions of the damaged o-rings, since these damage probabilities are of main interest.

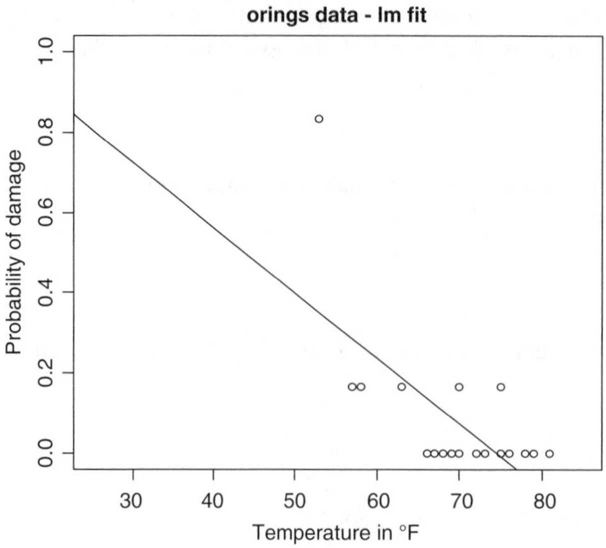

Figure 11.1 Data fitted with the `lm-function`.

Although the minimum value for the temperature given in the data is 53 °F, the x-axis starts at 25 °F, due to the fact that, firstly, the temperature of interest is 31 °F and, secondly, the forecast temperature for the day of the Challenger catastrophe was between 26 and 29 °F. Hence, these temperature values may also be of interest. The result of this first try with the `lm` function is shown in Figure 11.1 and, as expected, gives a terrible fit.

Now, logit-, probit-, and complementary log–log-link functions are used to fit a GLM to the data. This is done by the `glm` command, where our response is given to the function as a two-column matrix with the number of 'successes', i.e. damages, in the first column, and the number of 'fails' (`nodam`) in the second one:

```
> nodam<- 6- damage
> cbind(damage,nodam)
          damage  nodam
 [1,]       5       1
 [2,]       1       5
 [3,]       1       5
 ------------------------
 [20,]      0       6
 [21,]      0       6
 [22,]      0       6
 [23,]      0       6
> logit.mod<-glm(cbind(damage,nodam)~temp,family=binomial(link="logit"))
> probit.mod<-glm(cbind(damage,nodam)~temp,family=binomial(link="probit"))
> cloglog.mod<-glm(cbind(damage,nodam)~temp,
   family=binomial(link="cloglog"))
> summary(logit.mod)
Call:
glm(formula = cbind(damage, nodam) ~ temp, family = binomial(link =
"logit"))
Deviance Residuals:
  Min          1Q         Median        3Q          Max
-0.9529     -0.7345     -0.4393     -0.2079      1.9565
```

```
Coefficients:
            Estimate    Std. Error    z value     Pr(>|z|)
(Intercept)  11.66299    3.29626      3.538      0.000403 ***
temp         -0.21623    0.05318     -4.066      4.78e-05 ***
Signif. codes:  0 '***' 0.001 '**' 0.01 '*' 0.05 '.' 0.1 ' ' 1
(Dispersion parameter for binomial family taken to be 1)
Null deviance: 38.898  on 22  degrees of freedom
Residual deviance: 16.912  on 21  degrees of freedom
AIC: 33.675
Number of Fisher Scoring iterations: 6
> summary(probit.mod)
Call:
glm(formula = cbind(damage, nodam) ~ temp, family = binomial(link =
 "probit"))
Deviance Residuals:
   Min          1Q         Median          3Q         Max
-1.0134     - 0.7761      -0.4467      -0.1581      1.9983
Coefficients:
            Estimate    Std. Error   z value     Pr(>|z|)
(Intercept) 5.59145     1.71055       3.269      0.00108 **
temp        -0.10580    0.02656      -3.984      6.79e-05 ***
---
Signif. codes:  0 '***' 0.001 '**' 0.01 '*' 0.05 '.' 0.1 ' ' 1
(Dispersion parameter for binomial family taken to be 1)
    Null deviance: 38.898  on 22  degrees of freedom
Residual deviance: 18.131  on 21  degrees of freedom
AIC: 34.893
Number of Fisher Scoring iterations: 6
> summary(cloglog.mod)
Call:
glm(formula = cbind(damage, nodam) ~ temp, family = binomial(link =
 "cloglog"))
Deviance Residuals:
 Min          1Q          Median           3Q           Max
-0.9884     -0.7262     -0.4373        -0.2141       1.9520
Coefficients:
           Estimate     Std. Error   z value   Pr(>|z|)
(Intercept) 10.86388     2.73668     3.970     7.20e-05 ***
temp        -0.20552     0.04561     -4.506    6.59e-06 ***
Signif. codes:  0 '***' 0.001 '**' 0.01 '*' 0.05 '.' 0.1 ' ' 1
(Dispersion parameter for binomial family taken to be 1)
Null deviance: 38.898  on 22  degrees of freedom
Residual deviance: 16.029  on 21  degrees of freedom
AIC: 32.791
Number of Fisher Scoring iterations: 7
> inv.logit<-function(x){exp(x)/(1+exp(x))}
> inv.cloglog<-function(x){1-exp(-exp(x))}
> x<-seq(25,85, 0.5)
> inv.logit.val<-inv.logit(coef(logit.mod)[1]+coef(logit.mod)[2]*x)
> inv.probit.val<-pnorm(coef(probit.mod)[1]+coef(probit.mod)[2]*x)
> inv.cloglog.val<-inv.cloglog(coef(cloglog.mod)[1]+coef(cloglog.mod)[2]*x)
> plot(damage/6~temp, ylim=c(0,1), xlim=c(25,85), xlab="Temperature in °F",
   ylab="Probability of damage", main="orings data \n glm fits")
> lines(inv.logit.val, col="black", lt=2)
> lines(inv.probit.val, col="black", lt=1)
> lines(inv.cloglog.val, col="black", lt=3)
> legend(70,1, c("logit", "probit", "cloglog"),
     lt=c(2,1,3), lwd=c(1,1,1))
```

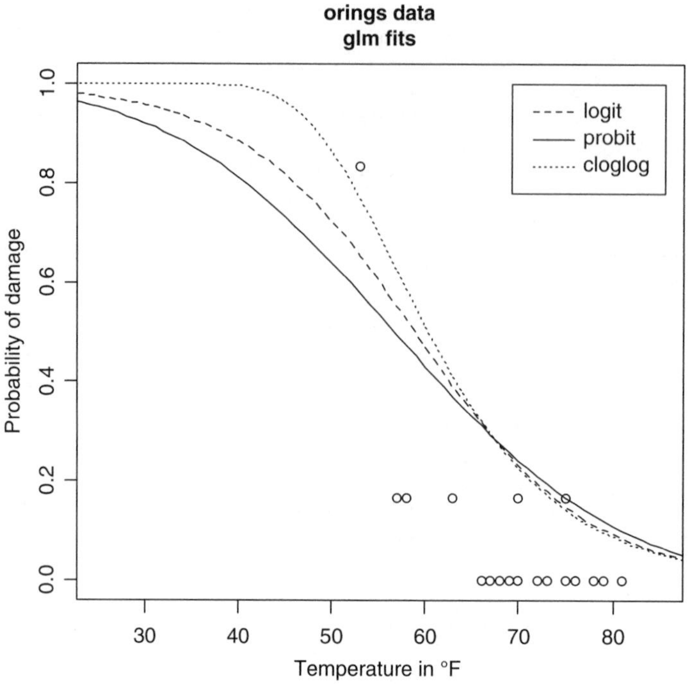

Figure 11.2 Data fitted with `glm-functions`.

To compare the results with the data, the response functions (inverse link functions) are needed, i.e. `inv.logit` and `inv.cloglog` as implemented in the code for logit and complementary log–log, respectively, and the function `pnorm` for the probit-link. The results are shown in Figure 11.2 and the following values for the intercept β_0, the first-order co-efficient of x, β_1, and the AIC values are obtained:

	β_0	β_1	AIC
Logit-link	11.66299	−0.21623	33.67
Probit-link	5.59145	−0.10580	34.89
Complementary log–log-link	10.86388	−0.20552	32.79

The AIC values stem from the Akaike criterion for model Choice. This and other criteria are explained in Rasch and Schott (2018, section 9.6.9).

The relative effects of the three different models are very similar, but the probit model, in particular, yields quite different coefficients. According to the AIC values, the complementary log–log-link has a slightly better, i.e. smaller, value. This may occur due to the fact that it fits the data point with the highest number of failures at 53 °F best, whereas in the region with most of the data points all three models give similar results.

This extreme jump in the data also becomes noticeable in the wide confidence intervals, calculated by

```
> library(MASS)
> confint(logit.mod)
Waiting for profiling to be done...
                 2.5 %          97.5 %
(Intercept)  5.575195    18.737598
temp         -0.332657      -0.120179
> confint(probit.mod)
Waiting for profiling to be done...
                          2.5 %              97.5 %
(Intercept)    2.4231051    9.04355971
temp           -0.1597757   -0.05742257
> confint(cloglog.mod)
Waiting for profiling to be done...
                 2.5 %           97.5 %
(Intercept)    5.4348673   16.8985804
temp           -0.3082858    -0.1184926
```

and resulting in:

		2.5%	97.5%
Logit	β_0	5.575195	18.737598
	β_1	−0.332657	−0.120179
Probit	β_0	2.4231051	9.043560
	β_1	−0.1597757	−0.05742257
Complementary log–log	β_0	5.4348673	16.898580
	β_1	−0.3082858	−0.1184926

Finally, the probabilities of failure at 31 °F for the different models, obtained via

```
> logit31<-inv.logit(coef(logit.mod)[1]+coef(logit.mod)[2]*31)
> probit31<-pnorm(coef(probit.mod)[1]+coef(probit.mod)[2]*31)
> cloglog31<-inv.cloglog(coef(cloglog.mod)[1]+coef(cloglog.mod)[2]*31)
```

are given as 0.9930342, 0.9895983 and 1 for logit, probit and complementary log–log, respectively, and therefore clearly agree with the horrible ending of the Challenger catastrophe.

Table 11.4 Values of $N_{ijk}(n_{ijk})$ of the block experiment.

Block	Genotype 1	Genotype 2
1	5(10), 4(10), 4(11), 5(11)	6(10), 7(10), 8(9), 9(10)
	5(10), 4(10), 3(10), 5(10)	10(11), 7(10), 8(11), 7(11)
	4(10), 3(9)	10(11), 7(11),
2	2(12), 2(9)	5(10), 5(10)
3	1(10), 1(10), 0(11), 0(10)	3(11), 2(9), 3(12), 3(10)
4	0(10), 0(10), 2(9), 1(8)	4(11), 2(9), 3(9), 4(11)
	3(12), 0(8), 1(10), 1(9)	3(10), 3(10), 3(9), 3(10)
	0(9), 2(10), 0(9), 1(11)	3(11), 5(10), 5(12)
5	4(11), 4(9), 4(11), 2(10)	5(9), 8(10), 7(8), 9(10)
	4(11), 3(9), 4(10), 0(10)	8(9), 7(9), 8(9), 6(9)
	3(11), 0(8)	6(10), 7(10)

Example 11.4 Table 11.4 shows the number N_{ijk} of infected peanut plants in block i, genotype j and plot k of an experiment with 5 blocks with at most 12 plots each in two genotypes. Using n_{ijk} the number of plants on block k is given. This example comes from Rasch et al. (1998, Procedure 4/51/0200, page 472–475).

p_{ijk} is the infection probability on plot (ijk). How does p_{ijk} depend on the fixed factors genotype and block? We assume that the $\boldsymbol{N}_{ijk}$ are binomially $B(n_{ijk}, p_{ijk})$ distributed and that no dependency between neighboured plots exist. Further we assume an additive model with the canonical link

$$\eta_{ijk} = \mu + b_i + g_j.$$

Problem 11.7 Analyse the infection probabilities of Example 11.4.

Solution

We use the command > `glm`.

Example

First we input the data in R:

```
> peanuts=read.table("C:/Rasch_applied_statistics/peanuts.txt",
        header=T)
> peanuts
   block G_type N_infected n_plants
1      1      1          5       10
2      1      1          4       10
3      1      1          4       11
-----------------------------------
73     5      2          6        9
74     5      2          6       10
75     5      2          7       10
```

Then we start the analysis:

```
> infected.plants <- sum(peanuts[,3])
> infected.plants
[1] 296
> number.plants <- sum(peanuts[,4])
> number.plants
[1] 749
> Block <- as.factor(peanuts[,1])
> Gtype=as.factor(peanuts[,2])
> peanuts1=as.data.frame(cbind(Block,Gtype,peanuts[,3],peanuts[,4]))
> peanuts1
   Block Gtype V3 V4
1      1     1  5 10
2      1     1  4 10
3      1     1  4 11
--------------------
73     5     2  6  9
74     5     2  6 10
75     5     2  7 10
> BLOCK <- factor(Block)
> GTYPE <- factor(Gtype)
> bin.add <- glm(V3/V4 ~ BLOCK + GTYPE, family = binomial,
            weights=V4, data=peanuts1)
> summary(bin.add)
Call:
glm(formula = V3/V4 ~ BLOCK + GTYPE, family = binomial, data = peanuts1,
    weights = V4)
Deviance Residuals:
     Min        1Q    Median        3Q       Max
-2.76222  -0.56956   0.05692   0.53349   1.67560

Coefficients:

            Estimate Std. Error z value Pr(>|z|)
(Intercept)  -0.4423     0.1728  -2.559  0.01049 *
BLOCK2       -1.1793     0.3903  -3.021  0.00252 **
BLOCK3       -2.3643     0.3559  -6.643 3.08e-11 ***
BLOCK4       -1.9122     0.2370  -8.069 7.09e-16 ***
BLOCK5       -0.3246     0.2211  -1.468  0.14201
GTYPE2        1.7287     0.1812   9.538  < 2e-16 ***
---
Signif. codes:  0 '***' 0.001 '**' 0.01 '*' 0.05 '.' 0.1 ' ' 1
(Dispersion parameter for binomial family taken to be 1)
    Null deviance: 257.808  on 74  degrees of freedom
Residual deviance:  55.719  on 69  degrees of freedom
AIC: 233.13
Number of Fisher Scoring iterations: 4
> anova(bin.add, test = "Chisq")
Analysis of Deviance Table

Model: binomial, link: logit
Response: V3/V4
Terms added sequentially (first to last)
      Df Deviance Resid. Df Resid. Dev  Pr(>Chi)
NULL                     74    257.808
```

```
BLOCK  4   98.772          70     159.037 < 2.2e-16 ***
GTYPE  1  103.317          69      55.719 < 2.2e-16 ***
---
Signif. codes:  0 '***' 0.001 '**' 0.01 '*' 0.05 '.' 0.1 ' ' 1
> bin.add1 <- glm(V3/V4 ~ GTYPE + BLOCK , family = binomial,
            weights=V4, data=peanuts1)
> anova(bin.add1, test = "Chisq")
Analysis of Deviance Table
Model: binomial, link: logit
Response: V3/V4
Terms added sequentially (first to last)
      Df Deviance Resid. Df Resid. Dev  Pr(>Chi)
NULL                       74     257.808
GTYPE  1   88.997          73     168.811 < 2.2e-16 ***
BLOCK  4  113.092          69      55.719 < 2.2e-16 ***
---
Signif. codes:  0 '***' 0.001 '**' 0.01 '*' 0.05 '.' 0.1 ' ' 1
```

Remarks

1) Number of observations = 75.
2) Number of infected plants = 296.
3) Number of plants = 749.
4) Coefficient estimate GTYPE2 = estimate (GTYPE2 – GTYPE1).
5) Estimate GTYPE1 = – (1.7287)/2 = – 0.8643 because $\sum_1^2 \text{GTYPE}_j = 0$.
6) Coefficient estimate BLOCK2 = estimate (BLOCK2 – BLOCK1).
7) Coefficient estimate BLOCK3 = estimate (BLOCK3 – BLOCK1).
8) Coefficient estimate BLOCK4 = estimate (BLOCK4 – BLOCK1).
9) Coefficient estimate BLOCK5 = estimate(BLOCK5 – BLOCK1).
10) Estimate BLOCK1 = –(–1.1793 – 2.3643 – 1.9122 – 0.3246)/5 = 1.15608 because $\sum_1^5 \text{BLOCK}_i = 0$.
11) Estimate MU = intercept – BLOCK1 – GTYPE1 = –0.4423 – 1.15608 – (–0.86435) = –0.7340.
12) Note that in the first analysis of deviance table the effect of GENOTYPE is tested and in the second analysis of deviance table the effect of BLOCK is tested. The command `glm` uses a hierarchical testing order.

11.5.2 Overdispersion

After fitting a GLM, it may happen that the estimated variance computed as the residual deviance exceeds the value expected – this we call overdispersion. We discuss this here for the binomial model and in Section 11.6.2 for the Poisson model. If the residual deviance exceeds the residual degrees of freedom overdispersion is present.

Possible sources of overdispersion are amongst others:

- A wrongly chosen systematic component of the GLM.
- Outliers are present.
- The data are otherwise not a realised random sample.

Let $\boldsymbol{p}$ be a random variable with $E(\boldsymbol{p}) = \mu$ and $\text{var}(\boldsymbol{p}) = \sigma^2$. Further let $\boldsymbol{k}$ be $B(n, p)$ distributed with a realisation p of $\boldsymbol{p}$. Then

$$E(\boldsymbol{k}) = \mu$$

Table 11.5 Number n of wasps per group and number k of these n wasps finding eggs.

Strain	$k(n)$
1	3(51), 8(68), 8(42), 21(96), 11(128), 4(49), 15(54), 3(84), 4(49), 5(63)
2	7(36), 20(79), 35(121), 31(121), 10(90), 12(70), 5(67), 22(81), 8(57), 14(62)
3	0(52), 5(111), 12(109), 10(87), 22(122), 0(65), 6(98), 3(71), 4(99), 7(97)

and

$$\text{var}(\boldsymbol{k}) = n\mu(1-\mu) + n(n-1)\sigma^2.$$

We see that var($\boldsymbol{k}$) exceeds for $n > 1$ the variance of a binomial distribution. If $n = 1$, overdispersion cannot be detected.

How can we reduce overdispersion?

First we try to correct the systematic component of the model. Further we choose a better link function and model the variance by

$$\text{var}(\boldsymbol{k}) = n\mu(1-\mu)\varphi. \tag{11.8}$$

φ is estimated by dividing the residual deviance by the corresponding degrees of freedom (see Example 11.5).

More detailed information about overdispersion can be found in Collett (1991). In an analogous way underdispersion can be handled.

Example 11.5 This example comes from Rasch et al. (1998, Procedure 4/51/0205, page 476–479).

In an experiment to investigate biological pest control the ability to find in a given time a nest with eggs of a pest in three strains of wasps was observed. The eggs have been hidden in a labyrinth and $y = 1$ means nest found and $y = 0$ means nest not found. In each of 10 groups in each strain of wasps the number of wasps was between 36 and 128. In Table 11.5 in each of the 30 groups in three strains n is the number of wasps per group and k the number of wasps finding eggs.

We use a GLM model with n_{ij} as number of wasps in strain i ($i = 1, 2, 3$) and group j ($j = 1, 2, \ldots, 10$). k_{ij} is the corresponding number of successes.

The systematic model component for k_{ij} is

$$\eta_{ij} = \mu + g_i$$

with g_i as the effect of strain i.

We use the logit-link function.

Problem 11.8 Analyse the infection probabilities of Example 11.5 where the dispersion parameter (scale-parameter) must be estimated.

Solution

We use the command `> glm from the package glm2.`

Example

```
> library(glm2)
> wasp <- read.table("D:/Rasch_Applied_Statistics/wasp.txt", header=T)
> wasp
   tribe  k   n
1      1  3  51
2      1  8  68
3      1  8  42
---------------
28     3  3  71
29     3  4  99
30     3  7  97
> wasp1 <- data.frame(wasp)
> TRIBE <- factor(wasp1[,1])
> number.events <- sum(wasp1[,2])
> number.events
[1] 315
> number.trials <- sum(wasp1[,3])
> number.trials
[1] 2379
> number.observations <- length(wasp[,3])
> number.observations
[1] 30
> k <- wasp1[,2]
> n <- wasp[,3]
> model1 <- glm(k/n ~ TRIBE, family=quasibinomial, data=wasp1, weight=n)
> summary(model1)
Call:
glm(formula = k/n ~ TRIBE, family = quasibinomial, data = wasp1,
    weights = n)
Deviance Residuals:
   Min      1Q  Median      3Q     Max
-3.200  -1.333  -0.678   1.267   3.762
Coefficients:
            Estimate Std. Error t value Pr(>|t|)
(Intercept)  -1.9935     0.2220  -8.981 1.35e-09 ***
TRIBE2        0.6637     0.2769   2.396   0.0237 *
TRIBE3       -0.5081     0.3241  -1.568   0.1286
---
Signif. codes:  0 '***' 0.001 '**' 0.01 '*' 0.05 '.' 0.1 ' ' 1
(Dispersion parameter for quasibinomial family taken to be 3.556208)
    Null deviance: 165.98  on 29  degrees of freedom
Residual deviance: 100.45  on 27  degrees of freedom
AIC: NA
Number of Fisher Scoring iterations: 5
> anova(model1, test = "Chisq")
Analysis of Deviance Table
Model: quasibinomial, link: logit
Response: k/n
Terms added sequentially (first to last)
      Df Deviance Resid. Df Resid. Dev  Pr(>Chi)
NULL                    29     165.98
TRIBE  2   65.535       27     100.45 9.962e-05 ***
---
Signif. codes:  0 '***' 0.001 '**' 0.01 '*' 0.05 '.' 0.1 ' ' 1
```

1) Number of observations = 30.
2) Number of estimate TRIBE2 = estimate (TRIBE2 – TRIBE1).
3) Coefficient estimate TRIBE3 = estimate (TRIBE3 – TRIBE1).
4) Estimate TRIBE1 = –(0.1556)/3 = –0.0519 because $\sum_1^3 \text{TRIBE}_{\text{j}} = 0$.
5) Estimate MU = intercept – TRIBE1 = –1.9935 – (–0.0519) = –1.9416.
6) Number of events = 315.
7) Number of trials = 2379.
8) After model fitting we find in the `glm` output in the deviance row a residual deviance 100.45 with 27 degrees of freedom. In SAS the scale parameter φ is estimated as the square root of the quotient of residual deviance divided by the corresponding degrees of freedom. In GLM the dispersion parameter is only the quotient of residual deviance divided by the corresponding degrees of freedom.

11.6 Poisson Regression

Poisson regression refers to a GLM model where the random component is specified by the Poisson distribution of the response variable $\boldsymbol{y}$, which is a count. However, we can also have $\boldsymbol{h} = \frac{\boldsymbol{y}}{\boldsymbol{t}}$, the rate as the response variable, where t is an interval representing time (hour, day), space (square meters) or some other grouping. The response variable $\boldsymbol{y}$ has the expectation λ. Because counts are non-negative the expectation λ is positive. The relation between the regressor variables $(x_0, x_1, \dots, x_k)$ and the parameter λ is described for $k = 1$ by

$$\ln \lambda_i = \beta_0 + \beta_1 x_i \tag{11.9}$$

and in the case of k regressors we have, analogous to (11.4),

$$\ln \lambda_i = \sum_{j=0}^{k} \beta_j x_{ij}.$$

We mainly handle in this section the case of one regressor variable. From (11.9) we receive the expectations (with equal variances)

$$\lambda_i = e^{\beta_0 + \beta_1 x_i}. \tag{11.10}$$

11.6.1 Analysis

We estimate the parameter λ by the maximum likelihood (ML) method minimising the logarithm of the likelihood function of $n > 0$ observations y_i, $i = 1, \dots, n$

$$\ln \prod_{i=1}^{n} L(y_i, \lambda) = \sum_{i=1}^{n} y_i \ln \lambda - n\lambda - \sum_{i=1}^{n} \ln y_i!, \lambda > 0, y = 0, 1, 2, \dots .$$

Derivation with respect to λ and zeroing this derivation gives the ML estimate

$$\hat{\lambda} = \bar{y}$$

i.e. the estimate is the arithmetic mean of the observed counts because the second derivative is negative and the solution gives a maximum.

Table 11.6 Number of soldiers dying from kicks by army mules.

Number of deaths	y = number of corp–year combinations
0	109
1	65
2	22
3	3
4	1
>5	0

Example 11.6 Ladislaus von Bortkiewicz was a Russian economist also dealing with 'probability theory' (including statistics) – see von Bortkiewicz (1893). He considered several data sets; one of the best known is from investigating 10 Prussian army corps over 20 years. He counted the number of soldiers per each of the 200 corps–year combinations dying from kicks by army mules. Table 11.6 shows his results.

Problem 11.9 Calculate the ML estimate for the Poisson parameter λ and calculate the chi-square-test statistic for the fit of the data of Example 11.6 for the Poisson distribution with $\hat{\lambda}$.

Solution

Calculate first $\hat{\lambda} = \bar{y}$ and calculate then the expected frequency E. Hereafter, calculate the Pearson chi-square-test statistic $X^2 = \sum(y - E)^2/E$. Let m be the number of classes of deaths and the significance probability (p-value) is $\Pr(CS(m-1) \geq X^2)$.

Example

```
> k <- c(0,1,2,3,4,5)
> y <- c(109, 65, 22, 3, 1, 0)
> m <- length(y)
> n <- sum(y)
> n
[1] 200
> y_bar <- sum(k*y)/n
> y_bar
[1] 0.61
> prob_k <- exp(- y_bar)* (y_bar^k)/factorial(k)
> prob_k
[1] 0.5433508691 0.3314440301 0.1010904292 0.0205550539 0.0031346457
[6] 0.0003824268
> E <- n*prob_k
> E
[1] 108.67017381  66.28880603  20.21808584   4.11101079   0.62692915
[6]   0.07648536
> result <- data.frame(k,y, E)
> result
```

```
  k   y       E
1 0 109 108.67017381
2 1  65  66.28880603
3 2  22  20.21808584
4 3   3   4.11101079
5 4   1   0.62692915
6 5   0   0.07648536
> diff <- (y-E)^2
> X2 <- sum(diff/E)
> X2
[1] 0.7818513
> p_value <- 1-pchisq(X2, df=m-1)
> p_value
[1] 0.9781765
```

Hence, because the p-value $= 0.9781765 > 0.05$ the null hypothesis that the horse kicks data are Poisson distributed is not rejected.

Example 11.7 This example comes from Rasch et al. (1998, Procedure 4/51/0300, Page 479–482).

In an experiment the influence of the kind of nutrient medium on the number of buds on a cucumber leaf was investigated. Observations refer to the number y_{ijk} of buds on a cucumber leaf in medium i ($i = 1, \ldots, 4$) of variety j ($j = 1, 2$) in replication k ($k = 1, \ldots, 6$).

We assume that

$\underline{y}_{ijk}$ is $P(\lambda_{ijk})$ distributed with the log-likelihood function $y_{ijk} \cdot \ln(\lambda_{ijk}) - \lambda_{ijk} + \text{const.}$

The systematic part of a GLM is

$\eta_{ijk} = \mu + m_i + g_j + (mg)_{ij}$ with

- m_i – effect of medium i
- g_j – effect of variety j
- $(mg)_{ij}$ – interaction effect.

We use the link function

$$\ln(\lambda_{ijk}) = \eta_{ijk}.$$

Table 11.7. Frequencies y_{ijk} from Example 11.7

Medium	Variety 1						Variety 2					
1	10	.	5	7	12	6	16	8	10	11	9	12
2	21	16	10	18	17	20	8	9	10	9	15	9
3	17	10	20	12	14	21	10	10	17	.	14	12
4	5	.	6	2	3	2	6	2	2	3	7	3

Problem 11.10 Analyse the data of Example 11.7 using the GLM for the Poisson distribution.

Table 11.7 Values of $k_{ijk}(n_{ijk})$ on plots $k = 1, \ldots, m_{ij}$ in block i and genotype j and the $m_{ij.}$

Block	Genotype 1	Genotype 2
1	5(10), 4(10), 4(11), 5(11)	6(10), 7(10), 8(9), 9(10)
	5(10), 4(10), 3(10), 5(10)	10(11), 7(10), 8(11), 7(11)
	4(10), 3(9)	10(11), 7(11)
	$m_{11} = 10$	$m_{12} = 10$
2	2(12), 2(9)	5(10), 5(10)
	$m_{21} = 2$	$m_{22} = 2$
3	1(10), 1(10), 0(11), 0(10)	3(11), 2(9), 3(12), 3(10)
	$m_{31} = 4$	$m_{32} = 4$
4	0(10), 0(10), 2(9), 1(8)	4(11), 2(9), 3(9), 4(11)
	3(12), 0(8), 1(10), 1(9)	3(10), 3(10), 3(9), 3(10)
	0(9), 2(10), 0(9), 1(11)	3(11), 5(10), 5(12)
	$m_{41} = 12$	$m_{42} = 11$
5	4(11), 4(9), 4(11), 2(10)	5(9), 8(10), 7(8), 9(10)
	4(11), 3(9), 4(10), 0(10)	8(9), 7(9), 8(9), 6(9)
	3(11), 0(8)	6(10), 7(10)
	$m_{51} = 10$	$m_{52} = 10$

Solution

We use the command > `glm with the package glm2.`

Example

```
>cucumber <-read.table("C:/Rasch_applied_statistics/cucumber.txt",header=T)
> cucumber
   Medium Cultivar  y
1       1        1 10
2       1        1  5
3       1        1  7
---------------------
43      4        2  3
44      4        2  7
45      4        2  3
> MEDIUM <- factor(cucumber[,1])
> CULTIVAR <- factor(cucumber[,2])
> model <- glm(y ~MEDIUM + CULTIVAR + MEDIUM:CULTIVAR, family=poisson,
           data=cucumber)
> summary(model)
Call:
glm(formula = y ~ MEDIUM + CULTIVAR + MEDIUM:CULTIVAR, family = poisson,
    data = cucumber)
Deviance Residuals:
    Min       1Q    Median       3Q      Max
-1.8405  -0.7401  -0.3063   0.6965   1.4710
Coefficients:
                  Estimate Std. Error z value Pr(>|z|)
(Intercept)         2.0794     0.1581  13.152  < 2e-16 ***
```

```
MEDIUM2               0.7538     0.1866   4.040 5.34e-05 ***
MEDIUM3               0.6721     0.1888   3.560 0.000371 ***
MEDIUM4              -0.7985     0.2838  -2.813 0.004902 **
CULTIVAR2             0.3185     0.2004   1.589 0.112001
MEDIUM2:CULTIVAR2    -0.8491     0.2581  -3.290 0.001003 **
MEDIUM3:CULTIVAR2    -0.5363     0.2582  -2.077 0.037791 *
MEDIUM4:CULTIVAR2    -0.2557     0.3731  -0.685 0.493179
---
Signif. codes:  0 '***' 0.001 '**' 0.01 '*' 0.05 '.' 0.1 ' ' 1
(Dispersion parameter for poisson family taken to be 1)
    Null deviance: 136.794  on 44  degrees of freedom
Residual deviance:  33.743  on 37  degrees of freedom
AIC: 230.53
Number of Fisher Scoring iterations: 4
> anova(model,test = "Chisq")
Analysis of Deviance Table
Model: poisson, link: log
Response: y
Terms added sequentially (first to last)
               Df Deviance Resid. Df Resid. Dev  Pr(>Chi)
NULL                              44    136.794
MEDIUM          3   87.603        41     49.192 < 2.2e-16 ***
CULTIVAR        1    3.648        40     45.544  0.056134 .
MEDIUM:CULTIVAR 3   11.801        37     33.743  0.008098 **
---
Signif. codes:  0 '***' 0.001 '**' 0.01 '*' 0.05 '.' 0.1 ' ' 1
> anova(model1, test=  "Chisq")
Analysis of Deviance Table
Model: poisson, link: log
Response: y
Terms added sequentially (first to last)
               Df Deviance Resid. Df Resid. Dev  Pr(>Chi)
NULL                              44    136.794
CULTIVAR        1    5.888        43    130.907  0.015247 *
MEDIUM          3   85.363        40     45.544 < 2.2e-16 ***
CULTIVAR:MEDIUM 3   11.801        37     33.743  0.008098 **
---
Signif. codes:  0 '***' 0.001 '**' 0.01 '*' 0.05 '.' 0.1 ' ' 1
```

As we can see the scale parameter is 1. The residual deviance is 33.74 with 37 degrees of freedom and nearly equal to the value 37 expected for the Poisson distribution; this means the model is adequate.

Example 11.5 ***Continued*** We extend the information received from the wasp experiment by counting the number m of eggs found by each of the g groups of wasps. Let m_{ij} be the number of eggs found by the wasps of strain $i = 1, 2, 3$ and group $j = 1, \ldots, 10$. We assume that $\boldsymbol{m}_{ij}$ is Poisson distributed with $E(\boldsymbol{m}_{ij}) = \theta_{ij} = k_{ij}\lambda_{ij}$; $i = 1, 2, 3$; $j = 1, \ldots, 10$ with λ_{ij} as the average of eggs found by wasps of strain i in group $j = 1, \ldots, 10$.

The systematic component of our GLM is

$$\ln(\theta_{ij}) = \ln(k_{ij}\lambda_{ij}) = \ln(k_{ij}) + \ln(\lambda_{ij}) = \ln(k_{ij}) + \mu + \gamma_i,$$

where now ln (k_{ij}) is a covariate with regression coefficient 1.

The observations are given in Table 11.8.

Table 11.8 Values k_{ij} and m_{ij} found in three strains.

Strain	$k_{ij}(m_{ij})$
1	3(46), 8(188), 8(137), 21(231), 11(93), 4(37), 15(63), 3(52), 4(48), 5(53)
2	7(45), 20(100), 35(181), 31(115), 10(32), 12(52), 5(37), 22(130), 8(24), 14(47)
3	0(0), 5(25), 12(57), 10(66), 22(150), 0(0), 6(36), 3(20), 4(27), 7(80)

Problem 11.11 Analyse the data of Table 11.8 using the GLM for the Poisson distribution with overdispersion.

Solution

We use the command > `glm with the package glm2` and we delete the data 0(0) twice for strain 3.

Example

```
>pest_eggs<-read.table("C:/Rasch_applied_statistics/pest_eggs.txt",
          header=T)
> pest_eggs
   Tribe  k   m
1      1  3  46
2      1  8 188
3      1  8 137
---------------
26     3  3  20
27     3  4  27
28     3  7  80
> lk <- log(pest_eggs[, 2])
> pest_eggs1 <- data.frame(pest_eggs, lk)
> TRIBE <- factor(pest_eggs1[,1])
> model <- glm(m ~TRIBE, family=quasipoisson, offset=lk, data=pest_eggs1)
> summary(model)

Call:
glm(formula = m ~ TRIBE, family = quasipoisson, data = pest_eggs1,
    offset = lk)

Deviance Residuals:
    Min       1Q   Median       3Q      Max
-9.6565  -1.7065  -0.0546   1.8865   8.7004

Coefficients:
            Estimate Std. Error t value Pr(>|t|)
(Intercept)   2.4476     0.1108  22.088  < 2e-16 ***
TRIBE2       -0.9102     0.1659  -5.485 1.07e-05 ***
TRIBE3       -0.5483     0.1937  -2.830  0.00904 **
---
Signif. codes:  0 '***' 0.001 '**' 0.01 '*' 0.05 '.' 0.1 ' ' 1

(Dispersion parameter for quasipoisson family taken to be 11.64097)
```

```
    Null deviance: 634.84  on 27  degrees of freedom
Residual deviance: 284.44  on 25  degrees of freedom
AIC: NA

Number of Fisher Scoring iterations: 4

> anova(model, test="Chisq")
Analysis of Deviance Table

Model: quasipoisson, link: log

Response: m

Terms added sequentially (first to last)


      Df Deviance Resid. Df Resid. Dev Pr(>Chi)
NULL                     27     634.84
TRIBE  2   350.39        25     284.44 2.91e-07 ***
---
Signif. codes:  0 '***' 0.001 '**' 0.01 '*' 0.05 '.' 0.1 ' ' 1
```

The deviance is 284.44 with 25 degrees of freedom, and we have an overdispersion. The GLM for rates has as link function the log of the rate.

11.6.2 Overdispersion

Because in the Poisson distribution the expectation equals the variance, overdispersion is often caused by some factors. Agresti (2018) mentioned that the negative binomial distribution may be better adapted to count data because it permits the variance to exceed the expectation.

11.7 The Gamma Regression

Gamma regression is a GLM model where the random component is specified by the Gamma distribution of the response variable $\boldsymbol{y}$, which is continuous. We use here the two-parametric Gamma distribution

$$f(y) = \begin{cases} \frac{\lambda^\nu}{\Gamma(\nu)} y^{\nu-1} e^{-\lambda y}, & \text{if } y \geq 0 \\ 0, & \text{if } y < 0 \end{cases}, \lambda > 0, \nu > 0.$$

The relation between the regressor variables $(x_0, x_1, \dots, x_k)$ and the parameters λ, ν is described by the link functions $\eta_1 = -\frac{\lambda}{\nu}$ and $\eta_2 = \frac{1}{\nu}$.

Assume that we have a random sample $\boldsymbol{Y}^T = (\boldsymbol{y}_1, \boldsymbol{y}_2, \dots, \boldsymbol{y}_n)$ of size n with components distributed like $\boldsymbol{y}$, i.e. they have the same parameters λ, ν. We further assume that $\boldsymbol{y}_i$ depends on regressor variables $(x_{i0}, x_{i1}, \dots, x_{ik})$ influencing the link function $g(\lambda_i, \nu_i)$, $i = 1, \dots, n$ via

$$g(\lambda_i, \nu_i \ ) = \sum_{j=1}^{k} \beta_j x_{ij}.$$

Without loss of generality we use the inverse link function $g(\lambda_i, v_i) = g(\eta_{1i}) = -\frac{1}{\eta_{1i}}$ in place of the canonical link function in the denominator. That there is no loss of generality stems from the fact that

$$g^*(\lambda_i, v_i) = -\frac{1}{\eta_{1i}} = \sum_{j=1}^{k} \beta_j^* x_{ij} \text{ leads with}$$

$$g(\lambda_i, v_i) = -g^*(\lambda_i, v_i) = \sum_{j=1}^{k} -\beta_j^* x_{ij} = \sum_{j=1}^{k} \beta_j x_{ij} \text{ and } -\beta_j^* = \beta_j \text{ to the form above.}$$

11.8 GLM for Gamma Regression

Example 11.8 Myers and Montgomery (1997) present data from a step in the manufacturing process of semiconductors. Four factors are believed to influence the resistivity of the wafer and so a full factorial experiment with two levels of each factor was run. Previous experience led to the expectation that resistivity would have a skewed distribution. The data are as follows in a data frame with 16 observations on the following five variables.

1) `x1` a factor with levels '–' '+'
2) `x2` a factor with levels '–' '+'
3) `x3` a factor with levels '–' '+'
4) `x4` a factor with levels '–' '+'
5) `resist` is resistivity of the wafer.

	x1	x2	x3	x4	resist
1	-	-	-	-	193.4
2	+	-	-	-	247.6
3	-	+	-	-	168.2
4	+	+	-	-	205.0
5	-	-	+	-	303.4
6	+	-	+	-	339.9
7	-	+	+	-	226.3
8	+	+	+	-	208.3
9	-	-	-	+	220.0
10	+	-	-	+	256.4
11	-	+	-	+	165.7
12	+	+	-	+	203.5
13	-	-	+	+	285.0
14	+	-	+	+	268.0
15	-	+	+	+	169.1
16	+	+	+	+	208.5

Problem 11.12 Analyse the data of Example 11.8 with a GLM with the gamma distribution and the log-link (multiplicative arithmetic mean model).

Solution

Use from the package glm2 the command > `glm(formula, family, data).`

Example

```
> library(faraway)
> data(wafer)
> plot(density(wafer$resist))
```

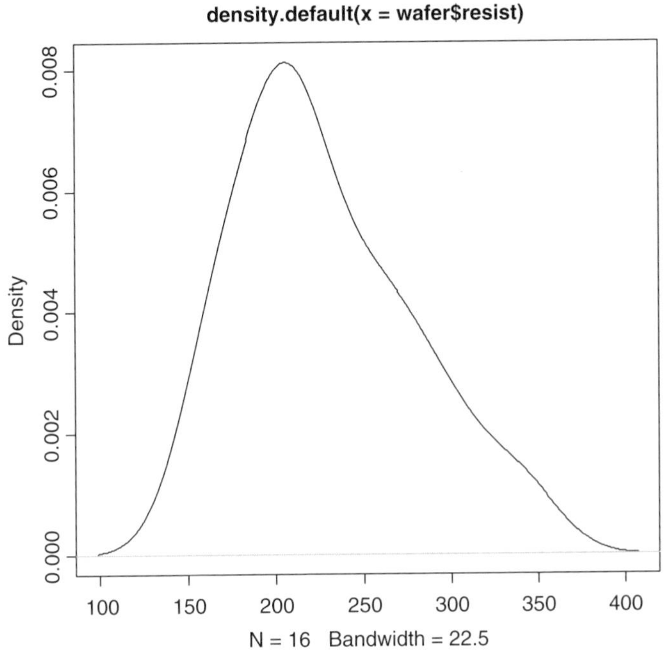


```
> resist.Gamma.log <- glm(resist ~ x1 + x2 + x3 + x4,
        family = Gamma(link="log"), data=wafer)
> summary(resist.Gamma.log)
Call:
glm(formula = resist ~ x1 + x2 + x3 + x4, family = Gamma(link = "log"),
    data = wafer)
Deviance Residuals:
     Min        1Q    Median        3Q       Max
-0.17548  -0.06486   0.01423   0.08399   0.10898
Coefficients:
            Estimate Std. Error t value Pr(>|t|)
(Intercept)  5.44552    0.05856  92.983  < 2e-16 ***
x1+          0.12115    0.05238   2.313 0.041090 *
x2+         -0.30049    0.05238  -5.736 0.000131 ***
x3+          0.17979    0.05238   3.432 0.005601 **
x4+         -0.05757    0.05238  -1.099 0.295248
---
Signif. codes:  0 '***' 0.001 '**' 0.01 '*' 0.05 '.' 0.1 ' ' 1
(Dispersion parameter for Gamma family taken to be 0.01097542)
```

```
    Null deviance: 0.69784  on 15  degrees of freedom
Residual deviance: 0.12418  on 11  degrees of freedom
AIC: 152.91
Number of Fisher Scoring iterations: 4
> anova(resist.Gamma.log, test = "Chisq")
Analysis of Deviance Table
Model: Gamma, link: log
Response: resist
Terms added sequentially (first to last)
     Df Deviance Resid. Df Resid. Dev  Pr(>Chi)
NULL                    15    0.69784
x1    1  0.05059        14    0.64725 0.0318037 *
x2    1  0.37740        13    0.26985 4.519e-09 ***
x3    1  0.13244        12    0.13741 0.0005132 ***
x4    1  0.01323        11    0.12418 0.2723316
---
Signif. codes:  0 '***' 0.001 '**' 0.01 '*' 0.05 '.' 0.1 ' ' 1
```

Note that it is possible to make inference about the arithmetic means on the original scale. For example, having `x1 = +` increases the log arithmetic mean by 0.12115. The exponential coefficient $\exp(0.12115) = 1.128794$ is the factor by which the arithmetic mean outcome in the original scale is multiplied, i.e. if `x1 = +`, the arithmetic mean on the original scale is 1.13 times higher compared to `x1 = –` within levels of other variables. The exponential intercept $\exp(5.44552) = 231.72$ is the arithmetic mean outcome for wafers which have – for all predictors. Therefore, this model assumes multiplicative effects on the original outcomes of the predictors.

Problem 11.13 Analyse the data of Example 11.8 with a GLM with the gamma distribution and the identity link (additive arithmetic mean model).

Solution

Use from the package glm2 the command > `glm(formula, family, data).`

Example

```
> library(faraway)
> data(wafer)
> resist.Gamma.identity <- glm(resist ~ x1 + x2 + x3 + x4,
        family = Gamma(link="identity"), data=wafer)
> summary(resist.Gamma.identity)
Call:
glm(formula = resist ~ x1 + x2 + x3 + x4, family =
      Gamma(link = "identity"), data = wafer)
Deviance Residuals:
      Min         1Q     Median         3Q        Max
-0.190394  -0.066533   0.005538   0.091549   0.126838
Coefficients:
            Estimate Std. Error t value Pr(>|t|)
(Intercept)   235.43      14.21  16.572 3.97e-09 ***
x1+            27.83      12.24   2.274 0.044017 *
x2+           -65.74      12.68  -5.184 0.000302 ***
```

```
x3+            36.94       12.32    2.999 0.012102 *
x4+           -11.88       12.15   -0.978 0.349294
---
Signif. codes:  0 '***' 0.001 '**' 0.01 '*' 0.05 '.' 0.1 ' ' 1
(Dispersion parameter for Gamma family taken to be 0.0124071)
    Null deviance: 0.69784  on 15  degrees of freedom
Residual deviance: 0.14009  on 11  degrees of freedom
AIC: 154.84
Number of Fisher Scoring iterations: 6
> anova(resist.Gamma.identity, test = "Chisq")
Analysis of Deviance Table
Model: Gamma, link: identity
Response: resist
Terms added sequentially (first to last)
     Df Deviance Resid. Df Resid. Dev  Pr(>Chi)
NULL                    15    0.69784
x1    1  0.05059        14    0.64725  0.043466 *
x2    1  0.37755        13    0.26970 3.461e-08 ***
x3    1  0.11782        12    0.15188  0.002059 **
x4    1  0.01178        11    0.14009  0.329797
---
Signif. codes:  0 '***' 0.001 '**' 0.01 '*' 0.05 '.' 0.1 ' ' 1
```

Note that if the effect of each predictor is considered additive on the original scale, the generalised model with the identity link function can be used. In this case, the raw coefficients are on the original scale. Having `x1 = +` adds 27.83 to the arithmetic mean outcome (additive effect). The raw intercept (235.43) is the arithmetic mean outcome for wafers that have – for all predictors.

Example 11.9 In McCullagh and Nelder (1989), on page 300–302, an example of clotting times of blood is described. The data are clotting time of blood, giving clotting times in seconds (y) for normal plasma diluted to nine different percentage concentrations with prothrombin-free plasma (u); clotting was induced by two lots of thromboplastin. The higher the dilution, the more the inference with the blood's ability to clot, because the blood's natural clotting capability has been weakened. For each sample, clotting was induced by introducing thromboplastin, a clotting agent, and the time until clotting occurred was recorded (in seconds). Five samples were measured at each of the nine percentage concentrations, and the mean clotting time was averaged; thus, the response is mean clotting time over the five samples. The data are shown in Table 11.9.

These data are positive with possible constant coefficient of variation. Thus, we consider the gamma probability model. Letting $\boldsymbol{y}_j$ be the clotting time at log(percentage concentration) x_j we, consider the model for the mean response with the reciprocal (or inverse):
$E(y_j) = 1/(\beta_0 + \beta_1 x_j)$.

Problem 11.14 Analyse the data of Example 11.9 with a GLM with the gamma distribution with $x = \log(u)$ and the inverse link for `Lot 1` and `Lot 2`.

Solution
Use from the package glm2 the command > `glm(formula, family, data)`.

Example

```
> # A Gamma example, from McCullagh & Nelder (1989, pp. 300-2)
> clotting <- data.frame(
    u = c(5,10,15,20,30,40,60,80,100),
    lot1 = c(118,58,42,35,27,25,21,19,18),
    lot2 = c(69,35,26,21,18,16,13,12,12))
> x <- log(clotting[,1])
> y1 <- clotting[,2]
> plot(x,y1, xlab="log(percentage conc of plasma)",
       ylab="mean clotting time", main="Lot1")
```

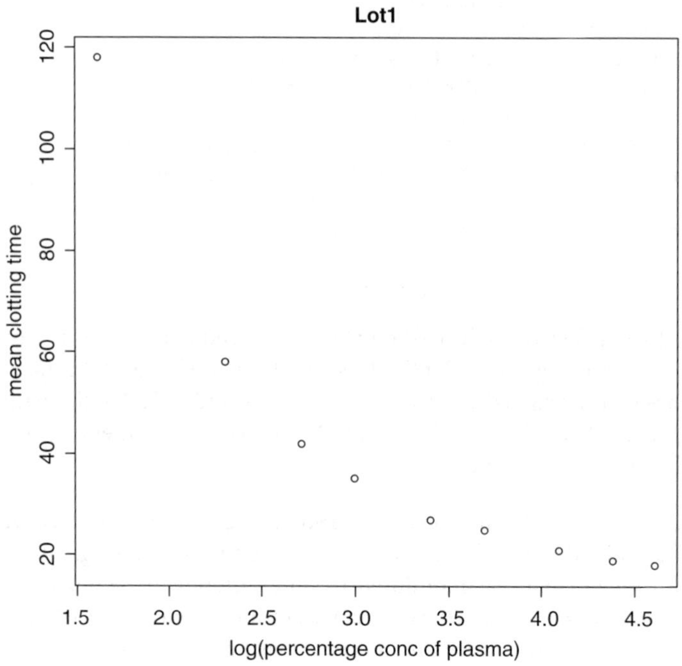


```
> lot1.log <- glm(lot1 ~ x, data = clotting,
     family = Gamma(link= "inverse" ))
> summary(lot1.log)

Call:
glm(formula = lot1 ~ x, family = Gamma(link = "inverse"), data = clotting)

Deviance Residuals:
     Min        1Q    Median        3Q       Max
-0.04008  -0.03756  -0.02637   0.02905   0.08641

Coefficients:
              Estimate Std. Error t value Pr(>|t|)
(Intercept) -0.0165544  0.0009275  -17.85 4.28e-07 ***
x            0.0153431  0.0004150   36.98 2.75e-09 ***
---
Signif. codes:  0 '***' 0.001 '**' 0.01 '*' 0.05 '.' 0.1 ' ' 1
```

```
(Dispersion parameter for Gamma family taken to be 0.002446059)

    Null deviance: 3.51283  on 8  degrees of freedom
Residual deviance: 0.01673  on 7  degrees of freedom
AIC: 37.99

Number of Fisher Scoring iterations: 3
> anova(lot1.log, test = "Chisq")
Analysis of Deviance Table

Model: Gamma, link: inverse

Response: lot1

Terms added sequentially (first to last)


     Df Deviance Resid. Df Resid. Dev  Pr(>Chi)
NULL                     8     3.5128
x     1   3.4961         7     0.0167 < 2.2e-16 ***
---
Signif. codes:  0 '***' 0.001 '**' 0.01 '*' 0.05 '.' 0.1 ' ' 1
> fit.lot1 <- fitted(lot1.log)
> data.fit1 <- data.frame(x,y1,fit.lot1)
> data.fit1
         x  y1  fit.lot1
1 1.609438 118 122.85904
2 2.302585  58  53.26389
3 2.708050  42  40.00713
4 2.995732  35  34.00264
5 3.401197  27  28.06578
6 3.688879  25  24.97221
7 4.094345  21  21.61432
8 4.382027  19  19.73182
9 4.605170  18  18.48317
```

Table 11.9 Clotting time of blood in seconds (y) for normal plasma diluted to nine different percentage concentrations with prothrombin-free plasma (u).

Clotting time		
u	y lot 1	y lot 2
5	118	69
10	58	35
15	42	26
20	35	21
30	27	18
40	25	16
60	21	12
80	19	12
100	18	12

Note that the reciprocal model fits the data very well and can be used for describing the percentage concentration of plasma-clotting relationship!

```
> y2 <- clotting[,3]
> plot(x,y2, xlab="log(percentage conc of plasma)",
       ylab="mean clotting time", main="Lot2")
```

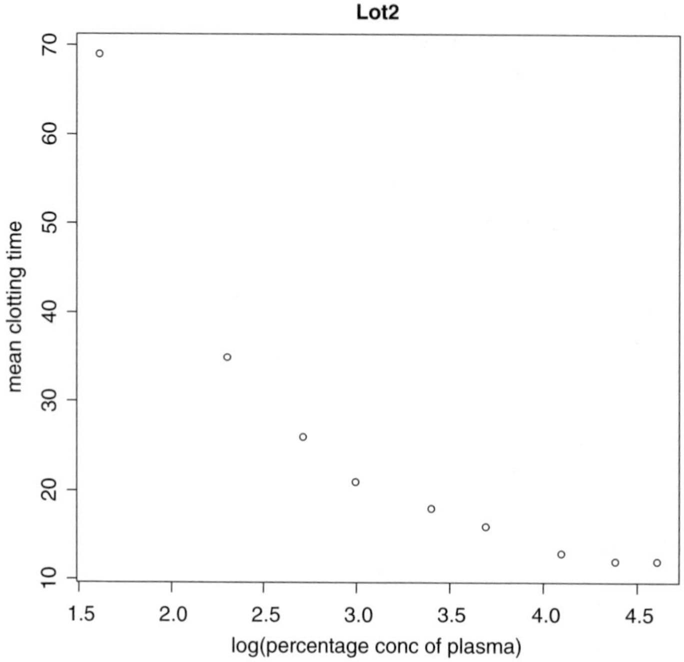


```
> lot2.log <- glm(lot2 ~ log(u), data = clotting,
  family = Gamma(link= "inverse"))
> summary(lot2.log)
Call:
glm(formula = lot2 ~ log(u), family = Gamma(link = "inverse"),
    data = clotting)

Deviance Residuals:
     Min        1Q    Median        3Q       Max
-0.05574  -0.02925   0.01030   0.01714   0.06371

Coefficients:
              Estimate Std. Error t value Pr(>|t|)
(Intercept) -0.0239085  0.0013265  -18.02 4.00e-07 ***
log(u)       0.0235992  0.0005768   40.91 1.36e-09 ***
---
Signif. codes:  0 '***' 0.001 '**' 0.01 '*' 0.05 '.'
                                             0.1 ' ' 1

(Dispersion parameter for Gamma family taken to be 0.001813354)

    Null deviance: 3.118557  on 8  degrees of freedom
Residual deviance: 0.012672  on 7  degrees of freedom
```

```
AIC: 27.032

Number of Fisher Scoring iterations: 3
> anova(lot2.log, test = "Chisq")
Analysis of Deviance Table

Model: Gamma, link: inverse

Response: lot2

Terms added sequentially (first to last)


       Df Deviance Resid. Df Resid. Dev  Pr(>Chi)
NULL                       8    3.11856
log(u)  1   3.1059         7    0.01267 < 2.2e-16 ***
---
Signif. codes:  0 '***' 0.001 '**' 0.01 '*' 0.05 '.' 0.1 ' ' 1
> fit.lot2 <- fitted(lot2.log)
> data.fit2 <- data.frame(x,y2,fit.lot2)
> data.fit2
         x y2 fit.lot2
1 1.609438 69 71.05806
2 2.302585 35 32.86152
3 2.708050 26 25.00038
4 2.995732 21 21.37279
5 3.401197 18 17.74399
6 3.688879 16 15.83627
7 4.094345 13 13.75235
8 4.382027 12 12.57800
9 4.605170 12 11.79664
```

Note that the reciprocal model fits the data very well and can be used for describing the percentage concentration of plasma-clotting relationship!

11.9 GLM for the Multinomial Distribution

The multinomial logit model is a generalisation of the binomial logit model. It describes an $(m-1)$-dimensional response variable $(\boldsymbol{y}_1, \boldsymbol{y}_2, \dots, \boldsymbol{y}_m)$, $\sum_{i=1}^{m} \boldsymbol{y}_i = n$ occurring with probabilities $p_i = P(\boldsymbol{y}_i = k_i); i = 1, \dots, m; \sum_{i=1}^{m} p_i = 1$. The probability (likelihood) function of the multinomial distribution is

$$P(\boldsymbol{y}_1 = k_1, \dots, \boldsymbol{y}_m = k_m) = \frac{n!}{y_1! \dots y_m!} p_1^{y_1} \cdots p_m^{y_m}; i = 1, \dots, m; 0 < p_i < 1; 0 < y_i < n.$$

$$P(\boldsymbol{y}_1 = k_1, \dots, \boldsymbol{y}_m = k_m) = \frac{n!}{y_1! \dots y_m!} e^{\sum_{j=2}^{m} y_j ln \frac{p_j}{p_1} + nlnp_1}.$$

The relation between the regressor variables $(x_0, x_1, \dots, x_k)$ and the probabilities p_i is described by the multinomial logit function as link function

$$\eta_j = \ln \frac{p_j}{p_1},$$

where p_1 is the probability of the reference category.

The model is fitted as described in Section 11.6.1 for the Poisson model because we have an equivalence between the multinomial distribution and a Poisson distribution with fixed sum of all counts.

Example 11.10 We use again the data of Example 11.4 but investigate now the character with the two outcomes infected–not infected as the regressor variable. We investigate the realisations of $\boldsymbol{k}_{ijk}$, the numbers of infected plants in block i, genotype j and on plot k.

The realisations k_{ijk} and the number of plants n_{ijk} in block i, genotype j and on plot k can be found in Table 11.7 as well as the number of plots m_{ij} in each block–genotype combination.

Let p_{ijk} be the infection probability of plants in plot (i, j, k); $i = 1, \dots, 5$; $j = 1, 2$; $k = 1, \dots, m_{ij}$ and assume that $\boldsymbol{k}_{ijk}$ is $B(n_{ijk}, p_{ijk})$-distributed.

Problem 11.15 Analyse the infection data of Example 11.10.

Solution

We use from the package glm2 the command > `glm`.

Example

```
> peanutsnew=read.table("D:/Rasch_Applied_Statistics/multinomial.txt",
        header=T)
> peanutsnew
    block G_type resp kn
1       1      1    1  5
2       1      1    1  4
3       1      1    1  4
------------------------
148     5      2    2  3
149     5      2    2  4
150     5      2    2  3
> results <- data.frame(peanutsnew)
> BLOCK <- factor(results[,1])
> GTYPE <- factor(results[,2])
> RESP <- factor(results[,3])
> y <- results[,4]
> model <- glm(y ~ RESP + BLOCK:GTYPE + RESP:BLOCK + RESP:GTYPE,
        family="poisson", data=results)
> summary(model)
Call:
glm(formula = y ~ RESP + BLOCK:GTYPE + RESP:BLOCK + RESP:GTYPE,
    family = "poisson", data = results)

Deviance Residuals:
     Min        1Q    Median        3Q       Max
-2.51857  -0.35922   0.06539   0.41226   1.83426

Coefficients: (1 not defined because of singularities)
             Estimate Std. Error z value Pr(>|z|)
(Intercept)    1.9063     0.1157  16.482  < 2e-16 ***
RESP2          0.4423     0.1728   2.559 0.010492 *
```

```
BLOCK1:GTYPE1  -0.5323    0.1920  -2.773 0.005562 **
BLOCK2:GTYPE1  -1.3568    0.4048  -3.352 0.000802 ***
BLOCK3:GTYPE1  -2.4443    0.3879  -6.302 2.94e-10 ***
BLOCK4:GTYPE1  -2.0915    0.2601  -8.041 8.89e-16 ***
BLOCK5:GTYPE1  -0.7521    0.1787  -4.209 2.57e-05 ***
BLOCK1:GTYPE2   0.1915    0.1530   1.252 0.210625
BLOCK2:GTYPE2  -0.2448    0.3018  -0.811 0.417386
BLOCK3:GTYPE2  -0.9257    0.3021  -3.064 0.002181 **
BLOCK4:GTYPE2  -0.6400    0.1841  -3.476 0.000508 ***
BLOCK5:GTYPE2       NA        NA      NA       NA
RESP2:BLOCK2    1.1793    0.3903   3.021 0.002517 **
RESP2:BLOCK3    2.3643    0.3559   6.643 3.08e-11 ***
RESP2:BLOCK4    1.9122    0.2370   8.069 7.10e-16 ***
RESP2:BLOCK5    0.3246    0.2211   1.468 0.142008
RESP2:GTYPE2   -1.7287    0.1812  -9.538  < 2e-16 ***
---
Signif. codes:  0 '***' 0.001 '**' 0.01 '*' 0.05 '.' 0.1 ' ' 1
(Dispersion parameter for poisson family taken to be 1)
    Null deviance: 297.692  on 149  degrees of freedom
Residual deviance:  61.361  on 134  degrees of freedom
AIC: 570.13
Number of Fisher Scoring iterations: 5
> anova(model, test = "Chisq")
Analysis of Deviance Table
Model: poisson, link: log
Response: y
Terms added sequentially (first to last)
            Df Deviance Resid. Df Resid. Dev  Pr(>Chi)
NULL                          149    297.692
RESP         1   33.155       148    264.538 8.512e-09 ***
BLOCK:GTYPE  9    1.088       139    263.450    0.9992
RESP:BLOCK   4   98.772       135    164.678 < 2.2e-16 ***
RESP:GTYPE   1  103.317       134     61.361 < 2.2e-16 ***
---
Signif. codes:  0 '***' 0.001 '**' 0.01 '*' 0.05 '.' 0.1 ' ' 1
> model1 <- glm(y ~ RESP + BLOCK:GTYPE + RESP:GTYPE + RESP:BLOCK,
         family="poisson", data=results)
> anova(model1, test = "Chisq")
Analysis of Deviance Table
Model: poisson, link: log
Response: y
Terms added sequentially (first to last)
            Df Deviance Resid. Df Resid. Dev  Pr(>Chi)
NULL                          149    297.692
RESP         1   33.155       148    264.538 8.512e-09 ***
BLOCK:GTYPE  9    1.088       139    263.450    0.9992
RESP:GTYPE   1   88.997       138    174.452 < 2.2e-16 ***
RESP:BLOCK   4  113.092       134     61.361 < 2.2e-16 ***
```

Compare these results with that of Example 11.4 in Problem 11.7.

Note that the estimates RESP2:BLOCK2,…, RESP2:GTYPE2 have the negative values of the estimates of BLOCK2, … , GTYPE2 in Problem 11.7. Note further that the deviance of RESP:GTYPE in the first analysis of deviance table is the same as the deviance of GTYPE in the first analysis of deviance table of Problem 11.7. Also, the deviance of RESP:BLOCK in the second analysis of deviance table is the same as the deviance of BLOCK in the second analysis of deviance table in Problem 11.7.

References

Agresti, A. (2018). *Categorical Data Analysis*. New York: Wiley.

von Bortkiewicz, L.J. (1893). *Die mittlere Lebensdauer. Die Methoden ihrer Bestimmung und ihr Verhältnis zur Sterblichkeitsmessung*. Jena: Gustav Fischer.

Collett, D. (1991). *Modelling Binary Data*. Boca Raton: Chapman & Hall.

Faraway, J.J. (2016). *Extending the Linear Model with R: Generalized Linear, Mixed Effects and Nonparametric Regression Models*, 2e. Boca Raton: Chapman & Hall/CRC Texts in Statistical Science.

McCullagh, P. and Nelder, J.A. (1989). *Generalized Linear Models*. New York: Springer.

Myers, R.H. and Montgomery, D.C. (1997). A tutorial on generalized linear models. *J. Qual. Technol.* 29: 274–291, (Published online: 21 Feb 2018).

Nelder, J.A. and Wedderburn, R.W.M. (1972). Generalized linear models. *J. R. Stat. Soc.* 135: 370–384.

Rasch, G. (1960). *Probabilistic Models for Some Intelligence and Attainment Tests*. Kopenhagen: Nissen & Lydicke.

Rasch, G. (1980). *Probabilistic Models for Some Intelligence and Attainment Tests*. Danish Institute for Educational Research, Copenhagen 1960, expanded edition with foreword and afterword by B.D. Wright. Chicago: The University of Chicago Press.

Rasch, D. and Schott, D. (2018). *Mathematical Statistics*. Oxford: Wiley.

Rasch, D., Herrendörfer, G., Bock, J. et al. (1998). *Verfahrensbibliothek Versuchsplanung und - auswertung, 2. verbesserte Auflage in einem Band mit CD*. München Wien: R. Oldenbourg Verlag.

12

Spatial Statistics

12.1 Introduction

Spatial statistics is a part of applied statistics and is concerned with modelling and analysis of spatial data. By spatial data we mean data where, in addition to the (primary) phenomenon of interest, the relative spatial locations of observations are also recorded because these may be important for the interpretation of data. This is of primary importance in earth-related sciences such as geography, geology, hydrology, ecology, and environmental sciences, but also in other scientific disciplines concerned with spatial variations and patterns such as astrophysics, economics, agriculture, forestry, and epidemiology, and, at a microscopic scale, medical and health research. Spatial statistics uses nearly all methods described in the first eleven chapters of this book and also multivariate analysis and Bayesian methods, neither of which are discussed in this book. We therefore restrict ourselves in this chapter to a few basic principles and give hints for further reading for other important methods. As a consequence of this, the list of references is relatively long.

We restrict our attention to continuous characteristics and to Gaussian distributions and analyse examples with the program package R as we did in other chapters of this book. Those who prefer SAS and understand a bit of German are referred to the procedures in 6/61 of Rasch et al. (2008).

6/61/0000 Spatial Statistics – Introduction
6/61/1010 Estimation of the covariance function of a random variable with constant trend
6/61/1020 Estimation of the semi-variogram of a random variable
6/61/1021 Estimation of the parameter of an exponential semi-variogram model
6/61/1022 Estimation of the parameter of a spherical semi-variogram model
6/61/1030 Definition of increments and of the generalised covariance function for non-steady state random variables
6/61/1031 Estimation of the generalised covariance function for non-steady state random variables
6/61/1100 Modelling spatial dependencies between two variables
6/61/2000 Spatial prediction – survey
6/61/2010 Prediction of stationary random variables using the covariance function
6/61/2020 Prediction of stationary random variables using the semi-variogram
6/61/2030 Prediction of non-steady state random variables

Applied Statistics: Theory and Problem Solutions with R, First Edition.
Dieter Rasch, Rob Verdooren, and Jürgen Pilz.

6/61/2040 Spatial prediction: Co-kriging
6/61/2050 Prediction of probabilities
6/61/2051 Prediction of exceedance probabilities
6/61/2052 Hermitean prediction.

We restrict ourselves to one- and two-dimensional regions $D \subset R^1$ and $D \subset R^2$, respectively. However, $D \subset R^3$ (oil and mineral prospection, 3D imaging) is also possible. In some fields such as Bayesian data analysis, design and simulation one even requires spaces D of dimension >3; this pertains, in particular, to the design and analysis of computer experiments with a moderate to large number of input variables.

Points in $D \subset R^2$ are written as $s^T = (x_1, x_2)$ $\left\{ s = \begin{pmatrix} x_1 \\ x_2 \end{pmatrix} \right\}$ and the coordinates x_1, x_2 are in geostatistics often the Gauss–Krüger coordinates on the earth based on the degrees (°) of the longitude meridional zone of the surface of the earth, see Krüger (1912). The surface is subdivided in meridional zones of a latitude of 3° running from the North Pole to the South Pole parallel to its central meridian. The degrees of the central meridian of each meridional zone counted from 0° eastwards are mapped to code numbers by dividing them by three as shown in Table 12.1.

The meridional zone is conformally mapped on a cylinder barrel with the axis in the equatorial plane and a radius equal to the curvature radius of the meridian. Its origin is the intersection of the central meridian and the equator. From the origin the coordinates of the points on the surface of the earth are defined like in a usual Cartesian coordinate system, positive to the east by the so-called easting (x_1), and to the north by the so-called grid north (x_2). The coordinates on the earth can be transformed to the Gauss–Krüger coordinates via https://www.koordinaten-umrechner.de. As an example, we give the Gauss–Krüger coordinates of the (first author's) house in Feldrain 73 in Rostock, Germany.

Degrees minutes seconds E 12° 11′ 41; N 54° 06′ 18
Easting: 4507572; grid north 59922353.

First we must select a spatial model for the observations (variables) at the points in $D \subset R^2$; their realisations are our observations.

Table 12.1 Gauss–Krüger code numbers.

Central meridian	Western longitude				Eastern longitude		
Degree	…	6°	3°	0°	3°	6°	…
Degree east from 0°		354°	357°	0°	3°	6°	
Code number	…	118	119	0	1	2	…

The general model is for observations $\boldsymbol{y}(s)$ at $s = \begin{pmatrix} x_1 \\ x_2 \end{pmatrix}$

$$\boldsymbol{y}(s) = \mu(s) + \boldsymbol{e}(s), s \in D \subset R^2; E[\boldsymbol{y}(s)] = \mu(s) \tag{12.1}$$

with side conditions for weak stationarity (second order stationarity):

$$E[\boldsymbol{y}(s)] = \mu(s) = \mu, \forall s \in D \subset R^2, \mu \in R^1, \tag{12.2}$$

$$\text{cov}(\boldsymbol{y}(s), \boldsymbol{y}(s+h)) = C(h), \forall s, s+h \in D \subset R^2. \tag{12.3}$$

Formula (12.3) leads for $h = 0$ to var($\boldsymbol{y}(s)$)= $C(0)$.

If we further assume that $C(h) = C(\|h\|)$ with the Euclidian norm $\|h\|$ of h, then C is called isotropic, otherwise anisotropic.

The positions of observation sites $s \in D$ can be fixed in advance (as, e.g. the position of wind power stations in an area) or may be random.

As examples where observation points occur randomly, we mention meteor strikes in a special area. To this situation also belongs the possibly oldest mapping of clusters of cholera cases in the London epidemic of 1854 (Snow 1855). Randomly means in this connection that we assume that a point is equally probable to occur at any location and that the position of a point is not affected by any other point.

Further, the observation points may not be fixed but determined by the scientist (monitoring). In this case, besides problems of analysis, design problems also exist by selecting the optimal observation points in an area. For this, readers are referred to Müller (2007).

Comprehensive treatments of the whole field of spatial statistics are given in Ripley (1988), Cressie (1993), and Gaetan and Guyon (2010).

Basically, there are four classes of problems which spatial statistics is concerned with: point pattern analysis, geostatistical data analysis, areal/lattice data analysis and spatial interaction analysis. These sub-problems are treated in overview papers such as: Pilz (2010), Mase (2010), Kazianka and Pilz (2010a), Diggle (2010), and Spöck and Pilz (2010).We discuss mainly geostatistical data analysis with some hints to areal data analysis.

For a good overview on software for different problem areas of spatial data analysis with R we recommend the book by Bivand et al. (2013), for the important issue of simulation of spatial models we refer to Lantuéjoul (2002) and Gaetan and Guyon (2010). An overview of methodology and software for interfacing spatial data analysis and geographic information system (GIS) for visualising spatial data is given in Pilz (2009).

Due to the fact that in some fields of application of spatial statistics special methods have been applied, and further because the theory and applications are still in development, we cannot give a closed presentation of the field. We describe basic methods used in geostatistics using Euclidean distances and then give examples. For point pattern analysis, areal/lattice data analysis, and spatial interaction analysis the reader is referred to the books mentioned above.

12.2 Geostatistics

In geostatistics D is a continuous subspace of R^2 or R^3 and the random variable (field) is observed at $n > 2$ fixed sites $s_1, s_2, \dots, s_n \in D$. Typical examples include rainfall data, data on soil characteristics (porosity, humidity, etc.), oil and mineral exploration data, air quality, and groundwater data. In this chapter only one characteristic observation variable is measured per observation point on a line or plane. Multivariate geostatistics dealing with observation vectors per observation point is described in detail in Wackernagel (2010).

The concept of stationarity is key in the analysis of spatial and/or temporal variation: roughly speaking, stationarity means that the statistical characteristics (e.g. mean and variance) of the random variable of interest do not change over the considered area. However, testing for stationarity is not possible. For spatial prediction the performance of a stationary and a non-stationary model could be compared through assessment of the accuracy of predictions.

In this chapter we assume that the random vectors $\boldsymbol{y}^T = [y(s_1), y(s_2), \dots, y(s_n)\,]$ follow an n-dimensional normal (Gaussian) distribution for any collection of spatial locations $\{s_1, s_2, \dots, s_n\} \subset D$ and any $n \geq 1$. In the literature the collection of random variables $\{y(s) : s \in D\}$ is then usually termed a Gaussian random field (GRF) over D. For other types of random fields and a detailed explanation of their mathematical structure and most important properties we refer to Cressie and Wikle (2011), where also extensions to so-called spatio-temporal random fields are considered. Often non-normal random variables may be transformed by a so-called Box–Cox transformation as a generalisation of the normal case by including an additional parameter λ. This transformation is given by:

$$y^* = \begin{cases} \dfrac{y^\lambda - 1}{\lambda} & \text{if } \lambda \neq 0 \\ \ln y & \text{if } \lambda = 0. \end{cases} \tag{12.4}$$

A GRF is completely determined by its expectation (trend function)

$$\mu(s) = E(y)$$

and covariance function $C(s_i - s_j) = \text{cov}[y(s_i), y(s_j)] : i, j = 1, \dots, n$.

Contrary to traditional statistics, in a geostatistical setting we usually observe only one realisation of y at a finite number of locations $s_i^T = (x_{1i}, x_{2i}); i = 1, \dots, n$. Therefore, the distribution underlying the random field cannot be inferred without imposing further assumptions. The simplest assumption is that of (strict) stationarity, which means that the normal distributions do not change when all positions are translated by the same (lag) vector h and this implies that (12.2) and (12.3) are valid.

Often (in areal/lattice data analysis) measurements are not related to points but to areas A_i, $i = 1, \dots n$. In such cases in place of distances between points measures of spatial proximity w_{ij} between two areas A_i and A_j are used and represented in a square $n \times n$ matrix $W = (w_{ij})$.

According to Bailey and Gatrell (1995) some possible criteria for determining proximities might be:

- $w_{ij} = 1$ if A_j shares a common boundary with A_i and $w_{ij} = 0$ else.
- $w_{ij} = 1$ if the centroid of A_j is one of the k nearest centroids to that of A_i and $w_{ij} = 0$ else.
- $w_{ij} = d_{ij}^\gamma$ if the inter-centroid distance $d_{ij} < \delta$ $(\delta > 0, \gamma < 0)$; and $w_{ij} = 0$ else.
- $w_{ij} = \frac{l_{ij}}{l_i}$, where l_{ij} is the length of the common boundary between $A_i$and A_j and l_i is the perimeter of A_i.

All diagonal elements w_{ii} are set to 0. Note that the spatial proximity matrix W must not necessarily be symmetric.

For more proximity measures we refer to Bailey and Gatrell (1995) and any other publications on areal spatial analysis like Anselin and Griffith (1988).

12.2.1 Semi-variogram Function

From now on, we focus on geostatistics and assume second order stationarity, i.e. (12.2) and (12.3) hold, and additionally, we assume isotropy. In geostatistics it is common to use the so-called semi-variogram function

$$\gamma(h) = \frac{1}{2}\text{var}[\boldsymbol{y}(s) - \boldsymbol{y}(s+h)] \text{ for all } s, s+h \in D. \tag{12.5}$$

Observing that under the assumption of stationarity it holds that

$$\gamma(h) = \frac{1}{2}\text{var}[\boldsymbol{y}(s) - \boldsymbol{y}(s+h)] = \frac{1}{2}E[\boldsymbol{y}(s) - \boldsymbol{y}(s+h)]^2 \tag{12.6}$$

we can simply use the empirical moment estimate to estimate the semi-variogram according to

$$\widehat{\gamma(\check{h})} = \frac{1}{2N(\check{h})}\sum_{i=1}^{N(\check{h})}[y(s_i) - y(s_i + h)]^2 \tag{12.7}$$

where $N(\check{h})$ denotes the number of sampling locations separated from each other by the distance $\check{h} = \| h \|$, also called the lag distance.

We call $\widehat{\gamma(.)}$ the estimated semi-variogram function or the sample semi-variogram function. Since a vector h is uniquely determined by its length $\| h \| = \check{h}$ and its direction, it is customary to form lag distance classes along given directions, usually this is done for the four main directions 0°, 45°, 90°, and 135°.

Then

$$\begin{aligned}\text{var}[\boldsymbol{y}(s) - \boldsymbol{y}(s+h)] &= \text{var}\,\boldsymbol{y}(s) + \text{var}[\boldsymbol{y}(s+h)] - 2\text{cov}[\boldsymbol{y}(s), \boldsymbol{y}(s+h)] \\ &= 2C(0) - 2C(\| h \|).\end{aligned} \tag{12.8}$$

This is called the variogram function. Dividing by two leads to the semi-variogram function

$$\gamma(\check{h}) = C(0) - C(\check{h}) \tag{12.9}$$

as a special case of (12.5).

For 'classical' estimation methods for the trend function and variogram parameters see Mase (2010), for Bayesian approaches we refer to Banerjee et al. (2014), Kazianka and Pilz (2010a) and Pilz et al. (2012). For non-stationary variogram modelling we refer to the review provided by Sampson et al. (2001) and Schabenberger and Gotway (2005).

Theoretically, at zero separation distance (lag = 0), the semi-variogram value $\gamma(0)$ is zero, because $\lim_{\check{h} \to 0} \gamma(\check{h}) = 0$. However, at an infinitesimally small separation distance, the semi-variogram function often exhibits a so-called nugget effect, which is some value greater than zero. For example, if the semi-variogram model intercepts the x_2-axis at 3, then the nugget effect is 3. Furthermore, often $\lim_{\check{h} \to \infty} \gamma(\check{h}) = c > 0$ is an asymptotic value of the semi-variogram model called sill. The nugget effect can be attributed to measurement errors or spatial sources of variation at distances smaller than the sampling interval (or both). Measurement error occurs because of the error inherent in measurement devices. Natural phenomena can vary spatially over a range of scales. Variation at microscales smaller than the sampling distances will appear as part of the nugget effect. Before collecting data, it is important to gain some understanding of the scales of spatial variation.

Problem 12.1 Calculate estimates of the semi-variogram function and show graphs.

Solution
We use the R package geoR.

Example 12.1
From the `library(geoR)` we use a simulated data set `s100` of 100 geodata in R^2. We use a part of the text of http://leg.ufpr.br/geoR/geoRdoc/geoRintro.html.

```
> library(geoR)
> data(s100)
> summary(s100)
Number of data points: 100
Coordinates summary
        Coord.X     Coord.Y
min 0.005638006 0.01091027
max 0.983920544 0.99124979
Distance summary
        min         max
0.007640962 1.278175109
Data summary
      Min.     1st Qu.      Median        Mean     3rd Qu.        Max.
-1.1676955  0.2729882  1.1045936  0.9307179  1.6101707  2.8678969
Other elements in the geodata object are
[1] "cov.model" "nugget"    "cov.pars"  "kappa"     "lambda"
```

The function `plot.geodata` shows a 2×2 display (as given in Figure 12.1) with data locations (top plots) and data versus coordinates (bottom plots). For an object of the class `"geodata"` the plot is produced by the command: `plot(s100)`

```
> plot(s100)
```

Empirical semi-variograms are calculated using the function `variog`. Theoretical and empirical semi-variograms can be plotted and visually compared. The most important theoretical variogram models will be given in the sequel. For example, Figure 12.2 shows the theoretical variogram model (exponential model) used to simulate the data `s100` and the estimated variogram values at different lags.

```
> bin1 <- variog(s100, uvec = seq(0,1,l=11))
variog: computing omnidirectional variogram
> plot(bin1)
> lines.variomodel(cov.model = "exp", cov.pars = c(1,0.3),nugget = 0,
    max.dist = 1,  lwd = 3)
> legend(0.4,0.3,legend=c("exponential model"),lty=1,lwd = 3)
```

Directional variograms can also be computed by the function `variog` using the arguments direction and tolerance. For example, to compute a variogram for the direction 60° with the default tolerance angle (22.5°) the command would be:

```
> vario60 <- variog(s100, max.dist = 1, direction=pi/3)
> plot(vario60)
> title(main = expression(paste("directional, angle = ", 60 * degree)))
```

and the plot is shown on the left panel of Figure 12.3.

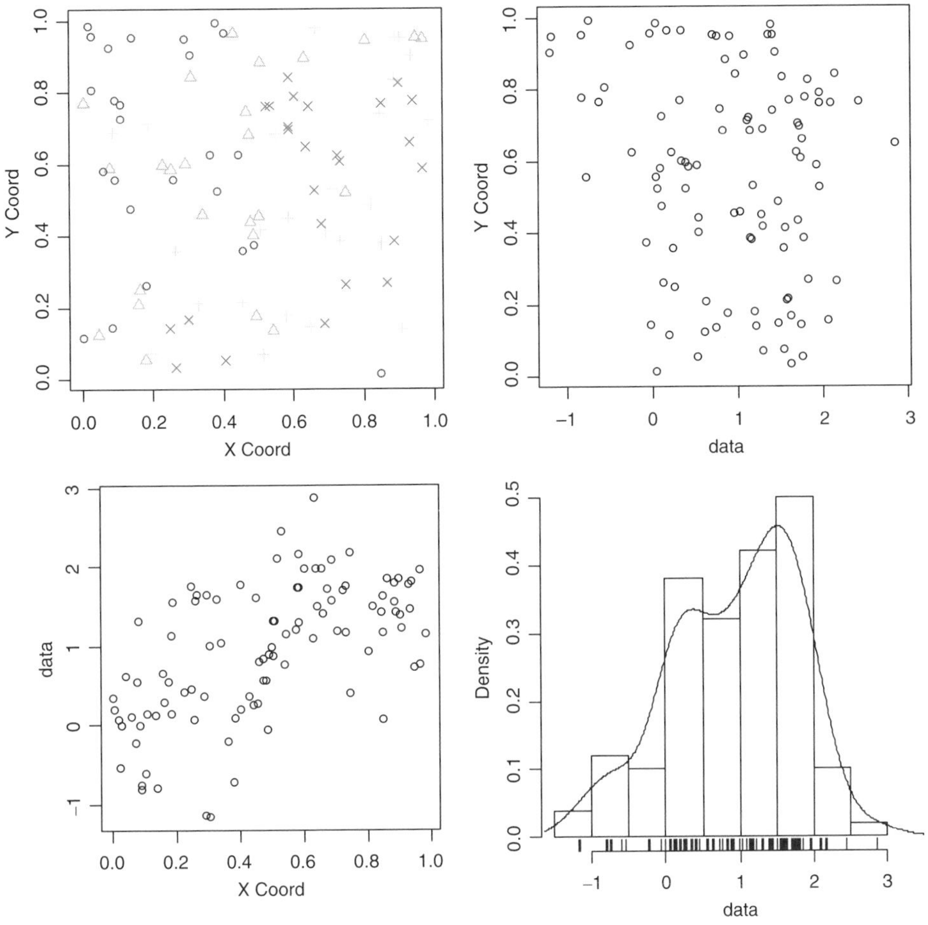

Figure 12.1 Exploratory spatial data analysis of dataset s100.

For a quick computation in the four main directions we can use the function `variog4` and the corresponding plot is shown on the right panel of Figure 12.3.

```
> vario.4 <- variog4(s100, max.dist = 1)
> plot(vario.4, lwd=2)
```

The semi-variogram as the graph of the semi-variogram function is often non-linear and can be estimated by the methods of Section 8.1.2 using $||h||$ as regressor from the estimates of the semi-variogram function as regressands. For this we have to select an appropriate regression model. However, we may only use semi-variogram models, which are conditionally negative semidefinite, which means that

$$\sum_{i=1}^{n}\sum_{j=1}^{n} a_i a_j \gamma(s_i - s_j) \leq 0$$

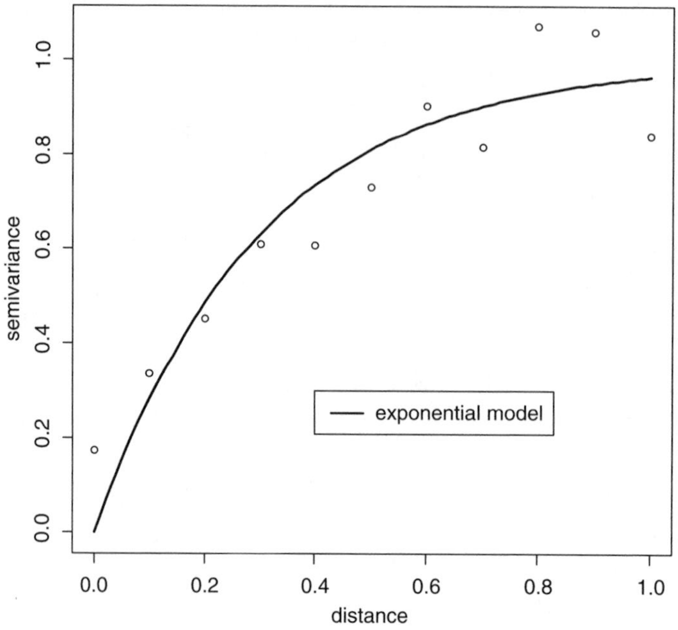

Figure 12.2 Exponential semi-variogram model underlying the dataset s100.

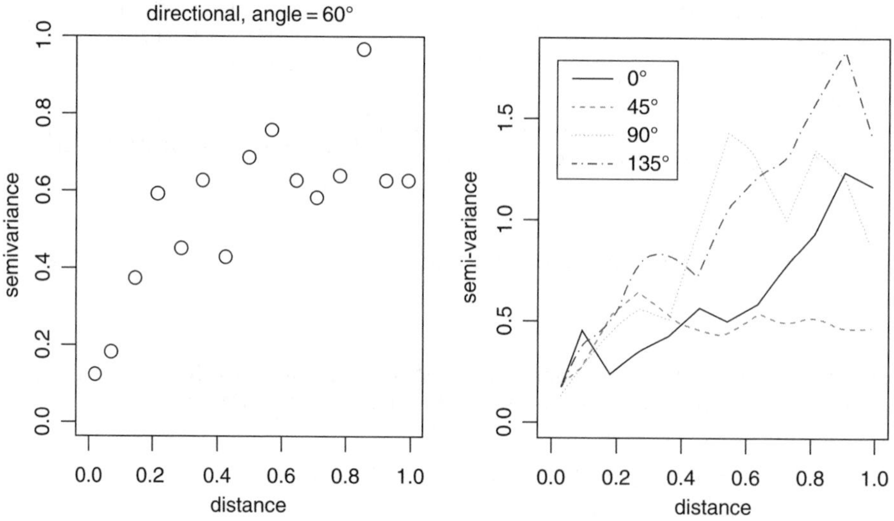

Figure 12.3 Empirical directional variograms for dataset s100.

must hold for any collection of locations $\{s_1, s_2, \ldots, s_n\} \subset D$, $n \geq 1$, and any vector $a^T = (a_1, \ldots, a_n) \in R^n$ such that $\sum_{i=1}^{n} a_i = 0$. This implies that

$$\text{var}[a_1 y(s_1) + \ldots + a_n y(s_n)] = \sum_{i=1}^{n} \sum_{j=1}^{n} a_i a_j C(s_i - s_j) = -\sum_{i=1}^{n} \sum_{j=1}^{n} a_i a_j \gamma(\|s_i - s_j\|) \geq 0$$

due to (12.9), for any linear combination of observations.

This non-negative definiteness is just what a covariance function $C(\cdot)$ is supposed to accomplish.

The following parametric semi-variogram models are in use (writing $\check{h} \geq 0$ for $\|h\|$).

Exponential model:

$$\gamma_E(\check{h}) = c_0(1 - \delta_0(\check{h})) + c\left(1 - e^{-\frac{3\check{h}}{a}}\right); a, c > 0; \check{h}, c_0 \geq 0. \tag{12.10}$$

Spherical model:

$$\gamma_S(\check{h}) = \begin{cases} c_0(1 - \delta_0(\check{h})) + c\left(\frac{3\check{h}}{2a} - \frac{1}{2}\left(\frac{\check{h}}{a}\right)^3\right) & \text{for } 0 \leq \check{h} \leq \text{a} \\ c_0 + c & \text{for } \check{h} > \text{a} \end{cases} \tag{12.11}$$

with $\delta_0(\check{h}) = \begin{cases} 1 \text{ for } \check{h} = 0 \\ 0, \text{ otherwise} \end{cases}.$

Gaussian model:

$$\gamma_G(\check{h}) = c_0(1 - \delta_0(\check{h})) + c\left(1 - e^{\left(-\frac{3\check{h}}{a}\right)^2}\right), \check{h}, c_0 \geq 0; a, c > 0 \tag{12.12}$$

Bessel model:

$$\gamma_B(\check{h}) = c_0(1 - \delta_0(\check{h})) + c[1 - b\check{h}K_1(b\check{h})]; \check{h}, c_0 \geq 0; b, c > 0. \tag{12.13}$$

The function $K_1(\cdot)$ is the modified Bessel function of the second kind and first order.

Matern model:

$$\gamma_M(\check{h}) = c_0(1 - \delta_0(\check{h})) + c\left[1 - \frac{2^{1-\nu}}{\Gamma(\nu)}\left(\sqrt{2\nu}\frac{\check{h}}{\rho}\right)^\nu K_\nu\left(\sqrt{2\nu}\frac{\check{h}}{\rho}\right)\right], \nu, c, \rho > 0; \check{h}, c_0 \geq 0. \tag{12.14}$$

$K_\nu(.)$ is the modified Bessel function of the second kind and order ν and $\Gamma(.)$ the gamma function.

Power model:

$$Y_P(\check{h}) = c_0(1 - \delta_0(\check{h})) + k\,\check{h}^\alpha \text{ with } 0 < \alpha < 2, k > 0, c_0 \geq 0 \text{ for } \check{h} \geq 0. \tag{12.15}$$

We remark that the exponential model and the Bessel model are special cases of the Matern model when we choose $\nu = 1/2$ and $\nu = 1$, respectively. Further, the Gaussian model appears as a limiting case of the Matern model when ν approaches infinity. This parameter ν, also called smoothness parameter, is crucial for determining the 'roughness' of the realisations of the underlying random field. The larger ν the smoother the observed realisations (paths for $D \subset R^1$ and surfaces for $D \subset R^2$, respectively) will be.

Problem 12.2 Calculate the estimate of the semi-variogram function.

Solution

We use the R package `geoR`.

Example 12.2
Surface elevation data are taken from Davis (2002). This is an object of the class geodata, which is a list with the following elements: `coords` x–y coordinates (multiples of 50 ft) and `data` elevations (multiples of 10 ft). The data are available in the R package `geoR`.

```
> library(geoR)
> data(elevation)
> summary(elevation)
Number of data points: 52

Coordinates summary
      x   y
min 0.2 0.0
max 6.3 6.2

Distance summary
     min      max
0.200000 8.275869

Data summary
    Min.  1st Qu.   Median     Mean  3rd Qu.     Max.
690.0000 787.5000 830.0000 827.0769 873.0000 960.0000

> par(mfrow=c(1,2))
> plot(variog(elevation, option="cloud"), xlab="h",
    ylab=expression(gamma(h)))
variog: computing omnidirectional variogram
> plot(variog(elevation, uvec=seq(0.8, by = 0.5)), xlab="h",
    ylab=expression(gamma(h)))
variog: computing omnidirectional variogram
```

On the left of Figure 12.4 is the empirical variogram of the elevation data given with the option '`cloud`'. This shows all the squared differences between the observed values

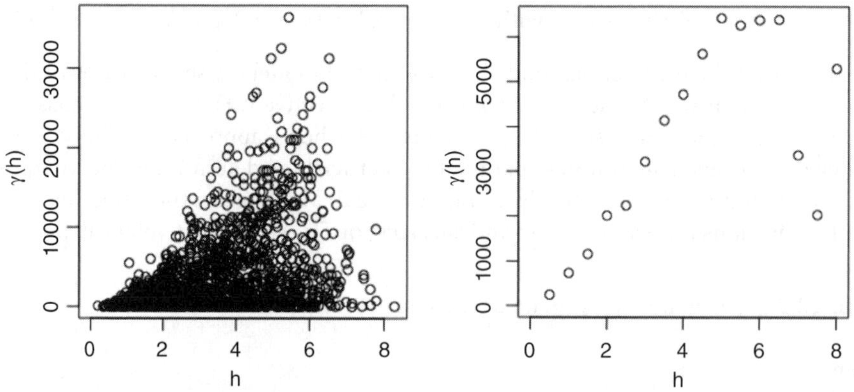

Figure 12.4 Variogram cloud and omnidirectional variogram of elevation data.

for all pairs of spatial locations. On the right side the sample variogram with $\check{h} = 0.5$ for the lag-class distance is given.

12.2.2 Semi-variogram Parameter Estimation

The unknown parameters occurring in the semi-variogram models (12.10)–(12.14) can be estimated by the least squares method of Section 8.2.1.1 or in the case of GRF the maximum likelihood method could be applied.

Problem 12.3 Show how the parameters of a Matern semi-variogram can be estimated.

Solution
Parameter estimation using semi-variogram based methods with the R package `geoR` proceeds in two steps. First the empirical (sample) semi-variogram is calculated and plotted.

Example 12.3
A data set `data(s100)` can be found in `geoR`.

```
> library(geoR)
> data(s100)
> v.s100<-variog(s100,max.dist=1)
variog: computing omnidirectional variogram
> plot(v.s100)
> lines.variomodel(seq(0,1,l=100),cov.pars=c(0.9,0.2),cov.model="mat",
     kap=1.5, nug=0.2)
```

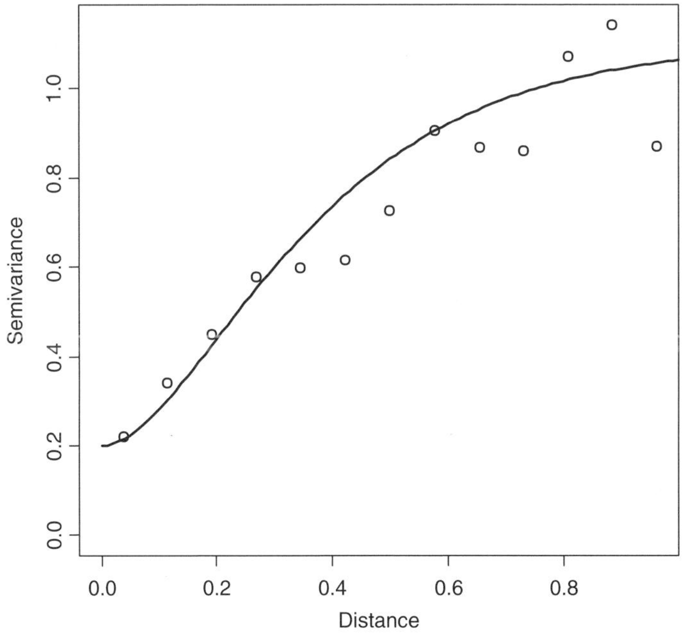

Figure 12.5 Empirical semi-variogram (circles) and the fitted theoretical semi-variogram.

Note: In "geoR", the smoothness parameter of the Matern variogram is denoted by kappa (kap, for short), the nugget variance by nug and the cov.pars stands for the vector of variogram parameters indicating the partial sill c and the range (correlation radius).

Then we use the function variofit with the empirical semi-variogram as argument and the Matern model (12.14) for the semi-variogram function (Figure 12.5).

```
> variofit(v.s100,ini=c(0.9,0.2),cov.model="mat",kap=1.5,nug=0.2)
variofit: covariance model used is matern
variofit: weights used: npairs
variofit: minimisation function used: optim
variofit: model parameters estimated by WLS (weighted least squares):
covariance model is: matern with fixed kappa = 1.5
parameter estimates:
tausq sigmasq phi
0.3036 0.9000 0.2942
Practical Range with cor=0.05 for asymptotic range: 1.395627
variofit: minimised weighted sum of squares = 23.6572
```

Note: In the "variofit" function the nugget variance c_0 is denoted by tausq (short for tausquared), sigmasq (short for sigmasquared) stands for the (partial) sill value c and phi denotes the correlation radius.

12.2.3 Kriging

In geostatistics kriging is a method of interpolation to predict values of the variable of interest in D at positions where no measurement has been made. It is based on a master thesis of Krige (1951) and the mathematical foundation by Matheron (1963). Observed values are modelled by a GRF. Under suitable assumptions, kriging gives the best linear unbiased prediction (BLUP) of values not observed in the corresponding area. The basic idea of kriging is to predict the value of a function at a given point by computing a weighted average or linear combination of the observed values of the function in the neighbourhood of the point. The methods used are a kind of regression analysis in two dimensions. Basically, we distinguish between ordinary and universal kriging: ordinary kriging assumes a constant trend function, $E(\boldsymbol{y}(s)) = \text{const}$ for all s in D, whereas universal kriging assumes a non-constant trend. Placing the problem in a stochastic framework permits precision-defining optimality for estimations of unknown parameters from the random variables for which the measurements are realisations. A criterion imposed is that the estimator be unbiased, or that in an average sense the difference between the predicted value and the actual value is zero. Another optimality criterion is that the prediction variance be minimised. This variance (kriging variance) is defined to be the expectation of the average squared difference between predicted and actual values. The kriging estimator minimises this variance. This minimisation is performed algebraically and results in a set of equations known as the kriging equations – in ordinary kriging we call them ordinary kriging equations (OK system).

For ordinary kriging, we must make the following assumptions already made above:

- $\boldsymbol{y}_i$; $i = 1, \dots, n$ are normally distributed.
- Second order stationarity, meaning, in particular, that the $\boldsymbol{y}_i$; $i = 1, \dots, n$ all have the same constant mean and variance.

We restrict ourselves to observation points in $D \subset R^2$ and the univariate case with one character measured at each of n points in D. In the case that in (12.1) $\mu(s)$ is known we speak about simple kriging. This is in many practical situations an unrealistic assumption. Therefore, we describe the so-called ordinary kriging where $\mu(s)$ is assumed to be unknown and constant.

The kriging prediction at an unobserved location $s_0 \in D$ is then given by a linear combination

$$y^* = y(s_0) = \sum_{i=1}^{n} \lambda_i y_i \tag{12.16}$$

where the weights in $\lambda^T = (\lambda_1, \dots, \lambda_n)$ are determined as solutions of the OK system of the observations $\begin{pmatrix} G & 1_n \\ 1_n^T & 0 \end{pmatrix} \begin{pmatrix} \lambda \\ \lambda_0 \end{pmatrix} = \begin{pmatrix} \gamma_0 \\ 1 \end{pmatrix}$.

Here $G = (\gamma(\|s_i - s_j\|))_{i,j=1,\dots,n}$ is the $(n \times n)$-semi-variogram matrix of the observations,

$$\gamma_0^T = (\gamma(\|s_0 - s_1\|), \gamma(\|s_0 - s_2\|), \dots, \gamma(\|s_0 - s_n\|))$$

is the vector of semi-variogram values, and λ_0 is the Lagrange multiplier, which is necessary due to the minimisation under the equality constraint $\sum_{i=1}^{n} \lambda_i = 1$, which, in turn, results from the unbiasedness condition $E(y^*) = \mu$. The kriging variance then becomes $\sigma^2(s_0) = \text{var}(y^*) = \lambda_0 + \lambda^T \gamma_0$. We illustrate this with the following toy example where $n = 4$.

Problem 12.4 Measurements $y(x_1, x_2)$ are taken at four locations:

$y(10; 20) = 40$
$y(30; 280) = 130$
$y(250; 130) = 90$
$y(360; 120) = 160.$

Predict the value at $y(180; 120)$ by ordinary kriging. Use as covariance function

$$C(\check{h}) = 2000\mathrm{e}^{-\frac{\check{h}}{250}}$$

and the distance $\check{h} = \sqrt{x_1^2 + x_2^2}$.

Solution and Example

First we initialise the vectors of coordinates, the observation vector and c_0

```
> x_1=c(10,30,250,360)
> x_2=c(20,280,130,120)
> Z=c(40,130,90,160)
> c_0=2000
```

Further we need some functions to calculate G, the semi-variogram matrix and γ_0:

```
> f=function(x,y){
  n=length(x)
  G=matrix(0,nrow=n,ncol=n)
  for(i in 1:n){
  for(j in 1:n){
  1
  G[i,j]=c_0-(2000*exp(-(sqrt((x[i]-x[j])^2+(y[i]-y[j])^2)/250)))
  }
```

```
}
G
}
> G=f(x_1,x_2)
> G
[,1] [,2] [,3] [,4]
[1,] 0.000 1295.259 1304.3323 1533.6761
[2,] 1295.259 0.000 1310.6009 1538.7534
[3,] 1304.332 1310.601 0.0000 714.2622
[4,] 1533.676 1538.753 714.2622 0.0000
> Gamma_0=function(x,y){
s=c(0,0,0,0)
for(i in 1:4){
s[i]=c_0-(2000*exp(-(sqrt((180-x[i])^2+(120-y[i])^2)/250)))
}
s
}
```

The two sides of the equations of the OK system are obtained by

```
> gamma_0=Gamma_0(x_1,x_2)
> rS=c(gamma_0,1)
> rS
[1] 1091.3326 1168.1651 492.7234 1026.4955 1.0000
> OK=matrix(1,nrow=5,ncol=5)
> OK[5,5]=0
> for(i in 1:4){
for(j in 1:4){
OK[i,j]=G[i,j]
}
}
> OK
 [,1] [,2] [,3] [,4] [,5]
[1,] 0.000 1295.259 1304.3323 1533.6761 1
[2,] 1295.259 0.000 1310.6009 1538.7534 1
[3,] 1304.332 1310.601 0.0000 714.2622 1
[4,] 1533.676 1538.753 714.2622 0.0000 1
[5,] 1.000 1.000 1.0000 1.0000 0
```

Now we solve the OK system for λ and λ_0:

```
> lambda=solve(OK)%*%rS
```

and the estimate becomes:

```
> estimate =sum(lambda [-5]%*%Z)
> estimate
 [1] 86.58756
```

the kriging variance is

```
> sigmasq<- t(lambda[-5])%*%gamma_0+lambda[5]
> sigmasq[,1]
 [,1]
[1,] 754.7532
```

Alternatively, the kriging prediction (12.16) can be written as

$$y^* = \hat{\mu} + k_0^T K^{-1}(y - \hat{\mu}1_n). \tag{12.17}$$

In (12.17) $k_0^T = [C(\|s_0 - s_1\|), C(\|s_0 - s_2\|), \dots, C(\|s_0 - s_n\|)]$, K is the covariance matrix $K = (C(\|s_i - s_j\|))_{i,j}$, and

$$\hat{\mu} = (1_n^T K^{-1} 1_n)^{-1} 1_n^T K^{-1} y \tag{12.18}$$

is the generalised least squares estimator of the unknown expectation μ. If $\mu(s)$ is no longer assumed to be an unknown constant then it can be more generally modelled as a linear regression setup

$$E(y(s)) = \beta_0 + \sum_{j=1}^{r} \beta_j f_j(s) \tag{12.19}$$

with given regression functions $f_1, \dots f_r$ and unknown regression coefficients β_0, $\beta_1, \dots, \beta_r$.

The case of ordinary kriging just considered corresponds to the special case where $\beta_1 = \dots = \beta_r = 0$. In the case of (12.17) with non-constant functions $f_1, \dots f_r$ we speak of universal kriging. In this case the OK predictor (12.17) is replaced by the universal kriging predictor (UK predictor)

$$y^*_{\mathrm{UK}} = f_0^T \hat{\beta} + k_0^T K^{-1}(y - F\hat{\beta}) \tag{12.20}$$

where $f_0^T = (1, f_1(s_0), \dots, f_r(s_0))$, F is the (spatial) $n \times (r+1)$ design matrix

$$F = \begin{pmatrix} 1 & f_1(s_1) & \dots & f_r(s_1) \\ 1 & f_1(s_2) & \dots & f_r(s_2) \\ \dots & \dots & \dots & \dots \\ 1 & f_1(s_n) & \dots & f_r(s_n) \end{pmatrix} \tag{12.21}$$

and $\hat{\beta}$ is the generalised least squares estimator

$$\hat{\beta} = (F^T K^{-1} F)^{-1} F^T K^{-1} y \tag{12.22}$$

with the covariance matrix K and the vector k_0 as defined before.

Since the GRF is completely defined by the trend function $\mu(s)$ and the covariance function $C(\check{h})$, we also can estimate the regression parameters $\beta^T = (\beta_0, \beta_1, \dots, \beta_r)$ and the covariance parameters $\theta^T = (c_0, c, \rho, \nu)$ jointly using a likelihood approach. Since the GRF assumption implies that the observation vector follows an n-dimensional normal distribution $N_n(F\beta, K)$, where $K = K(\theta)$, the log-likelihood function thus reads

$$\ln l(\beta, \theta, y) = -\frac{n}{2}\ln(2\pi) - \frac{1}{2}\ln|K(\theta)| - \frac{1}{2}(y - F\beta)^T K(\theta)^{-1}(y - F\beta).$$

For any given parameter vector θ this function is maximised by the realisation $\hat{\beta}$ of $\hat{\beta}$ in (12.22). Then the covariance parameter vector can be estimated by maximising the so-called profile log-likelihood function $\ln l(\widehat{\beta}(\theta), \theta, y)$ with respect to θ. This can be done using the R package geoR, which then also determines the predicted UK values according to (12.20).

The function `proflik(·)` in the R package geoR allows us to visualise profile likelihoods. This function requires a `likfit(·)` object and sequences for the variogram parameters to be plotted (sill or range sequences).

Problem 12.5 Show how to calculate (kriging)-predicted values.

Solution and Example

We use from Paulo J. Ribeiro Jr. a part of his 'geoR solution' article (https://www.stat.washington.edu/peter/591/geoR_sln.html).

Turning to the parameter estimation, we fit the model with a constant mean, isotropic exponential correlation function and allowing for estimating the variance of the conditionally independent observations (nugget variance).

```
> library(geoR)
> data(s100)
> s100.ml <- likfit(s100, ini = c(1, 0.15))
> s100.ml
likfit: estimated model parameters:
    beta     tausq  sigmasq       phi
"0.7766" "0.0000" "0.7517" "0.1827"
Practical Range with cor=0.05 for asymptotic range: 0.5473814
likfit: maximised log-likelihood = -83.57
> summary(s100.ml)
Summary of the parameter estimation
- - - - - - - - - - -  - - - - - - - - -
Estimation method: maximum likelihood
Parameters of the mean component (trend):
  beta
0.7766

Parameters of the spatial component:
   correlation function: exponential
      (estimated) variance parameter sigmasq (partial sill) =  0.7517
      (estimated) cor. fct. parameter phi (range parameter)  =  0.1827
   anisotropy parameters:
      (fixed) anisotropy angle = 0  ( 0 degrees )
      (fixed) anisotropy ratio = 1
Parameter of the error component:
      (estimated) nugget =  0
Transformation parameter:
      (fixed) Box-Cox parameter = 1 (no transformation)
Practical Range with cor=0.05 for asymptotic range: 0.5473814
Maximised Likelihood:
   log.L n.params      AIC      BIC
"-83.57"      "4"  "175.1"  "185.6"
non spatial model:
   log.L n.params      AIC      BIC
"-125.8"      "2"  "255.6"  "260.8"
Call:
likfit(geodata = s100, ini.cov.pars = c(1, 0.15)).
```

In the output above AIC stands for the Akaike-criterion (Akaike 1973) and BIC for the Schwarz criterion (Schwarz 1978), both are model choice criteria taking account of the number of parameters in a model. In the above program, the output `tausq` is the nugget variance, `sigmasq` the sill and `phi` the correlation radius.

Finally, we start the spatial prediction defining a grid of points. The kriging function by default performs ordinary kriging. It minimally requires the data, prediction locations and estimated model parameters.

```
> s100.gr <- expand.grid((0:100)/100, (0:100)/100)
> s100.kc <- krige.conv(s100, locations = s100.gr, krige =
  krige.control(obj.model = s100.ml))
krige.conv: model with constant mean
krige.conv: Kriging performed using global neighbourhood
> names(s100.kc)
[1] "predict"      "krige.var"     "beta.est"      "distribution" "message"
[6] "call"
```

If the locations form a grid the predictions can be visualised as image or contour. The plots in Figure 12.6 show the contours of the predicted values (left) and the image of the respective standard errors (right).

```
> par(mfrow = c(1, 2), mar = c(3.5, 3.5, 0.5, 0.5))
> image(s100.kc, col = gray(seq(1, 0.2, l = 21)))
> contour(s100.kc, nlevels = 11, add = TRUE)
> image(s100.kc, val = sqrt(krige.var), col = gray(seq(1, 0.2, l = 21)))
```

It is worth noticing that under the Gaussian model the prediction errors depend only on the coordinates of the data, but not on their values (except through model parameter estimates).

```
> image(s100.kc, val = sqrt(krige.var), col = gray(seq(1, 0.2, l = 21)),
   coords.data = TRUE)
```

The plot in Figure 12.7 shows the kriging standard deviations, the darker areas exhibit higher values. The black dots visualise the locations (coordinates) of the observations.

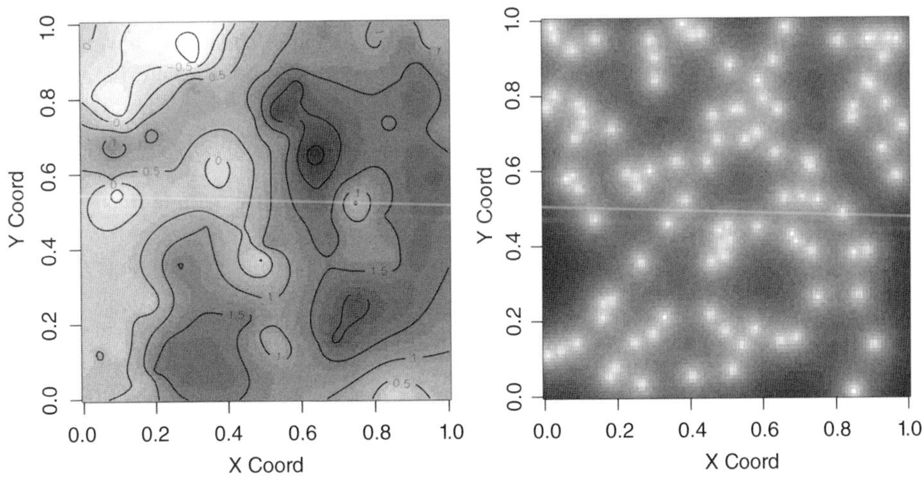

Figure 12.6 Predicted values and standard errors for s100.

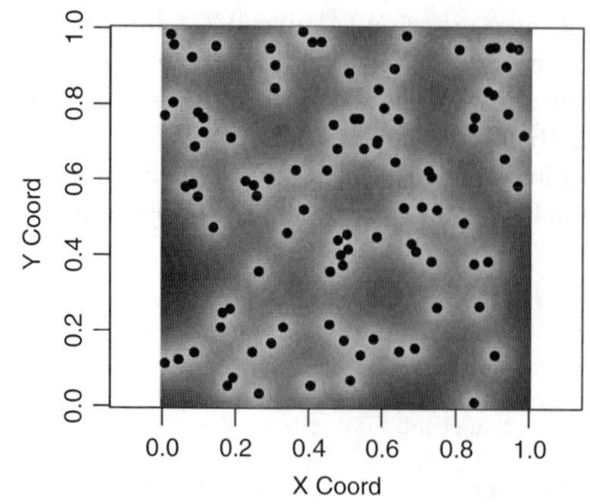

Figure 12.7 Association between standard deviations and data locations.

If, instead of the default 'ordinary kriging', we want to include a non-constant trend, then we have to specify this through the `trend.spatial` function. For more information on this see the section `details` in the documentation for `likfit(.)`.

12.2.4 *Trans*-Gaussian Kriging

Now we consider the case of non-normal random fields. It is then often possible to transform the observation such that the transformed data are realisations of a normal distribution. A striking example of this is to apply the Box–Cox transformation (12.4). This transformation depends on an additional parameter λ. We can find an appropriate value of λ with the function `boxcox(.)` in the package `geoR`. The function `boxcox(.)` gives a graphical summary of the log-likelihood function depending on λ and indicates a realised 95% confidence interval for λ. For λ we use an easy to interpret value from the set $\{\lambda = 1$ (no transformation), $\lambda = 0.5$ (square root transformation), $\lambda = 0$ (logarithmic transformation), $\lambda = -0.5$ (inverse square root transformation), $\lambda = -1$ (reciprocal transformation)$\}$. Kriging with Box–Cox transformed data is usually called trans-Gaussian kriging in the literature. This type of kriging and Bayesian extensions thereof were introduced by De Oliveira et al. (1997). Spöck and Pilz (2015) deal with the sampling design problem for optimal prediction with non-normal random fields and give an example of the optimal placement of monitoring stations for an existing rainfall monitoring network in Upper Austria.

The R package `gstat` also implements trans-Gaussian kriging using a `boxcox(.)` function, which may be found in the library (MASS).

For illustration we now use the Meuse Data `meuse.all` from the package `gstat`. This data set gives locations and top soil heavy metal concentrations (ppm), along with a number of soil and landscape variables, collected in a flood plain of the river Meuse, near the village Stein. Heavy metal concentrations are bulk sampled from an area of approximately 15×15 m.

Example 12.1 In this example we use `gstat` and data `meuse.all` from that package. This package and the data are well described in Bivand et al. (2013).

In this data set we use the zinc observations.

For nice graph representation (see Figure 12.8) we use also the R package sp.

```
> library(gstat)
> data (meuse.all)
> library(sp)
```

First, we look for an appropriate Box–Cox transformation.

```
> library(MASS)
> boxcox(zinc~1, data=meuse.all)
```

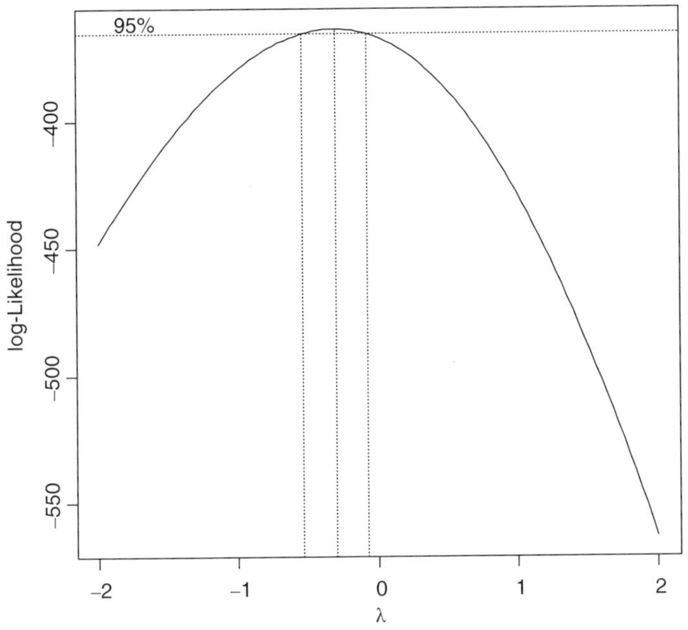

Figure 12.8 Determining a 95% confidence interval for Box–Cox parameter.

The plot suggests to choose $\lambda = 0$ for simplicity, i.e. to work with a logarithmic transformation of the observed data.

Before we can use the kriging function krige we call

```
> coordinates(meuse.all)<- c("x", "y" )
```

and calculate an empirical semi-variogram

```
> empv1<-variogram(log(zinc) ~ 1,meuse.all)
```

and fit it to the spherical semi-variogram model (12.11)

```
> v=vgm(0.6,"Sph",870,0.05)
> m1.zinc=fit.variogram(empv1,v)
> m1.zinc~
   model      psill       range
1  Nug        0.0506      0.000
2  Sph        0.5906      896.965
> plot(empv1,model=m1.zinc, main="spherical variogram fit of log(zinc)")
```

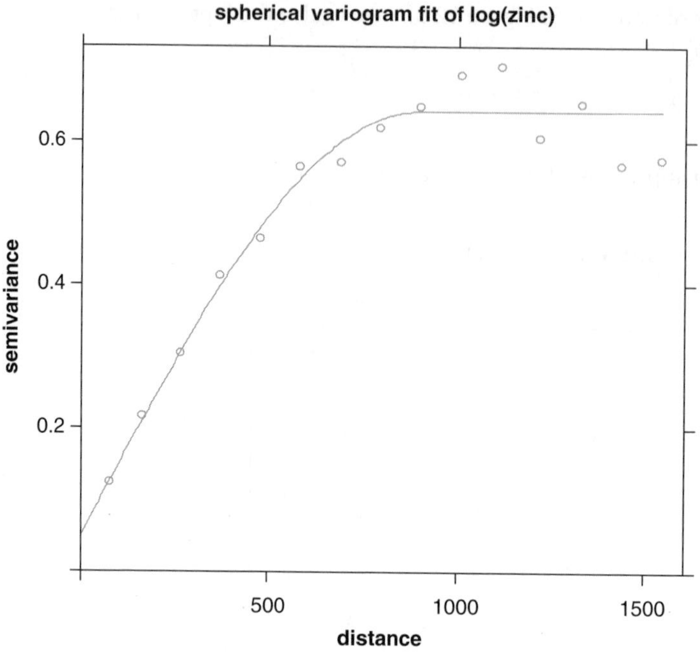

Figure 12.9 Empirical and fitted semi-variogram model for Meuse zinc data.

As can be seen from the output above Figure 12.8 and from the graph in Figure 12.9 this empirical semi-variogram has a nugget variance $c = 0.0506$; a (partial) sill $c = 0.5906$ (i.e. an overall sill 0.6412) and a range $a = 896.965$.

```
> gridded(meuse.grid)=~x+y
> summary(meuse.grid)
Object of class SpatialPixelsDataFrame
Coordinates:
     min      max
x  178440   181560
y  329600   333760
Is projected: NA
proj4string : [NA]
Number of points: 3103
Grid attributes:
  cellcentre.offset cellsize cells.dim
x            178460       40        78
y            329620       40       104
```

Next we prepare the predictions. The results are visualized in Figure 12.10.

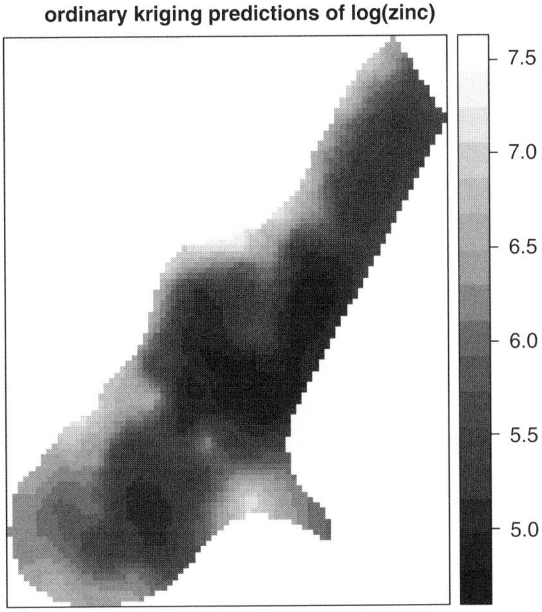

Figure 12.10 OK-predicted values of zinc data (on a log-scale).

```
> z=krige(log(zinc)~1, meuse.all, meuse.grid, model=m1.zinc)
[using ordinary kriging]
> z["var1.pred"]
Object of class SpatialPixelsDataFrame
Object of class SpatialPixels
Grid topology:
  cellcentre.offset cellsize cells.dim
x            178460       40        78
y            329620       40       104
SpatialPoints:
             x        y
1        181180   333740
2        181140   333700
...................................
3102   179180   329620
3103   179220   329620
Coordinate Reference System (CRS) arguments: NA
Data summary:
     var1.pred
 Min.    : 4.777
 1st Qu.: 5.238
 Median : 5.573
 Mean    : 5.707
 3rd Qu.: 6.172
 Max.    : 7.440
```

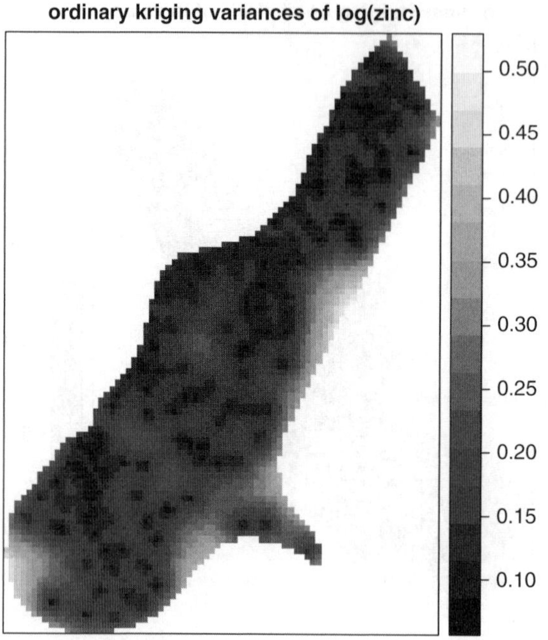

Figure 12.11 OK variances of predicted zinc values.

Next we find the predicted variances; these are plotted in Figure 12.11.

```
> z["var1.var"]
Object of class SpatialPixelsDataFrame
Object of class SpatialPixels
Data summary:
                    var1.var
 Min.    : 0.08549
 1st Qu.: 0.13728
 Median : 0.16218
 Mean    : 0.18533
 3rd Qu.: 0.21161
 Max.    : 0.50028
```

The commands for plotting the predicted values and variances, respectively, read as follows:

```
> spplot(z["var1.pred"], main = "ordinary kriging predictions of log(zinc)")
> spplot(z["var1.var"], main = "ordinary kriging variances of log(zinc)")
```

12.3 Special Problems and Outlook

We now briefly outline more recent developments which go beyond the traditional kriging and trans-Gaussian kriging considered in Section 12.2.3.

12.3.1 Generalised Linear Models in Geostatistics

In the same way as we had extended linear regression models to generalised linear regression models in Chapter 11, we may take the step from spatial linear models to so-called generalised linear geostatistical models (GLGMs). This allows us to model spatial random variables [with observations (realisations)] following a distribution from the exponential family. Spatial modelling and prediction for such observations is implemented in the R package `geoRglm` and is well described in Diggle and Ribeiro (2007), where in particular worked-through examples of binomially and Poisson distributed environmental and public health data are considered. Extensions to hierarchical Bayesian GLGMs can be found in Banerjee et al. (2014).

12.3.2 Copula Based Geostatistical Prediction

The GLGM framework does not allow distributions outside the exponential family. In particular we cannot use it for heavy-tailed/extreme value distributions, which have a much slower decay of probability in the tails than the normal distribution. A prominent example of such a distribution is the generalised extreme value distribution with distribution function

$$F(y;\mu,\sigma,\tau) = \exp\left\{-\left[1-\tau\frac{y-\mu}{\sigma}\right]^{-\frac{1}{\tau}}\right\}$$

with location, scale, and shape parameter μ, σ, and τ, respectively. Observations from non-Gaussian, skewed or heavy-tailed distributions are dealt with in the library (intamap). A general framework of handling such data are copulas which are distribution functions on the unit cube $[0, 1]^n$ with uniformly distributed margins, introduced by Sklar (1959). Copulas are invariant under strictly increasing transformations of the marginals; thus, frequently applied data transformations (e.g. square root and log transformations) do not change the copula. The relation between two locations separated by the lag-vector h is characterised by the bivariate distribution with distribution function

$$P[\boldsymbol{y}(s) \leq y_1, \boldsymbol{y}(s+h) \leq y_2] = C_h[F_y(y_1), F_y(y_2)]. \tag{12.23}$$

The copula C_h thus becomes a function of the separating vector h. Spatial copulas have been introduced by Bardossy (2006) and Kazianka and Pilz (2010b). Spatial copulas describe the spatial dependence over the whole range of quantiles for a given separating vector h, not only the mean dependence, as the variogram does. The main difference to GLGMs is that these only model the means $\mu_i = E(\boldsymbol{y}(s_i))$, $i = 1, \ldots, n$ through some link function $g(\mu_i) = f_i^T\beta + Z(s_i)$, where $Z(\cdot)$ is a stationary GRF with mean zero and given covariance function. The spatial copula, however, allows us to build a complete multivariate distribution of the random variables whose realisations are the observed values $[y(s_1), \ldots, y(s_n)]$. Recently, so-called vine copula have been developed which extend the concept of the bivariate copula C_h in (12.23) to higher dimensions – see e.g. Gräler (2014). The corresponding R packages `copula`, `spcopula` and `VineCopula` allow flexible spatial data modelling. A Matlab toolbox for copula-based spatial analysis is given in Kazianka (2013).

References

Akaike, H. (1973). Information theory and an extension of the maximum likelihood principle. In: *2nd International Symposium on Information Theory*, Tsaghkadzor, Armenia, USSR, September 2–8, 1971 (ed. B.N. Petrov and F. Csáki), 267–281. Budapest: Akadémiai Kiadó.

Anselin, L. and Griffith, D.A. (1988). Do spatial effects really matter in regression analysis? *Reg. Sci.* 65: 11–34.
Bailey, T.C. and Gatrell, T. (1995). *Interactive Spatial Data Analysis*. London: Longman Scientific & Technical.
Banerjee, S., Carlin, B.P., and Gelfand, A.E. (2014). *Hierarchical Modeling and Analysis for Spatial Data*, 2e. Boca Raton, Florida: CRC Press/Chapman & Hall.
Bardossy, A. (2006). Copula-based geostatistical models for groundwater quality parameters. *Water Resour. Res.* 42: W11416.
Bivand, R.S., Pebesma, E.J., and Gomez-Rubio, V. (2013). *Applied Spatial Data Analysis with R*, 2e. Berlin: Springer.
Cressie, N.A.C. (1993). *Statistics for Spatial Data*. New York: Wiley.
Cressie, N.A.C. and Wikle, C.K. (2011). *Statistics of Spatio-Temporal Data*. New York: Wiley.
Davis, J.C. (2002). *Statistics and Data Analysis in Geology*, 3e. New York: Wiley.
De Oliveira, V., Kedem, B., and Short, D.A. (1997). Bayesian prediction of transformed Gaussian random fields. *J. Am. Stat. Assoc.* 92: 1422–1433.
Diggle, P.J. (2010). Spatial point pattern. In: *International Encyclopedia of Statistical Science, Volumes I, II, III* (ed. M. Lovric), 1361–1363. Berlin: Springer.
Diggle, P. and Ribeiro, P. (2007). *Model-Based Geostatistics*. New York: Springer.
Gaetan, C. and Guyon, H. (2010). *Spatial Statistics and Modeling*. New York: Springer.
Gräler, B. (2014). Modelling skewed spatial random fields through the spatial vine copula. *Spatial Stat.* 10: 87–102.
Kazianka, H. (2013). spatialCopula: a Matlab toolbox for copula-based spatial analysis. *Stochastic Environ. Res. Risk Assess.* 27: 121–135.
Kazianka, H. and Pilz, J. (2010a). Model-based Geostatistics. In: *International Encyclopedia of Statistical Science, Volumes I, II, III* (ed. M. Lovric), 833–836. Berlin: Springer.
Kazianka, H. and Pilz, J. (2010b). Copula-based geostatistical modeling of continuous and discrete data including covariates. *Stochastic Environ. Res. Risk Assess.* 24: 661–673.
Krige, D.G, (1951) *A statistical approach to some mine valuations and allied problems at the Witwatersrand*, Master's thesis of the University of Witwatersrand, Johannesburg, S.A..
Krüger, L., (1912) *Konforme Abbildung des Erdellipsoids in die Ebene*, In: *Veröff. Kgl. Preuß. Geod. Inst*. Nr. 51.
Lantuéjoul, C. (2002). *Geostatistical Simulation. Models and Algorithms*. Berlin: Springer.
Mase, S. (2010). Geostatistics and kriging predictors. In: *International Encyclopedia of Statistical Science, Volumes I, II, III* (ed. M. Lovric), 609–612. Berlin: Springer.
Matheron, G. (1963). Principles of Geostatistics. *Econ. Geol.* 58: 1246–1266.
Müller, W. (2007). *Collecting Spatial Data*, 3e. Heidelberg: Springer.
Pilz, J. (ed.) (2009). *Interfacing Geostatistics and GIS*. Berlin-Heidelberg: Springer.
Pilz, J. (2010). Spatial statistics. In: *International Encyclopedia of Statistical Science, Volumes I, II ,III* (ed. M. Lovric), 1363–1368. Berlin: Springer.
Pilz, J., Kazianka, H., and Spöck, G. (2012). Some advances in Bayesian spatial prediction and sampling design. *Spatial Stat.* 1: 65–81.
Rasch, D., Herrendörfer, G., Bock, J. et al. (eds.) (2008). *Verfahrensbibliothek Versuchsplanung und - auswertung*, 2e. München Wien: R. Oldenbourg.
Ripley, B.D. (1988). *Statistical Inference for Spatial Processes*. Cambridge, UK: Cambridge University Press.

Sampson, P.D., Damien, D., and Guttorp, P. (2001). *Advances in Modelling and Inference*. London: Academic Press.

Schabenberger, O. and Gotway, C.A. (2005). *Statistical Methods for Spatial Data Analysis*. Boca Raton: Chapman & Hall/CRC Press.

Schwarz, G.E. (1978). Estimating the dimension of a model. *Ann. Stat.* 6: 461–464.

Sklar, A. (1959). Fonctions de repartition á *n* dimensions et leurs marges. *Publ. Inst. Statistique Univ. Paris* 8: 229–231.

Snow, J. (1855). *On the Mode of Communication of Cholera*. London: John Churchill.

Spöck, G. and Pilz, J. (2010). Analysis of areal and spatial interaction data. In: *International Encyclopedia of Statistical Science, Volumes I, II, III* (ed. M. Lovric), 35–39. Berlin: Springer.

Spöck, G. and Pilz, J. (2015). Taking account of covariance estimation uncertainty in spatial sampling design for prediction with trans-Gaussian random fields. *Front. Environ. Sci.* 3: 39, 1–39, 22.

Wackernagel, H. (2010). *Multivariate Geostatistics*. Heidelberg-Berlin: Springer.

Appendix A

List of Problems

This part of the Appendix shows the problems formulated in the chapters of this book. If we write 'calculate' or 'determine' we mean that the operations should be done using R.

Problem 1.1

Calculate the value $\varphi(z)$ of the density function of the standard normal distribution for a given value z.

Problem 1.2

Calculate the value $\Phi(z)$ of the distribution function of the standard normal distribution for a given value z.

Problem 1.3

Calculate the value of the density function of the log-normal distribution whose logarithm has mean equal to `meanlog = 0` and standard deviation equal to `sdlog = 1` for a given value z.

Problem 1.4

Calculate the value of the distribution function of the lognormal distribution whose logarithm has mean equal to `meanlog = 0` and standard deviation equal to `sdlog = 1` for a given value z.

Problem 1.5

Calculate the P%-quantile of the t-distribution with df degrees of freedom and optional non-centrality parameter `ncp`.

Applied Statistics: Theory and Problem Solutions with R, First Edition.
Dieter Rasch, Rob Verdooren, and Jürgen Pilz.

Problem 1.6

Calculate the $P\%$-quantile of the χ^2-distribution with df degrees of freedom and optional non-centrality parameter `ncp`.

Problem 1.7

Calculate the $P\%$-quantile of the F-distribution with df_1 and df_2 degrees of freedom and optional non-centrality parameter `ncp`.

Problem 1.8

Draw a pure random sample without replacement of size $n < N$ from N given objects represented by numbers 1,..., N without replacing the drawn objects. There are $M = \binom{N}{n}$ possible unordered samples having the same probability $p = \frac{1}{M}$ to be selected.

Problem 1.9

Draw a pure random sample with replacement of size n from N given objects represented by numbers 1,…,N with replacing the drawn objects. There are $M_{\text{rep}} = \binom{N+n-1}{n}$ possible unordered samples having the same probability $\frac{1}{M_{\text{rep}}}$ to be selected.

Problem 1.10

From a set of N objects systematic sampling with random start should choose a random sample of size n.

Problem 1.11

By cluster sampling from a population of size N decomposed into s disjoint subpopulations, so-called clusters of sizes $N_1, N_2, \ldots, N_s$ a random sample should be drawn.

Problem 1.12

Draw a random sample of size n in a two-stage procedure by selecting first from the s primary units having sizes N_i ($i = 1,\ldots,s$) exactly r units.

Problem 2.1

Calculate the arithmetic mean of a sample.

Problem 2.2

Calculate the extreme values $y_{(1)} = \min(y)$ and $y_{(n)} = \max(y)$ of a sample.

Problem 2.3

Order a vector of numbers by magnitude.

Problem 2.4

Calculate the $\frac{1}{n}$-trimmed mean of a sample.

Problem 2.5

Calculate the $\frac{1}{n}$-Winsorised mean of a sample of size n.

Problem 2.6

Calculate the median of a sample.

Problem 2.7

Calculate the first and the third quartile of a sample.

Problem 2.8

Calculate the geometric mean of a sample.

Problem 2.9

Calculate the harmonic mean of a sample.

Problem 2.10

Calculate from n observations $(x_1, x_2, \ldots, x_n)$ of a lognormal distributed random variable the maximum-likelihood (ML) estimate of the expectation $\mu_x = e^{\mu + \frac{\sigma^2}{2}}$.

Problem 2.11

Estimate the parameter p of a binomial distribution

Problem 2.12

Estimate the parameter λ of a Poisson distribution.

Problem 2.13

Estimate the expectation and the variance of the initial $N(\mu, \sigma^2)$-distribution after an 'in a' left-sided and after an 'in b' right-sided truncation.

Problem 2.14

Estimate the expectation of a $N(\mu, \sigma^2)$-distribution based on a left sided or a right-sided censored sample of type I.

Problem 2.15

The expectation μ of a of finite population is to be estimated from the realisation of a pure random sample or a systematic sampling with random start. Give the estimates of the unbiased estimator for μ and of the estimator of the standard error of the estimator of μ.

Problem 2.16

A universe of size N is subdivided into s disjoint clusters of size N_i $(i = 1, 2, \dots, s)$. n_i sample units are drawn from stratum i by pure random sampling with a total sample size $n = \sum_{i=1}^{s} n_i$ in the universe. Estimate the expectation of the universe if the n_i/n are chosen proportional to N_i/N.

Problem 2.17

Estimate the expectation of a universe with N elements in s primary units (strata) having sizes N_i $(i = 1,\dots,s)$ by two-stage sampling, drawing at first $r < s$ strata and then from each of the selected strata m elements.

Problem 2.18

Calculate the range R of a sample.

Problem 2.19

Calculate the interquartile range $\mathrm{IQR}(y)$ of a sample Y.

Problem 2.20

Calculate for an observed sample the estimate (realisation) of the square root $\boldsymbol{s}$ of $\boldsymbol{s}^2$ in (2.21).
Further give the estimate $\hat{\sigma}$.

Problem 2.21

Calculate the sample skewness g_1 from a sample y for the data with weight 1.

Problem 2.22

Calculate the sample kurtosis from a sample Y.

Problem 2.23

Calculate the association measure Q in (2.26) for Example CT2x2.

Problem 2.24

Calculate the association measure Y in (2.27) for Example CT2x2.

Problem 2.25

Calculate the association measure H in (2.28) for Example CT2x2.

Problem 2.26

Calculate χ^2 and the association measure C in (2.30) for Example 2.2.

Problem 2.27

Calculate the association measure C_{adj} in (2.31) for Example 2.2.

Problem 2.28

Calculate the association measure T in (2.29) for Example 2.3.

Problem 2.29

Calculate the association measure V in (2.33) for Example 2.3.

Problem 3.1

Determine the P-quantile $Z(P)$ of the standard normal distribution.

Problem 3.2

Calculate the power function of the one-sided test for (3.3) case (a).

Problem 3.3

To test $H_0 : \mu = \mu_0$ for a given risk of the first kind α the sample size n has to be determined so that the second kind risk β is not larger than β_0 as long as $\mu_1 - \mu_0 \geq \delta$ with $\sigma = 1$ in the one- and two-sided case described above.

Problem 3.4

Calculate the minimal sample size for testing the null hypothesis with $\sigma = 1$:
$H_0 : \mu = \mu_0$ against one of the following alternative hypotheses:

(a) $H_A : \mu > \mu_0$ (one-sided alternative),
(b) $H_A : \mu \neq \mu_0$ (two-sided alternative).

Problem 3.5

Let $\boldsymbol{y}$ be binomially distributed (as $B(n,p)$). Describe an α-test for $H_0 : p = p_0$ against $H_A : p = p_A < p_0$ and for $H_0 : p = p_0$ against $H_A : p = p_A > p_0$.

Problem 3.6

Let P be the family of Poisson distributions and $Y = (y_{1,\ldots,}y_n)^T$ a realisation of a random sample $\boldsymbol{Y} = (\boldsymbol{y}_{1,\ldots,}\boldsymbol{y}_n)^T$. The likelihood function is

$$L(Y, \lambda) = \prod_{i=1}^{n} \frac{1}{y_i \cdot} e^{\left[\ln \lambda \sum_{i=1}^{n} y_i - \lambda n\right]}, \quad y_i = 0, 1, 2, \ldots \;; \quad \lambda \,\varepsilon\, R^+$$

How can we test the pair H_o: $\lambda = \lambda_o$, $H_A : \lambda \neq \lambda_o$ of hypotheses with a first kind risk α?

Problem 3.7

Perform a Wilcoxon's signed-ranks test for a sample y.

Problem 3.8

Perform a Wilcoxon's signed-ranks test for two paired samples X and Y.

Problem 3.9

Show how, based on sample values Y, the null hypothesis $H_0 \, : \, \sigma^2 = \sigma_0^2$ can be tested.

Problem 3.10

Show how a triangular sequential test can be calculated.

Problem 3.11

How can we calculate the minimum sample size n per sample for the two-sample t-test for a risk of the first kind α and a risk of the second kind not larger than β as long as $|\mu_1 - \mu_2| > \delta$ with $\sigma = 1$?

Problem 3.12

Use the Welch test based on the approximate test statistic to test
$H_0 : \mu_1 = \mu_2 = \mu$; against one of the alternative hypotheses:
(a) $H_A : \mu_1 < \mu_2$,
(b) $H_A : \mu_1 > \mu_2$,
(c) $H_A : \mu_1 \neq \mu_2$.

Problem 3.13

Test for two continuous distributions
$H_0 : m_1 = m_2 = m$; against one of the alternative hypotheses:
(a) $H_A : m_1 < m_2$,
(b) $H_A : m_1 > m_2$,
(c) $H_A : m_1 \neq m_2$..

Problem 3.14

Show how a triangular sequential two-sample test is calculated.

Problem 4.1

Construct a one-sided $(1 - \alpha)$-confidence interval for the expectation of a $N(\mu, \sigma^2)$-distribution if the variance is known.

Problem 4.2

Construct a two-sided $(1 - \alpha)$-confidence interval for the expectation of a $N(\mu, \sigma^2)$-distribution if the variance is known.

Problem 4.3

Determine the minimal sample size for constructing a two-sided $(1 - \alpha)$-confidence interval for the expectation μ of a normally distributed random variable with known variance σ^2 so that the expected length L is below 2δ.

Problem 4.4

Determine the minimal sample size for constructing a one-sided $(1 - \alpha)$-confidence interval for the expectation μ of a normally distributed random variable with known variance σ^2 so that the distance between the finite (random) bound of the interval and 0 is below δ.

Problem 4.5

Construct a realised left-sided 0.95-confidence interval of μ for the x-data in Table 3.4.

Problem 4.6

Construct a realised right-sided 0.95-confidence interval for the x-data in Table 3.4.

Problem 4.7

Construct a realised two-sided 0.95-confidence interval for the x-data in Table 3.4.

Problem 4.8

Construct a realised two-sided 0.95-confidence interval for σ^2 for the random sample of normally distributed x-data in Table 3.4.

Problem 4.9

Determine the sample size for constructing a $(1 - \alpha)$-confidence interval for the variance σ^2 of a normally distribution so that

(a) $\frac{n-1}{2}\sigma^2 \left\{ \frac{1}{CS\left(\frac{\alpha}{2}, n-1\right)} - \frac{1}{CS\left(1-\frac{\alpha}{2}, n-1\right)} \right\} < \delta$ or

(b) $\frac{CS\left(1-\frac{\alpha}{2}, n-1\right) - CS\left(\frac{\alpha}{2}, n-1\right)}{CS\left(1-\frac{\alpha}{2}, n-1\right) + CS\left(\frac{\alpha}{2}, n-1\right)} < \delta_{\text{relativ}}.$

Problem 4.10

Compare the realised exact interval with the realised bounds (4.13) and (4.14) with the realised approximate interval (4.15).

Problem 4.11

Determine the sample size for the approximate interval (4.15).

Problem 4.12

Calculate the confidence interval (4.16).

Problem 4.13

Calculate n with (4.17).

Problem 4.14

Derive the sample size formula for the construction of a one-sided confidence interval for the difference between two expectations for equal variances.

Problem 4.15

We would like to find a two-sided 99% confidence interval for the difference of the expectations of two normal distributions with unequal variances using independent samples from each population with power = 0.90 and variances $\sigma_1^2/\sigma_2^2 = 4$. Given the minimum size of an experiment, we would like to find a two-sided 99% confidence interval for the difference of the expectations of two normal distributions with unequal variances using independent samples from each population and define the precision by $\delta = 0.4\sigma_x$. If we know that $\frac{\sigma_x^2}{\sigma_y^2} = 4$, we obtain $n_y = \left\lceil \frac{1}{2} n_x \right\rceil$.

Problem 5.1

Determine in a balanced design the sub-class number n in a one-way ANOVA for a precision determined by $\alpha = 0.05$, $\beta = 0.05$ and $\delta = 2\sigma$, and a test with (5.14).

Problem 5.2

Calculate the entries in the ANOVA Table 5.3 and calculate estimates of (5.9) and (5.10).

Problem 5.3

Test the null hypothesis $H_0 : a_1 = a_2 = a_3 = 0$ with significance level $\alpha = 0.05$.

Problem 5.4

Determine the $(1 - \alpha)$-quantile of the central F-distribution with df_1 and df_2 degrees of freedom.

Problem 5.5

Determine the sample size for testing $H_{A0} : a_1 = a_2 = \ldots = a_a = 0$.

Problem 5.6

Calculate the entries of Table 5.11 and give the commands for Table 5.12.

Problem 5.7

Calculate the sample size for testing $H_{AA0} : (a, b)_{11} = (a, b)_{12} = \ldots = (a, b)_{ab} = 0$.

Problem 5.8

Calculate the ANOVA table with the realised sum of squares of Table 5.14.

Problem 5.9

Determine the sub-class number for fixed precision to test $H_{A0}: a_1 = a_2 = \ldots = a_a = 0$ and $H_{B0}: b_1 = b_2 = \ldots = b_b = 0$.

Problem 5.10

Calculate the ANOVA-table of a three-way cross-classification for model (5.30).

Problem 5.11

Calculate the minimal sub-class number to test hypotheses of main effect and interactions in a three-way cross classification under model (5.30).

Problem 5.12

Calculate the ANOVA table for the data of Table 5.20.

Problem 5.13

Determine the minimal sub-class numbers for the three tests of the main effects.

Problem 5.14

Calculate the empirical ANOVA-table and perform all possible F-tests for Example 5.20.

Problem 5.15

Determine the minimin and maximin sample sizes for testing $H_{A0}: a_i = 0$ (for all i).

Problem 5.16

Determine the minimin and maximin sample sizes for testing $H_{C0}: c_k = 0$ (for all k).

Problem 5.17

Determine the minimin and maximin sample sizes for testing $H_{AxC0}: (ac)_{ik} = 0$ (for all i and k).

Problem 5.18

Calculate the ANOVA table and the F-tests for Example 5.21.

Problem 5.19

Determine the minimin and maximin sample sizes for testing $H_{A0}: a_i = 0$ (for all i).

Problem 5.20

Determine the minimin and maximin sample sizes for testing $H_{AxC0}: (ac)_{ik} = 0$ (for all i and k).

Problem 6.1

Estimate the variance components with the ANOVA method.

Problem 6.2

Estimate the variance components with the ML method.

Problem 6.3

Estimate the variance components using the REML method.

Problem 6.4

Test the null hypothesis $H_0 : \sigma_a^2 = 0$ for the balanced case with significance level $\alpha = 0.05$.

Problem 6.5

Test the null hypothesis $H_0 : \sigma_a^2 = 0$ for the unbalanced case with significance level $\alpha = 0.05$.

Problem 6.6

Construct a $(1 - \alpha)$-confidence interval for σ_a^2 and a $(1 - \alpha)$-confidence interval for the intra-class correlation coefficient (ICC) $\frac{\sigma_a^2}{\sigma^2+\sigma_a^2}$ with $\alpha = 0.05$.

Problem 6.7

Estimate the variance of the ANOVA estimators of the variance components σ^2 and σ_a^2.

Problem 6.8

Test for the balanced case the hypotheses:

$$H_{A0} : \sigma_a^2 = 0, \quad H_{B0} : \sigma_b^2 = 0, \quad H_{AB0} : \sigma_{ab}^2 = 0$$

with significance level $\alpha = 0.05$ for each hypothesis.

Problem 6.9

Derive the estimates of the variance components with the ANOVA method and using the REML.

Problem 6.10

Derive the ANOVA estimates for all variance components; give also the REML estimates.

Problem 6.11

Test the hypotheses $H_{A0} : \sigma_a^2 = 0$ and $H_{B0} : \sigma_b^2 = 0$ with $\alpha = 0.05$ for each hypothesis.

Problem 6.12

Use the analysis of variance method to obtain the estimators for the variance components by solving

$$\begin{aligned}
\mathbf{MS}_A &= s^2 + ns_{abc}^2 + cns_{ab}^2 + bns_{ac}^2 + bcns_a^2 \\
\mathbf{MS}_B &= s^2 + ns_{abc}^2 + cns_{ab}^2 + ans_{bc}^2 + acns_b^2 \\
\mathbf{MS}_C &= s^2 + ns_{abc}^2 + ans_{bc}^2 + bns_{ac}^2 + abns_c^2 \\
\mathbf{MS}_{AB} &= s^2 + ns_{abc}^2 + cns_{ab}^2 \\
\mathbf{MS}_{AC} &= s^2 + ns_{abc}^2 + bns_{ac}^2 \\
\mathbf{MS}_{BC} &= s^2 + ns_{abc}^2 + ans_{bc}^2 \\
\mathbf{MS}_{ABC} &= s^2 + ns_{abc}^2 \\
\mathbf{MS}_{\text{rest}} &= s^2.
\end{aligned}$$

Problem 6.13

Derive the F-tests with significance level $\alpha = 0.05$ for testing each of the null hypotheses $H_{AB0} : \sigma_{ab}^2 = 0; H_{AC0} : \sigma_{ac}^2 = 0; H_{BC0} : \sigma_{bc}^2 = 0; H_{ABC0} : \sigma_{abc}^2 = 0.$

Problem 6.14

Derive the approximate F-tests with significance level $\alpha = 0.05$ for testing each of the null hypotheses $H_{A0} : \sigma_a^2 = 0; H_{B0} : \sigma_b^2 = 0; H_{C0} : \sigma_c^2 = 0.$

Problem 6.15

Derive by the analysis of variance method the ANOVA estimates of the variance components in Table 6.11 and also the REML estimates.

Problem 6.16

Derive by the analysis of variance method the ANOVA estimates of the variance components in Table 6.13 and also the REML estimates.

Problem 6.17

Derive by the analysis of variance method the ANOVA estimates of the variance components in Table 6.15 and also the REML estimates.

Problem 7.1

Derive the analysis of variance (ANOVA) estimators of all variance components using case I in the fourth column of Table 7.1.

Problem 7.2

Derive the ANOVA estimators of all variance components using Case II in the last column of Table 7.1.

Problem 7.3

Test H_{01} :"All a_i are equal" against H_{A1} : "Not all a_i are equal" with significance level $\alpha = 0.05$.

Problem 7.4

Show the F-statistics and their degrees of freedom to test H_{02}: '$\sigma_b^2 = 0$' and H_{03}: '$\sigma_{ab}^2 = 0$', each with significance level $\alpha = 0.05$.

Problem 7.5

Estimate the variance components in an unbalanced two-way mixed model for Example 7.2.

Problem 7.6

Determine the minimin and maximin number of levels of the random factor to test H_{01}: 'all a_i are zero' if no interactions are expected for given values of α, β, δ/σ, a and n.

Problem 7.7

Determine the minimin and maximin number of levels of the random factor to test H_{01}: "all a_i are equal" if interactions are expected.

Problem 7.8

Estimate the variance components if the nested factor $\boldsymbol{B}$ is random by the ANOVA method.

Problem 7.9

How can we test the null hypothesis that the effects of all the levels of factor A are equal, H_{0A}: "all a_i are equal" against H_{aA}: "not all a_i are equal", with significance level $\alpha = 0.05$?

Problem 7.10

How can we test the null hypothesis $H_{B0}: \sigma^2_{b \text{ in } a} = 0$ against H_{BA}: $\sigma^2_{b \text{ in } a} > 0$ with significance level $\alpha = 0.05$?

Problem 7.11

Find the minimum size of the experiment which for a given number a of levels of the factor A will satisfy the precision requirements given by α; β, δ, σ.

Problem 7.12

Estimate the variance components σ^2 and σ^2_a by the analysis of variance method.

Problem 7.13

Test the null hypothesis H_{B0}:"the effects of all the levels of factor B are equal" against H_{B0}: "the effects of all the levels of factor B are not equal" with significance level $\alpha = 0.05$.

Problem 7.14

Test the null hypothesis $H_{A0}: \sigma^2_A = 0$ against H_{AA}: $\sigma^2_A > 0$ with significance level $\alpha = 0.05$.

Problem 7.15

Show how we can determine the minimin and the maximin sample size to test the null hypothesis $H_{A0}: \sigma^2_A = 0$.

Problem 7.16

Use the algorithm above to find the test statistic for testing $H_{A0}: a_i = 0, \forall i$ against H_{AA}: "at least one $a_i \neq 0$" for model III.

Problem 7.17

Use the algorithm above to find the test statistic for testing $H_{B0}: b_j = 0, \forall j$ against H_{BA}: "at least one $b_j \neq 0$" for model III.

Problem 7.18

Use the algorithm above to find the test statistic for testing $H_{AB0}: (ab)_{ij} = 0, \forall\, i, j$ against H_{ABA}: "at least one $(ab)_{ij} \neq 0$" for model III.

Problem 7.19

The minimin and maximin number of levels of the random factor **C** to test $H_{A0}: a_i = 0, \forall\, i$ against H_{AA}: "at least one $a_i \neq 0$" has to be calculated for model III of the three-way analysis of variance – cross-classification.

Problem 7.20

Calculate the minimin and maximin number of levels of the random factor C to test $H_{B0}: b_j = 0, \forall\, j$ against H_{BA}: "at least one $b_j \neq 0$" for model III of the three-way analysis of variance – cross-classification.

Problem 7.21

The minimin and maximin number of levels of the random factor **C** to test $H_{AB0}: (ab)_{ij} = 0, \forall\, i,\ j$ against H_{ABA}: "at least one $(ab)_{ij} \neq 0$" has to be calculated for model III of the three-way analysis of variance – cross-classification.

Problem 7.22

Estimate the variance components for model III and model IV of the three-way analysis of variance – cross-classification.

Problem 7.23

Calculate the minimin and the maximin sub-class number n to test H_{B0}:
$bj(i) = 0 \forall\, j,\ i$; H_{BA}: "at least one $bj(i) \neq 0$" for model III of the three-way analysis of variance –nested classification.

Problem 7.24

Estimate the variance components for model III of the three-way analysis of variance – nested classification.

Problem 7.25

Calculate the minimin and the maximin sub-class number n to test $H_{A0}: a = 0 \forall\, i$ for model IV of the three-way analysis of variance – nested classification.

Problem 7.26

Calculate the minimin and the maximin sub-class number n to test $H_0: c_{k(ij)} = 0 \forall k, j, i$; H_A: "at least one $c_{k(ij)} \neq 0$" for model IV of the three-way analysis of variance – nested classification.

Problem 7.27

Estimate the variance components for model IV of the three-way analysis of variance – nested classification.

Problem 7.28

Calculate the minimin and maximin number c of C levels for testing the null hypotheses $H_0: a_i = 0 \forall i$; against H_A: "at least one $a_i \neq 0$" for model V of the three-way analysis of variance – nested classification.

Problem 7.29

Calculate the minimin and maximin number c of C levels for testing the null hypotheses $H_0: b_{j(i)} = 0 \forall j, i$; against H_A: "at least one $b_{j(i)} \neq 0$" for model V of the three-way analysis of variance – nested classification.

Problem 7.30

Estimate the variance components for model V of the three-way analysis of variance – nested classification.

Problem 7.31

Calculate the minimin and maximin number c of C -levels for testing the null hypotheses for Model VI of the three-way analysis of variance - nested classification.

Problem 7.32

Estimate the variance components for model VI of the three-way analysis of variance – nested classification.

Problem 7.33

Calculate the minimin and maximin number b of B levels for testing the null hypotheses $H_0: b_{j(i)} = 0 \forall j, i$; against H_A: "at least one $b_{j(i)} \neq 0$" for model VII of the three-way analysis of variance – nested classification.

Problem 7.34

Estimate the variance components for model VII of the three-way analysis of variance – nested classification.

Problem 7.35

Calculate the minimin and maximin sub-class numbers for testing the null hypotheses $H_0 : c_{k(ij)} = 0 \forall k, j, i$; against H_A: "at least one $c_{k(ij)} \neq 0$" for model VIII of the three-way analysis of variance – nested classification.

Problem 7.36

Estimate the variance components for model VIII of the three-way analysis of variance – nested classification.

Problem 7.37

Calculate the minimin and maximin sub-class numbers for testing the null hypotheses $H_0 : a_i = 0 \forall i$; against H_A: "at least one $a_i \neq 0$" for model III of the three-way analysis of variance – mixed classification $(AxB) \succ C$.

Problem 7.38

Calculate the minimin and maximin sub-class numbers for testing the null hypotheses $H_0 : c_{k(ij)} = 0 \forall k, j, i$; against H_A: "at least one $c_{k(ij)} \neq 0$" for model III of the three-way analysis of variance – mixed classification $(AxB) \succ C$.

Problem 7.39

Estimate the variance components for model III of the three-way analysis of variance – mixed classification $(AxB) \succ C$.

Problem 7.40

Calculate the minimin and maximin sub-class numbers for testing the null hypotheses $H_0 : c_{k(ij)} = 0 \forall k, j, i$; against H_{A}: "at least one $c_{k(ij)} \neq 0$" has to be calculated for model IV of the three-way analysis of variance – mixed classification $(AxB) \succ C$.

Problem 7.41

Estimate the variance components for model IV of the three-way analysis of variance – mixed classification $(AxB) \succ C$.

Problem 7.42

Calculate the minimin and maximin sub-class numbers for testing the null hypotheses $H_0 : a_i = 0 \forall i$; against H_A: "at least one $a_i \neq 0$" for model V of the three-way analysis of variance – mixed classification $(AxB) \succ C$.

Problem 7.43

Calculate the minimin and maximin sub-class numbers for testing the null hypotheses $H_{B0} : b_j = 0 \forall j$ against H_{BA}: "at least one $b_j \neq 0$" for model V of the three-way analysis of variance – mixed classification $(AxB) \succ C$.

Problem 7.44

Calculate the minimum and maximum number of levels of factor C for testing the null hypothesis H_{AB0}: $(a,b)_{ij} = 0$ for all j, against H_{ABA}: "at least one $(a,b)_{ij} \neq 0$" for model V of the three-way analysis of variance – mixed classification $(\boldsymbol{A \times B}) > \boldsymbol{C}$.

Problem 7.45

Estimate the variance components for model V of the three-way analysis of variance – mixed classification $(\boldsymbol{A \times B}) > \boldsymbol{C}$.

Problem 7.46

Calculate the minimin and maximin sub-class numbers for testing the null hypotheses $H_0 : b_j = 0 \forall j$; H_A "at least one $b_j \neq 0$" for model VI of the three-way analysis of variance – mixed classification $(AxB) \succ C$.

Problem 7.47

Estimate the variance components for model VI of the three-way analysis of variance – mixed classification $(AxB) \succ C$.

Problem 7.48

Calculate the minimin and maximin sub-class numbers for testing the null hypotheses $H_0 : b_{j(i)} = 0 \forall j, i$; against H_A: "at least one $b_{j(i)} \neq 0$" for model III of the three-way analysis of variance – mixed classification $(A \succ B)xC$.

Problem 7.49

Estimate the variance components for model VI of the three-way analysis of variance – mixed model classification $(A \times B) \succ C$.

Problem 7.50

Calculate the minimin and maximin number sub-class numbers for testing the null hypotheses $H_0: c_{k(ij)} = 0 \forall k, j, i$; against H_A: "at least one $c_{k(ij)} \neq 0$" for model III of the three-way analysis of variance – mixed classification $(A \succ B)xC$.

Problem 7.51

Estimate the variance components for model III of the three-way analysis of variance – mixed classification $(A \succ B)xC$.

Problem 7.52

Calculate the minimin and maximin number of levels of the factor B for testing the null hypotheses $H_0: a_i = 0 \forall i$ against H_A: "$a_i \neq 0$ for at least one i" calculated for model IV of the three-way analysis of variance – mixed classification $(A \succ B)xC$.

Problem 7.53

Calculate the minimin and maximin number of levels of the factor B for testing the null hypotheses $H_0: c_{k(ij)} = 0 \forall k, j, i$; against H_A: "at least one $c_{k(ij)} \neq 0$" for model IV of the three-way analysis of variance – mixed classification $(A \succ B)xC$.

Problem 7.54

Calculate the minimin and maximin number of levels of the factor B for testing the null hypotheses $H_{AC0}: (ac)_{jk} = 0 \forall j, k$ against H_{ACA}: "at least one $(ac)_{jk} \neq 0$" for model IV of the three-way analysis of variance – mixed classification $(A \succ B)xC$.

Problem 7.55

Estimate the variance components for model IV of the three-way analysis of variance – mixed classification $(A \succ B)xC$.

Problem 7.56

Calculate the minimin and maximin number of levels of the factor C for testing the null hypotheses $H_0: a_i \ \forall i$ against H_A: "at least one a_i is unequal 0" for model V of the three-way analysis of variance – mixed classification $(A \succ B)xC$.

Problem 7.57

Calculate the minimin and maximin number of levels of the factor C for testing the null hypotheses $H_0: b_{j(i)} = 0 \forall j, i$ against H_A: "at least one $b_{j(i)} \neq 0$" for model V of the three-way analysis of variance – mixed classification $(A \succ B)xC$.

Problem 7.58

Estimate the variance components for model V of the three-way analysis of variance – mixed classification $(A \succ B)xC$.

Problem 7.59

Calculate the minimin and maximin number of levels of the factor C for testing the null hypotheses $H_0: b_{j(i)} = 0 \, \forall j, i$ against H_A: "at least one $b_{j(i)} \neq 0$" for model VII of the three-way analysis of variance – mixed classification $(A \succ B)xC$.

Problem 7.60

Estimate the variance components for model VII of the three-way analysis of variance – mixed classification $(A \succ B)xC$.

Problem 7.61

Estimate the variance components for model VIII of the three-way analysis of variance – mixed classification $(A \succ B)xC$.

Problem 8.1

Draw the scatter plot of Example 8.3.

Problem 8.2

Write down the general model of quasilinear regression.

Problem 8.3

What are the conditions in the cases 1,…, 4 in the example of Problem 8.2 to make sure that $\text{rank}(X) = k + 1 \leq n$?

Problem 8.4

Determine the least squares estimator of the simple linear regression (we assume that at least two of the x_i are different). We call β_0 the intercept and β_1 the slope of the regression line.

Problem 8.5

Calculate the estimates of a linear, quadratic, and cubic regression.

Problem 8.6

Write down the normal equations of case 2 in the example of Problem 8.2.

Problem 8.7

Calculate the determinant D of the covariance matrix of $\begin{pmatrix} \boldsymbol{b}_0 \\ \boldsymbol{b}_1 \end{pmatrix}$.

Problem 8.8

Show that the matrix of second derivatives with respect to β_0 and β_1 in Problem 8.4 is positive definite.

Problem 8.9

Estimate the elements of the covariance matrix of the vector of estimators of all regression coefficients from data.

Problem 8.10

Show how parameters of a general linear regression function $y = f(x_i, \beta) = \beta_0 + \beta_1 x_1 + \ldots + \beta_k x_k$ can be estimated.

Problem 8.11

Draw a scatter plot with the regression line and a 95%-confidence belt.

Problem 8.12

Test for a simple linear regression function H_0: $\beta_1 = -0.1$ against H_A: $\beta_1 \neq -0.1$ with a significance level $\alpha = 0.05$.

Problem 8.13

To test the null hypothesis $H_{0,1} : \beta_1 = \beta_1^{(0)}$, the sample size n is to be determined so that for a given risk of the first kind α, the risk of the second kind has at least a given value β so long as, according to the alternative hypothesis, one of the following holds:

(a) $\beta_1 - \beta_1^{(0)} \leq \delta$,

(b) $\beta_1 - \beta_1^{(0)} \leq -\delta$, or

(c) $|\beta_1 - \beta_1^{(0)}| \leq \delta$.

Problem 8.14

Test the two regression coefficients in a simple linear regression against zero with one- and two-sided alternatives.

Problem 8.15

Test the hypothesis that the slopes in the two types of storage in Example 8.4 are equal with significance level $\alpha = 0.05$ using the Welch test.

Problem 8.16

Determine the estimates of the Michaelis–Menten regression and draw a plot of the estimated regression line.

Problem 8.17

Test the hypotheses for the parameters of $f_M(x)$.

Problem 8.18

Construct $(1 - \alpha^*)$-confidence intervals for the parameters of $f_M(x)$.

Problem 8.19

Show how the parameters of $f_E(x)$ can be estimated.

Problem 8.20

Determine the determinant (8.44) of the matrix $F^T F$.

Problem 8.21

Test the hypotheses for all three parameters of $f_E(x)$ with significance level $\alpha = 0.05$.

Problem 8.22

Construct 0.95-confidence intervals for all three parameters of $f_E(x)$.

Problem 8.23

Show how the parameters of $f_L(x)$ can be estimated.

Problem 8.24

Derive the asymptotic covariance matrix of the logistic regression function.

Problem 8.25

Test the hypotheses with significance $\alpha^* = 0.05$ for each of the three parameters of $f_L(x)$ with R. The null-hypotheses values are respectively $\alpha_0 = 15$, $\beta_0 = 7$, and $\gamma_0 = -0.05$.

Problem 8.26

Estimate the asymptotic covariance matrix for this example.

Problem 8.27

Construct $(1 - \alpha^*)$ confidence intervals for all three parameters of $f_L(x)$.

Problem 8.28

Show how the parameters of $f_B(x)$ can be estimated.

Problem 8.29

Test the hypotheses with significance level $\alpha = 0.05$ for each of the three parameters of $f_B(x)$ with R. The null-hypotheses values are respectively $\alpha_0 = 3$, $\beta_0 = -1$, and $\gamma_0 = -0.1$.

Problem 8.30

Construct $(1 - \alpha^*)$ confidence intervals for all three parameters of $f_B(x)$.

Problem 8.31

Show how the parameters of $f_G(x)$ can be estimated with the least squares method.

Problem 8.32

Test the hypotheses with significance level $\alpha = 0.05$ for each of the three parameters of $f_G(x)$ using R. The null-hypothesis values are respectively $\alpha_0 = 15$, $\beta_0 = -2$ and $\gamma_0 = -0.3$.

Problem 8.33

Construct $(1 - \alpha^*)$-confidence intervals for all three parameters of $f_G(x)$.

Problem 8.34

Determine the exact *D*- and *G*-optimal design in $[x_l = 1,\ x_u = 303]$ with $n = 5$.

Problem 8.35

Estimate the parameters of a linear and a quadratic regression for the data of Example 8.2 and test the hypothesis that the factor of the quadratic term is zero.

Problem 9.1

Test the null hypothesis H_{B0}: $\beta_1 = \ldots = \beta_a$ against the alternative hypothesis H_{BA}: "there is at least one β_i different from another β_j with $i \neq j$" with a significance level $\alpha = 0.05$ using model (9.7) in Example 9.1.

Problem 9.2

Test in Example 9.1 with the ANCOVA model (9.1) the null hypothesis $H_{\beta 0}$: $\beta = 0$ against the alternative hypothesis $H_{\beta A}$: $\beta \neq 0$ with significance level $\alpha = 0.05$.

Problem 9.3

Test in Example 9.1 with the ANCOVA model (9.1) the null hypothesis H_{A0}: "all a_i are equal" against the alternative hypothesis H_{AA}: "at least one a_i is different from the other" with significance level $\alpha = 0.05$.

Problem 9.4

Estimate the adjusted machine means at the overall mean of the diameter.

Problem 9.5

Draw the estimated regression lines for the machines M_1, M_2, and M_3.

Problem 9.6

Give the difference of the adjusted means $M_1 - M_2$, $M_1 - M_3$, and $M_2 - M_3$. Give for these expected differences the $(1 - 0.05)$-confidence limits.

Problem 9.7

Test in Example 9.2 the null hypothesis H_{B0}: $\beta_1 = \ldots = \beta_a$ against the alternative hypothesis H_{BA}: "there is at least one β_i different from another β_j with $i \neq j$" with a significance level $\alpha = 0.05$ using the model of (9.8).

Problem 9.8

Test in Example 9.2 with the ANCOVA model (9.8) the null hypothesis H_B: $\beta = 0$ against the alternative hypothesis H_{BA}: $\beta \neq 0$ with significance level $\alpha = 0.05$.

Problem 9.9

Test in Example 9.2 with the ANCOVA model (9.8) the null hypothesis H_{A0}: "all a_i are equal" against the alternative hypothesis H_{AA}: "at least one a_i is different from the other" with significance level $\alpha = 0.05$.

Problem 9.10

Estimate the adjusted group means at the overall mean of x.

Problem 9.11

Give the difference of the adjusted means at the overall mean of x of group 1 – group 2. Give for this expected difference the (0.95)-confidence limits.

Problem 9.12

Give the estimated regression lines for the groups Fibralo = group 1 and Gemfibrozil = group 2 and make a scatter plot of the data with the two regression lines.

Problem 9.13

Test in Example 9.3 the null hypothesis H_{BO}: $\beta_1 = \ldots = \beta_a$ against the alternative hypothesis H_{BA}: "there is at least one β_i different from another β_j with $i \neq j$" with a significance level $\alpha = 0.05$ using the model (9.16).

Problem 9.14

Test in Example 9.3 with the ANCOVA model (9.15) the null hypothesis H_{BO}: $\beta = 0$ against the alternative hypothesis H_{BA}: $\beta \neq 0$ with significance level $\alpha = 0.05$.

Problem 9.15

Test in Example 9.3 with the ANCOVA model (9.15) the null hypothesis H_{AO}: "all a_i are equal" against the alternative hypothesis H_{AA}: "at least one a_i is different from the other" with significance level $\alpha = 0.05$.

Problem 9.16

Estimate the adjusted variety means at the overall mean of x.

Problem 9.17

Give the difference of the adjusted means at the overall mean of x of $V_1 - V_2$. Give for the expected difference of the adjusted means at the overall mean of x of $V_1 - V_2$ the (0.95) -Confidence Limits.

Problem 10.1

Determine the minimal sample size n so that for given a and δ the probability of a correct selection of the largest expectation P_C in (10.5) is at least $1 - \beta$.

Problem 10.2

Determine the minimal sample size n so that for given a and $\delta = r\sigma$ the probability of a correct selection of the t largest expectation P_C in (10.5) is at least $1 - \beta$.

Problem 10.3

Determine in a balanced design the sub-class number n for the multiple comparison (MC) Problem 10.1 for a precision determined by $\alpha_e = 0.05$, $\beta = 0.1$, and $\delta/\sigma = 2$.

Problem 10.4

Use Scheffé's method to test the null hypothesis of MC Problem 10.1. We consider all linear contrasts L_r in the μ_i.

Problem 10.5

Write the confidence interval down for the $\binom{a}{2}$ multiple comparisons of the expectations of a normal distributions with equal variances.

Problem 10.6

Construct confidence intervals for differences $\mu_i - \mu_j$ of a expectations of random variables y_i; $i = 1, \ldots, a$, which are independent of each other normally distributed with expectations μ_i and common, but unknown, variance σ^2. From a populations independent random samples $Y_i^T = (y_{i1}, \ldots, y_{in})$ $(i = 1, \ldots, a)$ of equal size n are drawn.

Problem 10.7

Construct confidence intervals for differences $\mu_i - \mu_j$ of a expectations of random variables y_i; $i = 1, \ldots, a$ which are independently from each other normally distributed with expectations μ_i and common, but unknown, variance σ^2. From a populations independent random samples of size n are drawn. $Y_i^T = (y_{i1}, \ldots, y_{in})$ $(i = 1, \ldots, a)$

Problem 10.8

Determine the minimal sample size for testing $H_{0,\,ij}$; $\mu_i = \mu_j\ i \neq j,\ i, j = 1, \ldots, a$ for a first kind risk α, a second kind risk β, an effect size δ and a standard deviation sd.

Problem 10.9

Determine the minimal sample size for testing $H_{0,\,ia}$; $\mu_i = \mu_a$, $i = 1, \ldots, a - 1$ with an approximate overall significance level α^* [hence we use for the test a first kind risk $\alpha = \alpha^*/(a-1)$], a second kind risk β, an effect size δ and a standard deviation sd.

Problem 10.10

Determine the minimal sample size for simultaneously testing $H_{0,\,i}$; $\mu_i = \mu_a$; $i = 1, \ldots, a - 1$ for a first kind risk α, a second kind risk β, an effect size δ and a standard deviation sd.

Problem 11.1

Show that the binomial distributions are an exponential family.

Problem 11.2

Show that the Poisson distributions are an exponential family.

Problem 11.3

Show that the gamma distributions are an exponential family.

Problem 11.4

Fit a generalised linear model (GLM) with an identical link to the data of Example 5.9.

Problem 11.5

Show the stochastic and systematic component and the link function for the binary logistic regression.

Problem 11.6

Analyse the proportion of damaged o-rings from the space shuttle Challenger data. The data can be found in Faraway (2016).

Problem 11.7

Analyse the infection probabilities of Example 11.4.

Problem 11.8

Analyse the infection probabilities of Example 11.5 where the dispersion parameter (scale-parameter) must be estimated.

Problem 11.9

Calculate the ML estimate for the Poisson parameter λ and calculate the chi-square test statistic for the fit of the data of Example 11.6 for the Poisson distribution with $\hat{\lambda}$.

Problem 11.10

Analyse the data of Example 11.7 using the GLM for the Poisson distribution.

Problem 11.11

Analyse the data of Table 11.8 using the GLM for the Poisson distribution with overdispersion.

Problem 11.12

Analyse the data of Example 11.8 with a GLM with the gamma distribution and the log link (multiplicative arithmetic mean model).

Problem 11.13

Analyse the data of Example 11.8 with a GLM with the gamma distribution and the identity link (additive arithmetic mean model).

Problem 11.14

Analyse the data of Example 11.9 with a GLM with the gamma distribution with $x = \log(u)$ and the inverse link for lot 1 and lot 2.

Problem 11.15

Analyse the infection data of Example 11.10.

Problem 12.1

Calculate estimates of the semi-variogram function and show graphs.

Problem 12.2

Calculate the estimate of the semi-variogram function.

Problem 12.3

Show how the parameters of a Matern semi-variogram can be estimated.

Problem 12.4

Measurements $y(x_1, x_2)$ are taken at four locations:
1) $y(10; 20) = 40$
2) $y(30; 280) = 130$
3) $y(250; 130) = 90$
4) $y(360; 120) = 160$.

Predict the value at $y(180; 120)$ by ordinary kriging. Use as covariance function

$$C(h) = 2000e^{-\frac{h}{250}}$$

and the distance $\| h \| = h = \sqrt{x_1^2 + x_2^2}$.

Problem 12.5

Show how to calculate predicted values.

Appendix B

Symbolism

In part, we use different notation to other mathematical disciplines. We do not use capital letters as in probability theory to denote random variables but denote them by bold printing. We do this not only to distinguish between a random variable $\boldsymbol{F}$ with F-distribution and its realisation F but mainly because linear models are important in this book. In a mixed model in the two-way cross classification of the analysis of variance with a fixed factor A and a random factor $\boldsymbol{B}$ is the model equation with capital letters written as:

$$Y_{ijk} = \mu + a_i + B_j + (aB)_{ij} + E_{ijk}.$$

This looks strange and is unusual. We use instead

$$\boldsymbol{y}_{ijk} = \mu + a_i + \boldsymbol{b}_j + (\boldsymbol{ab})_{ij} + \boldsymbol{e}_{ijk}.$$

Functions are never written without an argument to avoid confusion. So $p(y)$ is often a probability function but p a probability. Further is $f(y)$ a density function but f the symbol for degrees of freedom.

Sense	Symbol
Rounding up function	$\lceil x \rceil$ = smallest integer $\geq x$
Binomial distribution with parameters n, p	$B(n,p)$
Chi-squared (χ^2)-distribution with f degrees of freedom	CS (f)
Determinant of the matrix A	$\lvert A \rvert$, det(A)
Diagonal matrix of order n	$D(a_1.\dots.a_n)$
Direct sum of the sets A and B	$A \oplus B$
Identity matrix of order n	I_n
(n x m)-matrix with only zeros	0_{nm}
(n x m)-matrix with only ones	1_{nm}
Euclidian space of dimension n and 1 respectively (real axis), positive real axis	R^n; $R^1 = R$; R^+
y is distributed as	$\boldsymbol{y}\sim$

Applied Statistics: Theory and Problem Solutions with R, First Edition.
Dieter Rasch, Rob Verdooren, and Jürgen Pilz.

Sense	Symbol
Indicator function	Is A a set and $x \in A$, then $I_A(x) = \begin{cases} 1, & \text{if } x \in A \\ 0, & \text{if } x \notin A \end{cases}$
Kronecker delta	$\delta_{ij} = \begin{cases} 1 & \text{for } i = j \\ 0 & \text{for } i \neq j \end{cases}$
Interval on the x-axis	
Open	$(a, b) : a < x < b$
Half-open	$[a, b)$: $a \leq x < b$
	$(a, b]$: $a < x \leq b$
Closed	$[a, b]$: $a \leq x \leq b$
ith order statistic	$\boldsymbol{y}_{(i)}$
Cardinality (number) of elements in S	card (S); $\lvert S \rvert$
Constant in formulae	const
Empty set	$\emptyset$
Multivariate normal distribution with expectation vector μ and covariance matrix Σ	$N(\mu,\Sigma)$
Normal distribution with expectation μ and variance σ^2	$N(\mu,\sigma^2)$
Null vector with n elements	0_n
Vector with n ones	1_n
Parameter space	Ω
Poisson distribution with parameter λ	$P(\lambda)$
P-quantile of the $N(0, 1)$-distribution	$z(P)$ or z_P
P-quantile of the χ^2-distribution with f degrees of freedom	$CS(f, P)$
P-quantile of the t-distribution with f degrees of freedom	$t(f, P)$
P-quantile of the F-distribution with f_1 and f_2 degrees of freedom	$F(f_1,f_2, P) = F_P(f_1,f_2)$
Rank of matrix A	rk(A)
Rank space of matrix A	$R(A)$
Standard normal distribution with expectation 0; variance 1	$N(0,1)$
Trace of matrix A	$tr(A)$
Transposed vector of Y	$\boldsymbol{Y}^T$
Vector (column vector)	Y
Distribution function of a $N(0,1)$-distribution	$\Phi(x)$
Distribution Function	$F(y) = P(\boldsymbol{y} \leq y)$
Density function of a $N(0,1)$-distribution	$\phi(x)$
Random variable (bold print)	$\boldsymbol{y}$, $\boldsymbol{Y}$

Appendix C

Abbreviations

ASN	Average sample number
BLUE	Best linear unbiased estimator
BLUP	Best linear unbiased prediction
BQUE	Best quadratic unbiased estimator
df	Degrees of freedom
iff	If and only if
LSE	Least squares estimator
LSM	Least squares method
MINQUE	Minimum quadratic unbiased estimator
ML	Maximum likelihood
MLE	Maximum likelihood estimator
MS	Mean squares
MSD	Mean square deviation
ncp	non centrality parameter
REML	Restricted maximum-likelihood
SLRT	Sequential likelihood ratio test
SS	Sum of squares
UMP	Uniformly most powerful (test)
UMPU	Uniformly most powerful unbiased (test)
UMVUE	Uniformly minimum variance unbiased estimator
W.l.o.g.	Without loss of generality

Applied Statistics: Theory and Problem Solutions with R, First Edition.
Dieter Rasch, Rob Verdooren, and Jürgen Pilz.

Appendix D

Probability and Density Functions

Bernoulli distribution	$p(y,p) = p^y(1-p)^{1-y}, 0<p<1, y=0,1$
Binomial distribution	$p(y,p) = \binom{n}{y} p^y(1-p)^{n-y}; 0<p<1; y=0,\ldots,n$
Multinomial distribution	$P(\boldsymbol{y_1} = k_1, \ldots, \boldsymbol{y_m} = k_m) = \frac{n!}{y_1! \cdots y_m!} p_1^{y_1} \cdots p_m^{y_m};$ $i = 1, \ldots, m; 0 < p_i < 1; 0 < y_i < n$
Exponential family (canonical form)	$f(y,\eta) = h(y)e^{\sum_{i=1}^{k} \eta_i \cdot T_i(y) - A(\eta)}$
Uniform distribution in [*a*,*b*]	$f(y,a,b) = \frac{1}{b-a}, a<b, a \le y \le b$
Hypergeometric distribution	$p(y,M,N,n) = \frac{\binom{M}{y}\binom{N-M}{n-y}}{\binom{N}{n}}, n \in \{1,\ldots,N\},$ $y \in \{0,\ldots,N\}; M \le N$ integer
Normal distribution	$f(y,\mu,\sigma^2) = \frac{1}{\sigma\sqrt{2\pi}} e^{-\frac{(y-\mu)^2}{2\sigma^2}};$ $-\infty < \mu, y < \infty, \sigma > 0$
Poisson distribution	$p(y,\lambda) = \frac{\lambda^y}{y!} e^{-\lambda}, \lambda > 0, y = 0, 1, 2, \ldots$
Gamma distribution	$f(y,\lambda,\nu) = \begin{cases} \frac{\lambda^\nu}{\Gamma(\nu)} y^{\nu-1} e^{-\lambda y}, & if\ y \ge 0 \\ 0, & if\ y < 0 \end{cases}, \lambda > 0, \nu > 0$
Box-Cox transformation	$\boldsymbol{y}^* = \begin{cases} \frac{\boldsymbol{y}^\lambda - 1}{\lambda} & \text{if } \lambda \ne 0 \\ ln\boldsymbol{y} & \textbf{if } \lambda = 0 \end{cases}$

Applied Statistics: Theory and Problem Solutions with R, First Edition.
Dieter Rasch, Rob Verdooren, and Jürgen Pilz.

Exponential distribution	$f(y, \lambda) = \lambda e^{-\lambda y}; \lambda \in R^{+}; y \geq 0$
Beta-distribution	$f(y \cdot \theta) = \dfrac{1}{B(a, b)} y^{a-1}(1 - y)^{b-1}$ $0 < y < 1; 0 < a, b < \infty$
Matern-variogram	$\gamma_M(\check{h}) = c_0(1 - \delta_0(\check{h})) + c\left[1 - \frac{2^{1-\nu}}{\Gamma(\nu)}\left(\sqrt{2\nu}\frac{\check{h}}{\rho}\right)^{\nu} K_{\nu}\left(\sqrt{2\nu}\frac{\check{h}}{\rho}\right)\right]$, $\nu, c, \rho > 0; \check{h}, c > 0$

Index

Applied Statistics: Theory and Problem Solutions with R, First Edition.
Dieter Rasch, Rob Verdooren, and Jürgen Pilz.